Geology of the Bristol district

The country covered by the Bristol district special sheet includes the whole of the Bristol and Somerset Coalfield and the smaller basins on the south-east margin of the Severn Estuary. Mining in the area, active at the time of survey, has now ceased and the memoir presents a unique synthesis of the stratigraphy of this structurally complex region.

A comprehensive account is given of the stratigraphy and palaeontology of the underlying Lower Palaeozoic, Old Red Sandstone, Carboniferous Limestone and Millstone Grit rocks. The succession in the Coal Measures is described at length, with reference to their structure where necessary. The nature of the sub-Mesozoic unconformity and the relationships to it of the facies of the onlapping Triassic and Jurassic strata are examined in some detail. There is a summary of the Jurassic stratigraphy, and of recent developments in the interpretation of the Pleistocene and Recent deposits. The main economic products of the district are reviewed.

Cover photograph
View down Cheddar Gorge showing cliffs cut in Carboniferous Limestone. Photo: R W Gallois.

BRITISH GEOLOGICAL SURVEY

G. A. KELLAWAY and
F. B. A. WELCH

Geology of the Bristol district

Memoir for 1:63 360 geological special sheet (England and Wales)

CONTRIBUTORS

Biostratigraphy
H. C. Ivimey-Cook
M. Mitchell
B. Owens
A. W. A. Rushton
G. Warrington
D. E. White

Petrography
R. Dearnley

LONDON: HMSO 1993

© *Crown copyright 1993*

First published 1993

ISBN 0 11 884466 0

Bibliographical reference

KELLAWAY, G. A. and WELCH, F. B. A. 1993. Geology of the Bristol district. *Memoir of the British Geological Survey.*

Authors

G. A. Kellaway, DSc, CGeol
F. B. A. Welch, BSc, PhD
formerly *British Geological Survey*

Contributors

H. C. Ivimey-Cook, BSc, PhD, B. Owens, BSc, PhD, A. W. A. Rushton, BA, PhD, G. Warrington, BSc, PhD and D. E. White, MSc, PhD
British Geological Survey, Keyworth

R. Dearnley, BSc, PhD
British Geological Survey, Edinburgh

M. Mitchell, MA
formerly *British Geological Survey*

Other publications of the Survey dealing with this district and adjoining districts

BOOKS

British Regional Geology
Bristol and Gloucester, 3rd edition, in press

Memoirs
Geology of the Bristol district: the Lower Jurassic rocks (Bristol geological special sheet), 1984
Geology of the country around Monmouth and Chepstow (233, 250), 1961
Geology of the Malmesbury district (251), 1977
Geology of the country around Wells and Cheddar (280), 1965; second impression with amendments), 1977

Mineral Assessment Report: Celestite Resources
No. 25   North-east of Bristol (ST 68 and parts of ST 59, 69, 79, 58, 78, 67 and 77), 1976

MAPS

1:625 000
Geological map of the United Kingdom, South Sheet, 3rd edition, 1979
Quaternary map of the United Kingdom, South Sheet, 1977
Aeromagnetic map of Great Britain, Sheet 2, 1965
Bouguer gravity anomaly map of the British Isles, Southern Sheet, 1986

1:250 000
Bristol Channel, solid geology, 1987
Bristol Channel, Sea-bed sediments and Quaternary geology, 1986
Bristol Channel, Bouguer gravity anomaly, 1986
Bristol Channel, Aeromagnetic anomaly, 1980

1:50 000 and 1:63 000

Bristol district special sheet		1962
Sheet 250	(Chepstow)	Solid and Drift, 1972
Sheet 251	(Malmesbury)	Solid and Drift, 1970
Sheet 264	(Bristol)	Solid and Drift, 1974
Sheet 265	(Bath)	Solid and Drift, 1965
Sheet 280	(Wells)	Solid and Drift, 1984
Sheet 281	(Frome)	Solid and Drift, 1965

1:25 000

ST 45	Cheddar	Solid and Drift, 1969
ST 47	Clevedon and Portishead	Solid and Drift, 1968

Printed in the UK for HMSO
Dd 240439   C8   8/93   531\3   12521

CONTENTS

1 **Chapter 1 Introduction**
  Surface relief and drainage 2
  The Severn Estuary 5
  Outline of the history of earlier geological research 6

10 **Chapter 2 Pre-Carboniferous rocks**
  Lower Palaeozoic 10
    Cambrian 10
      Tremadoc 10
    Silurian 10
      Llandovery 11
      Wenlock 12
      Ludlow 13
  Old Red Sandstone 14
    Lower Old Red Sandstone 16
    Upper Old Red Sandstone 16

19 **Chapter 3 Carboniferous Limestone (Dinantian)**
  Lower Limestone Shale Group 26
  Black Rock Limestone Group 28
    Stratigraphical palaeontology 29
  Clifton Down Group 31
    Stratigraphical palaeontology 41
  Hotwells Group 42
    Stratigraphical palaeontology 45
  Conglomerates of doubtful age bordering the Severn Estuary 45
  Petrography of the volcanic rocks of Broadfield Down 46
  Details of stratigraphy 46
    Area west of Bristol 46
    Broadfield Down 58

62 **Chapter 4 Millstone Grit (Namurian)**
  Details of stratigraphy 63

66 **Chapter 5 Coal Measures (Westphalian)**
  Distribution and thickness 68
  History of research, and development of classification 71
  Principal coal seams 73
  Details of stratigraphy 73
    Lower and Middle Coal Measures 73
    Pennant Measures 95
      Downend Formation 95
      Mangotsfield Formation 101
      Pennant Measures undifferentiated 103
    Supra-Pennant Measures 104
      Farrington Formation 104
      Barren Red Formation 116
      Radstock Formation 120
      Publow Formation 126

128 **Chapter 6 Triassic**
  General description 128
  Details of stratigraphy 131
    Redcliffe Sandstone Formation 131
    Keuper Marl 132
    Tea Green Marl 135
    Dolomitic Conglomerate 135
    Rhaetic 140
      Westbury Beds 140
      Cotham Beds 141
    White Lias 141
  Mineralisation and metasomatism affecting Triassic rocks 142
    Haematitisation 142
    Quartz veins and geodes 143
    Dolomitisation 143
  Palynology 144

147 **Chapter 7 Jurassic**
  Middle Jurassic 147
  Details 147
    Inferior Oolite 147
    Great Oolite 10
      Fuller's Earth 150
      Great Oolite 150
      Forest Marble 151
      Harptree Beds 152

153 **Chapter 8 Economic geology**
  Iron ore 153
  Manganese 153
  Earth pigments (ochres) 153
  Lead and zinc 154
  Copper 155
  Celestite 155
  Roadstone and aggregate 156
  Building stone 157
  Sand and gravel 157
  Glass sand 157
  Pottery and brick clay 158
  Coal 158
  Fuller's earth 160

161 **Chapter 9 Pleistocene and Recent**
  Early to mid-Pleistocene (before about 600 000 years BP) 161
  Mid- to Late Pleistocene (about 600 000 to 10 000 years BP) 162
    Lower Severn terraces 162
    Bristol Avon terraces 162
    Head or Colluvium 163
  Flandrian (about 10 000 years BP to present) 164

166 **Appendices**
  1 Abstracts of selected shafts, wells and boreholes 166
  2 Glossary of local mining terms 177
  3 Biostratigraphy of the Black Rock Limestone in the Portway tunnel, Bristol 178

180 **References**

188 **Fossil index**

196 **General index**

FIGURES

1 Generalised geological map of the Bristol and Somerset Coalfield xii
2 Physiography of the Bristol district 2
3 Section through the Coal Measures at Stowey 7
4 Sections of a Coal Country in Somersetshire 8
5 Section showing the probable relationships of the Palaeozoic rocks between Thornbury and Mark 13
6 Comparative vertical sections of the Upper Old Red Sandstone and the Lower Limestone Shale Group of Burrington Combe, Bristol, Tortworth and the Forest of Dean 15
7 Dolomitisation of the Black Rock Limestone and Basal Viséan rocks on shelf areas and around the margins of the West Mendips Basin 21
8 Comparative vertical sections of the Carboniferous Limestone in the Avon Gorge area, Bristol and at Burrington Combe, Mendip Hills, to illustrate the relationship between lithostratigraphical and chronostratigraphical classifications of the rocks 22
9 Generalised horizontal section of the Upper Old Red Sandstone, Carboniferous Limestone and Millstone Grit of the western side of the Bristol and Somerset Coalfield 24
10 Comparative vertical sections in the Black Rock Group and in the lower part of the Clifton Down Group of Bristol and Burrington Combe 34
11 Section along the line of the Foul Water Tunnel between Roman Way, Sea Mills and Gully Quarry, Avon Gorge, Bristol 35
12 Isopachytes in (a) Upper Old Red Sandstone and Lower Limestone Shale Group; (b) Black Rock Group; (c) Clifton Down Group; (d) Hotwells Group 36
13 Comparative vertical sections of the Viséan rocks of the Bristol area 38
14 Comparative vertical section of the Viséan and Namurian strata of the Bristol and Somerset Coalfield. Key map shows the sites of the boreholes and measured sections 44
15 Sketch map showing the geology of the Avon Gorge Bristol 47
16 Section through the Carboniferous and Triassic rocks at the southern end of Falcondale Road, Westbury-on-Trym 50
17 Geological map of the Trym valley and Blaise Castle showing the structure of the Carboniferous rocks 53
18 Structure of the southern end of Kings Weston Hill, Bristol 54
19 Horizontal section through the western part of the Severn Tunnel:
   (a) Detailed section based on C. Richardson (1887)
   (b) Generalised section showing the relationship of the Coal Measures to the underlying formations 64
20 Generalised vertical sections of the Coal Measures of the Bristol and Somerset Coalfield 66
21 Map showing limits of area underlain by Westphalian rocks, and sites of principal shafts and boreholes included in Appendix 1 69
22 Comparative vertical sections of the Coal Measures between Yate and Winford 74
23 Horizontal section through the Kingswood Anticline between Fishponds and St George's Park, Bristol 77
24 Comparative vertical sections of the Middle and Upper Coal Measures in the Yate Deep and Westerleigh boreholes 89
25 Horizontal section through the southern part of the Radstock Basin between Luckington and Soho 92
26 Comparative vertical sections of the lower part of the Downend Formation in the Winterbourne area 96
27 Correlation of the coal seams in the lower part of the Downend Formation and the upper part of the Middle Coal Measures on the southern side of the Kingswood Anticline 98
28 Sections through the productive Coal Measures of the southern Radstock Basin 100
29 Comparative vertical sections of the productive measures of the Farrington Formation in the Radstock Basin 106
30 Structure contours on the No. 5 (Middle) Vein of the Farrington Formation south of the Farmborough Fault 108
31 Comparative vertical sections showing the probable correlation of the coals formerly worked at Bishop Sutton, Bromley and Pensford collieries 109
32 Horizontal section through the Coal Measures of the Coalpit Heath Basin between Winterbourne and Yate 113
33 Structure contour map of the Hard Vein, Coalpit Heath Basin 115
34 East–west section through Pensford Colliery 116
35 Comparative vertical sections of the Radstock Formation between Hursley Hill and Radstock 120
36 (a) Key map of the Radstock Basin showing approximate extent of the Radstock Formation and the names of collieries;
   (b) Isopachytes of the Slyving Vein;
   (c) Isopachytes of the Radstock Middle Vein;
   (d) Isopachytes of the measures between the Bull Vein and the Great Vein 121
37 Diagram showing the probable correlation of the principal coals of the Radstock Formation south of the Farmborough Fault 122
38 Sections showing the stratigraphy and structure of the productive Upper Coal Measures in the Radstock Basin 123

39 Section through the Farmborough Fault Belt between Dunkerton and Tenley showing faulting in the Upper Coal Measures 124
40 Section through the Farmborough Fault Belt proved in the Northside workings of Camerton Colliery showing faulting in the Radstock Formation 125
41 Comparative vertical sections in the Triassic rocks of the Bristol district 130
42 Sketch map showing the distribution of the Redcliffe Sandstone and homotaxial formations in areas east of the Severn 131
43 (a) Generalised base contour map of the Triassic rocks (b) Reconstructed base contours of the Triassic rocks after removing the effect of post-Rhaetian folding 138
44 Palynomorphs from the Triassic rocks of the Dundry (Elton Farm) Borehole 145
45 Dundry Hill, showing
(a) the effect of cambering, faulting and landslipping on the structure of the Inferior Oolite limestone plateau
(b) Structural inversion and marginal attenuation produced by cambering 148
46 Probable distribution of Triassic celestite deposits prior to dissection 156
47 Range diagram showing the vertical distribution of the fossils of the Lower Limestone Shale and Black Rock Groups in the Foul Water Tunnel of the Avon Gorge between Sneyd Park and the Gully 179

TABLES

1 Stratigraphical classification of the Carboniferous Limestone (Dinantian) in the Bristol–Mendips region 27
2 Stratigraphical classification of the Coal Measures (Westphalian) in the Bristol and Somerset Coalfield 67
3 Stratigraphical classification of the Triassic rocks of the Bristol district 128
4 Stratigraphical classifications of the late Triassic and Lower and Middle Jurassic rocks of the Bristol district 129
5 Stratigraphical classification of the Inferior Oolite of the Bristol district 149
6 Stratigraphical classification of the Great Oolite of the Bristol district 151

PLATES

1 Avon gorge, looking downstream from the Suspension Bridge 4
2 Cornstone in Old Red Sandstone, Portishead 18
3 Black Rock and Gully quarries, Avon gorge 28
4 Great Quarry, Avon gorge 30
5 Right bank of the Avon gorge from Bridge Valley Road to the Suspension Bridge 32
6 Vertical Concretionary Beds, Henbury Hill Quarry 39
7 Middle Cromhall Sandstone, Wick Quarry 40
8 Wicks Rocks Thrust, Wick Quarry 62
9 Thrust in Middle Coal Measures, Crofts End Brick Pit 83
10 Triassic and Jurassic rocks in the section at Aust Cliff 133
11 Dolomitic Conglomerate in Bridge Valley Road, Clifton 136
12 Sub-Triassic unconformity at Kilkenny Bay, Portishead 137

SIX-INCH MAPS

The primary six-inch survey of the area covered by the Bristol district special sheet was made on County Series maps at a scale of six inches to one mile of Gloucestershire, Monmouthshire and Somersetshire. Since publication of the special sheet, most of the component maps have been reconstituted on the National Grid basis at a scale of six inches to one mile, in many cases incorporating more recent information. The National Grid six-inch maps included wholly or in part in the special sheet are listed below, together with the initials of the surveyors and the dates of survey. The officers involved were: W. J. Barclay, R. Beveridge, W. Bullerwell, R. Cave, E. E. L. Dixon, G. W. Green, I. H. S. Hall, B. Kelk, G. A. Kellaway, T. R. M. Lawrie, I. B. Paterson, R. W. Pocock, D. R. A. Ponsford, H. C. Squirrell, F. M. Trotter, F. B. A. Welch, A. Whittaker and R. J. Wyatt. Manuscript copies of the maps are deposited for public reference in the libraries of the British Geological Survey. Uncoloured dye-line or photographic copies of these maps are available for purchase. Certain sheets, marked with an asterisk in the list, are not yet available; for information in these areas it will be necessary to consult the original County Series based material.

No.	Name	Initials	Dates
ST 34 NE	Mark	FBAW,RB,AW	1950–68
ST 35 SE	Cossington–Edington Burtle	FBAW	1949–50
ST 35 NE*	Banwell–Loxton	FBAW	1949
ST 36 SE*	Worle–Rolstone	FBAW,GWG	1949–67
ST 36 NE	Woodspring Bay	FBAW,GWG	1948–67
ST 38 SE	Goldcliff	FBAW,BK	1938–61
ST 38 NE	Llanwern Hill	FBAW,BK	1938–60
ST 39 SE	Llanbeder	FBAW,RWP,BK,HCS	1938–61
ST 44 NW	Blackford–Wedmore	FBAW,GWG,RB	1950–54
ST 44 NE	Theale–Henton	FBAW,GWG	1950–54
ST 45 SW	Axbridge–Weare–Clewer	FBAW,GWG	1949–52
ST 45 SE	Cheddar–Draycott	FBAW,GWG	1948–52
ST 45 NW	Sandford–Winscombe–Shipham	FBAW,GWG	1948–50
ST 45 NE	Burrington–Rowberrow–Charterhouse	GWG	1948–50
ST 46 SW	Puxton–Congresbury	FBAW,GWG	1948–49
ST 46 SE	Lower Longford–Wrington	FBAW,GAK,GWG	1948–51
ST 46 NW	Yatton	FBAW	1947–49
ST 46 NE	Brockley	FBAW,GAK	1947–51
ST 47 SW	Clevedon	FBAW	1947–49
ST 47 SE	Nailsea	FBAW,GAK	1947–51
ST 47 NW	Black Nore	FBAW	1948
ST 47 NE	Portishead	FBAW,GAK	1948–50
ST 48 SW*	Redwick	FBAW	1938
ST 48 NW*	Magor	FBAW	1938
ST 48 NE	Caldicot	FBAW,IHSH,WJB	1938–78
ST 49 SW*	Penham	FBAW	1938
ST 49 SE	Caer-went	FBAW	1938–39
ST 54 NW	Westbury–Wookey–Wells	FBAW,GWG	1950–55
ST 54 NE	West Harrington–East Harrington	FBAW,DRAP,GWG	1945–55
ST 55 SW	Priddy	FBAW,GWG	1948–51
ST 55 SE	Litton–Chewton Mendip	FBAW,GWG	1943–51
ST 55 NW	Blagdon–Ubley–Compton Martin	GWG	1948–50
ST 55 NE	West Harptree–East Harptree–Bishop Sutton	FBAW,GWG	1943–50
ST 56 SW	Butcombe–Newpnett Thrubwell	FBAW,GAK,GWG	1947–51
ST 56 SE	Chew Stoke	GAK	1945–52
ST 56 NW	Barrow Gurney	GAK	1949–52
ST 56 NE	Dundry	GAK	1948–49
ST 57 SW	Long Ashton	GAK	1949–51
ST 57 SE	Bristol South-West	GAK	1948–52
ST 57 NW	Avonmouth	GAK	1939–50
ST 57 NE	Bristol North-West	GAK	1939–49
ST 58 SW	Vimpennys Common	FBAW,GAK,EELD	1924–55
ST 58 SE	Almondsbury	FBAW,GAK	1939–55
ST 58 NW	Severn Tunnel	FBAW,EELD,TRML,FMT	1924–47
ST 58 NE	Aust Cliff	FBAW,FMT	1938–47
ST 59 SW	Chepstow	FBAW,TRML	1938
ST 59 SE	Oldbury Sands	FBAW,TRML,FMT	1938–47
ST 64 NW	Binegar-Ashwick-Oakhill	FBAW,DRAP,GWG	1945–61
ST 64 NE	Stoke Lane-Leigh-upon-Mendip-Holcombe	FBAW,DRAP,GWG	1945–61
ST 65 SW	Ston Easton	FBAW,BK	1943–69
ST 65 SE	Midsomer Norton	FBAW	1943–71
ST 65 NW	Clutton-Farrington Gurney-High Littleton	FBAW,BK	1943–69
ST 65 NE	Timsbury-Camerton	FBAW,BK	1943–69
ST 66 SW	Pensford	FBAW,GAK	1943–49
ST 66 SE*	Marksbury	FBAW,GAK	1943–46
ST 66 NW	Whitchurch	GAK	1945–49
ST 66 NE*	Keynsham	GAK	1945–47
ST 67 SW	Bristol South-East	GAK	1946–52
ST 67 SE	Kingswood	GAK	1946–48
ST 67 NW	Bristol North-East	FBAW,GAK	1946–52
ST 67 NE	Mangotsfield	FBAW,GAK	1946–47
ST 68 SW	Almondsbury	FBAW,GAK	1946–49
ST 68 SE	Frampton Cotterell	FBAW	1946–47
ST 68 NW	Alveston	FBAW	1939–47
ST 68 NE	Tytherington	FBAW	1939–46
ST 69 SW*	Thornbury	FBAW,FMT,IBP,RJW	1938–39
ST 69 SE	Cromhall	FBAW,GAK,FMT,RC,IBP,RJW	1945–70
ST 74 NW	Mells	FBAW,DRAP	1945–56
ST 75 SW	Hemington	FBAW,DRAP,GWF,BK	1944–69
ST 75 NW	Shoscombe	FBAW,GWG,BK	1944–69
ST 76 SW	Bath South-West	FBAW,GAK,GWG,DRAP,RJW	1944–79
ST 76 NW*	Bath North-West	GAK,WB	1946–59
ST 77 SW*	Wick	GAK,WB	1947
ST 77 NW*	Dyrham	FBAW,GAK	1946–47
ST 78 SW	Chipping Sodbury	FBAW,GAK	1946–57
ST 78 NW	Wickwar	FBAW,GAK	1945–57
ST 79 SW	Charfield Green	FBAW,GAK,RWP,RJW	1946–70

PREFACE

The area of the Bristol district special sheet lies within Old Series one-inch Sheets 19 and 35. These sheets were originally surveyed, at the one-inch scale, by Sir Henry de la Beche, W. T. Aveline, H. W. Bristow, T. E. James, J. Phillips, A. C. Ramsay, W. Saunders and D. Williams, and published in 1845. Later editions incorporated work by J. H. Blake, H B. Woodward and W. A. E. Ussher. Illustrating the geology were Horizontal Sections, Sheets 12, 14–17, 20–22, 103–107, 111–112 and Vertical Sections, Sheets 7, 11–12 and 46–52. The geology of the area was outlined by H. B. Woodward in his memoirs on *Geology of east Somerset and the Bristol Coal-fields* (1876) and *The Lias of England and Wales (Yorkshire excepted)* (1893).

Resurvey on the six-inch scale of the Chepstow (250) sheet in the north-western part of the district by F. B. A. Welch, T. R. M. Lawrie and F. M. Trotter, and by G. A. Kellaway on the Chepstow (250) Sheet and the western part of the Bristol (264) sheet, was interrupted by the 1939–45 war. The remainder of the area was surveyed between 1942 and 1952 by F. B. A. Welch, G. A. Kellaway and, after 1950, by G. W. Green. The District Geologists were R. W. Pocock and H. G. Dines. The one-inch-to-one-mile special sheet was published in 1962 and includes the whole of the working area of the Bristol and Somerset Coalfield and the small coal basins to the west. The component one-inch and six-inch maps were separately published during the next few years. Some revision of the north-east of the district was made in 1960–62 by R. Cave and I. B. Paterson and the results were incorporated in the Malmesbury (251) sheet.

When mapping was resumed in 1942–43, attention was concentrated on the coalfields to help maximise production. It proved difficult to determine coal thicknesses by drilling because the sheared coals were not easy to recover. In 1949 and 1950, tests were carried out on two boreholes at Harry Stoke of the new methods of gamma logging and resistivity measurements, to see if they could record the coals more accurately. Using extended logs of two holes and slow logging speeds, a series of successful runs was carried out by W. Bullerwell, in cooperation with G. M. Lees and the Anglo-Iranian Oil Company. By 1951, detailed geophysical logging of exploratory boreholes had become standard practice in the coalfield.

Between 1943 and 1955, the authors and G. W. Green, D. R. A. Ponsford and W. H. C. Ramsbottom examined in detail eleven deep boreholes as well as many underground boreholes and cross-measure drifts, all proving details of the Coal Measures succession. The supporting palaeontological work on the Namurian and Westphalian rocks was carried out by Sir James Stubblefield, M. A. Calver and W. H. C. Ramsbottom. M. Mitchell was a joint contributor to the studies of the Dinantian faunas including those proved in the Geological Survey's Ashton Park Borehole sunk in 1952.

The Geological Survey's Dundry (Elton Farm) Borehole, drilled in 1962–63, provided a section through the Lower Jurassic rocks; their stratigraphy and palaeontology were described by H. C. Ivimey-Cook and D. T. Donovan (1978). The memoir describing the Lower Jurassic rocks of the district (Donovan and Kellaway) was published in 1984.

Determinations of fossils by the senior author and by H. C. Ivimey-Cook, B. Owens, A. W. A. Rushton, G. Warrington and D. E. White have contributed to the present account. The petrography of the Carboniferous volcanic rocks is described by R. Dearnley. Miss E. Pyatt did much to further the progress of the memoir in its early stages. The structure of the Bristol district has been described by the senior author and P. L. Hancock (1983) and is therefore not described in detail here.

Grateful acknowledgement is made to numerous organisations and individuals, including landowners, quarry operators, consulting engineers and public and local authories, for general assistance during the survey. A debt of gratitude is owed to the officers of the National Coal Board and in particular to the late Mr J. Smith (General Manager) and Mr Arthur Savage (Chief Surveyor) of the Bristol and Somerset area, with whom close cooperation was maintained throughout the mapping. Several experienced mining engineers gave valuable assistance in the interpretation of old records. Among these were the late Mr W. H. Monks, Mr F. C. Sadler and Mr E. H. Staples; the last presented his papers and drawings to the Geological Survey in 1944. We also acknowledge with pleasure the collection of mining records kindly presented by Major Hippisley on behalf of his father, the late Mr H. E. Hippisley.

Special mention must be made of the assistance given by the staff of Bristol City Engineer's Department, then under the direction of Mr J. B. Bennett, in recording the geological results of trial boreholes, shafts and many miles of tunnels in Bristol. To the Port of Bristol Authority, we are indebted for help in surveying Denny Island. Bristol Waterworks Company and later the Wessex Water Authority, afforded valuable assistance, notably during the construction of the Chew Stoke Reservoir and the Clapton Tunnel, while engineers of British Rail (Western Region) assisted in the examination of the Clifton Down Railway Tunnel. To the engineers and contractors who built the M4 and M5 motorways, and especially to Messrs Freeman Fox & Partners our indebtedness is great. Extensive limestone quarries and mineral workings have been investigated with the active cooperation of the companies concerned. Helpful comment about the local glass industry has come from Mr R. J. Charleston of the Victoria and Albert Museum. To Dr and Mrs Victor Eyles and the staffs of the Bristol Archives Office and County Record offices of Taunton and Trowbridge, we are indebted for help with historical records, while Professor E. A. Vincent kindly gave permission for the examination of William Smith's papers at Oxford University. Factual observations made in earlier years have been contributed by Dr F. S. Wallis (Bristol Museum), Sir Arthur Trueman and Professor W. F. Whittard. To Professor D. L. Dineley and past and present members of the staff of the Geology Department of Bristol University, the senior author is indebted for help and encouragement.

As the primary six-inch survey advanced from 1943 onwards the newly acquired information was immediately put to use. At the end of the war and in its aftermath when coal was desperately needed, assistance was given to the local coal industry which at that time possessed neither the basic data nor the scientific manpower to carry out urgently needed investigations. War damage and industrial expansion faced Bristol with serious problems of reconstruction, to the resolution of which the Geological Survey was able to contribute. Immediate advice on the very complex local geology was given for many major works, including the long tunnels conveying storm water directly to the Avon Gorge and the foul water tunnels connecting the city with Avonmouth.

The foul-water drainage systems east and west of the River Avon were linked by a tunnel beneath the Avon gorge. The Geological Survey advised driveage in the lower Clifton Down Limestone and the tunnel was completed successfully without the use of compressed air. Other operations for which advice was given include the construction of the Brabazon Runway at Filton, and many site investigations in the old mining areas of east Bristol prior to rebuilding. In these and other projects the results of the survey have been used by quarry companies, planning and water authorities, consulting engineers and site investigators who depend largely upon the six-inches-to-one-mile- (or 1:10 000) scale maps for their information. The stratigraphical terms used in the memoir therefore agree with those given on the published 1:10 000 scale maps covering the Bristol and Somerset Coalfield. It was originally intended to metricate all the Imperial measurements, but in the course of doing this problems of recognition arose in connection with the depths and thicknesses given in the many old well, shaft and borehole records, of which there is commonly more than one version. Computors and calculators are readily available at the present day and errors are less likely to occur if metrication is applied by the user at the point where it is required.

In conclusion it is a sad duty to record the death in 1987 of Dr F. B. A. Welch, one of the authors of this memoir. A pioneer in the study of British Variscan tectonics and an outstanding geological surveyor.

Peter J. Cook, DSc
Director

British Geological Survey
Keyworth
Nottingham NG12 5GG

October 1992

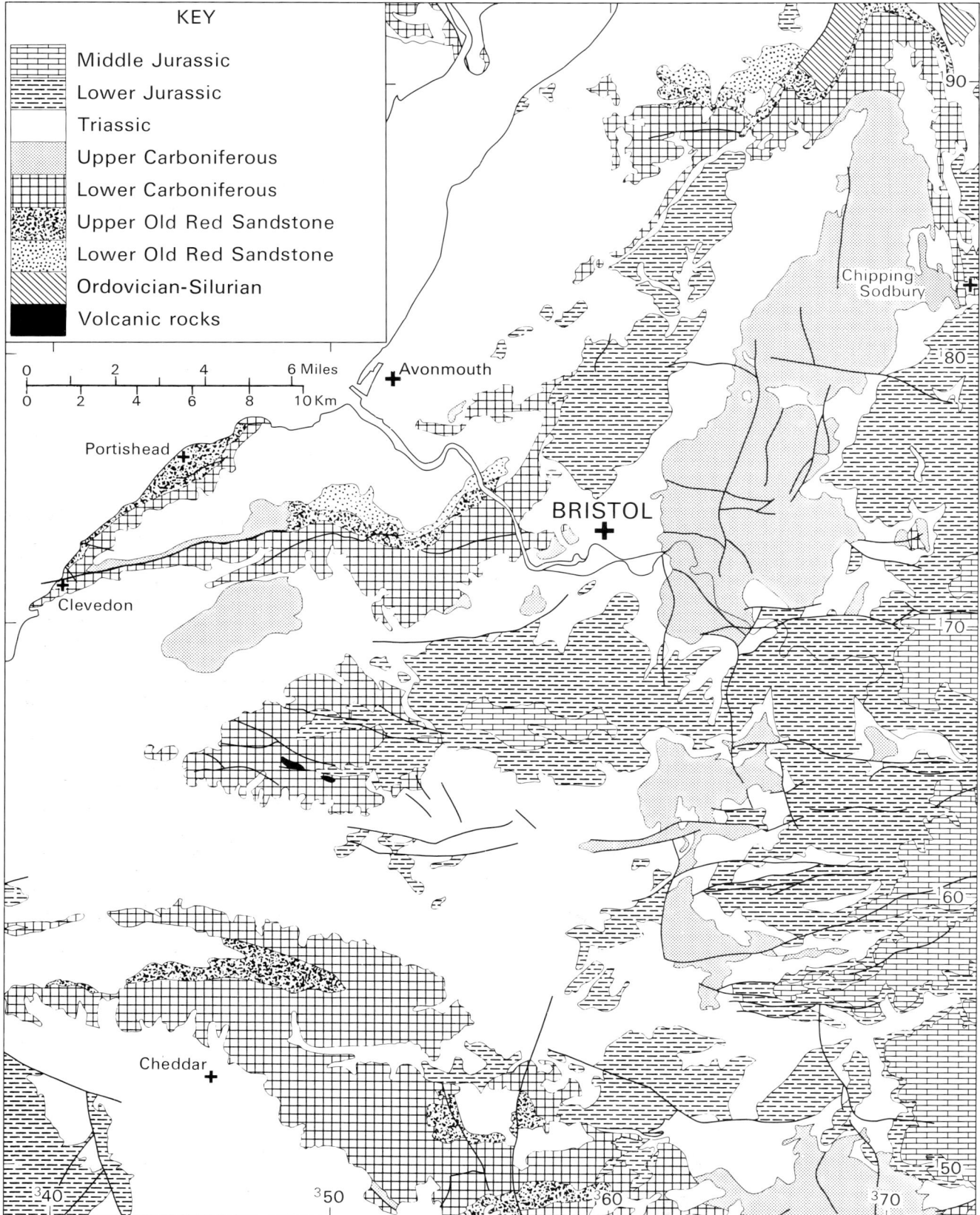

Figure 1 Generalised geological map of the Bristol and Somerset Coalfield

CHAPTER 1
Introduction

The Bristol and Somerset Coalfield may have been among the first to be worked as a source of fuel in the British Isles, for the Romans used coal to maintain the sacred fire on the altar of the Temple of Sul-Minerva at Bath. In his 'History of Kingswood Forest', Braine (1891, p.33) states that coal was worked in the Forest as early as the year 1200 and there were references to yearly charges, for the digging of sea-coal and earth for making pottery, in the Great Pipe Roll of Bristol Castle in 1223. From Tudor times onward to the late 18th century, the development of mining in the immediate proximity of Bristol and the Kingswood Anticline was largely influenced by the growth of the port of Bristol, but in the more southerly areas it was in the late 18th and early 19th centuries that improved transport led to a rapid expansion during the Industrial Revolution. Initially, much of the coal was used locally for domestic purposes and by blacksmiths, but it was also asssociated with other local industries based on indigenous resources of lead, zinc, iron ore, limestone, pottery clay and glass sand. Expansion continued throughout the 19th century, aided first by the development of canals and then by the railways, and this continued until about 1920, when some 10 000 persons were employed in the coal industry and its ancillary activities. After this, a gradual decline set in, due partly to the exhaustion of reserves in some of the older pits, and partly to the introduction of modern methods of working which were unsuited to conditions of complex geological structure; mechanisation proved to be uneconomic and the last working colliery, at Kilmersdon, closed in 1973.

The Bristol and Somerset Coalfield[1] includes a number of separate coal basins, some mainly exposed, others almost wholly concealed beneath Mesozoic rocks (Figure 1). Six geological sheets at a scale of six inches to one mile (250, 251, 264, 265, 280 and 281) are required to cover the whole of the exposed and concealed Upper Carboniferous rocks, and the Bristol district special sheet (published 1962) includes the greater part of these. Descriptive memoirs covering three of the one-inch sheets have already been published: Chepstow (250) in 1961, Malmesbury (251) in 1977, and Wells and Cheddar (280) in 1965.

Part of the working coalfield was included in William Smith's map of the country around Bath (1799), this being the first geological map 'worthy of the name' made in the British Isles. The primary one-inch-to-one-mile Geological Survey maps, to which Sir Henry de la Beche, the Survey's first Director, contributed, were published as sheets 19 and 35 in 1845. These were revised and published in 1873 and 1864 respectively, and a Memoir by H. B. Woodward on the entire coalfield area and the adjacent parts of Somerset, appeared in 1876. In July 1939 primary six-inch mapping began at Bristol, only to be suspended a few months later by the outbreak of war. It was resumed in north Somerset in 1942 and continued until 1970 when all the covering one-inch sheets had been published.

Most of the extensively worked coalfield areas are included within the Bristol district special sheet which covers a total area of about 615 square miles (Figure 1). This includes considerable tracts of country on either side of the Severn Estuary, and part of the Mendips bordering the coal basins, but excludes the concealed south-eastern part of the Radstock Basin about which less is known. In general, the coal basins are situated in areas of low relief, limited to the south by the high ground of the Mendip Hills and in the east by the southern Cotswolds. In the west, the Severn Coal Basin extends from Portishead beneath the Severn Estuary to Beachley. The northern margin of the Coalpit Heath Basin is marked by a low ridge of Old Red Sandstone and Carboniferous Limestone which also borders Ordovician and Silurian rocks exposed in the Tortworth Inlier to the north.

The three principal coal basins were originally a continuous deposit of Coal Measures extending from Cromhall in the north to the Mendips in the south; their present separation arises from the folding of the rocks. The most northerly is the Coalpit Heath Basin which is separated from the Pensford Basin by the Kingswood Anticline, a complex, transverse structural belt extending from east Bristol to Wick. South of the Kingswood Anticline is the Pensford Basin separated from the Radstock Basin by the Farmborough Fault Belt. On the western side of the coalfield are the two isolated basins of Nailsea and Clapton-in-Gordano. In addition, there are other isolated or partly concealed areas which may, in some cases, have buried connections with the basins listed above. They include the concealed Coal Measures of Westbury (Wilts), the exposed faulted outlier of Ebbor on the south side of the Mendips and a concealed area at Rodney Stoke (Green and Welch, 1965). A considerable area of concealed Coal Measures may exist beneath the Mesozoic rocks of the Vale of Wrington and in the Kenn–Puxton Moor area, and there is a small area of concealed Coal Measures at Barrow Gurney.

Between Yate and Wick, the eastern margin of the coalfield has a general north–south trend. South of the Avon valley at Bath, the margin swings sharply to an easterly direction but the Upper Carboniferous rocks are not exposed in the area between Bath and Frome.

Correlation between topographical relief and geology is poor in the coalfield areas, due to major unconformities below the Mesozoic cover (Figures 1 and 2). The widespread cover of younger rocks makes the task of elucidating both the stratigraphy and detailed structure of the older rocks immensely difficult. Large-scale horizontal and vertical displacements have taken place in the Palaeozoic rocks from

---

1 The greater part of the coalfield now lies within the newly constituted County of Avon which includes parts of the ancient counties of Gloucestershire, Somerset and the City and County of Bristol.

Figure 2 Physiography of the Bristol district

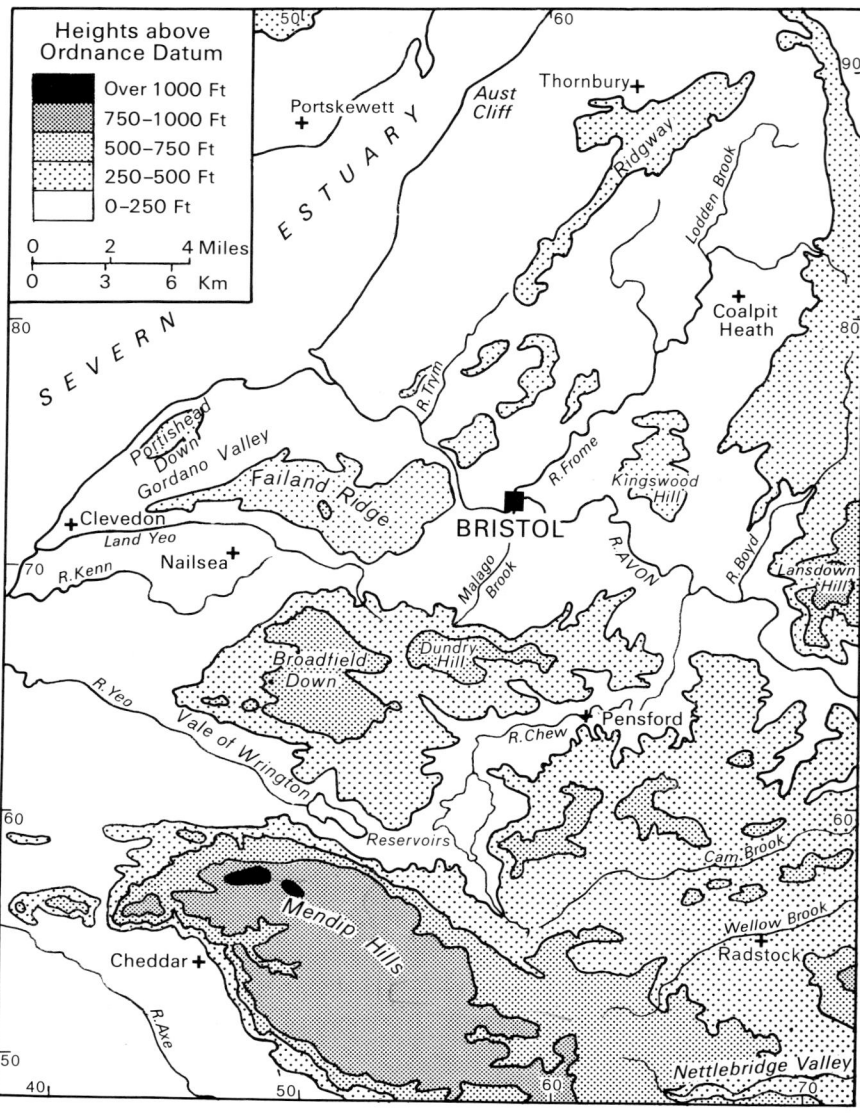

Silurian times onwards, the intense Carboniferous-Permian Variscan orogeny being succeeded by Permo-Triassic, Jurassic, Cretaceous and Tertiary movements. In particular, it has become clear that the long evolution of the north–south Malvern Fault Zone has to be considered in relation to the Variscan and later movements (Kellaway and Hancock, 1983). Recent investigations in the Bath area also suggest that Tertiary and Quaternary movements of tectonic origin may have played an important part in the development of the existing drainage and topography. Hotwells, in Bristol (Hawkins and Kellaway, in Kellaway, 1991, pp.179–203), and Bath are among the few places in Britain where thermal waters reach the surface; both these sources lie in the Avon valley and the waters rise through fissure systems which are of Quaternary age or which show Quaternary activation of older structures (Kellaway, 1991, pp.96–125).

SURFACE RELIEF AND DRAINAGE

At its southern margin, the Radstock coal basin is bounded by the Mendip Hills, extending some 30 miles from the Bristol Channel at Brean Down to the vicinity of Frome. Composed mainly of folded and faulted Carboniferous Limestone, but including considerable areas of Old Red Sandstone and some Silurian rocks, the Mendip Hills achieve their maximum altitude of 1068 ft on Black Down (Figure 2). Two of the best-known Mendip gorges, Cheddar and Burrington Combe, are incised into the southern and northern slopes of the Black Down Pericline. Three other periclinal structures, in which Old Red Sandstone forms high ground rising above the general level of the limestone plateau, are offset *en échelon* to the east-south-east. One of these, Pen Hill, is a prominent feature north of Wells. The most easterly one, Beacon Hill, has a core of Silurian rocks, mainly of volcanic origin. Many of the physical features of the Mendips, with its dry valleys, gorges, cavern systems, marginal overflow springs and underground rivers, can be likened to those of other Carboniferous Limestone uplands of Britain, and, as in other limestone environments on mainland Europe to the south, there is much evidence of the intermittent occupation of rock shelters and caves by ancient man from Late Palaeolithic times onwards.

At the eastern margin of the coalfield the Coal Measures are largely concealed by Mesozoic strata including the Middle Jurassic limestones which form the Cotswolds. Between Wotton-under-Edge and Bath, just east of the district, these limestones form a prominent escarpment, the Inferior Oolite being the dominant element in the north and the Great Oolite in the south. The Middle Jurassic escarpment enters the present district at Tog Hill east of Wick and continues southwards to Lansdown where the Great Oolite limestone plateau attains a height of 780 ft above OD. South of the Avon valley, the uplands are deeply dissected by the River Chew and other streams draining the southern part of the coalfield. Here, the Middle Jurassic rocks give rise to a number of broad ridges which merge farther south to form the high ground of the eastern Mendips around Mells and Nunney.

North of the Avon, the surface relief of the exposed Coal Measures in the Coalpit Heath Basin is very low. This is mainly due to erosion in Permo-Triassic times which produced a peneplain; although this surface has subsequently been exhumed, renewed dissection of the Palaeozoic rocks is still in its early stages. Post-Jurassic erosion of the Coal Measures of the Pensford and Radstock area to the south is even less advanced, the greater part of the Radstock Basin in particular being covered by Mesozoic rocks. The relief of the sub-Mesozoic surface, however, is very varied, both the Mesozoic and underlying Palaeozoic rocks having been affected by post-Variscan folding and faulting. This has resulted in the formation of some quite strong features, including the prominent Coal Measures sandstone ridge between Belluton and Stanton Wick and the wooded ridge of faulted Pennant sandstone at Temple Cloud.

Of the other structurally isolated or partly separated coal basins, that of Nailsea is situated on low ground between the Carboniferous Limestone uplands of Broadfield Down to the south and the Clevedon–Failand ridge to the north. It appears to have been isolated from the main basins to the east by Variscan folding and Permo-Triassic erosion, though the existing structure is partly of post-Triassic origin. The full extent of the Coal Measures beneath Kenn Moor and the alluvial flats at Kingston Seymour and Puxton on the western side of this basin, is unknown.

North of the Clevedon–Failand ridge is the low-lying Clapton Basin, almost entirely concealed beneath Triassic and Quaternary sediments. The strong ridges which separate these western coal basins are composed of structurally complex Old Red Sandstone and Carboniferous Limestone, and achieve altitudes of 300 to 500 ft above OD. They exhibit some notable limestone gorges, as for example, those of the River Avon at Clifton and the narrow defile occupied by the River Trym at Henbury and Combe Dingle.

Midway between Bristol and the Black Down Pericline in the Mendips lies the inlier of Carboniferous Limestone, called Broadwell Down by Strachley (1727) but since known to geologists as Broadfield Down (Figure 2). Although not formally recognised, the name appears on the Ordnance Survey maps in the form of Broadfield Farm, at the head of Goblin Combe near the south-western margin of Bristol (Lulsgate) Airport. Roughly triangular in plan, Broadfield Down extends from Yatton in the west for about six miles to Crown Hill, Winford in the east and for about four miles from Flax Bourton in the north to Redhill in the south. It is steep-sided on its north-western flank at Flax Bourton and Backwell and again in the south at Redhill and Wrington. The fertile red soils of the Vale of Wrington, derived from the weathering of Triassic marls, sandstones and conglomerates, contrast markedly with the barren moorlands of Black Down to the south and the windswept limestone uplands of Broadfield Down to the north.

The physical characteristics of Broadfield Down are similar to those of the central Mendips, though the uplands nowhere exceed 684 ft above OD. There are several limestone gorges, the most impressive being Goblin Combe and Brockley Combe, both dry and emerging westwards on to the broad Triassic terrain which separates the Carboniferous Limestone from the Coal Measures of the Nailsea Basin. Large parts of the central and eastern portion of Broadfield Down are formed of hard Lower Lias limestones and marginal conglomerates (Brockley Down Limestone), resting unconformably on the Carboniferous Limestone and locally very difficult to distinguish from it. This high-level cover of comparatively thin Upper Triassic and Lower Jurassic strata is continuous with the strata which underlie the main mass of Lower and Middle Jurassic rocks forming the Dundry outlier.

The eastern end of Broadfield Down terminates in a narrow valley draining to the River Chew at Winford where the ground rises steeply north-eastwards to the top of Dundry Hill, crowned by a plateau of cambered Inferior Oolite which achieves its highest level of 765 ft at its western end. Measured from east to west, the limestone plateau is about four miles long and presents an outstanding illustration of cambering and landslipping. On the east is the curious triangular promontory of Maes Knoll, with an Iron Age fort that coincides with the western termination of the Wansdyke. The shape of Maes Knoll has been determined by back-slip scars on the northern, south-eastern and south-western sides. Similar slips extend around the cambered limestone capping of Dundry Hill. At the foot of the steep, foundered slopes of Inferior Oolite, Lias clay and silt forming the upper part of the hill, lies a well-marked shelf of Blue Lias and White Lias limestone extending from Barrow Gurney in the west to Whitchurch in the east and thence along the southern side of the hill to Norton Malreward and Winford. East of Dundry, the Lias limestone plateau extends towards Queen Charlton, where its northerly inclination gives rise to a long dip-slope between the high ground at Hursley Hill and Charlton Field, and the Avon valley at Keynsham and Saltford. Farther north, clays replace the Lower Jurassic limestones and the associated surface features are less prominent. The more persistent basal Blue Lias limestones give rise to a strong plateau feature as far north as Itchington and Chipping Sodbury, but even this largely dies away north and north-east of Tortworth.

Large areas of dissected limestone plateaux formed by faulted Blue Lias and White Lias resting on Triassic mudstones and shales are present in the southern part of the Somerset Coalfield, east of a line from Whitchurch to East Harptree, and smaller areas occur west of Chew Valley Lake at Butcombe, Nempnett Thrubwell and Banwell. South of the Mendips, areas of low relief formed by the same rocks rise above the Somerset Levels at Wedmore and Weare.

Plate 1 Avon gorge, looking downstream from the Suspension Bridge.
The Clifton Down Limestone on the right is repeated, below the Avon Thrust, in Great Quarry within the trees. Black Rock Quarry is in the distance (A9763)

North of the Avon valley and extending eastwards from Eastville to Bridge Yate lies a broad ridge formed mainly of Middle Coal Measures rocks culminating in Kingswood Hill at over 370 ft above OD. This coincides roughly with a complex structure known as the Kingswood Anticline, though the Variscan structures are not directly responsible for the development of the present relief. Indeed, it would appear that the upheaval of the ground is largely post-Jurassic in age, and that the present land surface is almost coincident with the level of the sub-Mesozoic unconformity. Thus, post-Variscan warping produced an elevated mass of soft crushed shales and mudstones with the low ground to the north and south formed of hard massive Pennant sandstones, which were peneplained in Permo-Triassic times.

Of the rivers which drain the Bristol district, the Bristol Avon is the most important. Rising on the dip slope of the Middle Jurassic limestones in the south Cotswolds, it flows eastwards to Malmesbury and then southwards to Chippenham and Bradford-on-Avon. It then turns north-westwards to Bath cutting deep defiles in the Jurassic rocks, and entering the Bristol district at Twerton. From there to Bristol, its winding course is to the north-west crossing Lower Jurassic terrain and thence, by way of a gorge in Pennant sandstone at Hanham, to Bristol. The river then follows a westerly route across low-lying ground formed of Triassic rocks, but the course changes abruptly to north-north-west at Hotwells, where it enters the famous Avon Gorge in Carboniferous Limestone at Clifton (Plate 1), finally reaching the Severn Estuary at Avonmouth (Figures 1 and 2). The course of the river, though angular in pattern, shows little sign of control by major Variscan structures including the Avon thrust, though there is substantial evidence of adjustment to post-Jurassic faults and joint systems.

The confluence of the Wye and Severn, on the opposite bank at Chepstow lies only 7 miles north of that of the Bristol Avon and Severn Estuary at Avonmouth. Both the Wye and Avon have gorges incised in hard Devonian and Carboniferous rocks but there the similarity ends. The Wye is noted for its meanders, deeply incised and showing evidence of a long period of evolution on an easterly or south-easterly sloping peneplain. The terraces of the lower Wye are related to those of the lower Severn and possibly to the Avon terraces around Abbots Leigh and Shirehampton at the mouth of the river. Above Sea Mills, however, the Clifton and Hanham gorges of the Bristol Avon are narrow and steep sided, with sharply angled bends or elbows but no well-developed meanders like those of the Wye. In comparison with the Wye, the Avon is juvenile and its terrace sequence above Hanham suggests that a substantial part of its downcutting is likely to be middle–late Quaternary in age.

Several tributaries join the Avon between Bradford-on-Avon and Bristol. Among these are the Cam Brook and Willow Brook, both deeply incised into the Mesozoic cover of the Radstock basin. Like the Chew which joins the Avon at Keynsham, the Cam and Willow brooks rise on the western side of the Radstock Basin and flow north-eastwards, away from west-facing gaps which might have been expected to provide a more direct access to the Severn Estuary. This abnormal drainage pattern may have been due to thick ice in the Severn Estuary in early to mid-Pleistocene times, or it may have been due to differential uplift, or a combination of the two. Whatever the cause, the downcutting of these deeply incised valleys was accompanied by major cambering followed by landslipping of the Jurassic rocks on the steeper valley sides, a feature well seen on the Inferior Oolite limestone-capped hillsides of the Newton Brook between Priston and Newton St Loe. The strong cambering and slipping of the Inferior Oolite on the Dundry outlier (Figure 45) is likely to be of approximately the same age.

North of the Avon, the principal tributary streams include the River Boyd which has its confluence at Bitton, the Siston Brook with its headwaters near Siston and Staple Hill, and the Bristol Frome and its tributary, the Ladden Brook, which drain the Coalpit Heath Basin and join the Avon at Bristol. The Bristol Avon and all its tributaries run in gorges or deep, narrow defiles where they cross the outcrops of the more resistant formations. Thus, the River Boyd has cut a deep gorge in the Carboniferous Limestone of the Wick inlier, and the Bristol Frome flows through a substantial gorge in Pennant Sandstone between Hambrook and Stapleton. Farther west, below the Avon Gorge, the River Trym flows through a narrow gorge, about 250 ft deep, in Carboniferous Limestone and Dolomitic Conglomerate at Henbury and Coombe Dingle. Seen from the north, the notch is clearly cut in the highest part of the Carboniferous Limestone ridge and, but for the existence of a fault and shatter belt along which the stream has cut its channel, the feature could only be explained by superimposed drainage or by glaciation.

Tributaries joining the Avon on its south bank through Bristol show similar evidence of deep incision. An excellent example is seen in the deep gorge in Pennant Sandstone apparently cut by the tiny stream draining the Brislington area and joining the Avon at St Anne's Park. The development of these 'misfit' valleys and gorges has been widely attributed to superimposed drainage, but this process, though undoubtedly operative, cannot account for all the observed features of the extraordinary drainage pattern of the Bristol district. Many of the streams and rivers, including the Avon itself, show signs of structural control by geologically recent fault, joint or fissure patterns. Harmer (1907) attributed the formation of the gorges of the Avon to the presence of a glacially dammed lake in Wiltshire. However, there is unequivocal evidence to suggest that the emergence of the hot springs at Bath and Hotwells is due to Quaternary fissuring. This may well have been responsible for the severence of a much older drainage system whose formation predates the present course of the Lower Severn and the Bristol Avon. This decisive change may have commenced before or during the early glaciation (or glaciations) recorded at Kenn and Tickenham (p.161) and which antedate the last (Ipswichian) interglacial.

THE SEVERN ESTUARY

All the rivers and streams of the Bristol district, like those of the Chepstow area, drain to the Severn Estuary. Approximately 9 miles wide at Clevedon, it suffers a marked constriction upstream between Portskewett and Redwick, where the width is reduced to two miles. A further reduction to about one mile takes place between Purton and Sharpness beyond the confines of the present district, where the principal north–south faults of the Malvern Fault Zone cross the estuary.

The Severn Coal Basin, which underlies the estuary south of Aust, is largely concealed by Mesozoic rocks, mainly red marls and sandstones. These form the extensive intertidal reefs in the estuary known as the English Stones. The underlying Coal Measure sandstones are seen locally in the reef known as Lady Bench, which is situated on the line of the Severn Tunnel near Sudbrook. A small area of Upper Coal Measures is also exposed on the shore of the estuary at Portishead, marking the southern limit of the Basin; Carboniferous Limestone outcrops mark its northern limit between Beachley and Aust. These are the only Palaeozoic rocks known to be exposed in the bed of the estuary between Berkeley in the north and Denny Island in the south.

Denny Island occupies an isolated position three miles north-west of the entrance to Avonmouth docks. It is a flat-topped mass of Lower (Black Rock) Dolomite, a conspicuous feature of the estuary carrying an important navigational beacon. According to S. Smith and Willan (1937) the island is surrounded by reefs of Dolomitic Conglomerate but these were not visible at the time of survey. Apart from the cliffs on the north side of the Portishead promontory and along the coast extending from Portishead Point to Clevedon, there are few sections of geological interest in the lower reaches of the estuary. Low cliffs at Beachley are flanked by reefs of faulted Lower Dolomite with some Triassic marl, sandstone and conglomerate. On the east side of the river, however, the cliffs at Aust are well known for the excellent section showing faulted Triassic and Lower Jurassic strata resting on a platform of Carboniferous Limestone (Welch and Trotter, 1960). The Severn Bridge now dominates the river crossing at Aust and the bridge builders have utilised the reefs of hard rock in the river bed to provide a sound foundation for this impressive structure.

On the western bank of the Severn, above Beachley, a section comparable to that of Aust Cliff is also seen at Sedbury, and reefs of red Triassic marl and sandstone are exposed in the bed of the Severn at low tide, notably at Oldbury. Along the remainder of the coastline, from Portishead to Berkeley on the east bank and Redwick to Lydney on the west, the shores of the estuary are low, exposing only alluvial mud, occasionally with some peat or sand.

There is evidence of substantial changes in the distribution of land and water in historic times. Thus Denny Island is said to have been formerly accessible on foot from the Gwent shore at low tide, though it must now be approached by boat. Considerable changes have also taken place at the mouth of the River Avon, where the formation of the Swash Channel on the Somerset shore led to the isolation of Dung Ball Island (shown on Old Series Sheet 35) and, eventually, to the incorporation of the island in the northern or Gloucestershire bank of the Avon. Like other funnel-shaped estuaries the tidal portion of the Severn and its principal estuarine tributaries show very strong variations in water level. Thus, the rise and fall of the tide at Clevedon Pier is some 47 ft and at the mouth of the Wye near Beachley, the difference may be as much as 50 ft.

Under the influence of strong westerly or south-westerly winds, however, much higher water levels occur and lead to flooding. Coastal embankments and drains with self-acting tidal flaps have been built, to protect low-lying areas where the land is at or below ordinary high tide level, but these defences have not always proved adequate. In 1883, during the construction of the Severn Tunnel, one such sea flood flowed over the protective bank on the Gwent shore and entered the shafts and headings (Richardson, 1887). Another area, which is particularly vulnerable to sea floods generated by strong westerly gales, is the low-lying area at Kingston Seymour between Clevedon and Woodspring, where flooding took place as recently as 1981.

In addition to changes of level due to wind action in the lower reaches of the estuary, the Severn has an equinoctial tidal bore which becomes conspicuous in the upper reaches, notably between Newnham and Gloucester. Though less noticeable in the lower reaches, these tidal surges and the wind play an important part in the sedimentation and ecology of the huge, intertidal mudbanks and sandbanks and in the deposition and erosion of the muddy sediments. Locally, as at Aust and Woodhill Bay, Portishead the colonisation of intertidal mudflats by *Spartina* and saltmarsh vegetation stabilises them and extends the alluvial flats seawards. Elsewhere, as at the mouth of the River Avon, erosion predominates. The water of the Severn Estuary is extremely turbid and exhibits marked stratification (Mettam, 1979; Collins, 1987), the bulk of the material carried in suspension being of silt grade. The thick deposits of alluvial mud and silt which underlie the salt marshes and levels are of postglacial age but may be mainly periglacial or glacial in origin. The matter in suspension may have been brought down by tributary streams and rivers draining from adjacent inland areas but the estuary appears to form a trap in which finely divided solid matter, including foraminifera (Murray and Hawkins, 1976), may have been introduced by strong, wind-generated, westerly, tidal currents.

Stranded and floating seaweeds carrying small rocks and pebbles have been observed in the Bristol Avon, Wye and other tributaries of the Severn near the limits of tidal flow. At high tide, such material, as well as mud transported from the open estuary, tends to be deposited in slack water. Though vulnerable to erosion by strong flows of river water in time of flood, it is possible for much sediment of marine or estuarine origin to accumulate in this way, particularly when sea-floods take place.

OUTLINES OF THE HISTORY OF EARLIER GEOLOGICAL RESEARCH

It is beyond the scope of this introduction to review the vast and varied literature that over several centuries has touched upon the geology of the Bristol district. Significant contributions go back over four and a half centuries and the early years of the nineteenth century saw William Smith, a noted canal and mining engineer, develop his work on the Carboniferous, Triassic and Jurassic rocks that was to lay the foundations of modern stratigraphy and earn him the name of 'Father of English geology'.

Among the earliest local geological observations were those by Leland between 1535 and 1545. He saw Lower Lias ammonites at Keynsham which he described as 'stones figured like serpents' and visited St Vincent's Rocks at Clifton to see the quartz crystals (or 'Bristol Diamonds') which line some of the fissures in the Carboniferous Limestone and Dolomitic Conglomerate. More significant was the contribution of George Owen of Henllys (1570) whose *History of Pembrokeshire* was not published until 1796. In this, he described the distribution of the Carboniferous Limestone around the margins of the coal basins from Pembrokeshire in the west to Bristol in the east, presenting a picture which, as Buckland and Conybeare (1824) pointed out, indicated an orderly arrangement of the strata hitherto unrecognised. John Aubrey (1625–1697), born at Easton Piers near Malmesbury, made numerous observations on the geology of the country in and around Wiltshire. He speculated on the origin of the Avon Gorge and recognised that fossils are the remains of extinct animals; he also described the occurrence of the greywether sandstone boulders or sarsens which are scattered widely over west Wiltshire and the eastern Mendips.

In an account of early scientific activity in the Bristol region, Eyles (1955) has described the work of William Cole (died 1701), John Woodward (1665–1728) and others who investigated geological phenomena in the late 17th century. A distinguished foreign visitor was G. van Vallerius (1683–1744) who provided an interesting account of Aust Cliff on the Severn (Bradshaw and Matthews, 1970), comparing the striking bedding of the varicoloured Triassic rocks with that found in the Kupferschiefer in Germany.

Foremost among the early 18th century writers was John Strachey (1671–1743) of Sutton Court in Somersetshire, who published two papers (1719, 1725) in the *Philosophical Transactions*, followed by a pamphlet incorporating the contents of his earlier papers and including some additional observations in 1727. These were milestones in the history of geology, bringing to wide notice facts only previously known to practical miners (Eyles 1955; Fuller 1969). In his first two papers Strachey described and illustrated the succession of the coal seams (Figures 3 and 4), the presence of shells and fern-like plants in the roofs of seams, the effects of faulting, and the relations between the inclined strata of the Coal Measures and the flat or gently dipping 'red earth' and 'Lyas' which rest unconformably on the older strata[1]. Some of the information in his work is still of practical or scientific

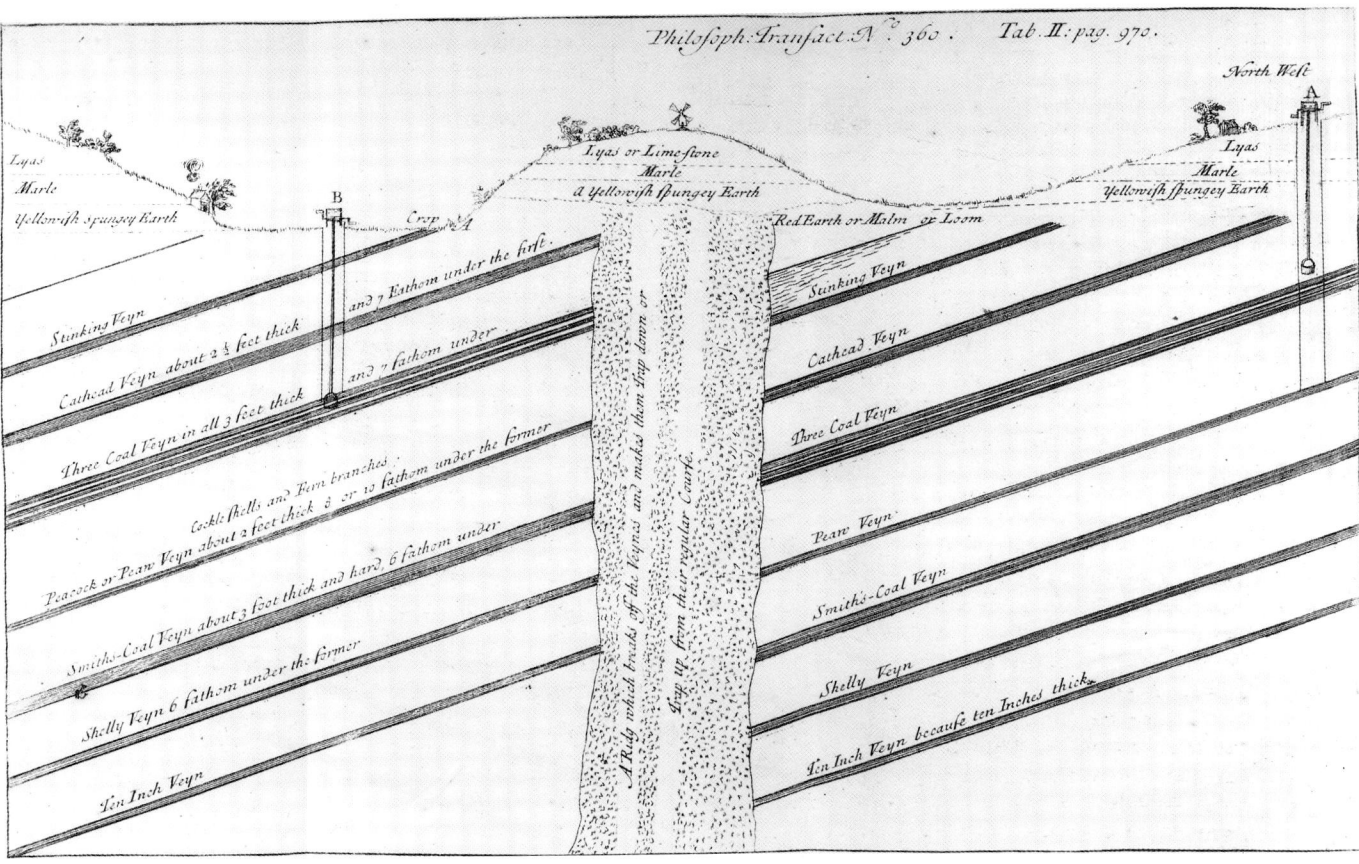

Figure 3 Section through the Coal Measures at Stowey (Strachey, 1719)

interest, since it relates to the extent of old workings or the condition of the coal seams. In one of his horizontal sections (1725, fig.1) Strachey showed steeply inclined veins of lead ore (associated with calamine and yellow-ochre) in the lead mines of Broadfield Down. This section (Figure 4) shows that he was uncertain as to the relationship of the Coal Measures to the Carboniferous Limestone, the contact being concealed by Mesozoic rocks. Many of Strachey's views on the origin of minerals and fossils were bizarre, in marked contrast to the rational approach of Aubrey and Hooke in earlier days. Nevertheless, the factual information which he presented was of great value and formed a signficant contribution to the geology of the coalfield as we know it today.

In 1754, there appeared a description of the geology of the Bristol area by Edward Owen. This included the first description of the Cotham or Landscape Marble, discovered beneath the White Lias in areas then being built up, on Cotham Hill in Bristol. Other important data, relating mainly to the lead mines of the Mendips and Bristol, notably Pen Park Hole at Henbury, were published in a work by the Reverend Alexander Catcott (1725–1779). All these publications added to the store of information which had been steadily accumulating during the 17th and 18th centuries and which now awaited a master hand to weld it into a coherent whole.

William Smith (1769–1839) started his working life as an apprentice to Mr Edward Webb, a surveyor of Stow-on-the-Wold. In 1791, Smith was sent to Somerset to survey an estate at Stowey near the ancestral home of John Strachey of Sutton Court. As a result of this work, he was asked to undertake an underground survey of coal mines at High Littenton and, subsequently, to act as Surveyor and Engineer to the Somerset Coal Canal. Existing accounts of his life and work include those by John Phillips (1844) and T. Sheppard (1920). A definitive biography by Mrs J.M. Eyles was in preparation at the time of her death (1986) and is being continued by Dr H. Torrens. Shorter works which should be mentioned are those of F.A. Bather (1926), L.R. Cox (1948) and V.A. Eyles (1955).

In the *Memoirs of William Smith, LLD*, published in 1844 by his nephew John Phillips, it is recorded that Smith was invited by the citizens of Bristol to attend the first meeting of the British Association in that city in 1836. Phillips goes on

---

1 The term Lias is of very great antiquity (Donovan & Kellaway, 1984). As used by Glanvil (1669) and Strachey (1719, 1725) it includes the White Lias and Blue Lias which together constitute the Lyas. White Lias and Blue Lias were both adopted by William Smith (1799 et seq.). They are lithostratigraphical terms. Phillip's definition (1829) of the base of the Lias is invalidated by the use of an imperfect section remote from the type area and Richardson's Langport Beds (1911), is a synonym of White Lias (Smith, 1799). The stratotype of the White Lias is situated at Rockhill near Radstock and was described by Tutcher and Trueman (1925, p.603). The base of the Lower Lias as defined on all the Geological Survey maps of the Bristol district and adjoining areas is taken at the base of the White Lias where this is present or at the base of the Blue Lias where it is not (see Kellaway, 1991, pp.39–50, for details).

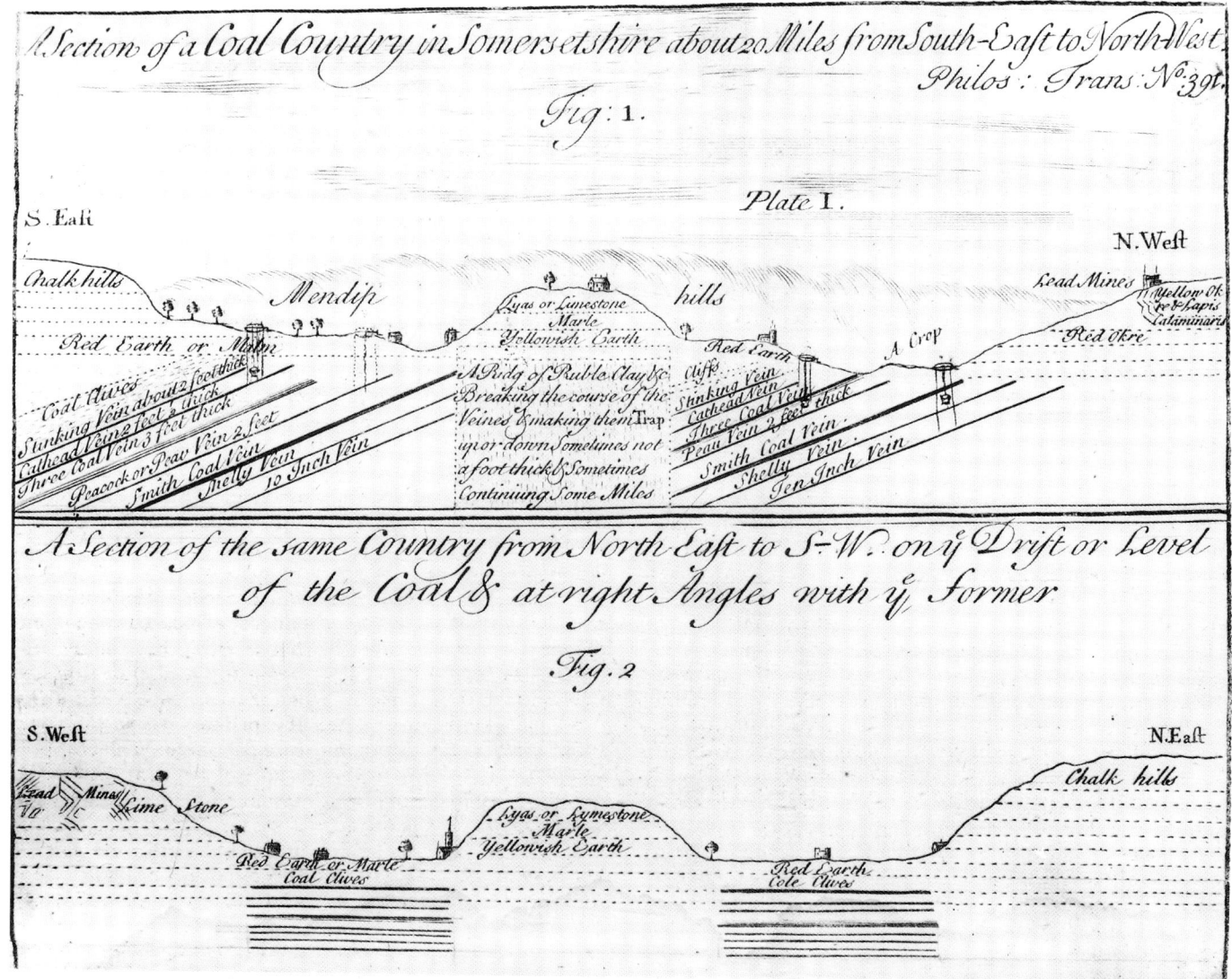

Figure 4 Sections of a Coal Country in Somersetshire (Strachey, 1725)

to remark that Smith made use of this opportunity to survey part of the coal district of Kingswood 'and coloured the results of his observations on the sheets of the Ordnance Survey'. This was not, however, the earliest geological map to be made of the Bristol district. William Smith had already published a geological map of the country extending for five miles around Bath in 1799. This map covers part of the eastern side of the Bristol and Somerset Coalfield, including the area around Corston and Newton St Loe in the Avon Valley between Bath and Saltford. Harrington's Coal Works at Newton St Loe, working seams in the lower part of the Pennant Grit, are shown on the base map, as are the Wellow and Dunkerton branches of the Somerset Coal Canal. The survey of the route of the canal, linking the Radstock basin and the Kennet and Avon Canal, had only just been completed under Smith's direction.

In 1801, there appeared Smith's first geological map of England and Wales on which the geology of the Mendips, the Bristol and Somerset Coalfield, the Cotswolds and Salisbury Plain is represented. This was followed by a map of Somersetshire (1805) which has since been lost (Sheppard, 1920, p.112) and by Smith's remarkable map of England and Wales drawn on a scale of five miles to one inch (1815).

Owing to the presence of a Mesozoic cover concealing the contact of the Carboniferous Limestone with the Millstone Grit and the Coal Measures, John Strachey and the early geologists had great difficulty in placing them in the correct order. Both in the Mendips and on Broadfield Down the Blue Lias consists of coarse fragmental limestone largely composed of recycled Carboniferous crinoids, corals and shell debris resting unconformably on the Carboniferous Limestone and projecting outwards over the adjacent Triassic rocks which conceal the Coal Measures. One such prominent bed of reconstituted Carboniferous material forms a prominent shelf at Ston Easton projecting above the worked coal seams of Farrington Gurney. Both Strachey and

Smith are likely to have been familiar with it and it may be because of this and other similar examples that Smith excluded the Mountain Limestone from his Table in 1799. In 1811 he made what he himself described as his 'mistake' by placing the Carboniferous Limestone above the Coal Measures but in 1817 he corrected the error.

Like John Strachey, William Smith was in direct contact with coal miners during his stay at High Littleton. In the later development of his ideas, however, he turned his attention to the Mesozoic rocks, and it was during this time that he began to formulate the general principles on which his pioneering stratigraphical work was subsequently based. Many accounts of his early geological work concentrate on his investigations of the Mesozoic rocks, but his manuscript notes and drawings include an excellent section through the Upper Old Red Sandstone, Lower Limestone Shales and Derbyshire (or Mountain) Limestone of Abbots Leigh near Bristol, made while he was residing at Leigh Court.

In 1824, nine years after the production of the final version of Smith's map of England and Wales, Buckland and Conybeare published their classic paper on the *South-western coal district of England and Wales*, and in the same year there appeared Thomas Weaver's account of the Silurian rocks of Tortworth.

The first editions of the engraved and hand coloured one-inch Geological Survey Old Series sheets 19 and 35 appeared in 1845, and showed the main outlines of the stratigraphy and structure of the district as described in Volume 1 of the Memoirs of the Geological Survey (De la Beche, 1846). These were substantially revised and new editions published: sheet 35 in 1866 and 1872 and sheet 19 in 1873. A third edition of sheet 19 was published in 1899, incorporating information collected during the preparation of J. Prestwich's report to the Royal Coal Commission of 1871. An explanatory memoir by H. B. Woodward entitled *The geology of the east Somerset and Bristol coalfields* appeared in 1876.

These, however, were not the only geological maps of the Bristol district to be published during the 19th century. In 1862, William Sanders (1799–1875) brought out a geological map on the scale of one inch to five miles covering an area of about 720 square miles centred on Bristol. One earlier map on the scale of 20 chains to one inch, together with a description, was published in 1836 in the *West of England Journal of Science and Literature*, and depicted the geology of the area around Portishead. The authorship of this work has never been established, but it may have been by Sir Henry de la Beche.

Palaeontological studies formed an important element of research in and around the Bristol and Somerset Coalfield from the days of William Smith onwards. Of those whose work has had implications of wider significance, Charles Moore (1815–1881) was a pioneer in the field of micropalaeontology. His work on the contents of Mesozoic fissure-fillings in the Carboniferous Limestone of the Mendips and the Bristol area yielded teeth of early mammals described as *Microlestes* (now renamed *Haramiya*); he also studied a wide range of microfossils including diminutive brachiopods, ostracods, foraminifera and conodonts (Winwood, 1892).

Biostratigraphical studies in the Bristol and Somerset Coalfield included zonation of the Jurassic rocks by ammonites (Buckman and Wilson, 1896; Vaughan and Tutcher, 1903). However it was the application of fossil zonal methods to the classification of the Lower Carboniferous strata that was to focus both national and European attention on the Bristol district. In 1905, A. Vaughan used the occurrence of corals and brachiopods to establish a standard sequence of zones in the Lower Carboniferous succession of the Avon Gorge.

Extrapolation of the Avonian zonal nomenclature and classification based primarily on a single section, the Avon Gorge, gave rise to problems of increasing complexity as the area of investigation extended beyond Bristol. Nevertheless A. Vaughan, T. F. Sibly, S. H. Reynolds and others developed and extended the classification to adjacent areas, and it was subsequently taken up by other workers and applied to most of the Lower Carboniferous areas of Britain, although with varying degrees of success.

Vaughan's work provided a much needed stimulus to the investigation of Lower Carboniferous palaeontology and biostratigraphy. In practice, however, it had many unsatisfactory features, not least being the deliberate replacement of long-established lithological terms by others having a 'zonal' significance. When the primary six-inch survey commenced in 1938, it soon became abundantly clear that it would be essential to revert to an independent lithostratigraphical nomenclature based on the mapping of distinctive rock types. Such a proposal was set out by G. A. Kellaway and F. B. A. Welch in 1955 and is used in this memoir. A new system of classification based on stages has since been introduced for the Dinantian rocks (George et al., 1976) and this is referred to in Chapter 3.

Structural features connected with the Variscan orogeny attracted the attention of geologists during the early stages of the investigation of the Bristol and Somerset Coalfield. Among these were the Radstock Slide studied by McMurtrie (1869) and the Carboniferous Limestone 'klippen' resting on the Coal Measures at Vobster and Luckington (Sibly, 1912; Welch, 1933). Important work on the structure of the Kingswood Anticline was carried out as early as 1865 by H. Cossham and later developed by Staples (1917) to reveal the extent of repetition by folding and overthrusting. In the period following the First World War, however, much of this earlier structural work was lost sight of, and palaeontological discoveries of considerable potential importance to stratigraphy were devalued by structural misinterpretation. The primary six-inch survey of 1938–70 was carried out against a background of advances in stratigraphical and palaeontological knowledge on an international scale and this has enabled some long-standing problems, of correlation and structural interpretation to be solved.

CHAPTER 2
Pre-Carboniferous rocks

LOWER PALAEOZOIC

Rocks belonging to the earliest Ordovician and to the Silurian crop out in the extreme north-east of the Bristol district. The main outcrops are the two small areas surrounding the northern tip of the Coalpit Heath Basin. The westerly one extends from Whitfield to Tortworth, east of which it is separated from the smaller one at Charfield Green by a strip of concealing Triassic rocks little more than half a mile wide. The exposed rocks which continue beneath this cover, form the southern limits of the Tortworth Inlier of Lower Palaeozoic rocks, which extends northwards to Tites Point on the River Severn and is bounded on the west by the large Whitfield and Berkeley faults.

The Lower Palaeozoic rocks form areas of low relief with subdued discontinuous ridges formed by limestones and lavas within the Silurian. Exposures are uncommon, except in these harder rocks which were once quarried for agricultural lime, building- and road-stones.

Early investigation of the rocks (Buckland and Conybeare, 1824; Weaver, 1824) classified them as Transition Series or Limestone, but Murchison (1839) and those after, such as Phillips (1848), considered all the rocks of the inlier to be Silurian. It was not until 1933 when S. Smith obtained fossils from an abortive coal shaft at Breadstone House, north of the district, that it was realised that an outcrop of rocks of Tremadoc age occupied a large part of the inlier (Smith and Stubblefield, 1933). More detailed accounts, particularly of the Silurian rocks, were made by Woodward (1876), Lloyd Morgan and Reynolds (1901), Reed and Reynolds (1908a and b) and Curtis (1972). The area now under consideration is described more fully by Cave (1977) in the Geological Survey's memoir on the Malmesbury (251) sheet.

Cambrian

TREMADOC

The main outcrop of Tremadoc rocks occurs beyond the northern limit of the district, between Damery and Purton. It is bounded on the south by unconformable Llandovery rocks, on the west by the large Whitfield and Berkeley faults, and on the east and north by overlapping Triassic marls, and it is from here that virtually all our knowledge of the sequence is derived. The Tremadoc has traditionally been regarded in Britain as the youngest part of the Cambrian, but international opinion is now moving decisively towards placing the base of the Ordovician at or near the lower limit of the Tremadoc. Pending a final decision they are retained in the Cambrian.

The Tremadoc rocks of the Tortworth Inlier consist of thinly bedded, slightly micaceous blue-grey mudstones which weather to a khaki colour. Thin beds of micaceous siltstone and very fine-grained sandstone are common in the upper part of the sequence, and a few thin calcareous bands with cone-in-cone structure occur. The greater development of arenaceous beds in the upper part is used as the basis of an imprecise division of the Tremadoc into Breadstone Shales below and Micklewood Beds above (Curtis, 1955, 1968). The thickness of the Breadstone Shales, the base of which is not seen, has been estimated at up to 5000 ft and of the Micklewood Beds, the top of which is not seen, as up to 2500 ft (Cave, 1977, p.11).

The fauna is very sparse. That of the Breadstone Shales indicates the zone of *Dictyonema flabelliforme* (Smith and Stubblefield, 1933), and a record of *Clonograptus sp.* from Mobley near Berkeley (Curtis, 1955) suggests that the zone of *C. tenellus* might also be present. Fragments of *Lingulella sp.* are common in some layers within the Micklewood Beds, which contain also *Schmidites sp., Angelina sedgwickii?* and *Peltocare olenoides?*. The age of the Micklewood Beds is in part at least Upper Tremadoc.

The only outcrop of the Tremadoc within the district is detached from the main outcrop and occurs southward of Charfield Mills [723 929], around Charfield Green [727 924]. Ditch exposures and field brash reveal clay and sandstone debris, the latter containing horny brachiopod fragments. Another small inlier occurs in the Little Avon River for a few hundred yards north and south of Bursall Bridge [727 898] where red shale and thin bands of pink grit are exposed.

Silurian

The marine Silurian rocks that crop out within the district are confined within the Llandovery and the Wenlock. These rocks are cut off to the west by the bounding Whitfield Fault, which is probably of Middle Devonian age. Ludlow rocks are absent east of this fault, following very strong pre-Upper Old Red Sandstone erosion and the latter rests unconformably upon Wenlock strata. Ludlow strata are, however, almost certainly present at depth, in conformity with the Lower Old Red Sandstone, to the west of the Whitfield Fault and brief mention is made of these subsequently. The younger Silurian rocks may be represented within the Thornbury Beds; but this formation is of Old Red Sandstone facies, and is considered under that heading, below.

Within the district, the main outcrop of the Silurian rocks is separated into two parts. The western part, consisting largely of Wenlock strata, occurs north of Horseshoe Farm [6734 9025] and around Brinkmarsh [677 910]. The eastern part, which consists mainly of Llandovery rocks, is situated around Charfield. Three small, more southerly, outcrops of Wenlock rocks occur in the course of the Little Avon River east of Wickwar.

The classification of the Silurian rocks of the Tortworth Inlier has evolved from the work of Lloyd Morgan and Reynolds (1901) and Reed and Reynolds (1908a and b) and this has been further developed by Curtis (1972).

The present classification is set out as follows:

	Formations		Thicknesses (ft)
Ludlow	Undivided	(mudstones and limestones)	(?700 concealed)
Wenlock	Brinkmarsh Beds	mudstones and sandstones	seen 100
		upper limestone	0 to 35
		mudstones	c. 400
		middle limestone	0 to 50
		mudstones and sandstones	c. 320
		lower limestones	40 to 90
	Tortworth Beds	(mudstones and sandstones)	350 to 1000
Llandovery (upper)	Upper Trap	(basalt)	0 to 210
	Damery Beds	(mudstones and sandstones)	up to 700
	Lower Trap	(basalt)	80 to 110

The Llandovery–Wenlock succession comprises shallow-water arenaceous and argillaceous marine sediments with some limestones, particularly in the Wenlock, and two developments of igneous rocks in the Llandovery. Shelly faunas confirm the shallow-water environment.

LLANDOVERY

In terms of O. T. Jones's (1925) classification of the type shelly Llandovery deposits at Llandovery in Dyfed, the rocks at Tortworth belong to substages ?C4-C5-C6 (for a recent review see Cocks et al., 1984). They represent only the top half of the late Llandovery sequence, all, or practically all, falling within the Telychian Stage, and thus the stratigraphical similarities of the Tortworth outcrops lie with those occurring to the east of the Malvern Hills rather than with those to the west.

The **Lower Trap** was described by Reynolds (1924, p.107) as an olivine-basalt; he was also of the opinion that the rock was intrusive. His evidence for this lay in the penetration by the basalt of fine cracks in xenoliths of sandstone, which led to the belief that high pressure was involved. Later views, such as that of Curtis (1955) favour the rock being extrusive and this is also the view resulting from the survey for the Malmesbury (251) sheet. The petrography of both this basalt and the Upper Trap (see below) is described in detail by R. W. Sanderson in the memoir of the Malmesbury sheet (Cave, 1977, pp.27–33).

The basalt thins consistently from south-east to north-west from about 110 ft near Charfield Green to 80 ft at Damery. It is amygdaloidal in places and clearly not transgressive. Although it was quarried in the past for roadstone, exposures are now few and contacts with sedimentary rocks above and below are not visible. Red micaceous sandstone was reported to overlie the igneous rock just south-east of Charfield Green (Reed and Reynolds, 1908b, p.514) and lenticles of red clay have been seen within the Trap indicating that lateritic weathering has taken place.

The **Damery Beds** consist of fine-grained sandstones, siltstones and mudstones. Limestone is present in subordinate amounts though some thin impure layers do occur and the sediments tend to be calcareous below the zone of weathering. The colour of the rocks is commonly dull red or green.

During the survey at a scale of six inches to one mile for the Malmesbury (251) sheet four subdivisions were mapped in the area north and north-west of Charfield Green, each consisting largely of either sandstone or mudstone, and these subdivisions assisted greatly in the understanding of the structure and stratigraphy. Unfortunately the same subdivisions could not be established in the outcrop south-south-west of Falfield where the ground is poorly exposed, and it may be that the lithological distinctions are less clearly defined in this area.

The lowest of the subdivisions consists predominantly of sandstone and is some 120–180 ft thick. In the middle of the outcrop of the Damery Beds there is a marked slack representing an argillaceous subdivision, about 120 ft thick near Charfield Green but becoming slightly thinner to the north-west. Above is another sandstone-with-mudstone group which also seems to vary slightly in thickness along the outcrop. Near Charfield Green it is estimated to be 290 ft thick, near Michael Wood [701 948] about 200 ft, while farther north-west the thickness appears to be greater again. Overlying this subdivision and beneath the Upper Trap is a mudstone, about 70 ft thick. This uppermost subdivision is not everywhere recognisable even along the outcrop from Charfield Green to Woodford.

The Damery Beds have yielded *Eocoelia curtisi*. Other members of the fauna include: *Atrypa reticularis, Brachyprion arenaceus, Costistricklandia lirata, Leptostrophia compressa, Mendacella* cf. *phiala, Stegerhynchus? weaveri, Gyronema octavium multicarinatum, Actinoceras nummularium, Tentaculites anglicus, Dalmanites weaveri, Encrinurus onniensis* and *Craspedobolbina (Mitrobeyrichia) clavata*. Corals are usually represented and colonies of *Favosites spp.* are locally common.

The **Upper Trap** is a basaltic lava (Sanderson in Cave, 1977, pp.27–33). It was described petrographically by Reynolds (1924, pp.108–110) and is distinguishable from the Lower Trap in containing small xenocrysts of quartz and feldspar. There seems to be no doubt about its effusive nature. It is amygdaloidal and it has been recorded that some exposures reveal irregular, thin deposits of limestone resting on the trap, and that this limestone is 'ashy', containing well-marked lapillae of the lava (Lloyd Morgan and Reynolds, 1901, p.271). Whether this deposit is a genuine tuff or merely the decomposition product of an irregular lava surface is immaterial; it indicates that the igneous rock was extrusive. The margin of the marine Silurian basin may therefore have lain at no great distance to the east.

The lava of the Upper Trap is about 210 ft thick near Charfield Green. This is its greatest known thickness; to the south and north-west it thins along crop, being only 100 ft thick at Avening Green [709 938].

The **Tortworth Beds** are fine-grained sandstones and mudstones. The sandstones are confined largely to the lower half of the sequence, whereas the upper half is predominantly argillaceous, sandstone beds being developed only subordinately. The mapping of the groups was largely based upon the topographical features produced by the rocks. The mudstones occupy tracts of lower ground whilst the sandstone forms the slightly higher ridges bounding these tracts. In the basal Tortworth Beds is the *Palaeocyclus* Band (Curtis, 1955) characterised by the coral *P. porpita*. This fossil is present above the Upper Trap in the outcrop north-west of Charfield Green, but it also occurs in the outcrop between Brinkmarsh and Falfield where, it seems likely, only Tortworth Beds are exposed. *Eocoelia sulcata* and *Costistricklandia lirata typica* have also been recorded.

Curtis (1955), introducing the name Tortworth Beds, recorded their thickness as about 200 ft but this may be slightly underestimated. The outcrop around Charfield is the only one which exhibits both top and bottom of the formation and thicknesses calculated along here are about 350 ft near St John's Church, Charfield Hill and 780 ft at Avening Green, becoming even greater beyond the district to the north-west.

WENLOCK

The **Brinkmarsh Beds** crop out mainly in the tract northeast of Buckover and Horseshoe Farm [6734 9025] to Daniel's Wood [696 941] and a strip also occurs on the west side of the Charfield Green Inlier. The strata consist predominantly of mudstones with thin layers of siltstone and fine-grained sandstone and some limestone. The sandstones are calcareous as are the mudstones which contain impersistent layers of hard calcareous nodules. Some beds are highly fossiliferous. Sedimentary structures which have been observed, particularly in the fine-grained sandstones, include ripple marking, current bedding and drag marks. Curved bedding and small rounded masses also occur within the sandstones suggestive of contemporaneous disturbance of the sediments. The total thickness of the exposed Wenlock succession appears to approach 1000 ft.

Within the Brinkmarsh Beds there are three prominent limestones which, however, are discontinuous and probably lenticular. They are associated with beds of sandstone and sandy limestone. This is especially true of the bottom limestone at Charfield Green which passes northwards into sandstone. The limestones are usually crinoidal and very fossiliferous, but impure and argillaceous. They form comparatively prominent ridges in the topography and yield a limestone brash to the soil. These factors and the small-scale quarrying that has taken place in the limestones have enabled them to be mapped in an otherwise poorly exposed area. The beds of limestone are thin, the one most commonly quarried being the lowest one which is about 100 ft thick. The thinnest and most difficult to trace is the top limestone, which proved to be 12 ft thick in the Buckover road cutting [667 907]. Overlying the bottom limestone in the Brinkmarsh area is a bed of mudstone characterised by an abundance of the coral *Pycnactis mitratus*.

The junction of the Wenlock with the Llandovery is ill-defined and has, for convenience, been taken at the base of the Brinkmarsh Beds. This junction appears to be conformable and transitional and has been identified near Little Whitfield Farm [6724 9132], where the bottom bed of limestone of the Brinkmarsh Beds contains a fauna, which is regarded as Wenlock. A subjacent sandstone has yielded Llandovery fossils. One anomalous occurrence here, is that of *Palaeocyclus porpita* found just beneath the bottom limestone, in the lane side [6751 9138] at Whitfield. This fossil is usually considered to be associated exclusively with the basal Tortworth Beds. In the Charfield Green area, beneath the bottom limestone, is a thin mudstone, from which it is presumed that Reed and Reynolds (1908b, pp.533,541) obtained a Wenlock-type fauna. This occurrence, which contrasts with the conditions observed at Whitfield, indicates that the faunal facies is controlled to some extent by the onset of conditions favouring calcareous deposition.

The highest exposed Brinkmarsh Beds occur in the south around Horseshoe Farm, where they are succeeded unconformably by the Upper Old Red Sandstone (Quartz Conglomerate). The stratigraphical break is clearly significant, and is greater in some places than others, though mapping of this particular area did not reveal much angular discordance. The magnitude of the easterly overstep of the Old Red Sandstone and Silurian rocks by the transgressive Upper Old Red Sandstone appears to be controlled by the presence of north-north-easterly trending faults, cutting the older rocks but not the unconformable cover of Upper Old Red Sandstone. In consequence high Wenlock rocks appear below the Quartz Conglomerate between Buckover, Horseshoe Farm and Little Daniel's Wood (Curtis and Cave, 1964, p.439), but east of the fault near Little Daniel's Wood, as far as Tortworth, the upper part of the Wenlock succession has suffered greater erosion, so that only the lowest part remains beneath the Upper Old Red Sandstone. This condition is thought to prevail also in the Charfield Green area though Triassic rocks obscure the junction.

West of the Whitfield Fault Wenlock rocks are not exposed, but the succession in the rocks overlain by the Upper Old Red Sandstone is more complete and a borehole would be expected to prove the whole of the Wenlock succession.

The three small outcrops of Wenlock rocks in the Little Avon River [730 881], [729 884] and [726 890] east of Wickwar were described by Whittard and Smith (1944, p.65). Mudstones, siltstones and limestones are exposed in juxtaposition with Downtonian rocks. The contact is probably faulted.

The bottom limestone of the Brinkmarsh Beds has yielded many fossils, particularly from the old quarries in the Brinkmarsh area. They include *Serpulites (Campylites?) perversus, Atrypa reticularis, Brachyprion waltonii, Stegerhynchus borealis, S. diodonta, Microsphaeridiorhynchus nucula, Striispirifer plicatellus, Leptaena depressa, Resserella elegantula, Rhynchotreta cuneata, Sphaerirhynchia davidsoni, Cypricardinia subplanulata, Acaste downingiae* and *Dalmanites caudatus. Favosites sp.*, crinoid columnals and bryozoa are also common.

The mudstone overlying this limestone has yielded crinoid columnals, *Brachyprion waltonii, Microsphaeridiorynchus nucula, Craniops implicatus, Cypricardinia subplanulata, Glassia sp., Howellella sp., Phaulactis* cf. *angusta, Pycnactis mitrata, Whitfieldella sp., Loxonema sp., Tentaculites sp.* and *Proetus sp.*

The middle limestone has yielded crinoid columnals, *Atrypa?, 'Camarotoechia' sp., Clorinda?, Eospirifer plicatellus, Leptaena depressa* and *Plectodonta sp.*

The mudstone overlying this contains crinoid columnals, *Clorinda?, Cyrtia exporrecta, Glassia obovata,* cf. *Leangella segmentum,* and *Whitfieldella sp.*

The fauna from the top limestone and higher beds is recorded in Curtis and Cave (1964). To that list can be added *Salopina conservatrix* and a small chonetoid, *Strophochonetes?.* A roadside exposure [6736 9021] at Horseshoe Farm has yielded crinoid columnals, bryozoa, *Amphistrophia funiculata, Fardenia?, Howellella sp.* [*Delthyris elevata*], *Meristina obtusa, Pentamerus?, Skenidioides lewisii, Loxonema?, Pterinea sp.*,

Tolmaia?, *Tentaculites ornatus*, *Dalmanites* cf. *aculeatus* and *Beyrichia sp.*

Ludlow

Rocks of Ludlow age crop out in a narrow north-to-south strip adjacent to the west side of the Berkeley Fault between Tites Point, Purton and Wanswell Green. This area lies north of the Bristol district, but rocks of the same sequence probably underlie the thick Thornbury Beds adjacent to the west side of the Whitfield Fault near Buckover.

The Ludlow rocks of Purton are in a normal Welsh Borderland shelf facies and probably comparable to those seen at May Hill (Lawson, 1955), Newnham and Tites Point (Cave and White, 1971). South of May Hill the Ludlow sequence becomes progressively thicker so that in the area of Milbury Heath and Thornbury it might be expected to be over 700 ft thick and consist predominantly of mudstones with subordinate limestones in thin and concretionary layers, the top 100 ft or so being rather silty or sandy, and equivalent to the Whitcliffe Formation.

The absence of exposed Ludlow rocks at Tortworth and Charfield makes it difficult to draw conclusions about the date of the earth movements which took place prior to the deposition of the continental Old Red Sandstone; it is reasonable to suppose that they are present in some form beneath the Lower Old Red Sandstone at Thornbury (Figure 5). On the other hand Ludlow rocks have not been found in the eastern Mendips where the marine Silurian sediments are thin.

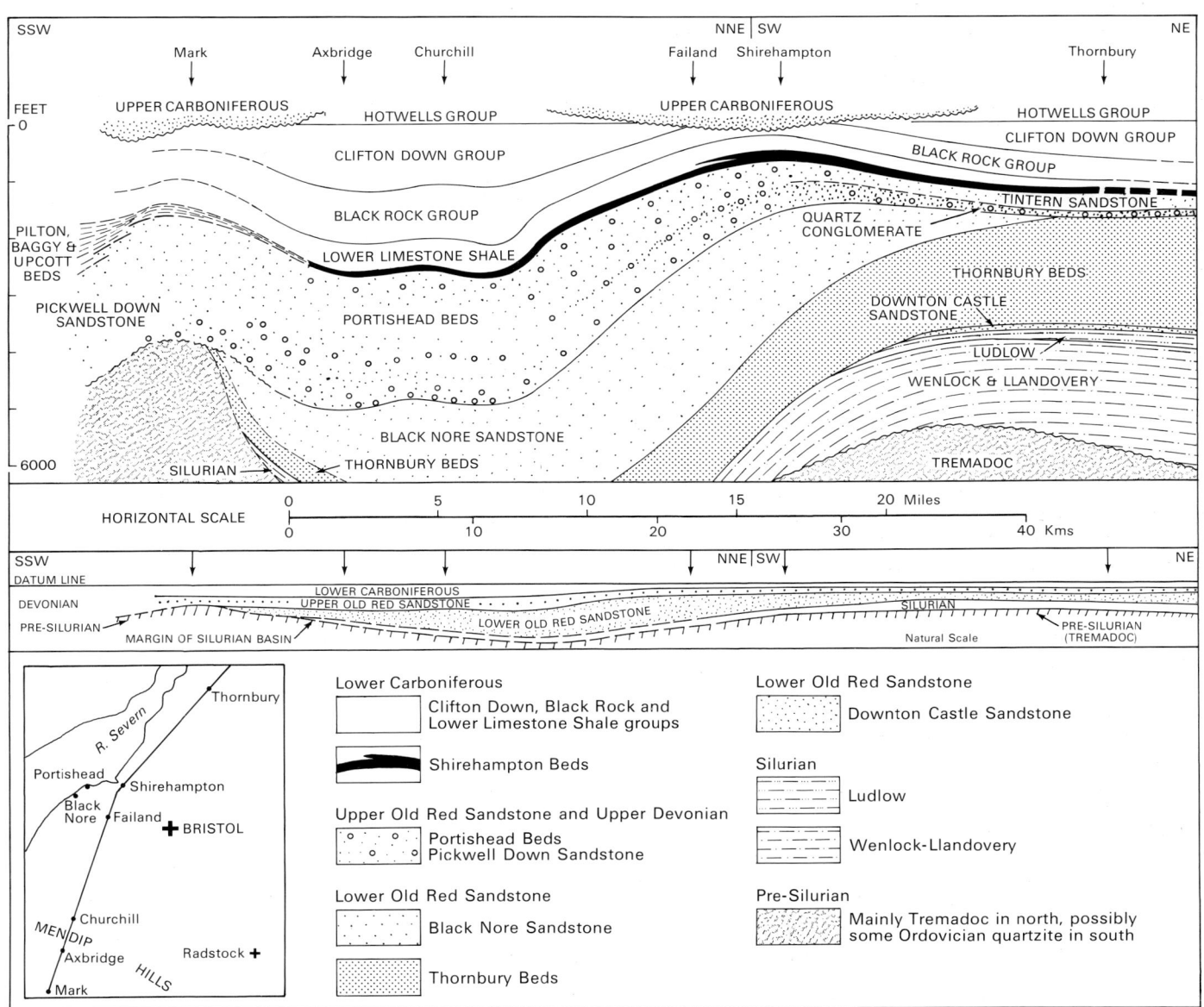

Figure 5 Section showing the probable relationships of the Palaeozoic rocks between Thornbury and Mark. No allowance has been made for shortening due to Variscan folding

OLD RED SANDSTONE (SILURIAN/DEVONIAN)

The Old Red Sandstone of the region between Gwent and the Mendips consists of a great thickness of sandstones, conglomerates, marls and mudstones, generally agreed to have been deposited under arid continental conditions. Neither the precise position of the Silurian–Devonian boundary (which probably lies somewhere within the Thornbury Beds), nor the stages defined in marine Devonian rocks can be recognised. The classification employed below is largely lithostratigraphical.

Old Red Sandstone rocks crop out in the following widely separated areas:

i   West of the Severn, on the south-east flank of the Usk Anticline;
ii The Thornbury area, extending eastwards around the northern rim of the Coalpit Heath Basin from Milbury Heath through Tortworth and Charfield to Wickwar;
iii West of Bristol, in the core of the Westbury-on-Trym Anticline, from Westbury-on-Trym to Clapton in Gordano;
iv The Clevedon–Portishead ridge;
v   The Mendips; in the cores of the Black Down, North Hill and Pen Hill periclines.

West of the Severn, the oldest Lower Old Red Sandstone exposed in the district under review belongs to the upper part of the St Maughans Group, which consists mainly of rhythmically interbedded red mudstones and sandstones. Intraformational conglomerates (conglomeratic cornstones), commonly fish-bearing, are present at many rhythm bases.

The succeeding Brownstones (2000 ft) are predominantly purple-grey, fluviatile, micaceous sandstones, with subordinate red and green mudstones. Intraformational conglomerates are present towards the base of the formation, and exotic pebbles occur near the top.

The overlying Quartz Conglomerate at the base of the Upper Old Red Sandstone rests unconformably on Lower Old Red Sandstone strata. Ranging from 10 to 50 ft thick, it comprises alternations of hard conglomerate and softer pebbly sandstone. The succeeding Tintern Sandstone Group consists of some 350 ft of soft, buff and greenish grey, micaceous sandstones with subordinate mudstones and a few pebbly lenses. Most of the sediments are fluviatile, but at the top these interdigitate with marine carbonates giving a transition into the overlying Carboniferous Limestone (Figure 6).

The area west of the Severn forms only a small part of the large outcrop of Old Red Sandstone rocks surrounding the Forest of Dean syncline. These rocks are described in the Monmouth and Chepstow Memoir (Welch and Trotter, 1961).

To the north of Bristol, the sequence present is similar to that seen on the west side of the Severn, but in the areas to the south detailed correlation is less certain and the classification and nomenclature based on the coastal outcrops between Clevedon and Portishead (Kellaway and Welch, 1955) is as follows:

West Bristol and Clevedon – Portishead ridge	Thornbury – Tortworth area
UPPER OLD RED SANDSTONE	
Portishead Beds — red and purple sandstones and subordinate, interbedded red and green mudstones, siltstones and conglomerates, 900 ft	Tintern Sandstone — brown, grey and green sandstones with subordinate red and green silty mudstones and cornstones, 400 ft
	Quartz Conglomerate — mainly conglomerates and pebbly sandstones with interbedded sandstones and silty mudstones, 50 to 100 ft
— Unconformity —	
LOWER OLD RED SANDSTONE	
Black Nore Sandstone — red and brown sandstones with green mottling, 2000 ft	Thornbury Beds — red and brown and purplish silty marls with impersistent beds of red and purple sandstones, 2000 ft

Probably because of its unfossiliferous nature and the somewhat monotonous character of its lithology, the Old Red Sandstone has received far less attention than that devoted to most of the other formations of the district. Lloyd Morgan (1886) recognised a lower and an upper division in the Portbury–Failand area, corresponding more or less with the present Black Nore Sandstone and Portishead Beds. A major contribution was made by Wallis (1928) who investigated both the stratigraphy and petrology of the Old Red Sandstone rocks of the Bristol district. One interesting fact to emerge from this work was that much of the pebbly material in the conglomerates has been derived from rocks similar to those of the Mona Complex of Anglesey. Since then further accounts have been given of the northern area in the memoir on the Malmesbury (251) sheet (Cave, 1977), on the Bristol district by Kellaway and Welch (1955) and on the Mendips in the memoir on the Wells and Cheddar (280) sheet (Green and Welch, 1965).

Fossils are rare, but two horizons with fish remains have been recognised in the Portishead Beds of the Bristol district. The fauna of the lower or Woodhill Bay Fish Bed was described by Wallis (1928) and Scott Simpson (1951), and the upper Sneyd Park Fish Bed by Martyn (1875). A third lenticular band with fish remains, the Buckover Fish Bed (Cave, 1977) was found near Milbury Heath, somewhat below the position of the Sneyd Park Fish Bed (Figure 6).

The deposits of the Lower and Upper Old Red Sandstone are the products of two distinct periods of sedimentation, separated by a major unconformity. Late Silurian times witnessed the regression of the Silurian seas and the onset of deposition of great thicknesses of the red and brown marls and sandstones of the Lower Old Red Sandstone over a subsiding continental basin stretching northwards through Gwent and the Welsh Borders. In what is generally considered to be Middle Devonian times, this period of deposition came to an end with uplift and erosion, particularly marked within the present district along a faulted ridge or horst passing northwards through Tortworth towards the Malverns. In turn, this period of emergence came to an end with renewed subsidence and the creation of a new continental basin of deposition covering a rather more extensive area

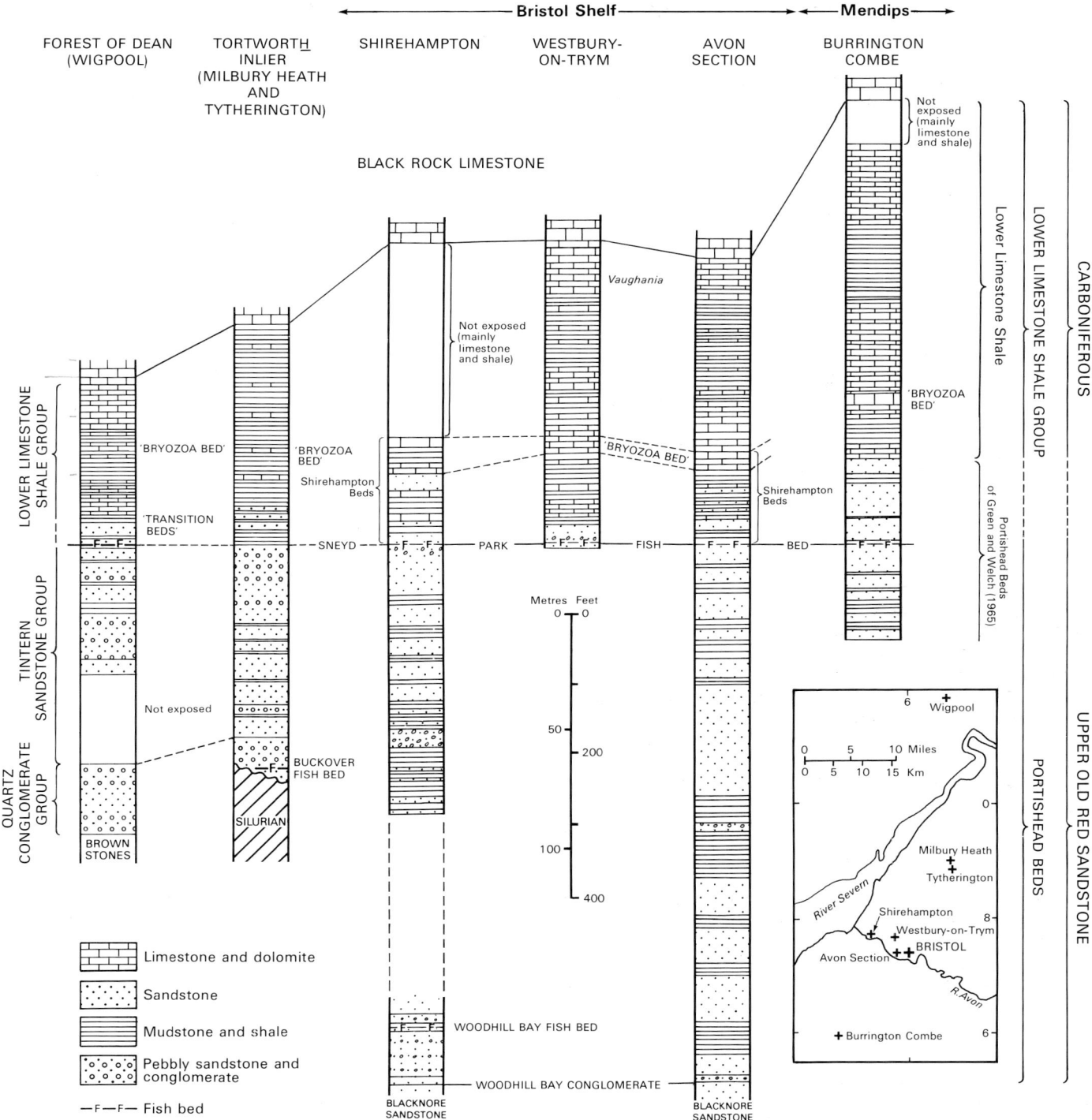

Figure 6 Comparative vertical sections of the Upper Old Red Sandstone and the Lower Limestone Shale Group of Burrington Combe, Bristol, Tortworth and the Forest of Dean

than that of the Lower Old Red Sandstone. This heralded a new major cycle of sedimentation in Upper Devonian times leading to deposition of Upper Old Red Sandstone conglomerates, sandstones and subordinate mudstones, and in the late Devonian to the transgression of the open seas from the south, culminating in the deposition of the Carboniferous Limestone. The probable relationship of the continental Old Red Sandstone to the marine Silurian and Devonian basins and to the transgressive Upper Old Red Sandstone is illustrated in Figure 5. Conditions east of the Malvern Fault Zone, however, differed in some respects from those on the western side.

Lower Old Red Sandstone

In the Thornbury–Tortworth–Wickwar area the Lower Old Red Sandstone consists of a sequence, over 2000 ft thick in some areas, of finely micaceous red, reddish brown and purplish silty marls, commonly showing pale green spheres and streaks. Impersistent layers of purple and red flaggy, very micaceous sandstones occur throughout, though they tend to be thicker and more numerous in the higher parts of the succession (Cave, 1977, p.42). Irregular dolomitic concretions are common in the marls, and in places they become so numerous as to form bands of impure limestone. The only fossils found within this area came from a small exposure east of Wickwar, where Whittard and Smith (1944, p.69) found fragments of the fish *Phialaspis*, suggesting a correlation with the Downtonian Stage of Wickham King (1934), the lowest division of the Old Red Sandstone of the Welsh Borderland. Correlation with the much fuller sequence found west of Severn is uncertain and for this reason the rocks have been named **Thornbury Beds** (Kellaway and Welch, 1955); they are thought to equate broadly with the Raglan Marl Group (Welch and Trotter, 1961, p.32) which is also regarded as being of Downtonian age[1].

Around Thornbury the exposed beds certainly exceed 1000 ft in thickness; at Tortworth they are absent, which suggests that the Malvern ridge at Tortworth was actually growing or in existence in late Silurian or early Devonian times. West of the ridge lay a depression, in which very thick mudstones accumulated and which terminates against the ridge, but the precise nature of the contact has not been ascertained. With the possible exception of one locality at Easton in Gordano [5139 7528) the base of the Thornbury Beds has nowhere been seen. If, however, the Malvernian ridge continues southwards beneath a cover of Upper Old Red Sandstone and Carboniferous rocks, the Thornbury Beds may be expected to occur on its western flank, possibly as far south as the Mendips (Figure 5).

The Thornbury Beds are overlain unconformably by the Upper Old Red Sandstone. The marls and sandstones of the St Maughans Group and the sandstones of the Brownstones, both well developed in the ground west of the Severn, are missing. The 'Thornbury Beds' shown on the Bristol special sheet in the Milbury Heath area resting directly upon Wenlock strata were proved in the A38 road cutting (Curtis and Cave, 1964) to belong to the Upper Old Red Sandstone.

West of Bristol there is no incontrovertible evidence of the presence of the Thornbury Beds, though red mudstones overlain by Dolomite Conglomerate, seen in a temporary exposure at Easton in Gordano [5139 7528] may represent the top of this formation. Bands of red mudstone also occur in the lower part of the Black Nore Sandstone, however, and the section may have exposed this (higher) part of the succession. The **Black Nore Sandstone** consists of at least 2000 ft of dingy, purplish red, current-bedded sandstones with much green mottling. When fresh, the rock is slightly calcareous, and bands of conglomeratic cornstone are sporadically developed in the lower part of the formation. Thin wisps and strings of dark cherty pebbles are locally present. The largest outcrop of the Black Nore Sandstone is in the Portbury–Failand area, but the locality in which the beds can best be examined is in the coastal section between Redcliffe Bay and Kilkenny Bay, near Portishead. This includes the Black Nore promontory, after which the formation is named. The Black Nore Sandstone was also proved in the Portway Foul Water Tunnel (see Figure 11). It is believed to correlate with the Brownstone Group of the Forest of Dean and is succeeded unconformably by conglomerates of the Upper Old Red Sandstone.

Rocks of Lower Old Red Sandstone age are not known to crop out in the Mendips. Indeed the only place where the base of the Upper Old Red Sandstone is seen, is about $1^{1/4}$ miles south-west of Oakhill (in the Beacon Hill Pericline just south of the Bristol district) where the Portishead Beds rest unconformably upon Silurian volcanic rocks (Green and Welch, 1965, p.10).

Upper Old Red Sandstone

Following the regional uplift and erosion the of Lower Old Red Sandstone in mid-Devonian times, deposition of the Upper Old Red Sandstone took place under less arid continental conditions when fluviatile and lacustrine deposits with crossopterygian fish remains were laid down. The Upper Old Red Sandstone is coarse and pebbly, and is strongly transgressive, particularly when traced eastwards from Gwent towards Tortworth. At Bristol the Upper Old Red Sandstone rests upon Brownstones or their correlatives, around Thornbury upon the Thornbury Beds, and at Tortworth on Silurian strata. Near Wickwar, the Thornbury Beds are present beneath Upper Old Red Sandstone and though the contact of the Lower and Upper Old Red Sandstone is concealed, it is known that Downtonian rocks persist at least as far south as Lansdown.

In general, the Upper Old Red Sandstone consists of reddish brown, feldspathic sandstones interbedded with bands of red, purple and green sandy shale and mudstone; quartz-conglomerates are widely developed, particularly in the lower part of the succession. Overall thicknesses range from about 450 ft in the Thornbury–Cromall area to 900 ft west of Bristol and at least 1600 ft in the Mendips (Figures 5 and 9).

In the northern outcrops, south and east of Thornbury, the succession closely resembles that in the Chepstow and Forest of Dean district and is divisible into the same two

---

1 The presence of rocks of Downtonian age at Hamswell, south-east of Freezing Hill has been described by Cave in Summary of Progress of the Geological Survey, 1963, p.35.

groups—the Quartz Conglomerate below and the Tintern Sandstone Group above (Figures 5 and 6).

The **Quartz Conglomerate** varies from about 50 to 100 ft in thickness. The best section was seen in the A38 road cutting at Buckover, east of Thornbury (Curtis and Cave, 1964) where 32½ ft of hard massive grey, green and purplish brown sandstone with beds of quartz-conglomerate were seen to overlie 15 ft of purplish brown and green mudstones, siltstones and fine sandstones with only a little conglomerate. The pebbles comprise well-rounded quartz, jasper and green mudstone. A thin sandstone about 4 ft above the base of the group has yielded *Bothriolepis sp.*, confirming the Upper Old Red Sandstone age of the beds.

The **Tintern Sandstone Group** consists of about 400 ft of purplish brown, grey and green flaggy and cross-bedded sandstones with subordinate layers of red and green silty mudstone. Scattered pebbles of quartz occur within the sandstones and there are also sporadic layers of yellowish or greenish cornstone, generally in the form of nodules or as continuous bands less than a foot thick. More than 300 ft of Tintern Sandstone were exposed in the road cutting at Buckover.

In the area between Westbury-on-Trym, Portishead and Clevedon the Quartz Conglomerate and Tintern Sandstone Groups cannot be separately identified and the Upper Old Red Sandstone sequence has been grouped as a single formation—the **Portishead Beds**. Here the Quartz Conglomerate has split up into a number of isolated lenticular masses of pebbly beds which occur at intervals throughout the succession. These lenticles form prominent features which, in the Failand area, are traceable over considerable distances. Apart from these conglomerates, the Portishead Beds are variable in lithology, ranging from red and green mudstones and marls to red siltstones and red, yellow and pale grey hard, fine-grained quartzose sandstones. The sandstones in particular are strongly lenticular and few individual beds can be traced very far; they nevertheless dominate the sequence (Figure 6). In thickness the sequence varies from about 800 to 900 ft. The lowest 130 ft are well exposed at Woodhill Bay, northwest of Portishead (Wallis, 1928, pp.768–769) but a much more complete, though less accessible, section is seen on the left bank of the Avon at Abbot's Leigh.

At the base of the sequence occurs the **Woodhill Bay Conglomerate** which consists of an unsorted mass of pebbles and cobbles, cemented by irregular masses of cornstone, the bed being about 10 ft in thickness (Plate 2). Whereas quartz pebbles characterise most of the Upper Old Red Sandstone conglomerates, the pebbles of the Woodhill Bay Conglomerate also include dark grey-green cherts and quartzites which are unlike any of the Lower Paleozoic rocks known to occur in the Bristol district. Some 100 ft above the base of the Portishead Beds at Woodhill Bay occurs the **Woodhill Bay Fish Bed**, a 33 ft band of siltstone, containing the remains of *Asterolepis*, *Bothriolepis*, *Coccosteus*, *Holoptychius*, *Glyptopomus* and eurypteroids. The bed is apparently lenticular and its value as a marker band is restricted.

However, the fish fauna, which includes *Coccosteus* (a form unknown in any of the higher fish beds) links it with the Bittadon 'Felsite', in fact a tuff, at the base of the Pickwell Down Beds and suggests a Middle Famennian age. The continental Portishead Beds may therefore pass laterally into the fluviatile and continental Pickwell Down Sandstone of north Devon and west Somerset (Figure 5).

In the Bristol area the Portishead Beds pass imperceptibly upwards into the wholly marine sequences of the Carboniferous Limestone (Figure 9). These passage beds—the Shirehampton Beds (p.26)—consist of interdigitations of continental sandstones and red mudstones with grey marine shales and limestones bearing a mixed Devonian-Carboniferous fauna, which make it difficult to identify any obvious boundary between the two systems. At the base of the Shirehampton Beds there occurs another locally well-developed conglomerate or pebbly sandstone, containing fish remains, including scales of *Holoptychius*. This is the **Sneyd Park Fish Bed** (Kellaway and Welch, 1955, p.6), which has been described by Martyn (1875) and Wallis (1928). It is typically developed between the Avon Gorge and Westbury-on-Trym and again in the Portway cutting at Shirehampton. The Sneyd Park Fish Bed has now been taken to mark the top of the Upper Old Red Sandstone in the Bristol area and, for the sake of convenience in mapping, this places the whole of the passage Shirehampton Beds within the Lower Carboniferous (Figure 6).

In the Mendips, only Upper Old Red Sandstone rocks are known. They crop out in the cores of the Black Down, North Hill, Pen Hill and Beacon Hill periclines. The base of the Portishead Beds rests unconformably on Silurian volcanic rocks in the Beacon Hill Pericline. Elsewhere in the Mendips the contact with the underlying formations is concealed. The beds are thickest in the Black Down Pericline, where more than 1600 ft are present (see Figure 12a). In the Beacon Hill Pericline, the succession between the top of the Silurian and the base of the Lower Limestone Shale has a thickness of about 1350 ft (Green and Welch, 1965, p.10). The beds have been mapped as Portishead Beds since they are of similar facies. They consist for the most part of dull red feldspathic and quartzitic sandstones, interbedded with fine-grained micaceous sandstones and sandy green, red and purplish shales and mudstones. In the lower part of the sequence pebbly sandstones and sandy quartz-conglomerates occur, somewhat reminiscent of the Woodhill Bay Conglomerate. Fossils are extremely rare, but a plant- and fish-bearing conglomerate has been found in the upper beds at Burrington on the north side of Black Down (Hepworth and Stride, 1950), which may correlate with the Sneyd Park Fish Bed at Bristol (Figure 6).

Plate 2 Cornstone in Old Red Sandstone [4579 7681] at Kilkenny Bay, Portishead.
Woodhill Bay Conglomerate at the base of the Upper Old Red Sandstone rests upon Black Nore Sandstone in the Lower Old Red Sandstone. A large cylindrical pipe of cornstone (immature calcrete) in the conglomerate extends down to form irregular nodules in the Sandstone (A10737)

CHAPTER 3
Carboniferous Limestone (Dinantian)

The term Carboniferous Limestone (Conybeare and Phillips, 1822) has long been used by British geologists for Lower Carboniferous rocks widely known in north-west Europe as Dinantian. The succession in the Bristol and Somerset area is predominantly marine and characterised by carbonate rocks, though there is a major development of shales and mudstones everywhere at the base. Deltaic sediments consisting of sandstones with some interbedded limestones and mudstones become dominant at higher stratigraphical levels in the northern half of the region.

Within the Bristol district the outcrop of the Carboniferous Limestone, though extensive, is separated into several well-defined areas (Figures 1 and 9). In the far north it forms the rim of the Coalpit Heath Basin between Over and Itchington north-eastwards to Cromhall and Charfield and thence southwards through Wickwar to Chipping Sodbury. Farther south it is largely concealed beneath Mesozoic strata, but it reappears in several small inliers at Codrington, Wick and Grandmother's Rock east of the coalfield.

West of the Coalpit Heath Basin Carboniferous Limestone forms low hills between Alveston and Olveston, and at Littleton-upon-Severn. Limestone reefs are seen at low tide at the foot of Aust Cliff and in the bed of the Severn, thus forming a link with occurrences on the west bank at Beachley. The Carboniferous Limestone at Olveston passes beneath the Coal Measures of the Avonmouth Basin, and south of Over the Carboniferous rocks are concealed by Mesozoic strata, but reappear at Upper Knole, north of Bristol.

Between Southmead and Westbury-on-Trym the Carboniferous Limestone outcrop divides to form two separate ridges; one branch extends westwards to Shirehampton terminating at Penpole Point, the other continues southwards to the spectacular Avon Gorge (Plate 1) at Clifton, thence westwards along the Failand Ridge to Tickenham, with the Clapton Coal Basin to the north and the Nailsea Basin on its southern side. The Tickenham–Clevedon ridge is cut by a major dry valley at East Clevedon. West of the gap the Carboniferous Limestone is confined to the eastern slopes of the north-easterly trending Clevedon–Portishead Ridge. Two miles north of Portishead Point, a flat-topped mass of Lower Dolomite, forming Denny Island, projects above the waters of the Severn Estuary, constituting another link with the Carboniferous Limestone uplands of Gwent.

Low ground occupied by the Nailsea Coal Basin is separated from the main Publow–Radstock Basin by the dome-like Carboniferous Limestone inlier of Broadfield Down, which has a width of about 6 miles between Yatton on the west and Winford on the east. South of Broadfield Down, the Vale of Wrington is underlain mainly by Triassic rocks which end abruptly at the foot of the steep northern face of the Mendips. Viewed from Broadfield Down, the Carboniferous Limestone terrain of the Mendips between Burrington and Churchill is dominated by bare moorland, marking the crop of the Old Red Sandstone in the core of the Black Down Pericline. The Carboniferous Limestone here has extensive outcrops which encircle the Old Red Sandstone and Silurian cores of several periclines that extend *en échelon* east-south-eastwards from Black Down through North Hill and Pen Hill to Beacon Hill. The northern flank of the Beacon Hill pericline is the only part of this structure lying within the area covered by the Bristol district special sheet.

It is in the central and western Mendips, in particular, that the limestones give rise to spectacular scenery, with deep gorges and towering cliffs as at Burrington and Cheddar, and extensive cave systems with clearwater springs forming the source of local streams and rivers. The water supply for the Bristol region depends primarily on three major reservoirs at Axbridge, Blagdon and Chew Valley, all of which are supplied by groundwater draining from the Carboniferous Limestone of the Mendip Hills.

The basal Dinantian rocks (Lower Limestone Shale *s.l.*) rest conformably upon the Old Red Sandstone in the Bristol area, and there is clear evidence of a gradation from the continental sandstones of the Portishead Beds, through the transitional Shirehampton Beds, into the increasingly marine sediments which form the Lower Limestone Shale. The fauna of the Shirehampton Beds has a mixed Devonian–Carboniferous aspect, but lacks the goniatites used by international agreement to define the *Gattendorfia subinvoluta* Zone at the base of the Carboniferous. Palynological evidence suggests that this may lie within the Upper Old Red Sandstone (Dolby and Neves, 1970; Utting and Neves, 1970). For practical purposes, however, the base of the Carboniferous Limestone has been taken at the Sneyd Park Fish Bed, which marks a change in lithology—at least in some areas—and the last local appearance of Upper Old Red Sandstone fish (Figure 6). In the Mendip basin the passage from Upper Old Red Sandstone to marine Lower Limestone Shale is usually more abrupt than on the Bristol shelf, although the basal sediments there also include several thin beds of 'alpha-limestone', similar to the Bryozoa Bed of the Avon Gorge.

After the establishment of the shallow marine, marginal environment of the Lower Limestone Shale the sea deepened and extended over the entire region, as far as the southern edge of 'St George's Land' (a Dinantian landmass which is said to have stretched across the British Isles from Ireland through mid-Wales and the English midlands to the Brabant massif of Belgium). For much of Dinantian times the southern part of the Bristol–Mendip area was largely free from terrigenous sediment and clear-water limestones of great variety were laid down. Depositional conditions varied locally from current-free lagoons, in which were deposited evenly bedded, fine-grained calcite- and dolomite-mudstones, with widespread stromatolitic growths, to deeper more open water in which massive current-bedded oolitic and bioclastic limestones accumulated in high energy environments. Corals, brachiopods, foraminifera, conodonts and other marine fauna abounded in the warm seas. The fossil remains of these marine organisms have proved to be of great value in correlation and biostratigraphical classification of the rocks.

Following the establishment of almost uniform shallow-water marine sedimentation in early Dinantian (Tournaisian) times a relatively deep open-water basin developed in the western and central Mendips. This was bounded on the north by a shallow-water shelf extending from Woodspring and Clevedon in the west to the southern part of Broadfield Down (Figure 5). A major depression extending down the eastern side of the region from Cromhall and Yate to Mells and Frome is marked by a general thickening of all the Dinantian sediments. This structural feature permitted the southern Tournaisian sea to penetrate at least as far north as Shropshire. After renewed uplift in Viséan times, the depression became a huge river and estuary system, draining southwards towards the present district, particularly into the Cromhall–East Mendip Depression, between the Bristol and Wickwar shelves (Figure 7).

Following the mid-Tournaisian marine transgression, deposition of the Black Rock Limestone was continuous in the Mendips, but slower, with much recycling of comminuted crinoid debris in late Tournaisian times on the shelf. The topmost beds of dolomitic limestone, including the so-called 'laminosa dolomite' (Vaughan, 1905), consist of finely divided shell and crinoid debris, contrasting markedly with the coarse crinoidal and shelly limestones found at lower levels and in the basinal sequences. The pale grey, early Viséan crinoidal limestones associated with the overlying Gully Oolite fill shallow depressions in the surface of the underlying Black Rock Limestone and Dolomite. Stylolites lined with red haematitic mud in the oosparites of the Gully Oolite indicate deposition in well-oxidised, shallow seawater, swept by tidal currents and presenting an environment which was hostile to most marine organisms.

On the surface of the very extensive marine shelf formed by the Gully Oolite, the lagoonal deposits of the Clifton Down Mudstone were laid down. These are composed of fine-grained dolomite-mudstone, algal limestones, and calcilutites with minor argillaceous deposits. Stylolitic seams in the calcilutites and in the fine-grained dolomitic rocks of the Clifton Down Mudstone are black, argillaceous and bituminous films commonly carrying tiny crystals of pyrite or galena. The rocks were laid down in stagnant water under reducing conditions, which contrasted markedly with those under which the Gully Oolite was formed, but corals, brachiopods and conodonts are again absent or poorly represented.

Although the stratigraphical break between the Gully Oolite and Clifton Down Mudstone gives rise to very striking sections (Plate 3), it is restricted to shelf areas. Between the top of the succeeding Clifton Down Limestone and the base of the Hotwells Limestone lies a much larger stratigraphical break, marked by bored surfaces, brecciated rocks, sandstone intercalations and even the development of thin coals. This hiatus is regional, not local, in extent.

The Carboniferous Limestone has a total thickness ranging from about 2500 ft in the northern part of the Coalpit Heath Basin to about 3700 ft in the eastern Mendips. The well-known sections of the Avon Gorge and Burrington Combe in the Mendips both expose about 2700 ft, (Figure 8); however at Burrington, especially, the Hotwells Group at the top of the local sequence has been subjected to pre-Namurian erosion, so the full succession is not present; the full sequence at Burrington was probably about 3360 ft thick before the erosion occurred. Pre-Namurian erosion removed some of the Dinantian sediments, mainly the Hotwells Group, not only in the West Mendip Basin, but also across parts of the Bristol–Wickwar and Chepstow shelves. Even though the full succession is not preserved, there is sufficient evidence to show that the (basinal) sequence in the Mendips was originally much thicker than that on the shelf areas (Figures 5, 8 and 9).

Along parts of the eastern margin of the Bristol Shelf, from Winford northwards to Cromhall, and farther east on the margin of the Wickwar Shelf, from Wick northwards to Yate, the upper Dinantian sequence (Hotwells Group) is more complete, but the presence of pebbly and phosphatic lag deposits at the base of the overlying Millstone Grit suggests that sedimentation may have been interrupted, even where there is no visible angular unconformity (Figures 9 and 14). At the western margin of the Bristol Shelf, however, between Shirehampton and Grovesend, the rocks of the Hotwells Group are attenuated and pass into arenaceous and pebbly sediments, with shales and thin limestones; they are in places violently contorted and show evidence of growth faulting and contemporaneous slumping. On the Chepstow Shelf to the west of the Lower Severn Axis (see Figure 7), which was active in Viséan times, the upper part of the Hotwells Group is also absent and the lower part is represented by sandstone.

Owing to the tectonic movements which took place along the Lower Severn Axis and Malvern Fault Zone in Tournasian–Viséan times the early Viséan shelf complex can, for purposes of description be conveniently divided into subordinate units (Figure 7). The Bristol Shelf extends southwards to Broadfield Down where the volcanic belt of Goblin Combe marks the hinge line forming its southern limit. Its eastern boundary is not clearly defined, but probably trends north-eastwards from a position east of Winford towards Cromhall. The margin of the Wickwar Shelf is evidently situated south of the Wick Inlier (where the Gully Oolite and Clifton Down Mudstone are shelf facies), but must swing northwards also towards Cromhall (Figure 7). The Cromhall–East Mendip Depression passes between the Bristol and Wickwar shelves; it may extend farther northwards towards the Forest of Dean, approximately along the line of the Malvern Fault Zone. Its margins are indicated mainly by facies and thickness changes at the edges of the shelf areas. It is only in the north-east, from Yate towards Cromhall, and in the south-east near Frome, that the thickness and character of the Dinantian sediments within the Depression can be seen.

In the northern part of the Coalpit Heath Basin the Viséan Cromhall Sandstone is up to 800 ft thick and the greater part of the Hotwells Group consists of sandstone, mudstone and fireclay, with only a few thin marine limestone intercalations. The sandstones show a northerly provenance, both in grain-size variation and in the nature of their clasts. Although even the most massive member, the Upper Cromhall Sandstone, passes southwards into calcareous

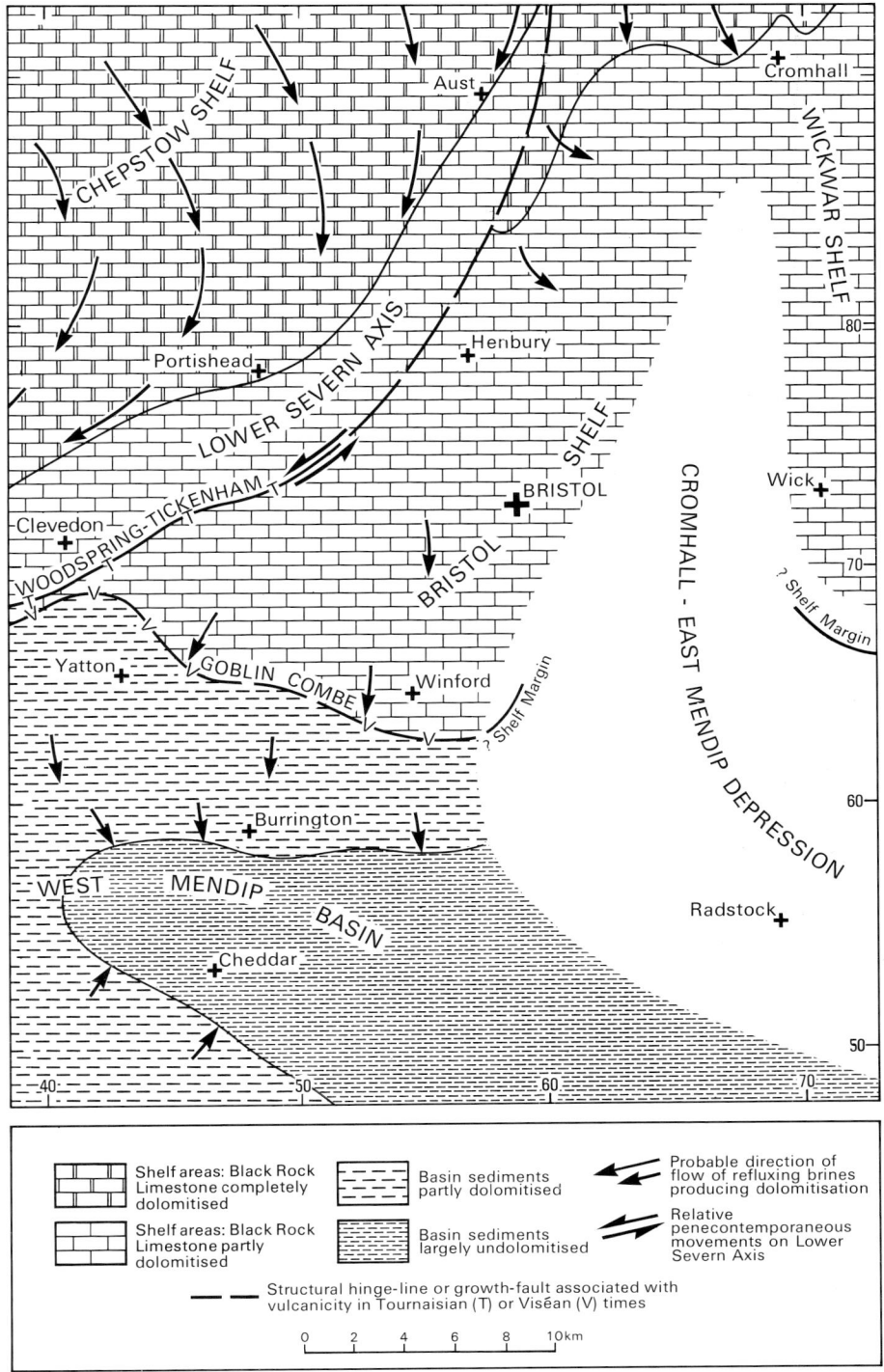

Figure 7 Dolomitisation of the Black Rock Limestone and basal Viséan rocks on shelf areas and around the margins of the West Mendip Basin

22 CHAPTER THREE CARBONIFEROUS LIMESTONE (DINANTIAN)

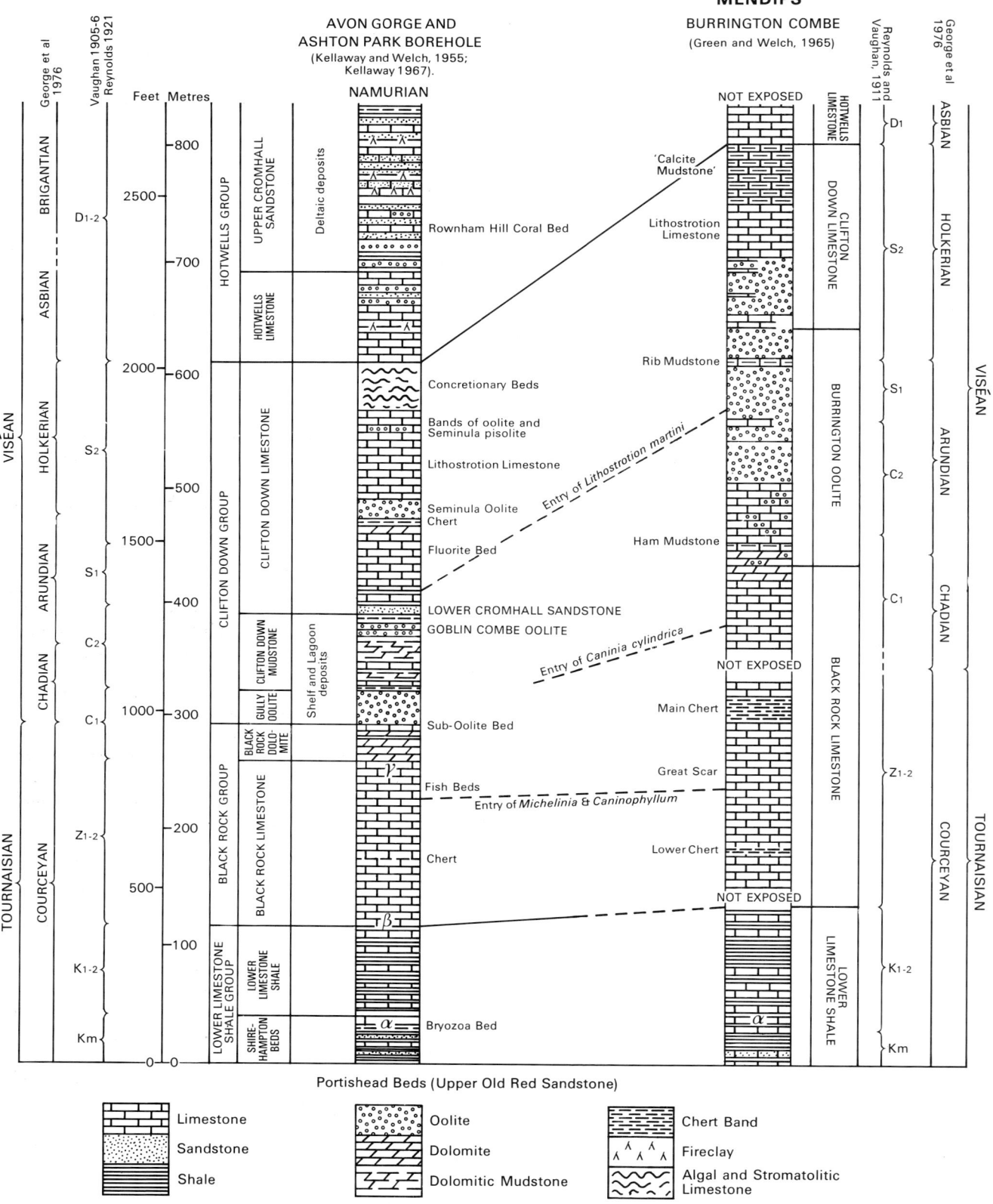

Figure 8 Comparative vertical sections of the Carboniferous Limestone in the Avon Gorge area, Bristol and at Burrington Combe, Mendip Hills, to illustrate the relationship between lithostratigraphical and chronostratigraphical classifications of the rocks

rocks in the Mendip Basin, it thickens considerably within the Cromhall–East Mendip Depression. For example, at Yate the upper sandstone member alone attains a thickness of 800 ft. As long ago as 1930 it was shown by Stanley Smith that at Wick some 670 ft of sandstone and shale (with a few thin limestone intercalations), which had been previously described as 'Millstone Grit' are of Viséan age. This prompted Dixon (in discussion on Smith, 1930, p.353) to suggest that these Viséan clastic sediments were deposited by a great river flowing from (central) Wales by way of Titterstone Clee and the Forest of Dean to a mouth situated near Bristol. (This imaginative suggestion was not followed up and was lost sight of in the postwar years.) However, the Dinantian strata along much of the East Mendip Depression are completely concealed beneath thick Upper Carboniferous and younger sediments. It is thought likely that the structures controlling the Depression had a general sinistral movement (along a southward extension of the Malvern Fault Zone), consistent with that of the Lower Severn Axis to the west (Figure 7).

Dolomitisation is an important feature of the Dinantian succession around Bristol. Excluding secondary Triassic dolomitisation, two main types of dolomite can be distinguished in the local Carboniferous Limestone. Fine-grained dolomite-mudstone, commonly associated with beds of calcilutite and argillaceous mudstone, is typically developed in the Clifton Down Mudstone. For example, the algal and stromatolitic limestones of the Clifton Down Group are accompanied by dense, fine-grained, unfossiliferous dolomites. These resemble the dolomite muds being deposited at the present day in coastal lakes and lagoons under semiarid climatic conditions in the Coorong of South Australia (Von der Borch et al., 1975; Von der Borch and Lock, 1979).

The second type of dolomitisation is characterised by changes produced in thick limestone sequences which have been penetrated by groundwaters moving seawards under pressure generated in supertidal areas. In the cases of the Chepstow, Bristol and Wickwar shelves the areas of active uplift marked by the Usk Anticline and other structures to the north and east may have provided the groundwater with sufficient hydrostatic pressure for deep seaward movement, causing it to combine with connate water to form strong refluxing brines. The converse situation is seen in the case of the Goblin Combe Oolite on the Bristol Shelf where a short-lived marine transgression had led to an influx of oceanic water which has temporarily suppressed dolomitisation.

There is a third type of dolomitisation (of Triassic age) which affects all the Dinantian rocks locally. It can usually be identified by its geological context and by the presence of red zoning in the dolomite crystals.

Early studies of dolomitisation and lagoon-phase deposits carried out by Dixon (1907) were published in the Geological Survey's Memoir on the Geology of the South Wales Coalfield, Part VIII, the country around Swansea (pp.13–20) and in a paper on Gower by Dixon and Vaughan (1911). This work was subsequently extended to Bristol by S. H. Reynolds (1921). Investigations at the Coorong mentioned above provide results which are in line with many of Dixon's conclusions. The Clifton Down Mudstone (*Caninia*-dolomite of Vaughan, 1905, 1906) is the most convincing example of a lagoonal deposit, with primary dolomite and magnesian limestones locally interbedded with mud of terrigenous and volcanic origin. The strong discontinuity at the base of the Clifton Down Mudstone in the Avon Gorge (p.33) is more clearly marked at Wick in the east and Kingsweston in the west. Farther north, in the Chepstow–Forest of Dean area it is even more pronounced. Yet the formation of the shelf on which the Gully Oolite and Clifton Down Mudstone were laid down had largely been accomplished before the Sub-Oolite Bed (p.31) filled the hollows on its surface.

The dimensions of this great lagoonal complex, which it is proposed to call the Dixon Lagoon are most surprising. Its western extremity lay at or beyond Milford Haven while in the east it continued to Bath or beyond. Its total length was in excess of 120 miles and its width, some 10–15 miles in Glamorgan, was more than 30 miles in extent between the Forest of Dean and Broadfield Down south of Bristol.

The closing stages in the life of the eastern part of the great lagoonal belt were presaged by a marine incursion from the south leading to the deposition of the Goblin Combe Oolite (p.41). The final breakdown was marked by sand with driftwood carried from the northern uplands down the channel marking the Malvern Fault Zone, and the deposition of the Lower Cromhall Sandstone (p.37).

Dolomitised, crinoidal or oolitic limestones which have been altered during late stage diagenesis or after consolidation has taken place have probably been affected by more than one phase of alteration, producing excessively hard, purplish grey rocks in which the MgO content rises to about 15 per cent. Weakly dolomitised limestones contain about 5 per cent MgO; below this figure the rock can only be described as a magnesian limestone, and is not usually mapped as a dolomite. In the Avon Gorge, for example, the highly dolomitised, purplish grey 'laminosa dolomite' in the Tournaisian Black Rock Group is overlain by the white, current-bedded Viséan Gully Oolite (Figure 8). The contact has been affected by dolomitisation and is not clearly defined. Palaeontological evidence shows that there is a major non-sequence at the base of the Viséan, and part of the dolomitisation may date from that period when a drop in sea level or a rise in the level of the landmass increased the hydrostatic pressure in the migrating groundwater. There is postdiagenetic dolomitisation, since in places both the Black Rock Limestone, below the break, and the Gully Oolite, above the break, are dolomitised. The diagenetic (or later) processes at the Tournaisian–Viséan boundary in the Avon Gorge are generally characteristic of the Bristol Shelf as a whole, although dolomitisation becomes more severe northwestwards.

Dolomitisation of the main mass of the Black Rock Limestone was largely by refluxion, that is by the deep circulation of Mg-rich brines (Figure 7), and may have been well advanced before Viséan subsidence took place. Correlatives of the Black Rock Group and the Gully Oolite are much more strongly dolomitised on the Chepstow Shelf and in the Forest of Dean, particularly in the Wigpool Syncline (Trotter, 1942, Welch and Trotter, 1961).

The northerly increase in the magnesium content of the dolomites of the Clifton Down Mudstone suggests deposition in shallow water with higher temperatures in a progressively

24 CHAPTER THREE CARBONIFEROUS LIMESTONE (DINANTIAN)

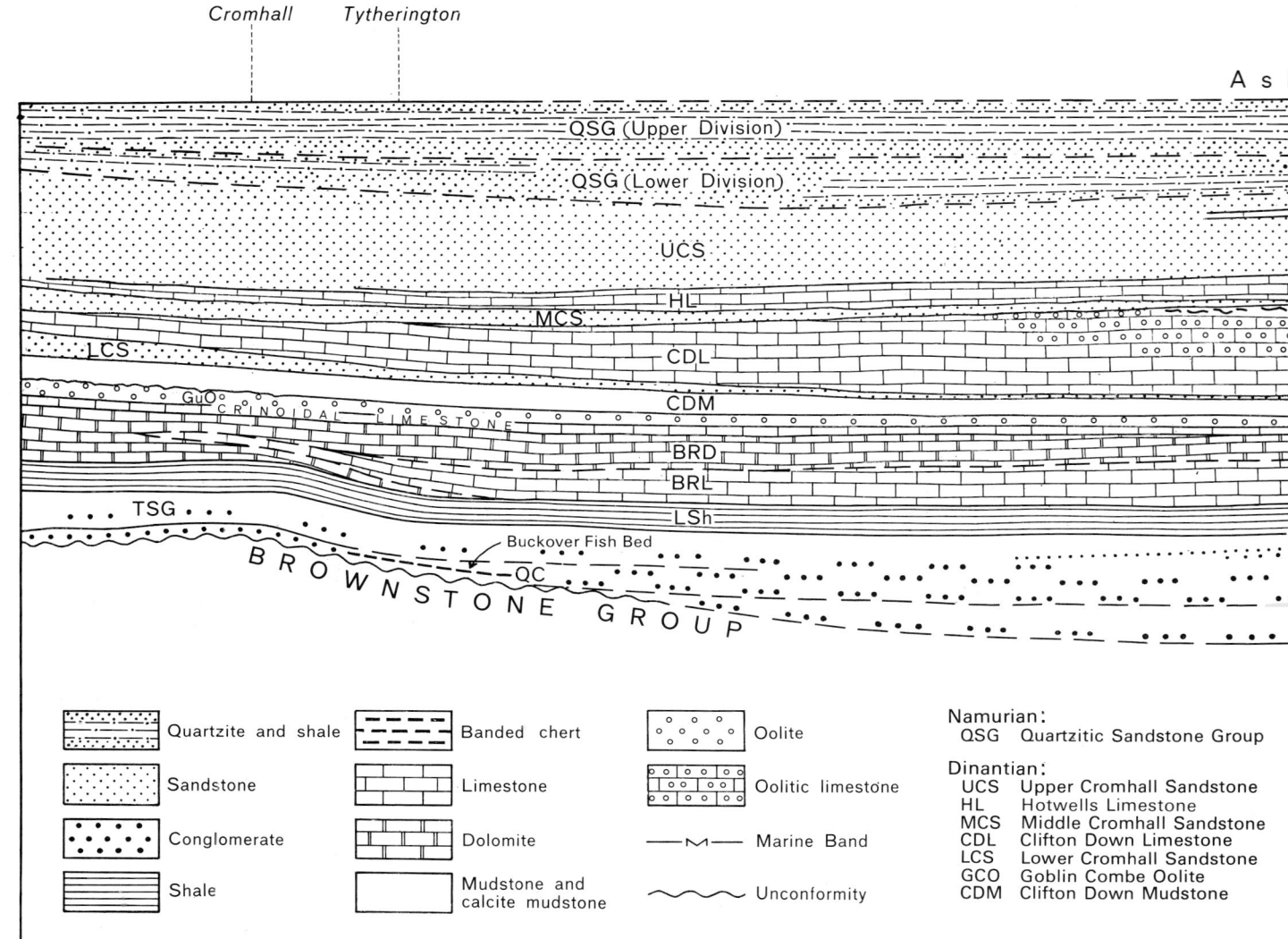

Figure 9 Generalised horizontal section of the Upper Old Red Sandstone, Carboniferous Limestone and Millstone Grit of the western side of the Bristol and Somerset Coalfield (including key map). No allowance has been made for shortening due to Variscan folding

more constricted environment. In the restricted, lagoonal conditions of the Clifton Down Group in the Bristol area (and even more in the Whitehead Limestone of Chepstow and the Forest of Dean) hard, dense, fine-grained beds of primary dolomite, virtually devoid of fossils, with black, stylolitic surfaces, are interbedded with shaly argillaceous mudstones.

One of the features considered to control the present day formation of dolomitic muds in the Coorong is the presence of barriers, which act as the interface between groundwater and sea water (Von der Borch et al., 1975, fig. 3). Such a barrier may have existed along the line of submarine fissure eruptions recorded by the Viséan volcanic rocks of Goblin Combe (Figures 7 and 9). The possibility that volcanic rocks may also have contributed to the supply of magnesium has also been suggested at Coorong (Von der Borch et al., 1975, p.285; Von der Borch, 1976, p.965). This may be important around Bristol, notably at Tickenham, and in the western Mendips at Weston-super-Mare, where there are two volcanic episodes, one in the upper part of the Black Rock Limestone, the other in the rocks equivalent to the Clifton Down Mudstone (Geological Survey 1:50 000 Sheet 279).

On the south side of the Mendips the incidence of dolomitisation points to the possibility that there was an emergent land mass formed in late Tournasian–Viséan times at no great distance to the south.

The present-day stratigraphical classification of the Carboniferous Limestone succession in the Bristol–Mendip region is summarised in Table 1, generalised lithologies and age relations in Figure 8.

The Geological Survey's nineteenth century one-inch maps of the region showed the Lower Carboniferous rocks

Table 1 Stratigraphical classifications of Carboniferous Limestone strata in the Bristol–Mendips region

		Regional stages		Bristol and ground to north	Central Mendips
Carboniferous Limestone Series Dinantian	Viséan	Brigantian	Hotwells Group	Upper Cromhall Sandstone (Tanhouse Limestone)	
		Asbian		Hotwells Limestone	Hotwells Limestone
		Holkerian	Clifton Down Group	Middle Cromhall Sandstone Clifton Down Limestone Lower Cromhall Sandstone Clifton Down Mudstone (upper) Goblin Coombe Oolite Clifton Down Mudstone (lower) Gully Oolite Sub-Oolite Bed	Clifton Down Limestone: Cheddar Oolite Cheddar Limestone
		Arundian			Burrington Oolite (Vallis Limestone)
		Chadian			
	Tournaisian	Courceyan	Black Rock Group	Black Rock Dolomite Black Rock Limestone	Black Rock Limestone — Upper Fauna / Middle Fauna / Lower Fauna
			Lower Limetone Shale Group	Lower Limestone Shale Shirehampton Beds	Lower Limestone Shales

The top of the Shirehampton Beds is marked by the Bryozoa Bed, a characteristic pink and grey limestone about 30 ft thick, described as 'Horizon α' by Vaughan (1905; 1906). The term 'α limestone' has been widely used for this rock type, which is developed locally, at more than one horizon, in the highest Devonian and lowest Carboniferous rocks of the South Wales, Bristol and Somerset coalfields and the Mendips. This fragmental crinoidal and bryozoan limestone facies is associated with late Devonian and Tournaisian molluscs in Pembrokeshire (Dixon, 1921) and with early Tournaisian species in the Bristol district. The Bryozoa Bed of the Avon Gorge is one of the thickest and most persistent red crinoidal and bryozoan limestone bands in the Shirehampton Beds but it is of only local value for correlation purposes. At Abbots Leigh [5495 7426], south of the Avon Gorge, it passes into a lenticular mass of oolite.

The fauna of the Shirehampton Beds is of a restricted marine or brackish type including *Lingula* and small calcareous brachiopods, the bivalves *Modiolus* and *Sanguinolites*, gastropods, ostracods, bryozoans, serpulids and calcareous algae, as well as fish fragments, including rhizodonts; rounded crinoid ossicles commonly make up the bulk of beds of bioclastic limestone. Plant remains are not uncommon in the Shirehampton Beds. They are usually very poorly preserved, but Utting and Neves (1970) have identified *Rhacophyton* from the lower part of the formation in the Avon section.

North of Bristol and west of the Severn, in the Clearwell area for example, a fairly massive but lenticular crinoidal limestone, up to 60 ft in thickness, is present at the base of the Lower Limestone Shale. Two lenticular bands of oolite, 10 to 15 ft thick and separated by 30 to 40 ft of shale are found near Coleford (Welch and Trotter, 1961, p.62). Jones and Lucy (1868) observed about 47 ft of 'transitional beds' near Drybrook [660 185], including sandstone and sandy crinoidal limestone with marine molluscs, *Ctenacanthus sp.* and a goniatite, *Imitoceras sp.* (Stubblefield, 1937). These rocks are overlain by typically argillaceous Lower Limestone Shale. The crinoidal limestones underlying the main mass of Lower Limestone Shale appear to equate with a bed of similar character in the Milbury Heath area east of the Severn. This includes bands of reddish crinoidal and bryozoa-rich limestone similar to the Bryozoa Bed of the Avon Gorge. Other indications of the presence of lenticular sandy and crinoidal limestones underlying the main mass of Lower Limestone Shale are found between Tortworth and Chipping Sodbury (Welch, *in* Cave, 1977, pp.52–57). It is therefore likely that the basal sandy and crinoidal limestones of the Chepstow–Tortworth–Chipping Sodbury area represent the upper part of the Shirehampton Beds of Bristol (Figure 6).

The upper and predominantly argillaceous part of the Lower Limestone Shale consists of grey-green shales with thin groups of dark grey, almost black, crinoidal limestone bands. These limestones are commonly very fossiliferous with a varied shelly fauna similar to that of the overlying Black Rock Limestone, though relatively poor in corals (Reynolds and S. Smith, 1925).

The best exposures of the Lower Limestone Shale occur in quarries in the lower part of the limestone succession. In the Mendips the railway cutting west of Maesbury Castle [607 475] provides a relatively continuous section through the upper two-thirds of the succession including the contact with the Black Rock Limestone (Green and Welch, 1965, p.45).

Resting on the Bryozoa Bed and filling slight hollows on its surface is a thin grey pebbly and phosphatic limestone with fish remains (Reynolds, 1908). This is known in the Avon Gorge as the Palate Bed (Stoddart, 1876; Vaughan, 1905, p.201). Its conglomeratic nature and its irregular contact with the Bryozoa Bed indicates a nonsequence, which is

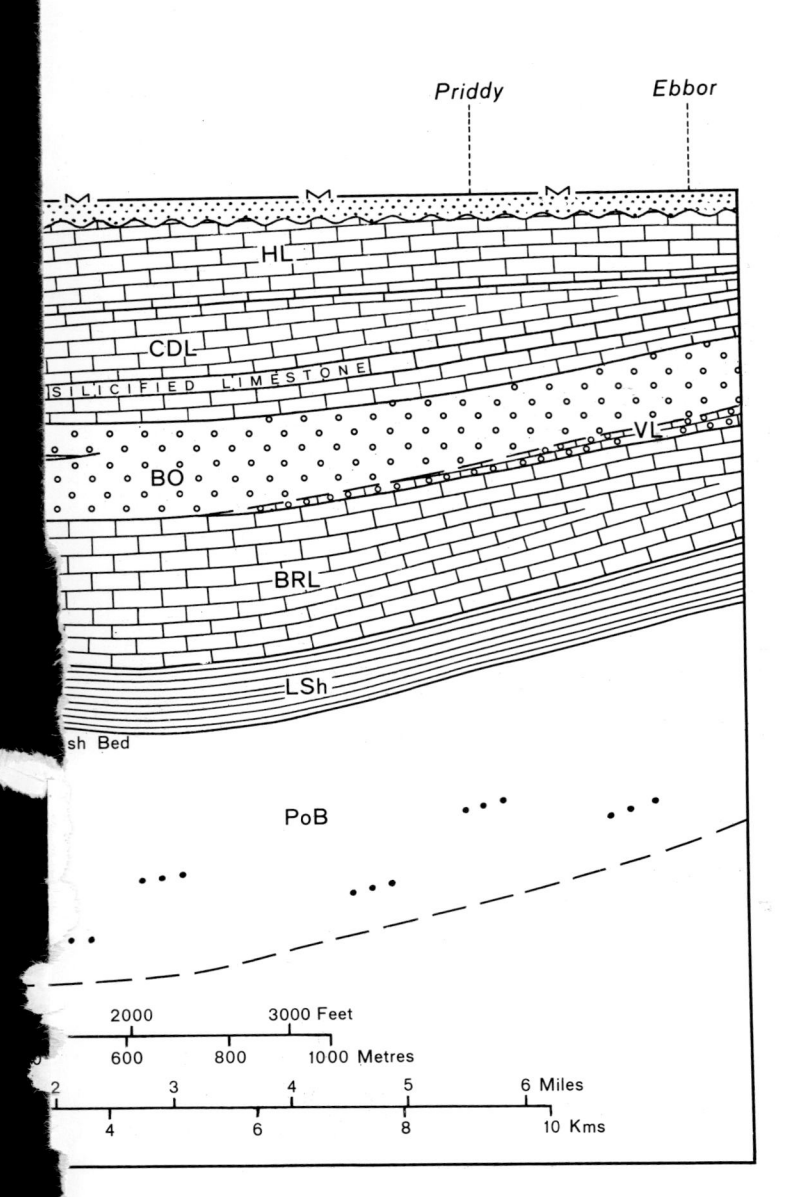

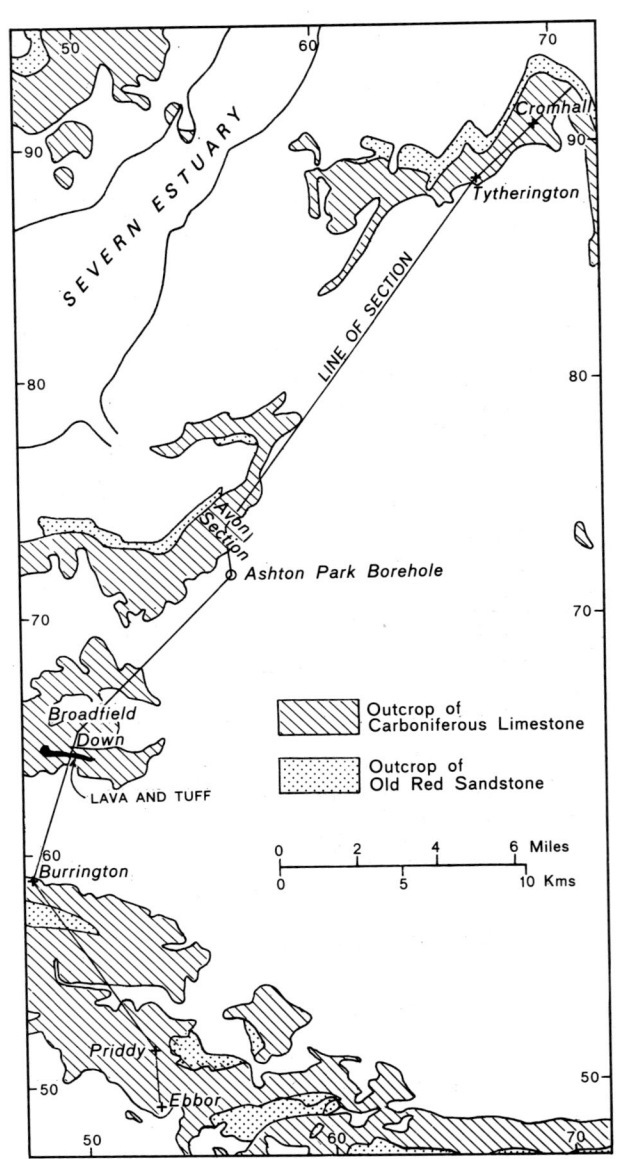

1955) applicable to the whole of the region is essential (Table 1).

LOWER LIMESTONE SHALE GROUP

As the name implies, this group is dominated by argillaceous rocks, though limestone bands are present in varying proportions throughout the succession (Figure 6). It represents a transition, both in lithological and faunal terms, between the essentially terrestrial facies of the Upper Old Red Sandstone and the fully marine deposits of the main Carboniferous Limestone above. Around Bristol two distinct lithological units can be recognised. The lower unit, distinguished as the Shirehampton Beds (Kellaway and Welch, 1955, p.8), is arenaceous, though with some thin limestone and shale. It constitutes a passage group of shallow-water brackish and marine strata separating the red mudstones, sandstones and conglomerates of the Portishead Beds from the dark marine shales and thin limestones which form the main, upper part of the Lower Limestone Shale Group. The Shirehampton Beds are about 160 ft thick in the Bristol area, but in the Mendips the equivalent strata are dominantly argillaceous and the Shirehampton Beds have not been mapped as a distinct formation. In the Chepstow area the passage beds, which include quite massive oolitic and crinoidal limestones, seldom exceed 60 ft in thickness though they attain a thickness of 95 ft in the Wigpool Syncline of the Forest of Dean (Sibly and Reynolds, 1937, pp.28–34).

The Shirehampton Beds take their name from Shirehampton, between Bristol and Avonmouth, where they are exposed in the deep Portway cutting (see Figure 6). They consist of about 160 ft of sandy limestones and shales with calcareous sandstones and thin red crinoidal limestones, interbedded with some sandstone and red and green mudstone of Old Red Sandstone type (Figure 6). They rest on a pebbly sandstone or conglomerate with fish remains, thought to correlate with the Sneyd Park Fish Bed of west Bristol (p.17).

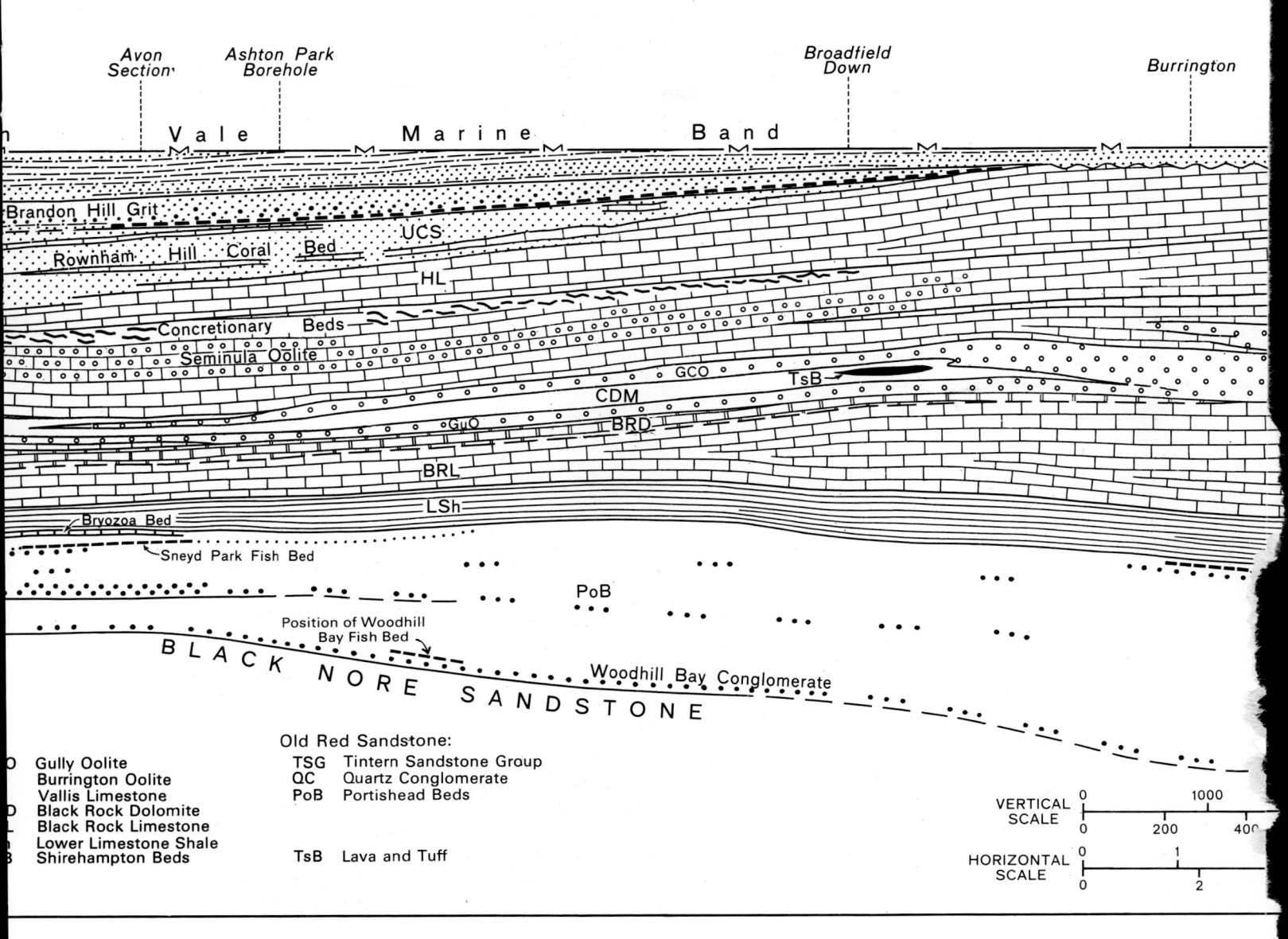

divided on lithological grounds into two groups: the Lower Limestone Shale and the overlying Mountain Limestone. However, the existence of distinctive suites of fossils (Dumont, 1832) in the Lower Carboniferous limestone sequence in Belgium led to two divisions, the Tournaisian below and the Viséan above, being widely applied throughout Britain, as well as on the continent. In 1905 Arthur Vaughan erected a series of coral-brachiopod zones for the Carboniferous Limestone, with the Avon Gorge as its type section (Plates 3, 4 and 5). This zonal scheme was modified (Reynolds, 1921) and extended to Lower Carboniferous rocks elsewhere in Britain. However, the application of Vaughan's zones to sections even as close to Bristol as the Mendips and Weston-super-Mare raised difficulties, and the use of the Avonian zonal terminology as a substitute for a normal lithostratigraphical nomenclature became a serious impediment to progress. Eventually, Rhodes and Austin (1971, p.341), Mitchell (1972, p.159, fig. 1) and Ramsbottom (1973, p.959, fig. 8) demonstrated the existence of several significant faunal nonsequences in the Avon Gorge succession, with important Dinantian faunas found elsewhere in Britain missing at Bristol. Ramsbottom (1973) proposed a new grouping based on six Major Cycles, thought to be characterised by major transgressive and regressive phases of sedimentation, the non-sequences in the succession occurring at the boundaries between cycles. George et al. (1976) erected six regional stages for the subdivision of the Dinantian in Britain (Table 1), defined in a series of stratotype sections and recognised by their contained faunas. A revision of the caninoid corals of the Black Rock Limestone has since been published (Mitchell, 1981); thus, although detailed revision of some of the stage boundaries can now be made, the general correlations given in George et al. (1976, Figure 4) are still viable. However, application of the Viséan stages to the local Carboniferous Limestone succession remains to be established in the thick arenaceous facies and a lithostratigraphical classification (Kellaway and Welch,

Plate 3 Panorama of the right bank of the Avon gorge showing the Black Rock and Gully quarries. Black Rock Limestone is overlain by Black Rock Dolomite, Gully Oolite and Clifton Down Mudstone

associated with a marked change in conditions of sedimentation, the coarse detrital and muddy sediments of the Shirehampton Beds being succeeded by fine-grained mudstones and shales with dark limestones. The Palate Bed is seldom more than 6 in thick and its chief interest derives from its value as a local marker.

The maximum known development (see Figures 5 and 12a) of the Lower Limestone Shale Group (including strata probably equivalent to, but not differentiated there as the Shirehampton Beds) is about 500 ft in the Black Down Pericline, north of Cheddar (Green and Welch, 1965, p.17). In the eastern Mendips about 400 ft occur. West of Bristol the total thickness appears to be about 350 ft, of which the Shirehampton Beds account for about one-third. The group is generally thinner on the north-west side of the Bristol Coalfield than it is to the east, being about 300 ft thick near Wickwar, but only about 190 ft near Milbury Heath (Kellaway and Welch, in Cave, 1977, p.52). Farther west, in the Chepstow area, the average thickness is about 260 ft. The northwards attenuation indicated by these figures is by no means regular, and sedimentation may have been influenced locally by penecontemporaneous earth movements. Thin fossiliferous limestones occur at the base of the upper shaly part of the group in the Avon section. These yield *Avonia bassa*, *Chonetes failandensis*, orthotetoids, and *Leptagonia* cf. *analoga* (see also Figure 47). At Bristol the encrusting coral *Vaughania vetus* (*'Cleistopora geometrica'*) is found in the top 60 ft of the Lower Limestone Shale. This fossil is generally rare or absent in the Chepstow area and the Mendips. The Lower Limestone Shale is placed in the lower part of the Courceyan Stage i.e. Lower Tournaisian.

At Clevedon thin beds of fine-grained siliceous sandstone are interbedded with the shales. Both here and in the eastern Mendips there is a broad similarity between the Lower Limestone Shale sediments and those of the Pilton Beds of north Devon (Butler, 1973).

BLACK ROCK LIMESTONE GROUP

The Black Rock Limestone is the lowest of three groups recognised within the main 'Mountain Limestone' facies of the Lower Carboniferous. It consists almost entirely of calcareous or dolomitic rocks and unlike the succession above, it contains only thin argillaceous beds and no arenaceous deposits. The name derives from the old quarrymen's name for the dark limestones so extensively quarried on the sides of the Avon Gorge between Clifton and Sneyd Park.

For the most part the sequence comprises dark grey to black, fine-grained, coarsely crinoidal limestone, which in places approaches a 'petit granit' in character. By careful examination of the abundant faunas, it has been shown that the upper part of the thick Mendip sequence is missing from the

Avon Gorge succession (Mitchell, 1981). This northwards attenuation has been accompanied by strong dolomitisation extending increasingly downwards from the surface of unconformity below the Clifton Down Group (Figures 9 and 10). From the western part of the Black Down Pericline through Broadfield Down to the western and northern fringes of the Bristol Coalfield, this dolomite facies is known as the Black Rock Dolomite. North-west of Cromhall, and over much of the Chepstow–Forest of Dean area to the west of the Severn, the rocks are strongly dolomitised and there the term Lower Dolomite is used. Irregular developments of dolomite, up to 300 ft thick, also occur locally to the south of the Mendips, between Cheddar and Croscombe, south-east of Wells.

In the Mendips two horizons, 100 ft and 500 ft above the base of the formation, are characterised by strong developments of chert bands and nodules (Figure 8). The 'Lower Chert' is present only in the Black Down Pericline, where it is about 13 ft thick in Burrington Combe. The 'Main Chert' is present everywhere in this southern region, and ranges from about 50 ft in the west to 90 ft in Burrington Combe and as much as 290 ft to the east of Leigh-on-Mendip (Butler, 1973). Silicified fossils are common in the limestones of the lower part of the succession.

The thickest, as well as most complete, succession of the Black Rock Limestone is to be found in the Beacon Hill Pericline of the eastern Mendips (Figure 12b); here upwards of 1200 ft of strata, mainly limestones, are present. In the Black Down Pericline of the western Mendips, the thickness ranges from 880 ft in Burrington Combe to 950 ft in Long Wood valley, north-east of Cheddar (Green and Welch, 1965, pp.25, 28). In the Bristol area, the thickness of the group is reduced in general to about 500 ft: it is about 520 ft in the Avon Gorge, and about 450 ft at Westbury-on-Trym and Southmead. This northward attenuation is less marked in the centre of the Coalpit Heath Basin (Figure 12b). At Wickwar the combined thickness of the Black Rock Limestone and the Black Rock Dolomite is about 400 ft, at Milbury Heath about 340 ft, and in the Chepstow and Forest of Dean area about 300 to 400 ft.

Stratigraphical palaeontology

A rich fauna of corals and brachiopods occurs throughout the strongly crinoidal Black Rock Limestone. The section in Burrington Combe has been described in detail by Mitchell and Green (*in* Green and Welch, 1965, pp.180–187, table 1). They recognised three distinct faunal assemblages referred to as the Lower Fauna, Middle Fauna and Upper Fauna (Figure 8), now equated with the *Zaphrentites delanouei*, *Caninophyllum patulum* and *Siphonophyllia cylindrica* assemblage biozones (Ramsbottom and Mitchell, 1980, p.62). The *Z. delanouei* Biozone (mid-Courceyan) is characterised by the corals *Fasciculophyllum omaliusi*, *Sychnoelasma clevedonensis* and *Zaphrentites delanouei*, and the brachiopods *Cleiothyridina glabristria*, *C.* cf. *glabristria* (Vaughan *non* Phillips), *Dictyoclostus multispiniferus*, *Pugilis vaughani*, *Rhipidomella michelini*, *Rugosochonetes vaughani* and *Syringothyris cuspidata cyrtorhyncha* are common. *Leptagonia analoga* and *Schellwienella aspis* range up into the *C. patulum* Biozone (late Courceyan) which is characterised by the first appearance of the corals *Caninophyllum patulum*, *Caninia cornucopiae*, *Cyathaxonia cornu*, *Cyathoclisia tabernaculum*, *Fasciculophyllum densum* and *Sychnoelasma konincki*. *Siphonophyllia cylindrica* is diagnostic of the *S. cylindrica* Biozone (early Chadian) together with such brachiopods as *Eomarginifera* aff. *derbiensis*, *Megachonetes magna*, *Pustula* cf. *pustuliformis*, *P. pyxidiformis*, *Schuchertella* cf. *wexfordensis* and *Syringothyris* aff. *elongata*.

In the Avon Gorge the succession between the Bryozoa Bed (see above) and the base of the Black Rock Limestone is poorly exposed and detailed correlation is difficult. However, in 1962–1963 a continuous section from the Upper Old Red Sandstone to the base of the Gully Oolite (see below) was proved in the Northern Foul Water or Portway Tunnel between Roman Way, Sea Mills and the Gully Quarry at the northern end of the Avon Gorge, west of Bristol (Figure 11). This was logged in detail by officers of the Geological Survey and the palaeontology of the Dinantian rocks was investigated by M. Mitchell (Appendix 3). He reported that the faunas found in the Black Rock Limestone of the tunnel agreed very closely with those found at Burrington Combe, with the important exception that the upper part of the *C. patulum* Biozone and the whole of the *S. cylindrica* Biozone were not represented among the fossils there collected. In particular the corals at the higher levels of the tunnel's Black Rock Limestone included *Caninophyllum patulum greeni* (= closely septate form) and *Caninia cornucopiae*, both characteristic of the lower levels of the *C. patulum* Biozone of Burrington and it would appear that the uppermost 400 ft of the Black Rock Limestone at Burrington are missing in the Avon Gorge west of Bristol, due to non-deposition or to sub-Gully Oolite erosion (Mitchell, 1971, p.97; 1972, p.159, fig.1; 1981). Both in character and thickness what remains of the Black Rock Limestone in the Portway Tunnel is closely similar to its stratigraphical counterpart at Burrington, and so it has been suggested that northward attenuation of the formation referred to at the beginning of this section is related to the development of this interformational unconformity.

North of Bristol and in the Chepstow area, the Black Rock Limestone is so strongly dolomitised that identifiable fossils are hard to find. However, the absence of any forms which can be referred to *Caninophyllum patulum* suggests that only the lower part of the Black Rock Limestone is preserved beneath Viséan strata.

Correlation of the Black Rock Limestone of Broadfield Down is not altogether clear. The base is not exposed but the lower part in general is well represented. The middle part, with the characteristic varieties of *C. patulum*, is well developed and it is possible that higher horizons, belonging to the Chadian, may also be present, though in an attenuated form, south of Brockley Combe.

Plate 4 Panorama of the right bank of the Avon gorge showing Great Quarry. The Clifton Down Limestone is overlain by the Hotwells Limestone.

The three biozones can also be generally identified in the central and eastern Mendips. Here the presence of *Delepinea carinata* at the base of the Vallis Limestone and of *Siphonophyllia garwoodi*, about 70 ft or so below the top of the Black Rock Limestone at Hale Combe quarry [431 566] indicates that the upper part of the formation extends upwards into the Arundian Stage in the Frome area (George, 1976, fig.4, section C).

CLIFTON DOWN GROUP

The Clifton Down Group comprises strata showing a wide range of lithologies, but largely characterised by their fine grain or oolitic nature and an impoverished fauna contrasting with the fossiliferous, crinoidal limestones of the Black Rock Limestone below and the richly fossiliferous Hotwells Limestone above (Figure 13). In the Bristol area the rocks are mainly calcareous, ranging from fine-grained calcite- and dolomite-mudstones (chinastones or calcilutites), some with algae, through coarser-grained limestones (calcarenites), some crinoidal and fossiliferous with corals, brachiopods and molluscs, to fine and coarse oolites. Argillaceous bands — mudstones and shales — are mainly restricted to the lower part of the group. Despite their variety, all the sediments are of shallow water origin, some having been laid down in fringing lagoons. To the north and west of Bristol the more massive limestones of the group are progressively replaced by the arenaceous Cromhall Sandstone (see below). Locally, in the Broadfield Down area, there are intercalations of lava and tuff. In thickness the group ranges from 900 to 1000 ft in the Bristol area; it thins rapidly westwards and north-westwards, but southwards it thickens to 1200 to 1500 ft in the Mendips (Figure 12c), where the sequence is dominated by oolites and bioclastic granular limestones.

The Clifton Down Group is best known from its development in the classic sections of the Avon Gorge (Vaughan, 1905; Reynolds, 1921). In the Bristol area generally the basal member of the group mapped is the highly distinctive Gully Oolite. It comprises about 90 ft of pinkish to pale grey medium- to rather fine-grained current-bedded oolite, which weathers white. It is massively bedded and is traversed by strong vertical joints. It is only sparsely fossiliferous and Vaughan's name of 'Caninia-Oolite' referred to its position in his zonal scheme and not to its fossil content. In the Avon Gorge, the base of the Gully Oolite is dolomitised, as is the underlying Black Rock Limestone, and the boundary between the two divisions is obscure. Elsewhere in the Bristol area, as at Southmead, the oolite itself is underlain by about 20 ft of pale grey well-sorted crinoidal limestones, described by Vaughan (1905, p.221) as the 'Sub-Oolite Bed'. This bed

Plate 5 Panorama of the right bank of the Avon gorge from Bridge Valley Road to the Suspension Bridge.

The upper Cromhall Sandstone is folded below the Avon Thrust, with the Clifton Down Limestone above. The latter is sheared and folded between the Thrust and the St Vincent's Rocks Fault. Dolomitic Conglomerate fills a small hollow in the Limestone surface near the eastern abutment of the Suspension Bridge

contains abundant brachiopods, including papilionaceous chonetoids and large orthotetoids, the fauna being of mixed Tournaisian and Viséan aspect.

The Clifton Down Mudstone overlies the Gully Oolite. It crops out in the Avon Gorge between the Gully Quarry and the Great Quarry, where it was described as early as 1885 by Lloyd Morgan as 'Middle Limestone Shales', and later known as 'Caninia shales and dolomites' (Vaughan, 1906, p.112) and 'Caninia Dolomite' (Reynolds, 1921, p.224). The lowest 120 ft of the Mudstone consists mainly of thinly bedded or lenticular pale grey calcite- and dolomite-mudstones or chinastones, commonly with stromatolitic algae, interbedded with dark grey or brown mudstones and shales. Apart from the algae and a few serpulids the limestones are virtually unfossiliferous. Three beds of massive cross-bedded crinoidal and oolitic limestones also occur in the top half of this succession, the lowest with probable late Chadian foraminifera and the highest with Arundian foraminifera (Whittaker and Green, 1983, p.26). These beds are succeeded by about 50 ft of hard grey oolitic and crinoidal limestones, which have yielded the gastropod *Bellerophon*, as well as productoids, spiriferoids and small rhynchonelloids. These beds were first described by Vaughan (1905, pp.212–214) as the 'Bellerophon Beds' of Failand. They are now regarded as being equivalent to the Goblin Combe Oolite of Broadfield Down (see below) and this term is now generally used where the beds are recognised in the Bristol area and to the east at Wick. The uppermost 20 ft of the Clifton Down Mudstone consist of thin limestones and shaly mudstones and these in turn are overlain by a hard sandy limestone marking the base of the Clifton Down Limestone.

The change from the massive current-bedded white Gully Oolite to the darker, thinly bedded fine-grained rocks of the Clifton Down Mudstone represents a remarkable change in conditions of sedimentation. In the Bristol area, as well as farther north in the Forest of Dean, the contact of the two formations is commonly very irregular and there is evidence of a pause in sedimentation with slight erosion of the upper surface of the Gully Oolite.

The Clifton Down Mudstone is overlain by the Clifton Down Limestone, a variable sequence of rocks, well exposed in the Great Quarry, on the right bank of the Avon Gorge and on the roadside extending northwards along the Portway. The sequence begins with a limestone bearing sand grains, marking the onset of the Cromhall Sandstone sedimentation (Kellaway and Welch, 1955, p.17), which became so widespread later in the ground to the north and north-west (see below). This limestone is succeeded by relatively unfossiliferous, uniform, splintery calcite- and dolomite-mudstones, some of them stromatolitic, generally similar to the underlying Clifton Down Mudstone, but more regularly bedded; they represent a continuation of shallow-water lagoonal conditions gradually giving way to more open-water conditions. These beds pass upwards through fossiliferous limestones with shale partings into massive

Figure 10 Comparative vertical sections in the Black Rock Group and in the lower part of the Clifton Down Group of Bristol and Burrington Combe

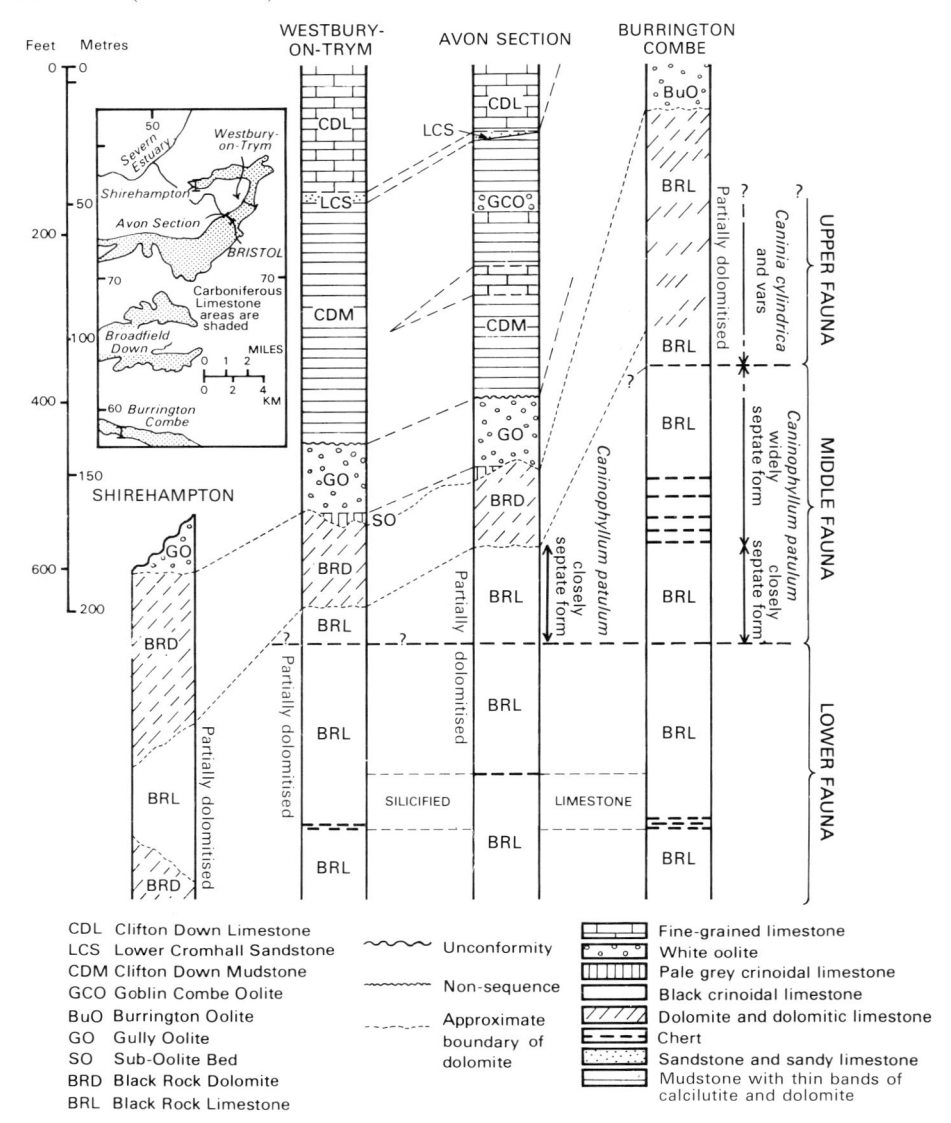

medium to dark grey limestones with colonies of *Lithostrotion martini* and bands of *Composita ficoidea* and *Linoprotonia*. Several locally identifiable bands are noteworthy in this part of the sequence. At the northern end of the Great Quarry a bed containing a diphyphylloid *Lithostrotion* was termed the 'Diphyphyllum Band'. Some distance above, four other local markers occur in close succession — the Lithostrotion basaltiforme Bed, the Trilobite Bed, the Fluorite Bed and the Caninia bristolensis (*Caninophyllum archiaci* var. *bristolensis*) Bed (Reynolds, 1921, p.241) (see Figure 13). These beds, however, have little persistence and they are of only limited use for correlation. There is some evidence that the Fluorite Bed, situated some 26 in above the Trilobite Bed in the Avon Gorge (Loupekine, 1951, 1953), may extend westwards as far as Lawrence Weston.

Not far above there is another sharp change in lithology, and oolites and oolitic limestones again become prominent. In the Avon Gorge two such developments are well marked and have been named the Seminula Pisolite and the Seminula Oolite (Reynolds, 1921, pp.228–229). Silicified fossils occur in the Seminula Pisolite, including *Composita* and *Lithostrotion* and near the top of the bed there are three bands of chert, which are separated from the Seminula Oolite by about 5 ft of cavernous dolomite. The Seminula Oolite, in particular, has a very sharply defined base, and Reynolds also noted a 'considerable lateral variability' in the

CLIFTON DOWN GROUP 35

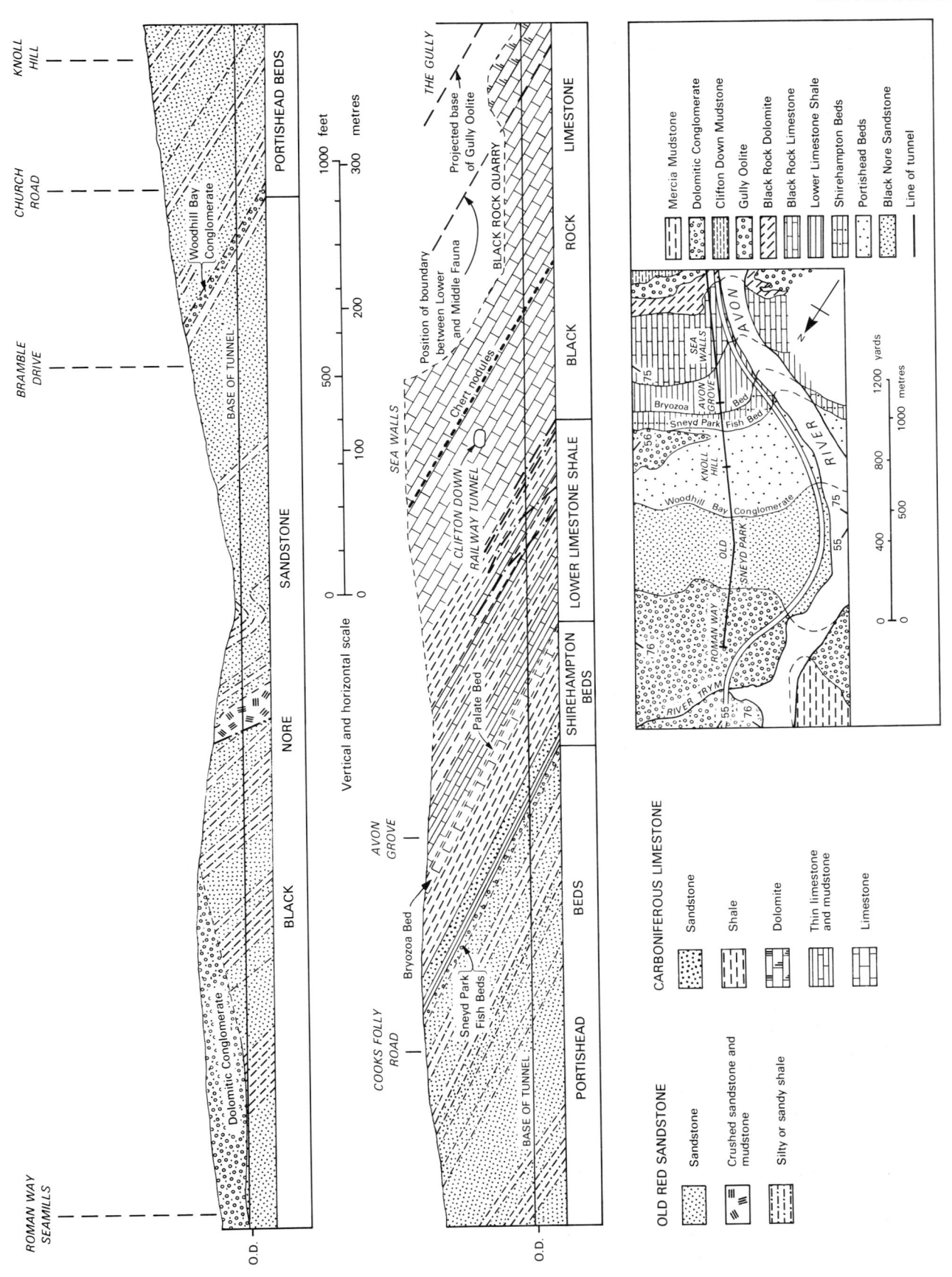

Figure 11 Section along the line of the Foul Water Tunnel between Roman Way, Sea Mills and Gully Quarry, Avon Gorge, Bristol

36 CHAPTER THREE CARBONIFEROUS LIMESTONE (DINANTIAN)

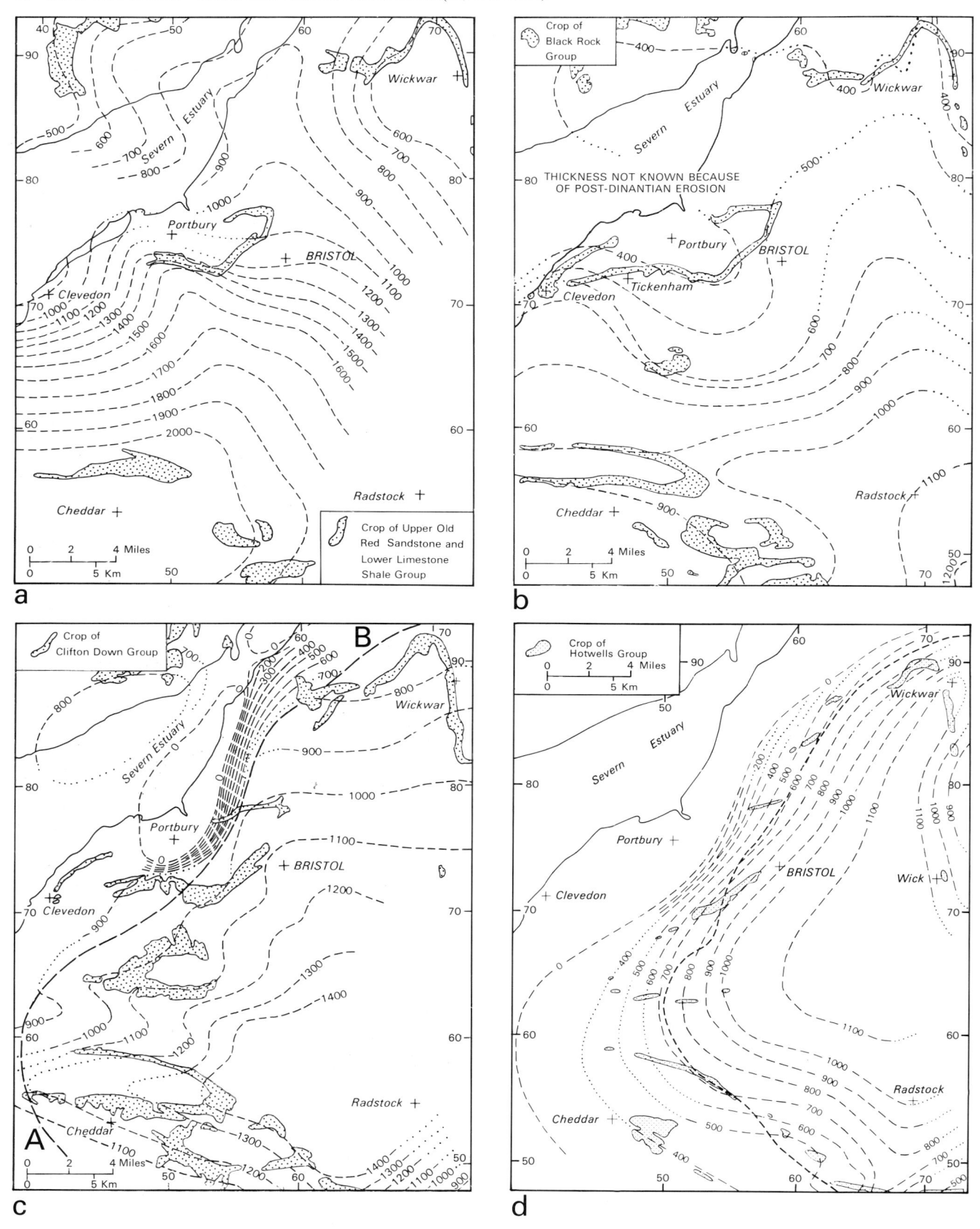

Figure 12 Isopachytes in (a) Upper Old Red Sandstone and Lower Limestone Shale Group; (b) Black Rock Group; (c) Clifton Down Group; (d) Hotwells Group. In areas west of the thick broken line A–B in (c) and (d), thicknesses have been affected by Dinantian and Namurian erosion

underlying beds. It seems probable, therefore, that there is a stratigraphical hiatus at this level, developed on the Bristol shelf. Traced laterally, it is seen that these oolites are part of a series of such developments, which may collectively be called the Clifton Down Oolite. The individual beds can seldom be mapped over any great distance, as they tend to pass laterally and vertically into nonoolitic limestone. In the Avon Gorge the Seminula Oolite grades upwards through oolitic limestones into limestones with *Lithostrotion* and then a thick sequence of calcite-mudstones, which mark a return yet again to the shallow lagoon conditions characteristic of the group as a whole. Finally these calcite-mudstones become markedly algal and stromatolitic in character and are interbedded with thin shales and brecciated limestones. These Concretionary Beds (Figures 8 and 13) have their maximum development in the area around Bristol, but they extend southwards to the northern and western sides of Broadfield Down and to Long Ashton, Belmont Hill and Cambridge Batch. They cannot be identified on the eastern side of the Coal Pit Heath Basin and they may be replaced by the Middle Cromhall Sandstone which persists as far south as Wick.

Northwards from Bristol the Gully Oolite and the basal crinoidal limestones associated with it (the 'Sub-Oolite Bed') can be traced to Henbury and the northern rim of the Bristol Coalfield, where they are again about 90 ft thick (Welch and Trotter, 1961, p.63). Between Tytherington and Wickwar on the northern rim of the Carboniferous basin, the Gully Oolite comprises 140 ft of almost unfossiliferous pale oolite, overlying about 70 ft of grey coarsely crinoidal highly fossiliferous limestone, presumably representing the 'Sub-Oolite Bed' (Welch *in* Cave, 1977, p.53). This is the maximum known thickness of these rocks (210 ft). At Bristol and Wickwar the comparable figures are about 110 and 150 ft respectively but the difference may be due in part to dolomitisation of the basal beds. To the west of the River Severn in the Forest of Dean and Chepstow districts, the equivalent rocks are known as the Crease Limestone. This consists of greyish white oolite resting on a basal development of pale grey crinoidal limestone, together ranging from 30 to 100 ft in thickness.

Traced northwards from the Avon Gorge, the hard grey oolitic and crinoidal bands of the Bellerophon Beds become finer in grain as they are followed across Durdham Down to Henbury and Southmead. Further north they become indistinguishable from the more or less uniform mass of pale grey splintery calcite- and dolomite-mudstone with shaly partings, about 100 to 200 ft thick, representing the Clifton Down Mudstone. In the Forest of Dean the equivalent Whitehead Limestone presents a similar succession of rocks, ranging in thickness from 50 to 150 ft and notable for the presence of concentrically ringed algae including *Garwoodia* [*Mitcheldeania*] *gregaria*, *Girvanella ducii*, *G. nicholsoni* and *Ortonella kershopensis* (Welch and Trotter, 1961, p.76).

North of Bristol the detailed succession in the Clifton Down Limestone is not well known. It consists of upwards of 450 ft of limestone, much of it oolitic, though massive limestones with corals are notable in some of the more southerly outcrops. At Henbury, limestones with algal nodules and polygonal desiccation cracks replace the lower part of the Concretionary Beds (Plate 6), the upper part being represented by the Middle Cromhall Sandstone.

North of a line between Elberton and Chipping Sodbury, sandstone beds are developed between the Clifton Down Mudstone and the Clifton Down Limestone. These are known as the Lower Cromhall Sandstone (Kellaway and Welch, 1955, p.17), the most southerly known development of this facies being seen in the Avon Gorge where limestone at the base of the Clifton Down Limestone contains scattered sand grains (see above). To the north, the sandy facies thickens rapidly: it is 20 ft at Chipping Sodbury, 60 ft at Tytherington, and it reaches its maximum at Cromhall, where it is 140 ft thick (Cave, 1977, p.54). West of the Avon Gorge thin sandstone lenses are present in the base of the Clifton Down Limestone at Lawrence Weston. The Lower Cromhall Sandstone consists dominantly of hard brownish and reddish, fine- to coarse-grained quartzitic sandstones with subordinate bands of shale and mudstone and rare thin limestones. Poorly preserved fossil drift wood occurs at Chipping Sodbury and other localities in the northern part of the Coalpit Heath Basin, this is clearly derived from a land mass to the north (St Georges Land), no evidence of an arenaceous facies at this stratigraphical level having been found at Wick. Locally the base is conglomeratic, with included fragments of the underlying calcite-mudstones.

West of the Severn, the Clifton Down Limestone is known as the Drybrook Limestone and the arenaceous beds beneath it as the Lower Drybrook Sandstone. The Drybrook Limestone consists of massive, predominantly oolitic, limestones with brachiopods and corals and in the southern parts of the Forest of Dean and Chepstow districts it reaches a thickness of 400 ft, though it thins northwards by passage into underlying and overlying sandstones.

In the north and north-east of the district, thinly bedded quartzitic sandstones with shales are again developed at the top of the Clifton Down Limestone. These constitute the Middle Cromhall Sandstone (Plate 7) and are now included within the Clifton Down Group. They are not represented in the Avon Gorge nor in the Ashton Park Borehole (p.42), where the Hotwells Limestone rests directly on the Concretionary Beds. At Henbury, however, on the north-west outskirts of Bristol, a thin development of quartzitic sandstones and shales has been observed directly beneath the Hotwells Limestone. Northwards these beds thicken markedly. Between Itchington and Tytherington, two bands of sandstone and shale, 40 and 60 ft thick, are separated by some 75 ft of rather coarse oolitic-crinoidal limestone. The sandstones are for the most part brown and white and fine grained. Between Wickwar and Chipping Sodbury, the Middle Cromhall Sandstone contains a thin median parting of limestone but traced southwards the upper layer of sandstone appears to die out. In Arnold's Quarry at Chipping Sodbury, about 90 ft of rippled sandstone and shales have been measured, with a 15 ft band of dolomitised limestone with thin shales interbedded about 12 ft above the base (Murray and Wright, 1971, p.263). Southwards the sandstones become sharply attenuated and only some 18 ft of ripple-marked sandstone and thin shales overlie 7 ft of nodular chinastones and thin shales at Wick (Murray and Wright, 1971, p.262). Here the shales

38 CHAPTER THREE CARBONIFEROUS LIMESTONE (DINANTIAN)

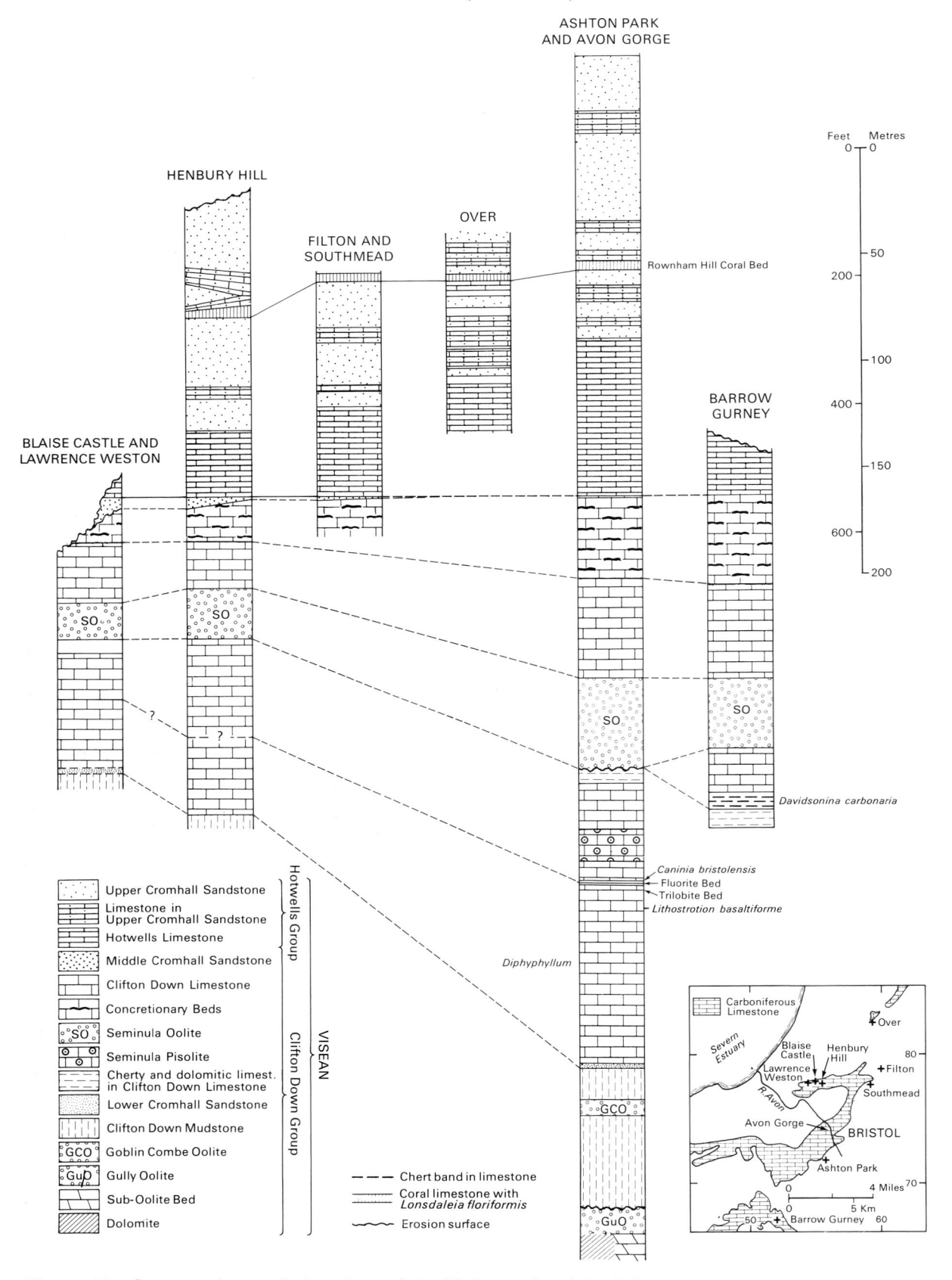

Figure 13 Comparative vertical sections of the Viséan rocks of the Bristol area

Plate 6 Vertical bedding plane in the Concretionary Beds at the top of the Clifton Down Limestone.

Polygonal dessication cracks occur in the top of the face which is also marked by numerous impressions of algal nodules. Henbury Hill Quarry [5655 7820] near Westbury-on-Trym (A10714). Height of section about 100 ft.

are associated with carbonaceous bands and thin coaly layers (S. Smith, 1930, pp.349–350).

On the east side of the Bristol Coalfield, the Clifton Down Group crops out as a narrow band extending south of Wickwar to Chipping Sodbury. To the south the Lower Carboniferous rocks are concealed beneath Mesozoic strata, but there is a small inlier at Tyning Farm, Codrington, and a larger, more important one at Wick. South of Wickwar the succession is similar to that at Tytherington though probably thicker: it comprises dark grey, massive compact limestones with a few interbedded shales and algal limestones. At Chipping Sodbury about 425 ft of Clifton Down Limestone was measured in detail in Arnold's Quarry by Murray and Wright (1971, pp.263–267). For the most part the sequence consists of dark fine-grained chinastones, oolites and oolitic limestones with a few shale bands near the top. Bioclastic debris is common throughout and corals, particularly *Lithostrotion*, are prominant in the lower middle part of the succession.

In the main part of the Wick Inlier, to the west of the Wick Fault, a virtually complete section of the Clifton Down Group is exposed in Messrs Wotton Bros quarries:

	Thickness ft
Middle Cromhall Sandstone:	18
Clifton Down Limestone: Massive grey limestone, oolitic in upper part,	

Plate 7 Red-stained Hotwells Limestone on the left rests on Middle Cromhall Sandstone including beds of carbonaceous shale, which in turn overlies Clifton Down Limestone. Quarry [7087 7301] at Wick (A10656)

passing down into hard grey bedded limestone and calcite-mudstone. The bottom 200 ft contains algal limestone bands and the limestones are splintery with sporadic shaly partings	800
Clifton Down Mudstone:	
Calcite-mudstone and fine-grained dolomitic limestone, some shale	60
Goblin Combe Oolite: hard grey crinoidal and oolitic limestone, with *Bellerophon*	28
Calcite-mudstone and shale (disturbed)	44
Brecciated and conglomeratic limestone with fragments of white oolite (Gully Oolite), fine-grained limestone, and greenish grey mudstone set in a matrix of calcite-mudstone; included fragments up to 1 ft across	1 to 12
Gully Oolite:	
White oolite with eroded upper surface; hollows, up to 20 ft across and 10 ft deep, filled by overlying conglomerates	90
Sub-Oolite Bed: hard pale grey crinoidal limestone, recrystallised, against Wick Fault	seen to 10

Traced south-westwards from Bristol into the extensive outcrops of Broadfield Down between Barrow Gurney and Congresbury, the pale current-bedded oolites of the Gully Oolite are indistinguishable from those in the Avon Gorge. They have increased slightly in thickness and in Goblin Combe, in the centre of the area, they are as much as 120 ft thick. The upper calcite-mudstones of the Clifton Down Mudstone have thinned southwards and are not separable from the basal beds of the Clifton Down Limestone, but the lower calcite- and dolomite-mudstones persist over the northern and western outcrops, and at Goblin Combe. In Goblin Combe and for about 2 miles eastward, they are in part replaced by volcanic rocks, including basalts, olivine basalts and their altered representatives, interbedded with tuffs consisting of fragments of these lavas set in a limestone matrix (see p.46). The volcanic rocks are about 40 ft thick and lie at a single horizon within the lower Clifton Down Mudstone; They are approximately the same age as the younger of the two volcanic episodes recorded near Weston-super-Mare—that in the Birnbeck Limestone at Spring Cove (Whittaker and Green, 1983, pp.12–13). There are two other small occurrences of volcanic rocks at approximately the same horizon within the district, one at Tickenham and the other at Cadbury Camp east of Clevedon.

The Goblin Combe Oolite (Bellerophon Beds of the Avon Gorge) can be traced westwards to Failand and Clevedon and southwards to Broadfield Down, where it has thickened to about 125 ft in its type locality in the Goblin Combe ravine. It consists largely of grey oolites and oolitic limestones and it can be generally distinguished from the Gully Oolite by virtue of its darker colour and coarser texture, and by the presence of lenses of crinoidal limestone with fossils, including *Palaeosmilia murchisoni* and bellerophontoids.

The Clifton Down Limestone has extensive outcrops around the northern, western and southern margins of the Broadfield Down Inlier. Although there are changes in the detailed succession that make precise correlation with the Avon section difficult, it is overall very similar. The Clifton Down Oolite facies is well developed above some massive thickly bedded limestones, particularly on the west and south, and it is here given the name Brockley Oolite after its type-locality Brockley Combe, just east of that village. The middle part of the sequence is marked by an increase in the development of silicified masses of *Lithostrotion* and bands and nodules of chert are seen in the massive limestones below the Brockley Oolite (Wallis, 1922, p.212). The uppermost calcite-mudstones however exhibit fewer algal growths and the Concretionary Beds lose their identity southwards across Broadfield Down.

Over most of the western and central Mendips the Gully Oolite and the Goblin Combe Oolite come together, and the combined limestone succession so formed is known as the Burrington Oolite, after its type locality in Burrington Combe. This important section on the northern limb of the Black Down Pericline has been described in detail by Mitchell and Green (*in* Green and Welch, 1965, pp.187–190). The sequence consists predominantly of grey massive fine- to coarse-grained oolites and oolitic-crinoidal limestones, current bedded and dolomitised in part, with a few sporadic layers of grey calcite and dolomite-mudstone. The lower part of the succession, particularly, includes much coarse crinoidal debris, and in the uppermost part thin developments of pale grey calcite- and dolomite-mudstone occur. One such band, 125 ft below the top of the Burrington Oolite, has been called the Rib Mudstone (George et al., 1976, p.16). Oolitic limestone pebbles and pale chinastone pellets occur throughout, particularly in the top half of the succession. The thickness of the Burrington Oolite ranges from 600 to 700 ft (it was measured as 680 ft in Burrington Combe) over most of the western Mendips. Traced eastwards across the central and eastern Mendips, the non-oolitic crinoidal facies in the lower part of the Burrington Oolite becomes increasingly marked and over much of the eastern part of the North Hill Pericline beyond Priddy and in the ground to the east and south it has been separately mapped as the Vallis Limestone, replacing the lower part of the Burrington Oolite sequence. The Vallis Limestone, named after Vallis Vale, near Frome, just to the south-east of the present district, consists of pale grey coarse-grained crinoidal limestones, and bears comparison with the crinoidal limestones of the base of the Gully Oolite in the Bristol area and to similar rocks on Broadfield Down. It reaches a thickness of 450 to 500 ft in the Binegar–Ashwick area and farther east near Mells and Whatley the Burrington Oolite cannot be separately recognised. Thus in the east Mendips, south of the structural depression marking the Malvern Fault Zone the massive Black Rock Limestone is succeeded by Vallis Limestone, Clifton Down Limestone and Hotwells Limestone in an apparently conformable Viséan sequence totalling about 2750 ft in thickness.

In the western and central Mendips the Clifton Down Limestone is divisible into three broad units of overall differing lithologies (Green and Welch, 1965, p.19). The base is ill defined, but is usually taken, as in Burrington Combe, at the onset of the lowest significant development of calcite-mudstones. The lowest division, about 150 to 300 ft thick, consists of alternations of calcite-mudstones, white oolites and dark splintery limestones with scattered *Lithostrotion*; these are succeeded by grey to black fine-grained cherty limestones with sheets of *Lithostrotion*, about 150 to 250 ft thick (Lithostrotion Limestone); and the highest division comprises grey to black calcite-mudstones or chinastones, about 170 to 300 ft thick (Figure 8). On the south side of the Black Down Pericline, in the Cheddar area, the lowest division is exceptionally thick and can be separated into two distinct parts: a lower dark limestone, 100 to 120 ft thick, called the Cheddar Limestone, overlain by 120 to 190 ft of white oolite, separately mapped as the Cheddar Oolite. The boundary with the Burrington Oolite is again ill defined, and when traced eastwards this difficulty is still further emphasised by the apparent passage of the lowermost calcite mudstones into oolites with the consequent merging of the Cheddar Oolite with the Burrington Oolite, which in the Wells area is of considerable thickness. The middle division is characterised by an abundance of silicified masses of the coral *Lithostrotion* and the presence of chert layers and nodules. The silicification is much more strongly marked than that noted in Broadfield Down. The upper division consists dominantly of calcite-mudstones with some algal mudstones, oolites, pisolites and mudstone-pellet beds. They are usually well bedded, but the 'concretionary' structures of the upper part of the Clifton Down Limestones farther north are not so evident.

Stratigraphical palaeontology

For the most part fossils are neither abundant nor varied in the Clifton Down Group, though they do occur prolifically at certain horizons. Corals and brachiopods are the most significant forms for correlation, though they show little real variation throughout the group itself. Molluscs are not common, excepting *Bellerophon* which is abundant locally in the crinoidal limestone facies of the Goblin Combe Oolite. Some poorly preserved bivalves, including *Edmondia*-type shells occur in the Concretionary Beds at the top of the Clifton Down Limestone at Filton and these are associated with *Aviculopecten* and *Parallelodon* (Whittard and Smith, 1943, p.442) and with *Spirorbis* at Brentry nearby. Nautiloids are rare and goniatites have yet to be recorded. The few arthropods that occur are restricted to a few localised marker bands such as the Trilobite Bed noted in the Clifton Down Limestone of the Avon Gorge (see above).

The 'Sub-Oolite Bed' of the northern areas, as already indicated, contains abundant large orthotetoids, papilionaceous chonetoids and other brachiopods of mixed Tournaisian and Viséan aspect, while the Vallis Limestone of the south yields *Haplolasma* aff. *subibicina*, *Palaeosmilia murchisoni*, *Siphonophyllia* cf. *caninoides*, *Sychnoelasma* aff. *kentensis* and the large chonetoid *Megachonetes magna*, all of which indicate a lower Viséan age. The Gully Oolite seldom yields any identifiable corals or brachiopods. Conodants are said to include forms referable to the *Mestognathus beckmanni-Polygnathus bischoffi* Zone (Rhodes et al., 1969).

Palaeosmilia murchisoni and *Lithostrotion martini* are the commonest corals in the Burrington Oolite, the former in the lower part and the latter in the upper part. Brachiopods include *Gigantoproductus* (maximus), *Delepinea* cf. *carinata* and *Davidsonina carbonaria* as well as *Megachonetes papilionaceus* which is common in the lower beds.

The chinastone facies of the Clifton Down Mudstone has few macrofossils, but the interbedded Bellerophon Beds or Goblin Combe Oolite contain *Palaeosmilia murchisoni* as well as abundant bellerophontoids.

The faunas of the Clifton Down Limestone are in general monotonous. The splintery calcite-mudstones at the base yield *Composita ficoidea*, and this form ranges through the formation, being locally abundant. Among other brachiopods which have been used to characterise the Clifton Down Limestone is *Davidsonina carbonaria*. This fossil is found in basinal areas such as the central Mendips and in the northern part of the Coalpit Heath Basin, but it is unknown at Bristol and is not a reliable indicator for correlation purposes. *Linoprotonia corrugatohemispherica* is more widely distributed as are chonetoids of the papilionaceous group. Of the common Viséan corals *Palaeosmilia murchisoni* is normally conspicuous by its absence in the Clifton Down Limestone. *P. murchisoni* first appears at the top of the Black Rock Limestone in the Goblin Combe area of Broadfield Down and is present in the Goblin Combe Oolite, and in the Vallis Limestone and Burrington Oolite of the Mendips. There are few coarse bioclastic rocks comparable with those of the Black Rock Limestone and Goblin Combe Oolite in the Clifton Down Limestone, but the genus is well represented again in the coarse oolitic and crinoidal limestones of the Hotwells Limestone above. By contrast, Lithostrotiont corals are associated with medium- to fine-grained limestones and large unbroken colonies are seen in the lower part of the Clifton Down Limestone of Bristol. In the Mendips, these *Lithostrotion* limestones can be mapped as a consistent and conspicuous marker member, 140 to 150 ft thick, in the middle of the Clifton Down Limestone (Green and Welch, 1965, p.19; Whittaker and Green, 1983, table 1, pp.6,12). *L. martini* is the commonest species and it ranges through the whole of the formation.

The Clifton Down Group ranges in age from the upper part of Chadian through the Arundian to the top of the Holkerian (George et al., 1976, fig.4). The Chadian age of the Sub-Oolite Bed and the Gully Oolite has been confirmed by the identification of the brachiopod *Levitusia humerosa* at the base of the Gully Oolite in the Middle Hope Peninsula, near Weston-super-Mare (Whittaker and Green, 1983, p.22) and of *Delepinea notata* at the top of the underlying dolomite. Since fossils of diagnostic value are generally lacking throughout the Gully Oolite and the Clifton Down Mudstone the base of the Arundian is generally taken at the top of the lower group of calcite-mudstones below the Goblin Combe Oolite. *Delepinea carinata* found in the Burrington Oolite about 150 to 200 ft above the base confirms an Arundian age. The Clifton Down Limestone is of late Arundian and Holkerian age in northern areas. However, the presence of *Davidsonina carbonaria* in the upper part of the Burrington Oolite in Burrington Combe (Mitchell and Green *in* Green and Welch, 1965, p.189) suggests that the base of the Stage should be drawn some distance below the base of the Clifton Down Limestone in the Mendips.

HOTWELLS GROUP

The Hotwells Group comprises the uppermost division of the Carboniferous Limestone in the Bristol–Mendips region, extending upwards from the top of the Clifton Down Limestone or the Middle Cromhall Sandstone (where that formation is present) to the sub-Namurian–Westphalian unconformity. It is made up of two very contrasting facies—an almost wholly calcareous one, the Hotwells Limestone, below and a predominantly arenaceous one, the Upper Cromhall Sandstone, above (Figures 13 and 14). In the south, along the Mendips and south of the Radstock Basin with the exception of the Mells area, the succession consists wholly of limestones, but on the south-east side of Broadfield Down sandstones and shales begin to replace the upper limestones and this process continues progressively northwards until around Tytherington, at the extreme north-western rim of the Bristol Coalfield, the Upper Cromhall Sandstone has virtually replaced the Hotwells Limestone in its entirety.

The best-documented and most complete section is that proved in the Ashton Park Borehole (Kellaway, 1967) where 268 ft of Hotwells Limestone is overlain by 458 ft of Upper Cromhall Sandstone, making a total thickness of 726 ft. This compares with about 750 ft at the southern margin of the Radstock Basin, 740 ft at Winford on the south-east side of Broadfield Down, 950 ft at Wick and 1100 ft in the area north of Yate (Figure 12d). However, these thicknesses are not strictly comparable because of the nonsequence and erosion at the base of the overlying Namurian Quartzitic Sandstone, a feature which becomes markedly accentuated as the rocks are traced westwards. On the western margins of the Bristol and Somerset coalfields, and in the western Mendips, the Hotwells Group is incompletely represented or even absent (Figure 13). This is due to uplift and erosion associated with two tectonic events, the first in pre-Namurian, the second in late Westphalian times, along the line of the Lower Severn Axis. These movements have been responsible for the rapid changes both of facies and thickness, observed in the surviving rocks of the Hotwells Group and for the major intra-Carboniferous unconformities seen in the Clapton and Severn coal basins. Thus the thin Namurian strata of the Severn Coal Basin rest on early Viséan or Tournaisian rocks but all the pre-Westphalian formations (including the Namurian) have been subjected to folding and faulting followed by erosion before the deposition of the unconformable cover of Upper Coal Measures.

Over most of the district there is a well-marked lithological distinction between the relatively unfossiliferous calcite-mudstones at the top of the Clifton Down Limestone and the more massive oolitic and crinoidal rocks of the Hotwells Limestone (Kellaway and Welch, 1955, p.16), which contain abundant corals and brachiopods. The sharpness of the transition does, however, vary. In the Avon Gorge it is somewhat ill defined but in the Long Ashton area, notably at Providence, the uppermost bed of the Clifton Down Limestone is pitted and bored, indicating some kind of hiatus. In the Mendips the junction is relatively sharp, marked by alternations of splintery chinastones and crinoidal limestones over a vertical distance of about 6 ft. Overall the quiet sheltered lagoon conditions represented by the Clifton Down Limestone appear to have given way to open shelf marine conditions of turbulence and high energy deposition.

The type locality of the Hotwells Limestone is in the Avon Gorge, where it is about 175 ft thick. The thickest sequences are, however, seen at Winford, where the dominantly limestone succession is about 700 ft thick, and along the northern margin of the Mendips, where a wholly calcareous sequence ranges from 600 to 750 ft. In the Ashton Park Borehole (Kellaway, 1967, pp.65–67) the Hotwells Limestone is 268 ft thick and consists of hard grey to dark grey foraminiferal crinoidal oolitic bioclastic limestones interbedded with sporadic bands of grey and dark grey shales and mudstones, some of them carbonaceous or calcareous, the whole sequence showing marked rhythmic sedimentation (Figure 14).

Only the basal 100 ft of the Hotwells Limestone is seen in Burrington Combe (Green and Welch, 1965, p.194); for the most part they consist of massive grey granular crinoidal limestones with scattered corals and gigantoproductoids. On the south side of the Black Down Pericline, in the Cheddar–Westbury–sub-Mendip area, limestones, with chert nodules and seams, about 10 ft thick overlying some 20 ft of flaggy, black, splintery limestone, form a useful marker band about 70 ft above the base of the Hotwells Limestone. The upper part of the formation is characterised by a diversity of rock types interbedded with the more typical grey limestones. These include 'rubbly beds' or pseudobreccias, limestones with chert, thin quartzitic sandstones, and, particularly on the north side of the Mendips, purplish shales with nodular, white, porcellaneous dolomitic limestones (Green and Welch, 1965, p.21).

At Wick, the Hotwells Limestone is about 240 ft thick. To the north the formation thins as the overlying Upper Cromhall Sandstone thickens. Indeed, in the area between Tytherington and Cromhall the Limestone appears to have disappeared almost entirely and the Upper and Middle Cromhall Sandstones are mapped together. Farther west towards Itchington, the Limestone reappears and crops out in some small outliers, only the basal beds being exposed. Farther still to the south-west, the Limestone becomes even thicker on the Ridgway at Almondsbury and this development is maintained towards Henbury and Filton. West of a line through Henbury to Olveston, however, the Hotwells Limestone undergoes another drastic change, both of thickness and facies, again with rapid attenuation of the limestone and passage into Upper Cromhall Sandstone.

The Upper Cromhall Sandstone (Kellaway and Welch, 1955, p.18) is by far the thickest and most extensive arenaceous deposit in the Carboniferous Limestone of the district although the exposed rocks are largely confined to areas north of the Vale of Wrington. Over most of the ground it is not well exposed and the best documented section is that of the Ashton Park Borehole (Kellaway, 1967, pp.106–115). Here the Upper Cromhall Sandstone is taken to extend from 1315 ft 1 in to 1773 ft 1 in, a thickness of 458 ft. The sequence consists of grey grits, quartzitic sandstones and calcareous sandstones, sandy crinoidal and oolitic limestones, dark grey shales and mudstones and grey fireclays. Most of the rocks contain pyrite, the oxidation of which is responsible for the strong red colours which characterise the formation at outcrop. The sequence is strongly cyclic, the typical cycle ranging upwards from calcareous beds through mudstone, siltstone and sandstone to seatearth, an order strongly reminiscent of the Yoredale facies of northern England. Such sediments represent a change of environment from the open shelf seas of the Hotwells Limestone to shallow water marginal and deltaic conditions, changing from shallow marine to brackish or estuarine and back to marine again. The presence of mudflats colonised by vegetation is indicated by the occurrence of rootlet-bearing seatearths and mudstones with plant remains. From Bristol the Upper Cromhall Sandstone can be traced south-westwards as far as Cambridge Batch, near Flax Bourton. It is seen again on the south side of Broadfield Down at Redhill, but it is only 60 ft thick at Stoke Lane (Welch, 1931) on the north side of the Mendips and is absent on the southern flanks at Ebbor. This change is due partly to attenuation and partly to pre-Namurian erosion.

To the east of Bristol, the Upper Cromhall Sandstone is well developed in the western part of the Wick inlier (S. Smith, 1930). Here the succession consists predominantly of about 700 ft of interbedded sandstones and mudstones with a 65 ft bed of crinoidal and oolitic limestone—the Castle Wood Limestone—about 175 ft from the base, and a 20 ft sandy limestone a further 200 ft higher. The Castle Wood Limestone contains *Orionastraea* and abundant *Lonsdaleia floriformis*, 10 ft above the base and correlates with the Rownham Hill Coral Bed of Bristol. The latter is named after Rownham Hill near the Avon Gorge (Vaughan, 1905; Kellaway, 1967) where the coral-bearing rock was formerly quarried to be cut and polished at Clifton. Some 50 ft from the top of the Upper Cromhall Sandstone at Wick occurs the Mollusca Bed, a useful marker in separating the Sandstone from the Quartzitic Sandstone (Millstone Grit Series) above. The succession proved on the west side of the Wick Fault appears to be thicker than that seen on the east and north of the inlier, where the Castle Wood Limestone also contains substantial intercalations of sandstone and is not quite so thick.

The boundaries of the Upper Cromhall Sandstone, Quartzitic Sandstone Group (Namurian) and basal Westphalian rocks shown on the Bristol district special sheet (1962) were later shown to be in need of revision (Kellaway, 1967), and

44 CHAPTER THREE CARBONIFEROUS LIMESTONE (DINANTIAN)

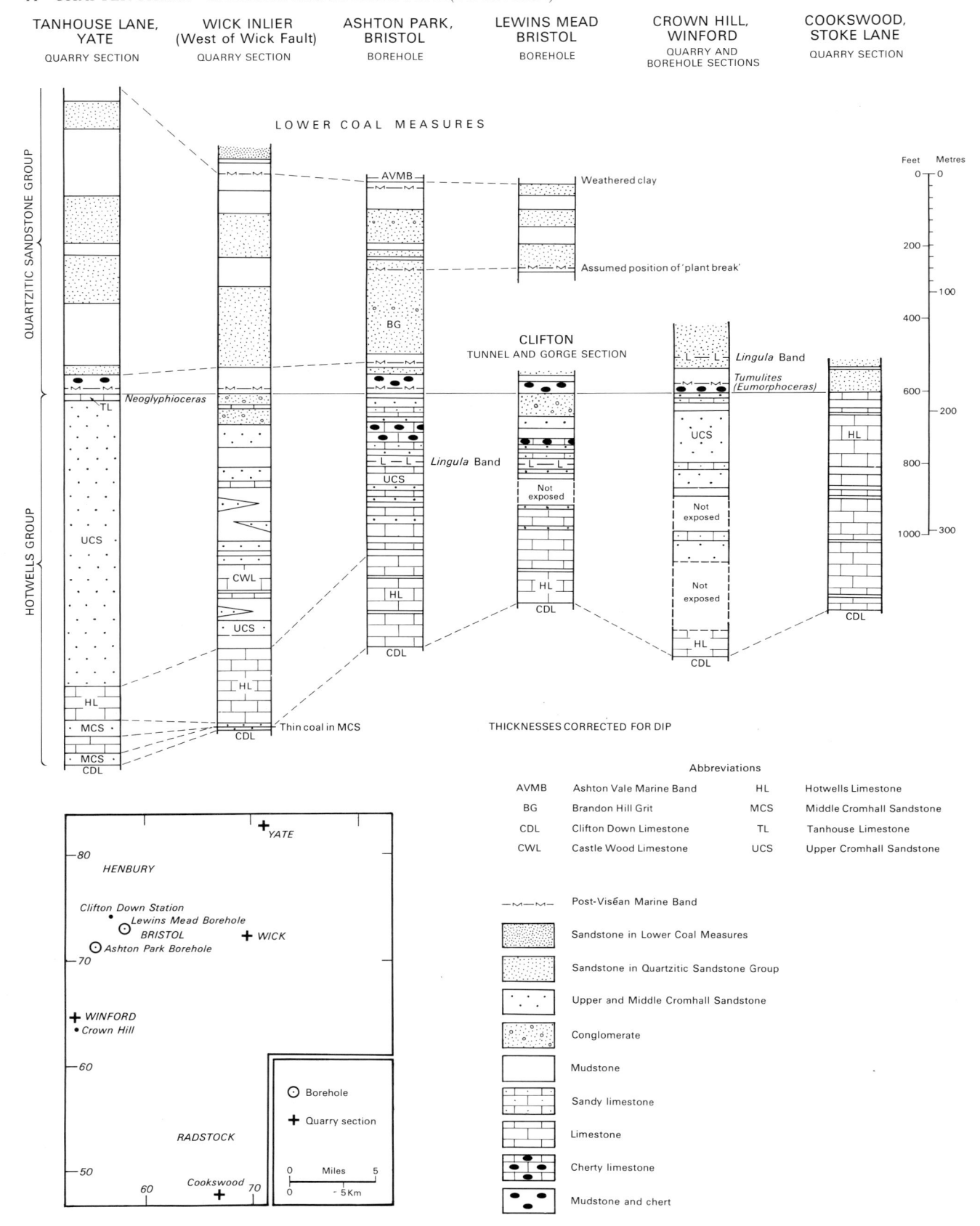

Figure 14 Comparative vertical section of the Viséan and Namurian strata of the Bristol and Somerset Coalfield. Key map shows the sites of the boreholes and measured sections

the revised positions were shown on the Malmesbury (251) sheet published in 1970.

Farther north, along the narrow crop between Yate and Cromhall on the eastern flank of the Dinantian structural depression, the Upper Cromhall Sandstone is nearly 800 ft thick. At the top occurs the Tanhouse Limestone, a thin but useful marker setting an upper limit to the Carboniferous Limestone (Figure 12); it yields *Neoglyphioceras*, *Fasciculophyllum carruthersi*, *Buxtonia*, *Lingula* and *Paladin* at Yate (S. Smith, 1942, pp.335–336).

Traced northwards from Bristol not only does the Hotwells Limestone thin steadily, but the proportion of calcareous strata in the Upper Cromhall Sandstone also diminishes. For about 2 miles along the crop in the Tytherington area, the Middle and Upper Cromhall Sandstones unite to form a single arenaceous unit about 750 ft thick. North-eastwards the Hotwells Limestone reappears and at Cromhall in the central region of the north–south depression the Upper Cromhall Sandstone is separated from the Middle Cromhall Sandstone by 100 ft of oolitic limestone. Here the Upper Sandstone is about 520 ft thick and is capped by a thin brown decalcified sandy limestone, crowded with impressions of *Buxtonia*, presumably the correlative of the Tanhouse Limestone of Yate.

Stratigraphical palaeontology

Of all the Dinantian strata the Hotwells Group is the most remarkable for its rich and varied fauna. Fossils occur abundantly throughout the Hotwells Limestone and in the calcareous beds of the Upper Cromhall Sandstone, particularly in the Bristol region. In general terms the group corresponds to the *Dibunophyllum* (D) Zone of Vaughan (1905), which was divided into a lower, D Subzone and an upper, D Subzone, divisions which correspond with the Asbian and Brigantian regional stages (George et al., 1976, pp.11–12). Because of the major facies change and the lack of faunas at critical levels, the boundary between those two stages cannot usually be drawn with precision in the Bristol district, but in the Bristol and Wick areas it has been taken at the junction of the Hotwells Limestone with the Upper Cromhall Sandstone (George et al., 1976, fig.4).

There is evidence for the existence of a nonsequence at the base of the Hotwells Limestone. The *Daviesiella llangollensis* fauna, which characterises the early part of the Asbian Stage in northern England, has not been found within the present district, and, furthermore, *Davidsonina septosa*, which is restricted to the later part of the stage in northern England and Wales has been found only at the base of the Hotwells Limestone in the Mendips (Green and Welch, 1965, p.21).

The Asbian Stage is characterised by a rich coral-brachiopod fauna. The corals include such typical forms as *Axophyllum vaughani*, *Clisiophyllum* aff. *delicatum*, *Dibunophyllum bourtonense*, *D. bipartitum*, *Koninckophyllum vaughani* (= ϕ), *Lithostrotion junceum*, *L. martini*, *L. pauciradiale*, *L.* cf. *sociale*, *Palaeosmilia murchisoni* and *Syringopora spp.* Among the brachiopods gigantoproductoids are common, including *Gigantoproductus maximus*; *Linoprotonia hemisphaerica* also occurs, as well as athyroids, chonetoids and spiriferoids.

The Brigantian Stage carries many corals in the limestones, including *Dibunophyllum bipartitum bipartitum*, *Diphyphyllum lateseptatum*, *L. junceum*, *L. martini*, *L. pauciradiale*, *Lonsdaleia floriformis*, *Nemistium edmondsi* and *Syringopora spp.* Brachiopods include species of *Athyris*, *Buxtonia*, chonetoids, gigantoproductoids (especially in the lower beds) and spiriferoids.

The mudstones within the Upper Cromhall Sandstone commonly carry brachiopods, including *Buxtonia* and *Lingula*, and bivalves including such forms as *Aviculopecten*, *Promytilus* and *Sanguinolites*.

The rich marine fauna in the upper part of the Hotwells Limestone at Compton Martin contains goniatites (Green and Welch, 1965, p.27) and a diversified gastropod fauna (Batten, 1966) indicating full open-sea conditions. The Brigantian sea penetrated the structural depression east of Bristol and extended beyond Wick in the east. In the west, uplift along the Lower Severn Axis (Kellaway and Welch, 1948) and the Mendips restricted the distribution of the Upper Cromhall Sandstone and the West Mendips Basin was converted into a barrier dividing west Somerset from the Radstock Basin. The connection with the open late Viséan seas lay to the east or south-east of Radstock and became increasingly restricted as time went by. Soon after the deposition of mudstones and cherts with *Tumulites* (*Eumorphoceras*) at Winford the connection with the sea was broken and the depression marking the Malvern Fault Zone gave rise to a series of enclosed basins in which nonmarine sediments were laid down.

CONGLOMERATES OF DOUBTFUL AGE BORDERING THE SEVERN ESTUARY

In 1959 several boreholes were drilled at Kings Weston on the north-western outskirts of Bristol, in connection with the Foul Water Trunk Sewer Tunnel. Borehole D [5463 7783], about 450 yds north-east of Kings Weston House, proved inverted strata with about 16 ft of breccia or conglomerate intervening between the Clifton Down Limestone and Pennant Sandstone. The beds comprised fragments of red and purple mudstone, limestone, dolomite and sandstone set in a mudstone matrix, interbedded with red and purplish grey shales and red and yellow sandstones with some thin limestone and sandy limestone. This Kings Weston Conglomerate is clearly older than Westphalian and younger than the Clifton Down Limestone, and being at least partly calcareous, could well be of late Viséan age. It is clearly a marginal facies which may have been contemporaneous with the thick limestones, sandstones and shales of the Hotwells Group to the east. The Kings Weston Conglomerate crops out over a small area near the site of Borehole D (Figure 18), though it is not now exposed. Where seen it consists of calcareous sandstone, shale and red and grey conglomerate or breccia, composed largely of fragments of dolomite and impure limestone set in a muddy matrix.

Coarse breccias and boulder beds resting on the upper surface of the Drybrook (or Clifton Down) Limestone in the Severn Tunnel (Richardson, 1887) may correlate with the Kings Weston Conglomerate and with other coarse conglomerates found between Dinantian and Westphalian rocks proved in the intake tunnels of Portishead Generating Station [473 773]. Conglomerates found filling hollows in the upper surface of the Drybrook Limestone may also be of late Viséan age.

PETROGRAPHY OF THE VOLCANIC ROCKS OF BROADFIELD DOWN

The following account is based on a detailed examination of thin sections by Dr R. Dearnley. The volcanic rocks which occur interbedded within the Clifton Down Mudstones of Goblin Combe and the ground to the east (see p.24) include basalt lavas and their altered representatives, together with interbedded tuffs consisting of fragments of the lavas set in a matrix of limestone. A previous account of these rocks was given by S. H. Reynolds (1916).

When fresh, the lavas are fine-grained porphyritic basalts and olivine-basalts (some amygdaloidal), which consist of phenocrysts of olivine, augite and pigeonite in a groundmass of andesine-labradorite laths with minor dark interstitial glass, and accessory prisms of apatite and irregular plates of ilmenite. A small amount of alteration is found in even the freshest rocks, but progressive stages of alteration can be traced which lead to alkali-feldspar-chlorite-calcite rocks in which the original igneous texture is still faintly discernible. Alteration of the ferromagnesian minerals is common resulting eventually in calcite-chlorite pseudomorphs after olivine or pyroxene. The feldspars lose their albite twinning during alteration and change in composition to albite associated with calcite. Potash-feldspar may sometimes occur as an outer rim to plagioclase, and in the most highly altered rocks alkali-feldspars are dominant. Patches of pale green chlorite are common and the amygdales have calcite-chlorite-chalcedony cores and chalcedony outer rims.

The change in the composition from andesine-labradorite to alkali-feldspar is a progressive one associated with increasing alteration of the ferromagnesian minerals to chlorite and calcite and with accompanying minor silicification. Since no fresh rocks examined contained orthoclase and no altered rock andesine-labradorite, it seems likely that, contrary to the view expressed by Reynolds (1916, p.40), the alkali-feldspars are not original constituents of the basalts, but are purely secondary, and presumably due to the metasomatic introduction of potash. These alterations must have taken place after the crystallisation of the rocks had been completed, the last phase of which was the formation of interstitial glass, and before the brecciation of the rocks during the subsequent deposition of the tuffs.

The tuffs consist predominantly of basalt fragments, altered (devitrified) shards and sporadic detrital quartz, microquartzite and limestone embedded in a limestone matrix. The basalt fragments are usually highly altered and replaced by calcite and chlorite, although the original texture is commonly preserved as ghost structures. A feature of some specimens is the flattening of the chloritised glassy fragments, but the haematitised lava fragments are not flattened. Scattered quartz and microquartzite fragments occur, as do limestone ooliths and crinoid fragments. The tuffs usually contain finely divided iron oxide, mainly in the form of haematite, which gives the red colouration to the rocks. The basal fragments are extremely irregular and embayed, indicating a close local derivation rather than any transportation by water. The matrix is usually fine-grained calcite-mud, though some dolomite material has been seen. The close association of these limestone tuffs with the altered potash-rich (?spilitic) lavas suggests accumulation under shallow lagoonal conditions.

DETAILS OF STRATIGRAPHY

Details of the outcrops of the Carboniferous Limestone in the area surrounding the northern half of the Coalpit Heath Basin of the Bristol Coalfield are given in the memoirs on the Monmouth and Chepstow (sheets 233 and 250) and Malmesbury (sheet 251) districts (Welch and Trotter, 1961, pp.66–85; Cave, 1977, pp.55–59). In the Mendip Hills in the south of the Bristol district detailed descriptions of the Carboniferous Limestone of the Black Down, North Hill, Pen Hill and Beacon Hill periclines are given in the memoir on the Wells and Cheddar (sheet 280) district (Green and Welch, 1965, pp.23–51). The following account is confined to the area lying west of Bristol, extending through Failand and Tickenham to Clevedon and thence north-westwards to Portishead, and to the extensive outcrops of Broadfield Down about six miles south-west of Bristol.

Since the publication of Vaughan's first account of the Avon section (1905) considerable changes have taken place in the physical condition of the Gorge at Clifton. Many of the best annotated illustrations of the sections described by Vaughan were published in his second paper (1906). The photographs taken there by S. H. Reynolds, before the construction of the Portway, show the gorge as it was when the old railway line from Sea Mills to Hotwells was still in existence. Some of the subsequent changes in the Avon section have been recorded by Reynolds (1920; 1921; 1926; 1936) though only a few of the later photographs (Reynolds, 1936, pl.1, 4–9) show the gorge after removal of the railway and the construction of the Portway. In recent years the construction of deep tunnels extending from the mouth of the Trym near Sea Mills [5509 7600] to Gully Quarry [5621 7441] and thence to Clifton Down Station [5771 7417] has supplied much new information regarding the succession between the Black Nore Sandstone (Lower Old Red Sandstone) and the Quartzitic Sandstone Group (Millstone Grit Series). Owing to structural disturbances and the incomplete nature of the exposures it is difficult to construct an accurate vertical section of the strata in the Avon Gorge. In particular the higher part of the Dinantian sequence has never been adequately exposed. For this reason a borehole was drilled by the Geological Survey at Ashton Park [5633 7146] in 1952–53 to determine the succession in the Millstone Grit and the higher part of the Carboniferous Limestone Series, and this has already been described at length (Kellaway, 1967).

Area west of Bristol

Right (east) bank of the Avon Gorge (Figure 15)

Over recent decades there has been some deterioriation in the condition of the Avon Gorge. Some sections are almost unapproachable and have become grown over. One such area is the old railway cutting [5362 7710] alongside the Portway north of Sea Walls, now largely obscured by debris and vegetation. The Bryozoa Bed and the underlying shale and sandstone in the upper part of the Shirehampton Beds can still be seen, but the main mass of the Lower Limestone Shale is obscured. The succession in the upper part of the Lower Limestone Shale was recorded by Reynolds (1926, pp.318–319) following the construction of the Portway. At Press's Quarry [5596 7469], the most northerly and somewhat inaccessible part of the cliff at Sea Walls, the succession in the Black Rock Limestone is broken by a number of small bedding-plane

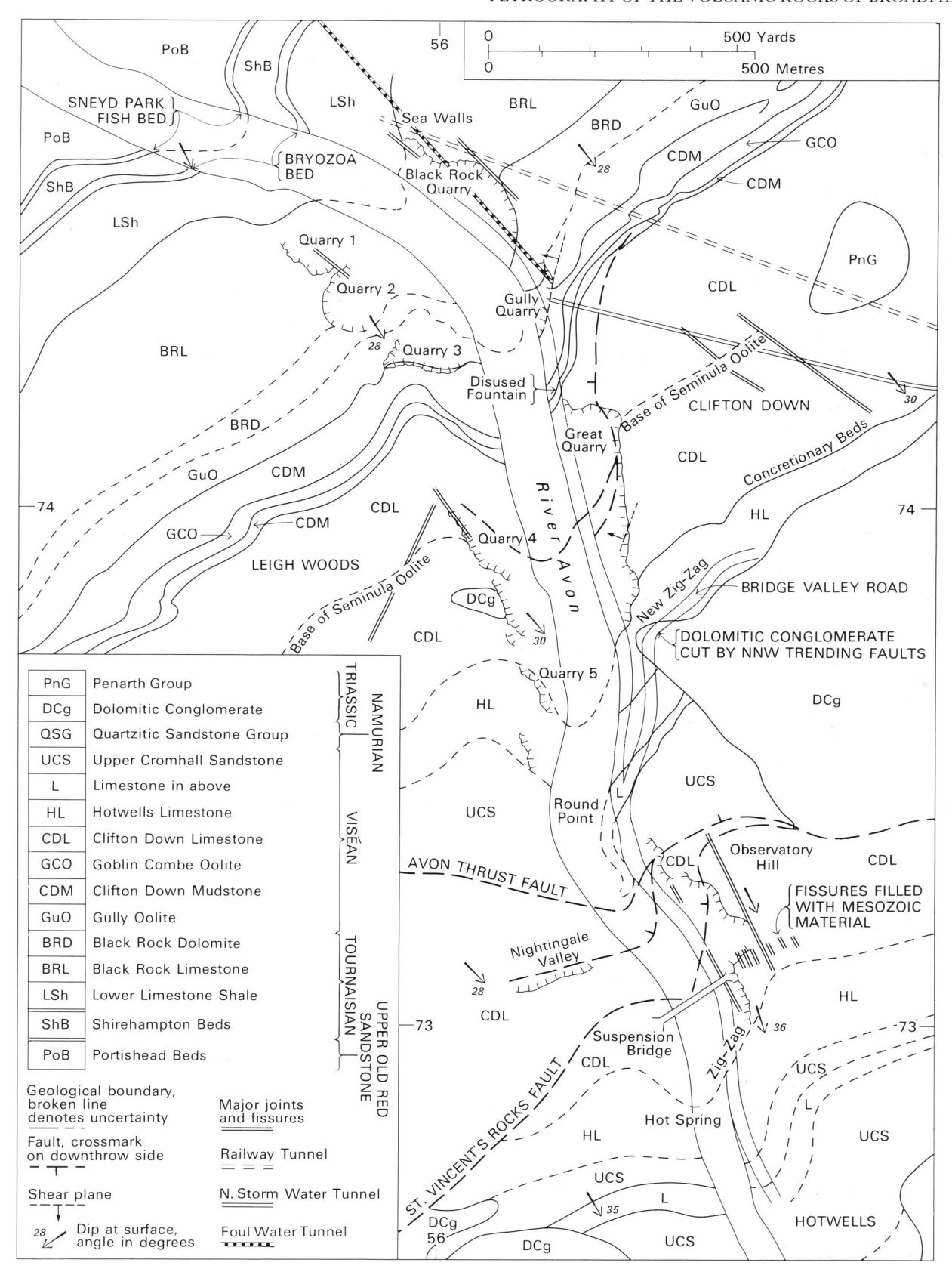

Figure 15 Sketch map showing the geology of the Avon Gorge Bristol. Thermal water is said to issue from the fissure belt between Hotwells and the disused Fountain north of Great Quarry

slides. In the adjacent road cutting thin fossiliferous limestones at the contact of the Lower Limestone Shale and the Black Rock Limestone can still be seen.

The main cliff at Sea Walls forms part of Black Rock Quarry in which the Black Rock Limestone was formerly worked. The base of the limestone known as 'Horizon β' in Vaughan's original accounts corresponds approximately with the dividing line between the Lower and Middle divisions of the Black Rock Limestone in the Portway Tunnel (Figure 11). Reynolds (1921, p.210) gives the thickness of 'Horizon β' as 60 ft, but the measured section in the tunnel appears to be nearer 100 ft, a discrepancy which may be partly explained by structural complications or perhaps by differences in the depth of dolomitisation in the upper part of the Black Rock Limestone.

Succeeding the Black Rock Limestone is a hard purplish grey dolomite, the so-called 'Laminosa'-dolomite, about 100 ft thick and locally including the basal part of the Gully Oolite which has also been dolomitised to varying degrees. On the east bank the Sub-oolite Bed cannot be seen though a few fossils occur in the base of the less strongly dolomitised part of the Gully Oolite (Reynolds, 1921, p.224).

South of Sea Walls, at the entrance to the ravine known as Walcombe Slade or, more familiarly, 'The Gully', massive white Gully Oolite is overlain by shale with bands of calcite mudstone, breccia and dolomite (Reynolds, 1921, pl.xi). The rocks are tolerably well exposed in the upper part of the Gully Quarry section though they are somewhat difficult of access. From this point southwards to the Great or Tennis Court Quarry the section in the Clifton Down Mudstone has deteriorated markedly and many of the features shown on the early photographs of the gorge (Vaughan, 1906, pls 4, 5 and 5a) are almost completely overgrown or degraded. The coarse crinoidal and oolitic limestone with *Bellerophon*, now correlated with the Goblin Coombe Oolite, once formed a prominent crag on the side of the Sea Mills–Hotwells railway (Vaughan, 1906, pls 5, 5a) and above it the upper part of the Clifton Down Mudstone is also poorly exposed.

Between the Gully and Great quarries the base of the Clifton Down Limestone is taken at a prominent band of hard sandy limestone which may mark the southern limit of the Lower Cromhall Sandstone (p.33). It is succeeded by thinly bedded calcite- and dolomite-mudstones with shaly partings seen in the roadside section extending to the northern boundary of the Great Quarry. Here the Diphyphyllum Band was formerly exposed (= Lithostrotion Band of Vaughan, 1906, pl.5). Reynolds (1921, p.227) described the lower part of the Clifton Down Limestone, which includes several characteristic fossil bands, for example the Lithostrotion basaltiforme Band, the Trilobite Bed and the Caninia bristolensis Bed, still to be seen at the time of survey. About 50 ft of strata including the Fluorite Bed are repeated by a low-angle thrust fault seen in the main quarry face (Loupekine, 1953). The corrections made by Loupekine are important for an understanding of the stratigraphy and structure of the lower part of the Clifton Down Limestone in the Great Quarry section. Many of the individual fossil bands cannot be traced for any great distances beyond the gorge, though there is evidence to suggest that the Fluorite Bed may be of fairly wide extent.

For the rocks above the Caninia bristolensis Bed Reynolds description (1921) supplies the best account of the Seminula Pisolite, the Seminula Oolite and the rocks up to and including the Concretionary Beds. This part of the succession in the Clifton Down Limestone was also proved in part of the Northern Storm Water Interception Tunnel lying between Gully Quarry and Clifton Down Station.

The upper part of the Clifton Down Limestone and much of the Hotwells Group are seen in the roadside section adjacent to the Portway. The exposures are not continuous however and even at the time of the construction of the Portway when the new cuttings were measured by S. H. Reynolds and F. W. Wallis (Reynolds, 1926) it was impossible to construct a complete section through the Hotwells Group north of the Avon Thrust. In the Avon Gorge section shown in Figure 13 all the evidence presented by Reynolds has been incorporated, although less than half the thickness of the Hotwells Group can be measured in detail. Apart from poor exposure, there are structural complications which no doubt affect the sequence. The Point Villa section, as it was formerly known (Reynolds, 1926, 1936), is well shown in two photographs published by Vaughan (1906, pls 8,9). Point Villa was demolished during the construction of the Portway, but the least disturbed part of the section, including the richly fossiliferous limestones with *Lonsdaleia* and *Palaeosmilia regia* are well seen in Vaughan's plate 9. This group of limestones, which correlates with the coral-rich horizon at Rhododendron Walk in Blaise Castle Woods (Figure 17), the Rownham Hill Coral Bed and the Castle Wood Limestone of Wick, was also proved in the Ashton Park Borehole (Kellaway, 1967). It is by far the most important of the marker bands in the Upper Cromhall Sandstone and can be traced to Henbury and Over in the north and north-west, to Wick in the east, and at least as far south as Wrington in Somerset.

Near the foot of Bridge Valley Road and immediately to the north of the Avon Thrust Fault the strata forming the footwall of the fault are too highly sheared and folded for accurate measurements. The thicknesses shown in Figure 13 are those measured by Wallis (*in* Reynolds, 1926, p.322) but the original descriptions make it clear that the rocks are highly disturbed.

South of the Avon Thrust the massive limestones of the Clifton Down Group are structurally thickened and distorted. Beyond the St Vincent's Rocks Fault, however, in the tall cliff beneath the Suspension Bridge the rocks have a fairly uniform south-easterly dip and can be seen in an almost continuous section as far as the Colonnade [5657 7286]. Here in the cliffs between the foot of the Zig Zag and the Colonnade the richly fossiliferous Hotwells Limestone can be inspected. *Dibunophyllum bourtonense, Palaeosmilia murchisoni, Lithostrotion martini* and gigantoproductoids are seen in section in the limestone which is one of the most characteristic and easily recognised formations in the Carboniferous Limestone Series. At low water a fissured mass of Hotwells Limestone can be seen projecting through the river bed near the northern end of the Hotwells Landing Stage. From this crag issues the thermal water spring (24°C) which formerly supplied Hotwells Spa. South of the Colonnade the higher parts of the Hotwells Group are mainly concealed by the urban areas of Clifton and the contact with the Namurian has not been seen. However a prominent ridge of hard quartzitic sandstone lying in the upper part of the Upper Cromhall Sandstone forms the foundation of Windsor Terrace overlooking the southern entrance to the Clifton gorge. A short distance to the north of Windsor Terrace lies the General Draper public house [567 727] behind which an exposure of a thin limestone is said to have yielded fossils ascribed by Vaughan (1905, p.199) to Horizon ε, and regarded by him as the top of the D Zone and the upper limit of the Carboniferous Limestone Series.

Clifton – Westbury-on-Trym

The succession in the Avon Gorge (Figures 8 and 15) may be taken as representative of the area of Clifton and Durdham Downs, Bristol. East of the gorge there is little to see at the surface. Much of the surface of the Downs was formerly covered by limestone boulders and griked bedrock but nearly all of this has been removed in historic times, either by lime burners or as ornamental stone for use in rockeries. Most of the quarries, lead workings and small craggy outcrops which formerly existed on the Downs have now been filled in or obliterated.

Between the railway section west of Sea Walls [5573 7479] and Westbury-on-Trym the crop of the Lower Limestone Shale is largely concealed by buildings, roads and gardens. Most of the important sections seen to date have been recorded from temporary exposures such as the one described by Reynolds and Smith (1925) at St Monica's Home [571 762]. Another, smaller section in the Lower Limestone Shale was recorded by Coysh (1926) at Wills Hall (formerly known as Downside), where the Bryozoa Bed was seen [568 757]. One useful section (Figure 16), hitherto undescribed, was recorded by G. A. Kellaway during the widening of the junction of Falcondale and Westbury roads [5722 7680] in 1931, and is summarised below. The succession is given in descending order, Bed 7 being the topmost limestone of the Shirehampton Beds. Thicknesses have been corrected for dip.

		Thickness	
		ft	in
LOWER LIMESTONE SHALE GROUP			
1	Massive crinoidal limestone, fossiliferous, with *Eumetria sp.*, *Leptagonia analoga*, *Avonia bassa*, *Schellwienella crenistria* and other brachiopods	20	0
2	Brown shaly limestone parting	1	0
3	Limestone, crinoidal, thinly bedded and very fossiliferous with *Macropotamorynchus mitcheldeanensis*, *Cleiothyridina roysii* and *Conularia quadrisulcata*	1	0
4	Shale	1	6
5	Limestone, sandy	1	6
6	Clay (weathered shale)	2	0
7	Hard grey crinoidal limestone divided roughly into 3 parts by shaly bands. Top part is barren but basal division is fossiliferous with phosphatic pellets and fish remains similar to the Palate Bed of the Avon Gorge. The phosphatic layer is about 2–4 inches in thickness and is situated about 5 ft above the top of the Bryozoa Bed	10	0
8	Sandy shale, ochre stained but greenish colour at base	2	3
9	Thinly bedded limestone, green with shaly partings and seamed with calcite veins	2	3
10	Bryozoa Bed (Horizon α of Vaughan, 1904). Limestone, part thinly bedded, part massive, sub-crystalline and crinoidal. Colour mainly red but with green and purple layers, becoming fine-grained and shaly towards the base. Fossiliferous with *Schellwienella crenistria*, *Fenestella plebeia*, *Orthoceras sp.* and *Helodus sp.*	9	6
11	Shale, green, purple and red	4	0
12	Limestone, grey when freshly fractured with much calcite veining and red-stained joints. Pseudobreccia at base. Fossils include *Macropotamorhynchus mitcheldeanensis*, *Modiola sp.* and *Bellerophon sp.*	3	0
13	Shale, grey, dark purple and red	3	0
14	Thinly bedded limestone	0	4
15	Thinly bedded sandy limestone	1	0
16	Mottled shale	0	8
17	Green and yellow shale	4	6
18	Limestone and pseudobreccia	0	6
19	Yellow friable calcareous sandstone	2	0

The thickness of the Lower Limestone Shale Group appears to be about 400 to 420 ft in the Sneyd Park–Westbury Park area, though structural deformation makes realistic estimation difficult.

Resting on the Lower Limestone Shale, the Black Rock Group forms the well-marked plateau extending from Sea Walls to Durdham Downs. Only the lower part of the Black Rock Limestone is seen at the surface in the area between the White Tree [5725 7581] and the Red Maids' School [574 770], the upper part of the Black Rock Limestone and the Black Rock Dolomite being concealed by Rhaetic and Lower Lias. Between Eastfield and Southmead the whole of the Black Rock Limestone and Black Rock Dolomite crop out on the plateau which is dissected by the upper reaches of the River Trym. The only exposures of consequence in this area are those seen in the long line of quarries opened along the strike of the Gully Oolite at Southmead (Vaughan, 1905, p.234; Reynolds, 1920, p.95). The southernmost of these quarries has now been filled with tipped material as has the one immediately north of Eastfield Road. The northern quarry, now flooded, and known as Henleaze Bathing Lake [582 777] still shows the top of the Black Rock Dolomite overlain by the Sub-Oolite Bed and Gully Oolite, as first described by Vaughan in 1905.

North of Henleaze Bathing Lake the crop of the Black Rock Limestone and Black Rock Dolomite is now largely built-up, and the surface geology of the Southmead area has been deduced from a limited number of observations made during trenching and building operations. The structure is very complex and the six-inch map (ST 57 NE) can only be a simplified interpretation. In general the Black Rock Group appears to be about 450 ft thick at Westbury-on-Trym and Southmead as compared with about 520 ft in the Avon Gorge (Figure 10). The rocks, however, appear to contain a higher proportion of dolomite, and the characteristic hard purplish grey dolomite ('Laminosa'-dolomite) at the top of the group may be a little thicker than it is in the Avon section.

Caninoid corals are comparatively rare in the Black Rock Limestone of Westbury-on-Trym and Southmead. Most of them are strongly recrystallised and occur in partly dolomitised crinoidal limestone within 20 to 30 ft of the overlying Black Rock Dolomite.

The Gully Oolite appears to have been quarried on Durdham Downs to the north-east of the clump of trees known as the Seven Sisters [5719 7536]. At Westbury Park both the Gully Oolite and the overlying Clifton Down Mudstone are concealed beneath Rhaetic sands and clays, through the position of the incrop has been revealed by shallow excavations, as for example near the junction of North View and Etloe Road [5736 7578] where hard grey dolomite was proved beneath the Rhaetic, and in an electricity substation at the northern end of Downs Park East, where Gully Oolite was encountered at a depth of 20 ft beneath Rhaetic clay and shale. At Eastfield the Gully Oolite emerges again from beneath Mesozoic cover and was worked in the half-mile long belt of quarries referred to above. These terminate in the highly faulted ground which crosses the gorge of the River Trym at Southmead.

The presence of typical Viséan fossils, including *Composita*, was first recognised by Vaughan (1905, p.234) in the Sub-Oolite Bed at a point on the River Trym marked by a dip arrow on six-inch Sheet ST 57 NE [5805 7776]. The Sub-Oolite Bed seldom exceeds 10 ft in thickness and the pale grey crinoidal limestone appears to pass up imperceptibly into the base of the Gully Oolite. Elsewhere in the area dolomitisation has affected not only the top of the Black Rock Limestone but also the base of the Gully Oolite, and it may also have affected the Sub-Oolite Bed resulting in its variable thickness.

In contrast with the Gully Oolite the Clifton Down Mudstone contains too much shaly waste to be attractive to quarrymen. Perhaps for this reason the crop of the Clifton Down Mudstone is traceable over the greater part of Clifton and Durdham Downs from the Avon Gorge to the Stoke Road. The hard crinoidal and oolitic bands which mark the position of the 'Bellerophon Beds' of Failand (Goblin Combe Oolite) can be identified on the surface of the ground and are visible on aerial photographs. Only the basal beds of the Clifton Brown Mudstone ('Caninia'-dolomites) were proved above the Gully Oolite in the Southmead Quarries (Vaughan, 1905, p.234). It seems likely that the Goblin Combe Oolite is too thin to be identifiable north of Durdham Down.

From the Great Quarry in the Avon Gorge the Clifton Down Limestone extends in a north-westerly direction to Durdham Park and St Alban's Church [577 754]. Many quarries and lead work-

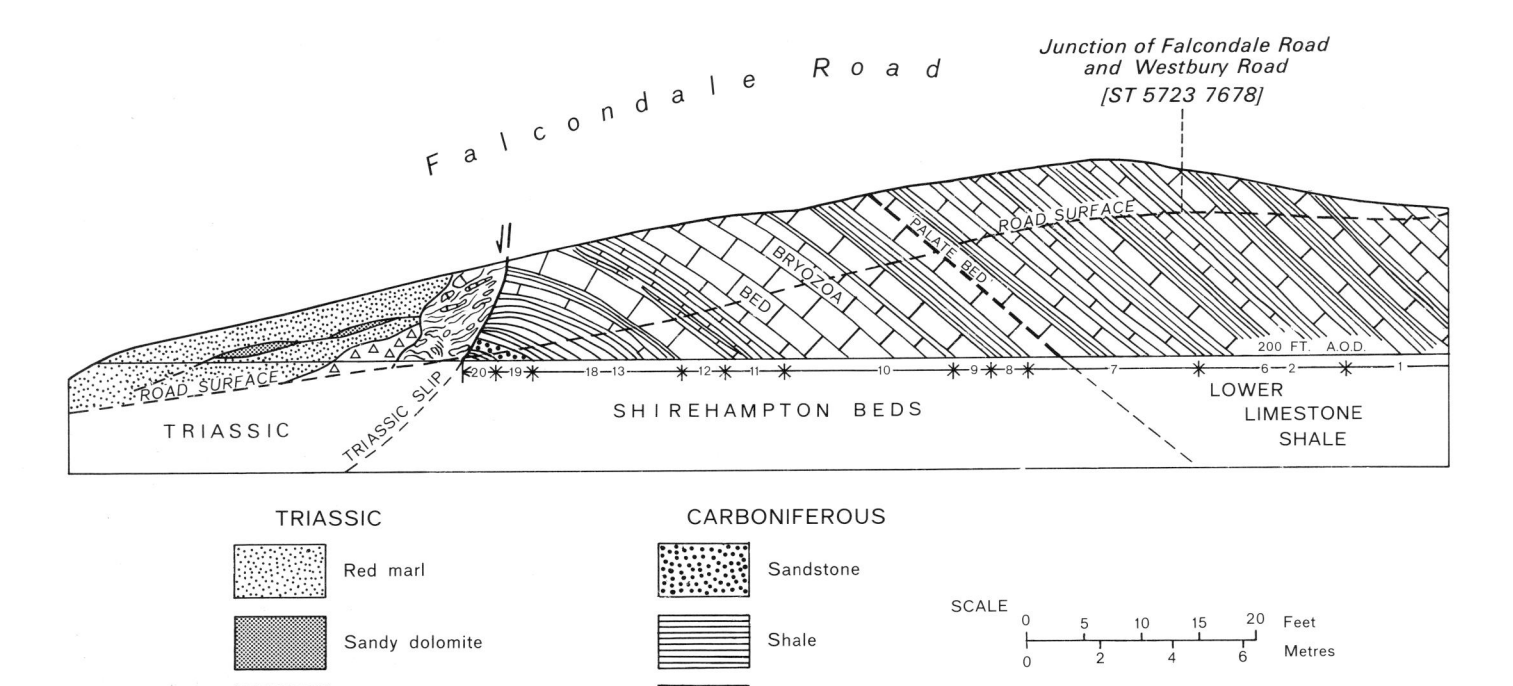

Figure 16 Section through the Carboniferous and Triassic rocks at the southern end of Falcondale Road, Westbury-on-Trym

ings were opened in the past on Clifton Down (Reynolds, 1920, pp.96–97] but most have now been filled in. One large quarry in the upper part of the Limestone was located near the northern end of Pembroke Road [568 743] and another was situated south of Stoke Road near the Reservoir on Durdham Down [5715 7500]. Other small quarries, possibly including some open trench-like excavations near Upper Belgrave Road [570 747] may be lead workings which were subsequently enlarged to supply limestone. One famous quarry was situated in the triangle of land known as 'The Quarry' between Upper Belgrave Road, Worral Road and Black Boy Hill [572 746]. Here the quarry was situated near the top of the Clifton Down Limestone but W. Sander's map (1865) shows the Dolomitic Conglomerate was also present and this rock is the source of the Triassic reptiles described by Etheridge (1870).

The best section recorded in this part of Bristol and still partly visible, is the one in the old quarry known as 'The Glen' [5740 7505]. Reynolds (1920, p.97) recorded the presence of the Lithostrotion basaltiforme Band in the northern part of the quarry, and like the same horizon in the Great Quarry in the Avon Gorge (Vaughan, 1906, pl.6, p.113) it underlies limestones with chert. These are succeeded by the Seminula Oolite (Figure 11) and Concretionary Beds occur at the extreme southern end of The Glen. *L. basaltiforme* auctt. occurs on Durdham Down in old workings [5705 7473], 300 yd west-south-west of St John's School. If this marks the L. basaltiforme Band of Great Quarry and The Glen it would appear that faulting may account for the apparent excessive width of outcrop of the Concretionary Beds in the vicinity of Redland Hill.

North of The Glen the Clifton Down Limestone appears to underlie Durdham Park, but is concealed by Rhaetic and Lower Lias at Westbury Park and Henleaze. A small valley inlier at Halsbury Road [580 758] was described by Vaughan (1905, p.234) as 'in Westbury Park, near Cold Harbour Farm'. According to Vaughan the rocks in the old quarry at Halsbury Road, long since infilled with waste material, are probably in 'the lower part' of the *Seminula* Zone. If correct, this would suggest that the concealed Carboniferous rocks of this area are structurally disturbed. Algal limestone forming the basal part of the Clifton Down Mudstone resting on Gully Oolite was formerly exposed in a small quarry [5826 7797] 100 yd north of the northern end of Henleaze Bathing Lake. This was briefly described by Reynolds (1920, p.95) but it has since been obliterated by tipping. Clifton Down Limestone, mainly hard grey splintery limestone with *Composita* and *Lithostrotion*, was formerly exposed in the bed of the Trym and was proved in many shallow excavations between Charfield Road and Coleford Road during building operations at Southmead. This area marks the northern end of the tract of exposed Carboniferous Limestone Series though the Lower Carboniferous rocks have been proved by boring to extend in a north-easterly direction towards Filton. At Pen Park E Borehole [5892 7889], 600 yd south-east of Pen Park Hole, Clifton Down Limestone was proved beneath about 30 ft of Rhaetic and Lower Lias strata. Greater thicknesses of Mesozoic rocks (including Keuper Marl) were encountered in the Bristol Aeroplane Co.'s Filton Nos. 1 and 2 boreholes (pp.169–170). In both these holes the Hotwells Group was proved, showing that the higher beds of the Carboniferous Limestone Series are present beneath the Mesozoic at Horfield and Filton.

The gully up which the New Zig Zag path leads from the Avon Gorge on to Clifton Down at the upper end of Bridge Valley Road [566 739] is cut in Hotwells Limestone. From here the crop extends north-eastwards towards The Avenue [5696 7423] where it is cut off by the overlap of the Dolomitic Conglomerate. From here northwards the Hotwells Group is entirely concealed by Triassic rocks though it is present in the Northern Storm Water Interceptor Tunnel, the contact of the Clifton Down and Hotwells groups being proved at mean sea level at a point 433 yd west-north-west of the eastern portal of Clifton Down Railway Tunnel. The base of the Quartzitic Sandstone Group was seen in the Northern Stormwater Tunnel at Clifton Down Station [8766 7417] but the Hotwells Group was incompletely exposed, being in part cut out by Triassic rocks.

South of the Avon Thrust Fault the Clifton Down Limestone forming Observatory Hill crops out on the broad ridge extending eastwards to Christ Church, Clifton [5703 7337]. Further east in the vicinity of Clifton Park the Carboniferous rocks are concealed by Dolomitic Conglomerate. The Hotwells Group has a more extensive crop in Clifton extending from the gorge in a north-easterly direction to the Victoria Rooms [5774 7346]. A large cirque-like depression at Hotwells marks the position of an embayment filled with Triassic sandstone and breccia dominated by the steep curving slopes on which much of 18th century Clifton, including Royal York Crescent, was built. Between Hope Chapel Hill [569 726] and The Triangle [579 732] the contact of the Hotwells Group and Quartzitic Sandstone Group has not been precisely located. However, it is thought to run from near Goldney House [5742 7284] across the valley at Jacobs Well Road and thence by way of the foot of the steep slope on the northern face of Brandon Hill to the vicinity of Berkeley Place [5778 7314].

Brentry – Henbury – Shirehampton

The two ridges of Carboniferous Limestone surrounding the Westbury Anticline meet in the tract of featureless plateau south of Brentry Hill and north-east of Westbury-on-Trym. Most of this area is now built up but over much of the hinge area at the centre of the 'anticline' the Carboniferous rocks are concealed by a thin unconformable cover of Rhaetic clay. The structure throughout this area is highly complex and the stratigraphy of the Carboniferous Limestone Series is imperfectly known. North of the extensive area of Rhaetic and Liassic rocks at Southmead a narrow ridge of exposed Carboniferous Limestone extends from Upper Knole [5860 7935] to Brentry. This consists of vertical or steeply dipping Clifton Down Limestone overlain by a thin representative of the Middle Cromhall Sandstone, succeeded in turn by the Hotwells Limestone.

A quarry, now obliterated by tipping, formerly existed at Upper Knole [583 591] on the south side of the Brentry–Charlton road; the section in the Concretionary Beds, was on the northern side of the quarry and not the southern side as stated by Reynolds (1920, p.971), and comprised limestone with *Spirorbis* and poorly preserved bivalves, algae and stromatolitic beds and limestone with *Chaetetes septosus*. Between the quarry in Hotwells Limestone north of the Brentry–Charlton road and the much larger quarry on the south side the Middle Cromhall Sandstone crops out. This sandstone gives rise to a slight ridge which can be traced from a point 150 yd north-west of Pen Park Hole in a south-westerly direction to Brentry Colony [5788 7896].

The general succession at Upper Knole is:

	Thickness ft
Hotwells Limestone, exposed	about 200
Middle Cromhall Sandstone	40 to 60
Clifton Down Limestone	
Concretionary Beds	about 100
Pale grey limestone with bands of oolite and abundant *Composita*	about 200
Hard grey limestone with *Lithostrotion*	about 50

About 150 yd south-west of the large quarry at Upper Knole a smaller quarry [5808 7895] described by Reynolds (1920, p.97) as 'east of Brentry Farm', appears to have been in the lower part of the Clifton Down Limestone. From this point the limestone ridge extends in a south-westerly direction towards Brentry Hill, where extensive sections in quarries, which are now mostly overgrown or filled in, were seen by Vaughan (1905, p.235) and Reynolds (1920, pp.97-98). A section in a cutting on the east side of the road at Brentry Hill [5739 7848] is still visible. This shows the upper part of the Clifton Down Limestone including the Seminula Oolite and Concretionary Beds, the latter being about 100 ft thick according to Reynolds (1921). Owing to structural disturbances the thickness of the Clifton Down Limestone hereabouts is difficult to determine but is likely to be of the order of 500 to 600 ft. The rocks beneath the Seminula Oolite have not been extensively worked, but an old quarry lying immediately west of the junction of Charlton Road and Passage Road [5730 7830] formerly showed hard splintery calcite-mudstone or chinastone characteristic of the base of the Clifton Down Limestone (Reynolds, 1920, p.98).

From Brentry Hill westwards to Henbury Hill the Clifton Down Limestone forms a well-marked limestone ridge bounded on the north by the steep wooded slope of Sheep Wood. Much of the area has now been built up and the structure of the ground to the south is so complex that it is impossible to map the various formations with any certainty. However it would appear that all the rocks including the Clifton Down Mudstone, Gully Oolite, Black Rock Group and Lower Limestone Shale are present.

At Henbury Hill it is once more possible to determine the succession in the Viséan rocks and the underlying Black Rock Group. On the steep wooded slopes on the south side of Coombe Hill the steeply dipping Black Rock Limestone appears to be about 400 ft thick overlain by about 100 ft of Black Rock Dolomite. The thickness of the overlying Gully Oolite and Clifton Down Mudstone is not precisely determinable but is probably of the order of 400 ft.

Figure 13 shows the generalised section of the Clifton Down Limestone and Hotwells Group at Henbury. Most of the measurements have been made in Henbury Hill Quarry [566 782] (Plate 6), now disused and partly tipped, or in the cutting alongside the path leading from Henbury Hill Lodge [5668 7818] to Blaise Castle House [5620 7873]. Vaughan (1905, p.235) recorded an exposure by the side of the Rhododendron Walk (Figure 17) which yielded *Lonsdaleia floriformis* and other corals characteristic of the Rownham Hill Coral Bed and the Castle Wood Limestone of Wick (p.39). This appears to be represented by a coarse crinoidal and oolitic limestone intercalated in the Upper Cromhall Sandstone.

Generalised succession at Henbury Hill (see Figure 13):

	Thickness ft
QUARTZITIC SANDSTONE GROUP (NAMURIAN)	
HOTWELLS GROUP	
Upper Cromhall Sandstone	
Hard quartzitic sandstone and quartz-conglomerate	about 150
Coarse-grained crinoidal and oolitic limestone (= Rownham Hill Coral Bed)	about 40
Sandstone	about 70
Sandy limestone	15
Sandstone	60
Hotwells Limestone	
Crinoidal and oolitic limestone with *Palaeosmilia murchisoni* and productoids	about 100
CLIFTON DOWN GROUP	
Middle Cromhall Sandstone	15
Clifton Down Limestone	
Concretionary Beds	50
Pale grey limestone	75
Seminula Oolite	80

Grey limestone with chert 100 ft from top	250
Coombe Hill Thrust	

No fossils have been found in the structurally disturbed quartzites and quartz-conglomerates which are developed above the limestones of Rhododendron Walk (Figure 17). Their age, therefore, is uncertain. It would seem that the conglomeratic beds are almost certainly Lower Carboniferous, while the highest quartzitic sandstones seen in the centre of the synclinal fold near Rhododendron Walk are thought to be of Namurian age though the Winford cherts have not yet been found at Blaise Castle Woods.

South of the Kings Weston Hill ridge the oldest Tournaisian rocks have been proved from time to time in excavations made during drainage and building operations, more particularly in that part of Coombe Dingle [c.552 777] bounded on the north by Southside Wood, on the east by Grove Road, on the west by Aldercombe Road and on the south by Westbury Lane. Over much of this area the Carboniferous Limestone rocks are concealed by a thin veneer of Dolomitic Conglomerate, and even where they reach the surface they are red-stained and weathered. However, it has been possible to identify the Lower Limestone Shale Group in a long trench section running from Westbury Lane [5532 7736] to the southern margin of Southside Wood. Here shales interbedded with thin limestones, including the Bryozoa Bed, were proved, all red-stained and overlain by patches of Triassic red marl and conglomerate. About 35 yd south of the southern boundary of Southside Wood crinoidal limestones of Black Rock Limestone type were proved. Several other excavations made in the vicinity of Grove Road [5537 7781] have shown beyond doubt that the basal Black Rock Limestone is inverted along an east–west strike, and in all probability is in sequence with vertical or slightly overfolded Lower Limestone Shale exposed to the south.

The Black Rock Limestone crops out on the southern flanks of the wooded ridge of Kings Weston, the higher parts of the ridge being formed of Black Rock Dolomite and Gully Oolite. In the disused Kings Weston Quarries [547 774] inverted Black Rock Dolomite, Gully Oolite and contorted Clifton Down Mudstone were all formerly well exposed (Figure 18). The top of the Black Rock Limestone was seen at the southern entrance to the quarry succeeded by about 100 ft of Black Rock Dolomite and 80 to 90 ft of Gully Oolite. The Clifton Down Mudstone consisted of 50 ft of shales with well marked breccia bands, all highly contorted and folded. At Kings Weston Hill the Black Rock Limestone, much of it strongly dolomitised, forms the southern slopes and farther west the Black Rock Dolomite extends as far as Penpole Wood. Gully Oolite and Clifton Down Mudstone crop out in a narrow belt on the north side of the Black Rock Dolomite about 250 yd south and south-east of Kings Weston House, but farther west these rocks are cut out by the overlap of Triassic rocks, mainly Dolomitic Conglomerate. In the Foulwater Tunnel beneath Kings Weston, Pennant sandstones (Upper Coal Measures) rest unconformably on Clifton Down Limestone, or locally on quartzitic sandstones of the Millstone Grit (Figure 18).

Left (west) bank of the Avon Gorge (Figure 15)

About 430 yd west-south-west of the western entrance to the Clifton Down Tunnel a band of conglomerate, thought to mark the top of the Old Red Sandstone, is overlain by about 85 ft of thick-bedded shale, sandstone and limestone, above which lies the Bryozoa Bed. The lowest bed in this sequence, a red and green mudstone, has yielded scales of a Rhizodont fish (?*Strepsodus*) according to Vaughan (1906, p.101). The Bryozoa Bed crops in the river bank and railway cutting [5557 7465] and is overlain by 50 ft of thinly bedded shales, mudstones and limestones (Reynolds, 1921, pp.219–220). *Vaughania vetus* has been found in the upper part of the sequence. Some distance above about 35 ft of fossiliferous limestone and shale, representing the highest part of the Lower Limestone Shale have also been recorded (Reynolds, 1921, p.220) but it is not until the Black Rock Limestone is reached that the section on the left bank becomes more continuous.

Ever since Vaughan first described the sequence, the principal quarries on the left bank have been known by numbers (Reynolds, 1936, pls.10–13, 15). Quarries 1 and 2 [5575 7450; 5585 7440], together with the nearby river-bank and railway sections, prove almost the whole of the Black Rock Group, including the Black Rock Dolomite or 'Laminosa'-dolomite. Quarry 3 [560 743], and the adjacent railway section, are in Gully Oolite, underlain by Black Rock Dolomite and overlain by the basal beds of the Clifton Down Mudstone, consisting mainly of calcite- and dolomite-mudstone and shale with algal limestones.

The greater part of the Clifton Down Mudstone is very poorly exposed, but a prominent band of hard grey crinoidal and oolitic limestone can be seen on the wooded side of the gorge and crossing the railway midway between Quarries 3 and 4. This represents the Bellerophon Beds or Goblin Combe Oolite and can be traced almost continuously to Failand and Tickenham and thence to Clevedon. Quarries 4 and 5 [561 739; 562 738] show the greater part of the succession from the Trilobite Bed to the lower part of the Hotwells Limestone, although that part of the sequence between the Seminula Oolite and the Concretionary Beds is poorly exposed. In Quarry 5 many of the sharply defined bedding planes are slides on which appreciable movement has taken place; some may be seen in the river cliff beneath the towpath (Vaughan, 1906, pl.12; Reynolds, 1936, pl.10). In this area a series of very strong joints or near-vertical shear-planes trending north-west–south-east cut the Clifton Down Limestone and the Hotwells Limestone, and controlled the form of the old working faces in Quarries 4 and 5. From Quarry 5 to the footwall of the Avon Thrust Fault at Stoneleigh Camp [562 733] the rocks are intensely sheared and distorted.

Beyond the hanging wall of the Avon Thrust Fault, folded, thrust-faulted and sheared Clifton Down Limestone forms the steep wooded cliffs of Nightingale Valley and Burgh Walls. About 140 yd south of Clifton Suspension Bridge [5645 7288] the base of the Hotwells Limestone crosses the towpath, but most of the section between this point and the foot of Rownham Hill has long been overgrown.

Abbots Leigh – Ashton Park – Long Ashton – Failand – Wraxall

Westwards from the Avon Gorge the shales, thin sandstones and limestones of the Shirehampton Beds form a belt of relatively low-lying land bounded on the south-east by a strong feature formed by the Bryozoa Bed, which can be traced for about 600 yd from the Gorge in a south-westerly direction. About 130 yd south-west of the Lodge [5508 7436] there is an old quarry with coarse oolitic and crinoidal limestone dipping 30° south-eastwards; this may be the quarry described by Coysh (1926, p.324) as being about 400 yd north-east of Abbots Leigh Church, though the main workings are somewhat farther away. Coysh pointed out that the Bryozoa Bed was overlain by oolite. In fact only 6 ft out of a total of 25 ft of limestone is made up of crinoidal and bryozoan limestone and it seems that here there is a passage of crinoidal into oolitic limestone. Reynolds (1918, pp.189–190) and Coysh (1926, p.324) noted that oolitic limestone occurs both above and below the crinoidal and bryozoan limestone in the Abbots Leigh – Failand area.

Above the Bryozoa Bed the main mass of the Lower Limestone Shale gives rise to the well-marked hollow dominated on the south-east by the scarp of the Black Rock Limestone. West of Yew Tree Cottage [5346 7263] the Lower Limestone Shale Group is so disturbed by folding and faulting that only the broad outlines of the stratigraphy can be made out. Isolated masses of disturbed limestone appear within the main body of shale, particularly in the area to the south of Failand Farm [5257 7255]. The lowest beds ap-

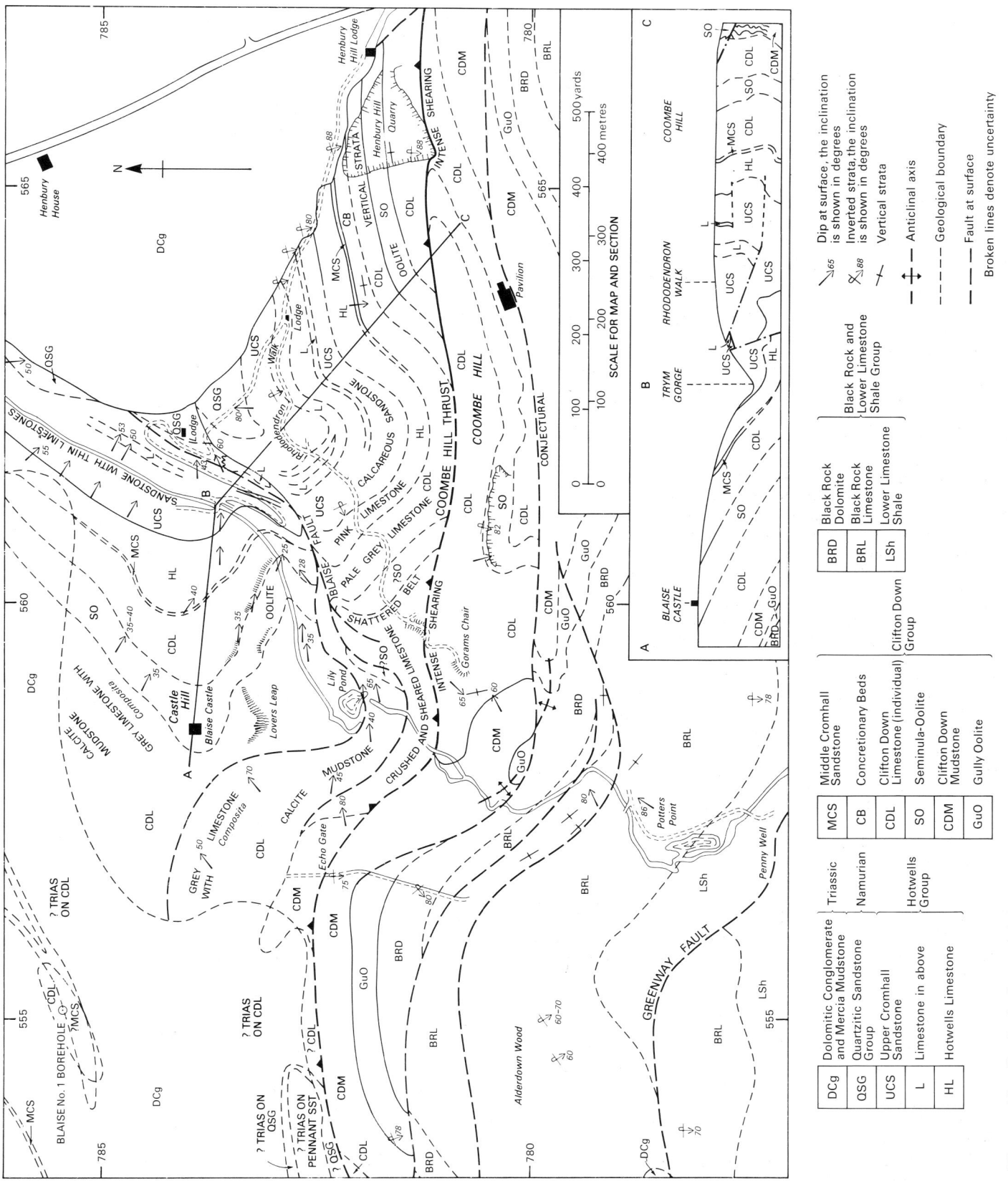

PETROGRAPHY OF THE VOLCANIC ROCKS OF BROADFIELD DOWN 53

Figure 17 Geological map of the Trym valley and Blaise Castle showing the structure of the Carboniferous rocks

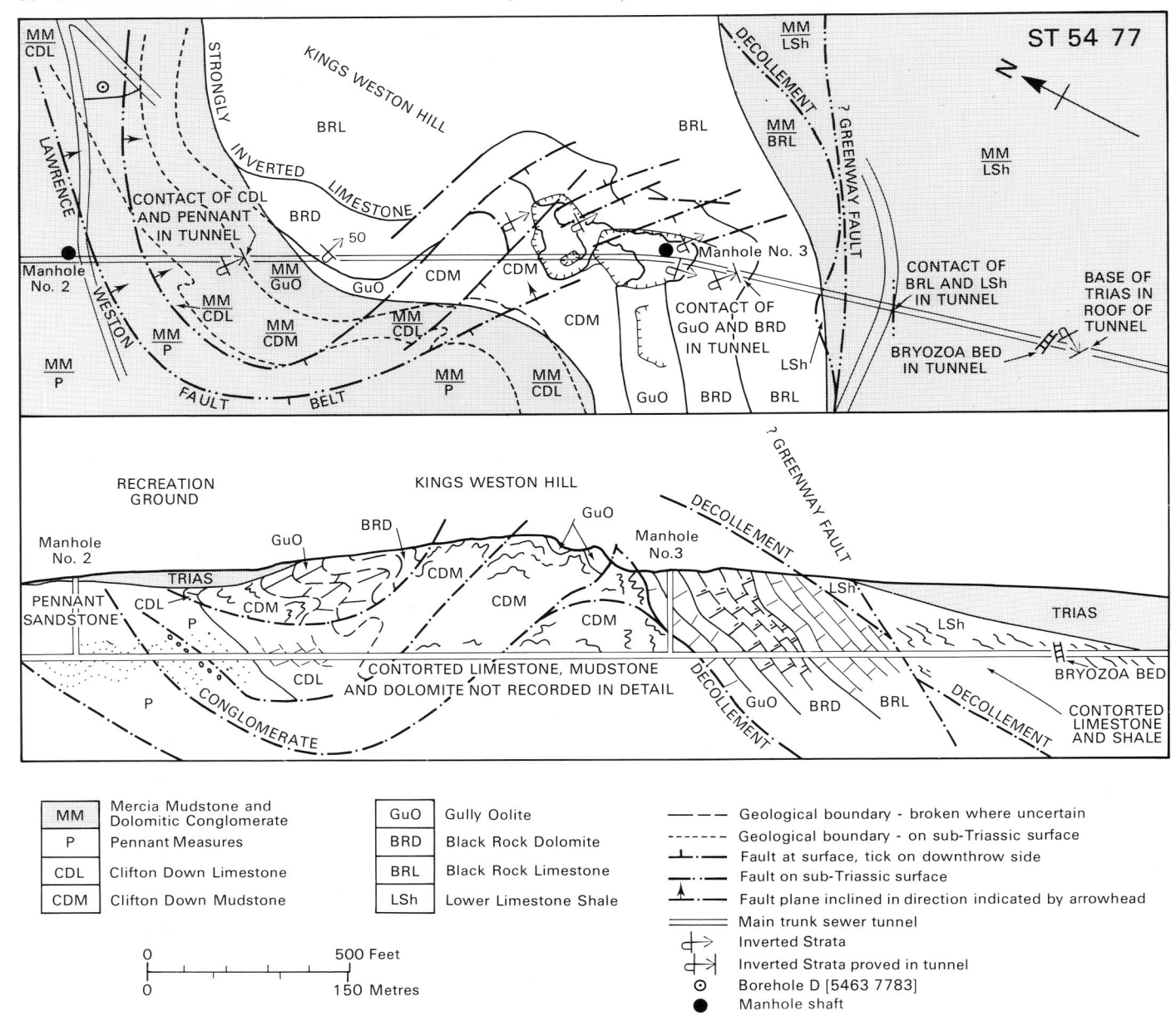

Figure 18 Structure of the southern end of Kings Weston Hill, Bristol. Small masses of Namurian quartzite sandstone are present between the Clifton Down Limestone and the Pennant Measures

pear to consist of flaggy sandy limestone, seen near Yew Tree Cottage. To the east of Failand Farm a band of shale separates crinoidal and shaly limestones from the lower flaggy sandy beds. At Ox House Bottom [520 727] the flaggy limestone has been worked, and from here westwards the shales are violently disturbed, though traces of the flaggy beds and the overlying crinoidal limestone can be seen. To the north of Failand Hill House the crop is displaced once more by the Avon Thrust Fault. West of Racecourse Farm [505 734] the basal sandy limestone and the overlying shale with thin crinoidal limestone can still be identified. Near Charlton House [494 737] and Moat House Farm [487 731] the outcrop of the Lower Limestone Shale Group is over 600 yd wide. This can only be due to incompetent folding and crushing due to movement on the Avon Thrust Fault and its associated structures.

Little is known of the Black Rock Limestone sequence south of Abbots Leigh but towards Round Hill Clump [538 726] it becomes increasingly disturbed in the vicinity of the Avon Thrust Fault. West of Fifty Acre Plantation the Black Rock Limestone forms a plateau which extends to Failand. The rocks are exposed intermittently, mainly in small workings on the scarp face, but they are tectonically disturbed. Between Failand and Wraxall they are again only poorly exposed and are so sheared and contorted that the sequence and thickness can only be conjectured.

From Quarry 3 in the Avon Gorge the Black Rock Dolomite and the overlying Gully Oolite can be traced south-westwards through Leigh Woods to the point where they are cut off by the Avon Thrust Fault [541 726] near Quarry Plantation, where the Gully Oolite was formerly quarried. At Wraxall Piece an old quarry in Gully Oolite

[528 718] was first described by Vaughan (1905, p.213). He also noticed a small quarry on the north side of Wraxall Piece where oolitic limestone yielded orthotetoids and chonetoids, presumably indicating the presence of the Sub-Oolite Bed. Evidence of the Sub-Oolite Bed has also been found on the north side of the Bristol-Clevedon road about 200 yd south-west of Failand Lodge [514 721].

The Bellerophon Beds or Goblin Combe Oolite makes a rib-like feature through Leigh Woods and across the open fields to 200 yd north of Upper Farm [544 726] where it is in turn cut off by the Avon Thrust Fault. The Clifton Down Limestone is poorly exposed in Leigh Woods but the base of the Seminula Oolite can be followed from Quarry 4 to the Avon Thrust Fault about 230 yd south-south-west of the Lodge [5502 7329] on the Bristol–Portishead road.

South of the Avon Thrust Fault a belt of highly disturbed strata runs from Stoneleigh Camp to Upper Farm bounded on the south by a parallel subsidiary fault. Between the two faults the disturbed sequence includes the higher beds of the Black Rock Limestone, the Black Rock Dolomite, Gully Oolite, Clifton Down Mudstone (with Goblin Combe Oolite) and the greater part of the Clifton Down Limestone, though individual thicknesses are difficult to determine. A well-defined ridge marks the crop of the Goblin Combe Oolite at Beggar's Bush on either side of the Bristol–Portishead road [5524 7310]. Reynolds (1918, p.193) refers to a quarry, now obliterated, in Leigh Woods 'opposite the end of Beggar's Bush Lane where dolomitised oolite contains pebbles of oolite apparently due to pene-contemporaneous erosion'. In the northern corner of Ashton Park the Goblin Combe Oolite is highly contorted and faulted, at least one of the faults being mineralised. Passing in a south-westerly direction the ridge formed by the oolite crosses Beggar's Bush Lane, and after being slightly displaced by the southern fault, extends along Beggar's Bush Lane to the crossroads at the south-western corner of Fifty Acre Wood, where it appears to split into two distinct beds of hard grey oolitic and crinoidal limestone, with a total thickness of about 50 ft. It can be followed by way of Long Wood and Wraxall Piece to one of two 'Bellerophon-Beds' quarries [523 717] mentioned by Vaughan (1905, pp.212–213), who estimated its position as being about 150 ft above the Gully Oolite. From here the upper part of the Goblin Combe Oolite can be traced as a ridge westwards to the vicinity of the old limekiln [5173 7193] and thence to the other of Vaughan's two 'Bellerophon-Bed' quarries south of Failand Inn [5142 7196]. The lower part of the Goblin Combe Oolite, thinner and more fine grained, can be traced as a subsidiary feature lying some 50 to 70 yd north of the main feature. About 250 yd south-west of Failand Inn a quarry in work in 1950 showed 40 ft of coarse current-bedded oolite with *Bellerophon* and *Euomphalus*, the beds having a southerly dip of 22°. This quarry is a continuation of the one mentioned by Vaughan.

To the west of Failand the thickness of the Goblin Combe Oolite is difficult to estimate partly because of structural complications and partly because of the thickening of the intervening mudstone. The greater width of outcrop near Sidelands Cottages [4968 7213] is due largely to folding and shearing along the Sidelands Fault.

Within the area bounded by Beggar's Bush Lane, Rownham Hill, Providence and Long Ashton lies the interesting ground of Ashton Park and Ashton Hill. Here the oldest formation is the Clifton Down Limestone which crops out on the plateau of the north-western part of the Park. The lower part of the formation consists mainly of calcite-mudstone, pseudobreccia and hard splintery limestone with *Composita sp.* and *Lithostrotion sp.*, with a cherty band near the top. This is overlain by grey oolitic limestone and pale grey oolite (Seminula Oolite) passing up into oolitic, fine-grained and stromatolitic limestones with the Concretionary Beds in characteristic form at the top.

The Clifton Down Limestone is succeeded by the Hotwells Limestone, partly crinoidal and oolitic and commonly richly fossiliferous. In a small exposure 470 yd west-south-west of the entrance lodge at the top of Rownham Hill pale crinoidal limestone with *Dibunophyllum* was observed. Pink and grey oolitic and crinoidal limestones with *Palaeosmilia murchisoni* and other corals and brachiopods can be traced in a south-westerly direction to New Barn [545 717] and thence to Pill Grove and Coombe Plantation where the crop meets the Providence Fault. An extensive cover of Triassic rocks hides much of the Hotwells Group in Ashton Park. The contact of the Hotwells Limestone and the overlying Upper Cromhall Sandstone is exposed here and there in Deer Park, but mostly it is concealed either by Triassic rocks or Pleistocene deposits. The full succession of the Hotwells Group was proved nearby in the Ashton Park Borehole (Kellaway, 1967).

West of the Ashton–Failand road lies the isolated mass of Ashton Hill, composed partly of limestone, shale and sandstone of the Hotwells Group, and partly of hard quartzitic sandstone and mudstone of the Quartzitic Sandstone Group. The rocks are very poorly exposed; Reynolds (1918, p.196) refers to 'a considerable quarry in massive grey, coarsely crinoidal and foraminiferal limestone … 300 yd W of Convalescent House'. Another section no longer visible was the old cricket pitch section described by Lloyd Morgan (1885, pp.163–165) consisting of 89 ft of alternating bands of limestone, shale and sandstone belonging to the Upper Cromhall Sandstone.

About 330 yd east of the Chapel at Providence crinoidal limestone with *Lithostrotion pauciradiale* was seen in a small exposure, and in another old quarry 85 yd east of the Miners Arms [5383 7063] crinoidal limestone with *L. pauciradiale* and *P. murchisoni* was seen. This may be the Rownham Hill Coral Bed and it is overlain by sandstone and shale.

There are few exposures in the Upper Cromhall Sandstone; perhaps the best was an old quarry in a plantation in the Deer Park, 430 yd west of Ashton Court. Here decalcified limestone with abundant *Buxtonia scabriculus* and other fossils marks one of the shelly bands so common in the north-east of the district.

Several quarries in the Clifton Down Limestone are located in the vicinity of the crossroads north of Providence. The largest is Longwood Quarry [535 715] near Round Plantation, connected by a tunnel under the Providence–Abbots Leigh Road with a smaller quarry in Ashton Park. In the larger quarry a well-marked chert band is overlain 100 ft higher in the sequence by paler grey Seminula Oolite. This chert band may not be the chert found immediately below the Seminula Oolite in Great Quarry in the Gorge, but the equivalent of silicified limestone found in the lower part of the Seminula Pisolite. Beneath the chert in Longwood Quarry about 175 ft of grey fine-grained limestone and calcite-mudstone can be referred to the Clifton Down Limestone.

Some of the best sections seen in the Concretionary Beds are in an old quarry about 100 yd north of Providence Place [5379 7094]. Here the top of the Concretionary Beds, dipping 60° WSW, contains beds crowded with U-shaped borings which show up well on the weathered surfaces. About 225 ft of the uppermost Clifton Down Limestone is present between the old quarry and the faulted ground to the east. Part of this succession was formerly to be seen in the western part of the old quarry near the crossroads [5377 7114]. The rocks are folded and strongly sheared.

The chert band about 100 ft below the Seminula Oolite (see above) can be traced across the plateau to south of Longwood House. Bands of white oolite which may equate with the Seminula Oolite have been noted in Bendle Combe [5064 7190], near the Battleaxes at Wraxall [495 716], and to the east and west of Ham Farm [4845 7187], but only in the southern parts of Tyntesfield Plantation and at Belmont [513 707] to the south is the upper part of the Clifton Down Limestone preserved. Elsewhere around Tyntesfield and in the area west of Wraxall village Millstone Grit quartzitic sandstones rest on, or fill pipes in the oolite and grey splintery limestones of the lower and middle portions of the Clifton Down Limestone. In the northern part of Tyntesfield Plantation, on the steep slopes near Tyntesfield House and in Bendle Combe the succession appears to be:

	Thickness ft
Quartzitic Sandstone Group	
Quartzitic sandstones resting unconformably on limestones down to and including the Seminula Oolite and filling pipes and solution cavities	
Hotwells Limestone	0 to 75
Clifton Down Limestone	
Hard grey limestone (including Concretionary Beds at Belmont)	0 to 175
Pale grey limestone and oolite (Seminula Oolite)	about 75
Hard grey tough splintery limestone and calcite-mudstone with bands rich in *Lithostrotion sp.* and *Composita sp.*	about 300
Clifton Down Mudstone (including Goblin Combe Oolite)	about 250
Gully Oolite	about 75

Probably the best exposure in the Clifton Down Limestone is the quarry at Limekiln Plantation, Wraxall [506 724]. Here calcite-mudstones and hard grey limestone with *Lithostrotion martini* and *Composita sp.* dip 30° south-west. They are next seen in The Sidelands, north-north-east of Wraxall village as a limestone promontory crossed by Sidelands Fault, surrounded by Dolomitic Conglomerate, and ending in an isolated inlier near the Battleaxes. The main hill consists of hard grey splintery limestone and calcite-mudstone belonging to the lower division of the Clifton Down Limestone, as does the smaller mass north of the fault, where Reynolds (1918, p.195) saw compact limestone with chinastones and *Seminula* bands. In the south-east corner of The Sidelands [4985 7170] a small quarry shows southwards-dipping hard grey limestone with chert, overlain by Dolomitic Conglomerate. This chert appears to lie in the same position as that seen in Longwood Quarry.

The great width of the crop of the Hotwells Limestone on the slopes west of Providence is due mainly to the relatively low dip of the rocks and the southerly slope of the ground. There is also some indication on the lower slopes near Fenswood Farm [5338 7006] and Gatcombe Court [526 698] that the calcareous facies has expanded westwards at the expense of the sandstones and shales of the Upper Cromhall Sandstone of Ashton Park. Reynolds (1918, p.197) noted 'grits and limestones of D_2 age' at the south-eastern corner of George's Hill Plantation. Red-stained rubbly limestones were seen during the present survey, but the presence of sandstone would support the view that the higher limestones on the lower slopes of the valley above Ashton Watering are indeed a more strongly calcareous development of the lower part of the Upper Cromhall Sandstone to the east. The total thickness of the main Hotwells Limestone throughout this area cannot be less than 300 ft and the 50 ft of overlying rubbly limestone and shale should also probably be included in that formation. With the exception of a small outcrop of coarse oolitic and crinoidal limestone of Hotwells type on the steep wooded slope on the south-west margin of Tyntesfield Plantation [511 711], and a still smaller but more doubtful occurrence beneath Quartzitic Sandstone near the Chapel at Tyntesfield House [5067 7155], the extensive tract of Hotwells Limestone between Providence and Belmont Hill is the formation's last outcrop of any consequence west of Ashton Park.

The highest Carboniferous Limestone Series strata form a number of low mound-like areas at the lower end of the dip slope of the Hotwells Limestone between Fenswood and Gatcombe. The rocks are not generally well exposed but the outlier at Gatcombe [526 699] consists of alternating beds of limestone, shale and sandstone characteristic of the Upper Cromhall Sandstone. In one small exposure about 170 yd east of Gatcombe Court a colony of *Lonsdaleia floriformis* in coarse red crinoidal oolite is suggestive of the Rownham Hill Coral Bed.

The most southerly exposures of the Carboniferous Limestone in this tract occur along the railway line in the vicinity of Cambridge Batch. Here a natural ridge of anticlinal fossiliferous Hotwells Limestone is continuous with the main mass on Belmont Hill and forms part of the folded mass of limestone at Kingcot Farm [5205 7026] and Cook's Wood from which it is separated only by a tract of Dolomitic Conglomerate flanking the east side of the ridge at Cambridge Batch. At the southern end of the ridge there are good sections in the approach cuttings to the tunnel: the Hotwells Limestone dips to the south-west on the western side and to the south-east on the eastern side. About one mile east of Cambridge Batch two inliers of Carboniferous Limestone are seen in the railway cutting. The more westerly [5350 6995] is about 100 yd in length and consists of southerly dipping crinoidal limestone with a small infolded mass of quartzitic sandstone on the north-west side. The second inlier [5380 6982] lies about 250 yd to the east and shows gently folded crinoidal and oolitic limestone with a variable but generally northerly dip. *Palaeosmilia murchisoni* and *Lithostrotion pauciradiale* occur on the north side. Some of the ooliths have sand-grain cores; it is therefore likely that the section lies in the higher part of the Hotwells Group, possibly the Rownham Hill Coral Bed horizon.

Resting on the Hotwells Group at Ashton Watering are hard quartzitic sandstones regarded as belonging to the Quartzitic Sandstone Group (Millstone Grit). Similar sandstones crop out at a number of localities in Tyntesfield Park and occur as pipes and 'sheets' associated with the Clifton Down Limestone as far west as Tickenham. Some of the occurrences, however, have been thought to be of Lower Carboniferous age. Reynolds (1918, p.195), for example, ascribed some quartzitic sandstone near Ham Farm [484 719] to a position within the S Zone, and he compared it with 'a similar strong band of grit in S_1 in the Olveston area' i.e. with the lower Cromhall Sandstone.

Clevedon area: Wain's Hill – Walton Park – East Clevedon

From time to time the Lower Limestone Shale succession is well exposed after storms in Clevedon Bay, though normally only the harder beds project through the sand and mud. At the landing stage 150 yd south of the Pier, the red crinoidal Bryozoa Bed, 9 ft thick, dips at 30° to the south-south-east. According to Bush (1928) some 45 ft of crinoidal limestone overlies the Bryozoa Bed. The rest of the formation consists of grey-green and red shale with thin bands of reddish crinoidal limestone particularly prominent in the upper part. In these beds the coral *Vaughania vetus* ('*Cleistopora geometrica*') occurs near the tank [4018 7162] on the beach, some 270 yd south-south-west of the Pier entrance. In Littleharp and Salthouse bays two small outcrops of Lower Limestone Shale occur bounded by thrust faults. On the north side of Dial Hill the Lower Limestone Shale, dipping 30°–40° to the south-east, strikes north-eastwards through the built-up district of Walton Park as far as the major fault which extends east-south-east from Ladye Bay.

Wain's Hill, Church Hill and much of Clevedon northwards to the fault along Hill Road on the south-west of Dial Hill is formed by Black Rock Limestone and Black Rock Dolomite. On the Blackstone Rocks reef, south-west of Wain's Hill, dark crinoidal limestone with silicified corals and brachiopods is seen dipping 30°–40° to the south-east. In the cliff section at the southern end of Wain's Hill a nearly complete succession of Black Rock Limestone is seen capped by Black Rock Dolomite, also dipping south-east. The rocks are mainly dark crinoidal limestones with silicified fossils and chert nodules near the base, and details of the succession here at Church Hill are given by Bush (1928). At Hangstone Quarry [405 711], 400 yd west-north-west of Clevedon station, the Black Rock Dolomite contains chonetoids and tylothyroids on the large bedding-planes dipping 40° to the south-east. At the side of Chapel Hill, ½ mile north of Clevedon Station, Black Rock Limestone dips 40° to the south-south-east, the overlying Black Rock Dolomite being exposed in the cutting at the junction of Chapel Hill and Highdale Avenue. On the eastern side of The Park, overlooking the valley at East Clevedon, Black Rock Limestone, Black Rock Dolomite and Gully Oolite are exposed along woodland tracks on the south side of the Clevedon Fault.

North of the Clevedon Fault, over the area extending north-eastwards from Dial Hill to the Ladye Bay Fault, the Gully Oolite is exposed in a number of overgrown exposures with dips of 20°–40° south-east. It was worked in the large quarry [4164 7257] on the south side of Holly Lane, ½ mile north of East Clevedon Church. The oolite in the quarry immediately to the north of Holly Lane is bounded to the north by the shatter-zone of the Ladye Bay Fault. About 100 yd south-east of this quarry thin purple and green clay pocketed into the top of the Gully Oolite and overlain by pink dolomite is either Clifton Down Mudstone or Triassic deposits similar to those seen in fissures at Nightingale Lane Quarry, Portishead (see below).

Clifton Down Mudstone occupies a narrow belt of low-ground south-east of Dial Hill, and is poorly exposed in crags [4136 7225] 520 yds north-west of East Clevedon Church, where it consists of greyish white and purple dolomite-mudstone dipping 30° south-east, overlain by Goblin Combe Oolite, coarse white oolite containing *Bellerophon*. The same formation is exposed in an old pit [4106 7186] 640 yd west of East Clevedon Church, and in crags above Hill Road 650 yd north by west of Clevedon Station. Traces of splintery Clifton Down Limestone occur ¼ mile west-north-west of East Clevedon Church, within 50 yd of an old quarry in unconformable Pennant Sandstone.

Walton-in-Gordano – Weston-in-Gordano

Throughout this belt of country the greater part of the exposed Carboniferous Limestone Series forms the north-west limb of a sharp syncline, the southern limb of which is seen only for a short distance south of Portishead Down. Between Black Hill and Walton Down to the north and Castle Hill and Canon's Wood to the south, the Lower Limestone Shale occupies a long narrow valley filled mainly with blue-grey clay, with the basal crinoidal limestone, including the Bryozoa Bed, forming a slight ridge at the base. The overlying Black Rock Limestone comprising dark crinoidal limestones with silicified fossils is exposed in the series of steep crags extending from the south-west end of Castle Hill to Canon's Wood. South of Portishead Down dolomite is strongly developed in the base of the Black Rock Limestone. This is well exposed in the large Nightingale Quarry [450 748], 1000 yd north-east of Weston-in-Gordano Church: at the base of the quarry pale grey fine-grained sporadically crinoidal dolomites with nodules of chert are seen, and over the remainder of the section, which extends for some 180 yd southwards, dark crinoidal limestones dip at 50° south-south-east. In the centre of the quarry an east–west fault is accompanied by shattering and steepened dips. This fault can be traced across the quarry on the east side of the road, where the dark crinoidal limestone is overlain by Black Rock Dolomite marking the top of the formation. At the top of the quarry face about 10 ft of pale yellow Dolomitic Conglomerate overlies the limestone unconformably. In the old quarry to the south [4512 7472] Gully Oolite dips at 45° south-south-east. At the junction of the Oolite and the underlying Black Rock Dolomite there are some curious vertical infillings of spherical dolomite surrounded by radiating red-brown calcite associated with traces of purple shale.

Some 500 yd east of these quarries the north-eastern tip of the syncline can be observed, the hitherto south-easterly dipping rocks bending sharply round to dip north-westwards. In the abandoned Black Rock Quarry [4571 7470] 600 yd west-south-west of Weston Road, vertical and inverted Black Rock Limestone and Black Rock Dolomite strike north-east.

Portishead Down – Portishead – Portishead Point

The greater part of this area is occupied by Lower Limestone Shale forming part of the northern limb of the same syncline as that just described. South of Portishead Down a second belt of crinoidal limestone is developed at a slightly higher level than the basal group, and the two bands form conspicuous dip-slopes as far north as Dry Hill, west of New Town. Here in the foundations for a reservoir a good section of the passage beds immediately below the Lower Limestone Shale was seen. At Woodhill the two limestones appear to have united to form a considerable thickness of limestone (Smith and Reynolds, 1935). Immediately west of the Power Station a narrow belt of shale occupies the axis of a sharp syncline, on the eastern limit of which the beds, striking north-east, are vertical.

This general north-easterly strike is abruptly terminated by a large fault-belt which extends westwards from the Pier, and to the north the rocks have dominant east–west strike. At the northern end of Woodhill Bay the uppermost beds of the Lower Limestone Shale are exposed on the foreshore, where they are folded into a series of shallow anticlines and synclines, also visible in the road-cutting at the north end of the marine lake. In the red crinoidal limestones on the foreshore *Vaughania vetus* is common.

The wooded ridge extending eastwards from Portishead Point on the north side of the fault-belt is formed almost entirely of Black Rock Limestone and Black Rock Dolomite, striking east–west, often sharply folded, and on the south side of Portishead Point showing local inversion.

About ¼ mile south-west of Portishead Church is the steep Fore Hill, on the top of which a large overgrown quarry [4640 7578] shows a 50 ft face of dolomitised basal Black Rock Limestone dipping 20° east-south-east. This limestone mass overlies almost vertical Portishead Beds (Upper Old Red Sandstone) exposed in the lanes on the north and south sides of the hill.

East Clevedon – Tickenham – Wraxall

Between the valley at East Clevedon and the Tickenham Fault, some five miles to the east, the Carboniferous Limestone outcrop is sharply bounded on the north by the Clevedon Fault and the Naish House Fault, with Pennant sandstones occurring on the north side. Most of this narrow limestone ridge is formed by Black Rock Limestone overlain to the south by Black Rock Dolomite. In places, however, particularly in the eastern half of the outcrop, higher formations appear through the Triassic cover. In all the general dip of the Carboniferous rocks is 30°–40° south-south-east.

To the east and south-east of Clevedon Church Black Rock Limestone and Black Rock Dolomite are exposed in the hillside paths dipping 40°–50° south-south-east. Some 300 yd south-east of the Church 20 ft high crags of Gully Oolite are seen. On the north side of the Warren, 700 yd north-east of Clevedon Court, a lenticular mass of shattered oolite, 370 yd in length, appears beneath the main Clevedon Fault between crags of Black Rock Limestone and Pennant Sandstone. A smaller mass of oolite occupies a like position 500 yd to the east, and both are similar in their relations to the wedge of broken oolite lying between Black Rock Limestone and Pennant Sandstone 200 yd west of East Clevedon Church (Reynolds and Greenly, 1924, fig.8, p.458).

Between Middletown and Cadbury Camp, the Gully Oolite, Clifton Down Mudstone, Goblin Combe Oolite and Clifton Down

Limestone dip 30°–40° south-east, but all are poorly exposed except for the Goblin Combe Oolite, of which 25 ft can be seen in an old quarry [4478 7200] 180 yd north of the main road at Middletown; the limestone contains abundant *Bellerophon*.

The Naish House Fault forms the boundary between Black Rock Limestone and Pennant Sandstone for nearly a mile eastwards from Westpark Wood, beyond which Lower Limestone Shale appears between the two formations. At Naish House the fault dies out and the Pennant Sandstone rests unconformably upon Lower Limestone Shale and then upon Portishead Beds.

The excavation in 1952–53 of a tunnel 1512 ft long to carry water mains through the hill afforded a section starting in Pennant Sandstone and passing through the Naish House Fault and continuing in the Black Rock Limestone. The main part of the section was kept under observation by Mr R. B. Wilson. The north portal [4586 7300] lies at a height of 178.75 ft above OD, that on the south [4576 7254] at 173.75 ft above OD. For the first 60 yd inwards from the north portal typical Pennant Sandstone with subordinate shale has a general dip of 40° in a southerly direction; between 60 and 80 yd the rocks are highly disturbed with numerous shear planes. At a distance of 80 yd from the entrance the Naish House Fault plane dips at 36° to the south. It is a clean fracture about 1 ft wide and the disturbance in the Carboniferous Limestone Series on the south side is very small and limited to a belt 10 yd wide. Immediately south of the fault plane Lower Limestone Shale, consisting of shale with thin limestones, extends inwards for a distance of 20yd and is succeeded by grey crinoidal Black Rock Limestone dipping at 25° to the south. At a distance of 176 yd from the entrance a small synclinal fold is developed, part of the low fold which widens the outcrop of the Black Rock Dolomite at Cadbury Camp to the south-west. The remainder of the tunnel is cut in southerly dipping Black Rock Limestone.

About ½ mile south-south-west of Naish House and immediately north-east of Hale's Farm, the Clifton Down Mudstone is cut out by unconformable Pennant Sandstone, and in the same area the Goblin Combe Oolite consists of two bands of oolite separated by mudstone.

In the area between Cadbury Camp and Lime Breach igneous rocks occur in the Clifton Down Mudstone. These consist of highly weathered basaltic lavas, and two small outcrops were mapped. The westerly exposure [4557 7214], situated 780 yd north-north-west of Tickenham Church is about 3 yd across; the eastern and larger exposure [4663 7245] is about 60 yd long and 15 yd wide and lies on the south side of Lime Breach Wood, 1440 yd north-east of Tickenham Church. This exposure was investigated by Reynolds (1916) who dug a number of trial holes.

Between Tickenham and West Hill a wedge-shaped outcrop is bounded on its north side by the Tickenham Fault and contains rocks ranging from the Black Rock Limestone to the Clifton Down Limestone. There is a general dip of 30° south-south-east, though in the proximity of the fault slight anticlinal folding is seen at Stone-edge Batch and on the north side of West Hill.

Gully Oolite is exposed in a line of wooded crags 400 yd north-east of Hale's Farm, and in an old quarry [4670 7181] 1070 yd north-east of Tickenham Church splintery limestone with pink-yellow shale and mudstone represents the Clifton Down Mudstone. The Goblin Combe Oolite is split by a narrow band of mudstone on the north side of West Hill and was worked in a line of old quarries about 700 yd north-west of Wraxall Court. The Clifton Down Limestone forms the remainder of the area, occupying a large dip slope extending southwards to the valley of the River Yeo. In the side of the stream at Tickenham [4662 7160] splintery chinastones with fasciculate *Lithostrotion* dip 30° south. Between Birdcombe and Wraxall Court fine-grained quartzite (? Millstone Grit) occurs as sheets and pipes in the Clifton Down Limestone, being prominent in an old quarry [4856 7183], 500 yd south-west of Wraxall Court.

Broadfield Down

Yatton – Wrington – Redhill

Between Yatton and the main Bristol road (A370) seven detached inliers of Carboniferous Limestone emerge through the Trias, the dips indicating a gentle anticlinal structure. Most of the rocks belong to the upper part of the Clifton Down Limestone, which is well exposed in a quarry [4410 6525], 1100 yd east-south-east of Yatton Church. Here splintery calcite-mudstones with *Lithostrotion* and productoids dip 14° west-north-west. Similar beds are seen in an old quarry [4350 6500] on Frost Hill, 600 yd south-east of the Church, dipping 12° south-south-west. Hotwells Limestone consisting of dolomitised oolitic and crinoidal limestone with *Dibunophyllum sp.* and *Palaeosmilia murchisoni* forms a thin cap on Cadbury Hill, the largest of the inliers. The junction between the two formations can be seen in the crags on the south side of the hill at a point [4400 6485], 1180 yd south-east of Yatton Church. In the small inlier on the north side of Congresbury, oolitic and crinoidal limestone with *Palaeosmilia murchisoni* dips 20° south-south-west.

Farther to the east much of the ground south of the Wrington Hill Fault is occupied by poorly exposed Brockley Oolite developed within the Clifton Down Limestone; it dips 8°–10° south and is in turn overlain by calcite-mudstones. On the southernmost margin of the Carboniferous Limestone a number of small disconnected patches of Hotwells Limestone are separated by tongues of Dolomitic Conglomerate extending northwards from the main Triassic outcrop. In an old quarry [4566 6383], 440 yd east-south-east of Prospect Farm, massive oolitic limestone is overlain by oolitic crinoidal limestone with red shale bands, full of *Lithostrotion* and productoids, dipping 20° south-south-west. Hotwells Limestone is exposed in the now overgrown quarry [4750 6317], 1000 yd north-east of Wrington Church. Here 25 to 30 ft of coarse grey oolite are overlain by rubbly limestone and shale full of corals. This quarry was noted by Vaughan (1905, p.242) as the best locality in which to collect corals from his Upper *Dibunophyllum* Sub-zone, and he recorded a formidable list of species.

To the north of Redhill the Brockley Oolite thins eastwards and most of the high ground through Butcombe Court as far as the Ridgehill Fault is formed of splintery calcite-mudstones of the Clifton Down Limestone. To the south occurs a fuller succession of the Hotwells Limestone than that seen to the west. About 400 ft above the base a band of gritty Upper Cromhall Sandstone is developed, first seen in a crag [6537 6283], 500 yd east of Scars Farm, whence a discontinuous outcrop strikes eastwards to a point 950 yd south-south-west of Butcombe Court. Still farther east in the small inlier ½ mile north-west of Ridgehill Church, a second grit band occurs separated from that just mentioned by about 50 ft of limestone.

In small crags 550 yd and 900 yd east of Scars Farm quartzitic grits with chert bands are regarded as lying at the base of the Millstone Grit.

Goblin Combe and adjacent areas

In the ground between the Wrington and North Hill Faults a fairly complete succession from the Black Rock Limestone to the Hotwells Limestone is exposed, the chief section being that of Goblin Combe. The broad structure is that of a shallow anticline broken along its axis by the Cleeve Fault, which follows the line of Goblin Combe for a distance of nearly a mile in its western part.

South of the North Hill Fault the Black Rock Limestone occupies a roughly triangular outcrop extending from a little east of Warren House to near Cornerpool Farm [504 645]. It is exposed in old quarries 500 yd south-east of Warren House and immediately north of the Cleeve Fault, where it dips 15°–18° north. Black crinoidal limestones, dipping 15° south-west, form crags 1100 yd south-east of Warren House. West of Cornerpool Farm the North Hill Fault passes into a belt of shattered ground which has been extensively mined for lead. The succeeding Black Rock Dolomite is relatively thin and poorly exposed.

Gully Oolite strikes in a broad belt south-westwards from the North Hill Fault through Warren House, beyond which it is exposed in a series of fine crags on the north side of Goblin Combe, where it dips 20°–25° north-north-west. Black Rock Dolomite and some Black Rock Limestone are seen in the base of these crags immediately north of the Cleeve Fault, on the south side of which the Gully Oolite has been shifted eastwards. It reappears some 200 yd south of Warren House, here dipping 30° south. It then follows a sinuous course south-eastwards, being affected by minor folds, and is well displayed in 30 ft crags, 650 yd south-south-east of Warren House, the dip here being 10° west-south-west.

North-west of Warren House the Clifton Down Mudstone occupies a belt some 200 yd wide, and near Goblin Combe this mudstone occupies a depression between the Gully Oolite and the Goblin Combe Oolite. An impersistent thin oolite forms a slight feature in the upper half of the Mudstone, and about the middle the first traces of volcanic ash are found, but this appears to die out eastwards.

The first important exposure of volcanic rocks occurs some 500 yd south of Warren House, around the meeting point of four tracks in the valley bottom. About 540 yd south-south-east of Warren House a low knoll—the Bethel Stone—comprises poorly exposed basaltic lava overlain by alternations of limestone and tuff (Reynolds, 1916, pp.24–25). The main outcrop of basaltic lava and ashy limestone extends eastwards for nearly a mile from a point [483 645] 970 yd south-east of Warren House to 550 yd south-west of Cornerpool Farm. Reynolds (1916, fig.1) depicted the outcrop of the lava and tuff following the sinking of numerous trial-pits. The survey of 1948–49, however, showed that the volcanic rocks continued eastwards beyond the limits set by Reynolds, and indicated that most of the rock is tuff and ashy limestone, though vesicular lava crops out from the road 200 yd east of Broadfield Farm to within 600 yd of the main Bristol road (A38) near High Wood [500 641]. It is uncertain whether ashy limestone, tuff or lava predominates here; the rocks crop out in the valley floor and it is likely that some of the material seen on the surface is derived from Drift deposits.

The highest crags of Goblin Combe are formed by the Goblin Combe Oolite, which reaches 120 ft in thickness. On the north side of the Cleeve Fault the Oolite forms the cliff which rises some 300 yd east-south-east of Cleeve Toot and extends for 900 yd eastwards on the north side of the combe, the dip being 20° north-north-west. The Cleeve Fault displaces the Goblin Combe Oolite eastwards so that on the south side of the fault it emerges at the base of the combe some 820 yd east-south-east of Cleeve Toot with low dips to the south-west. About ¼ mile south-west of Warren House the crags on the north side of the combe are formed of almost vertical Goblin Combe Oolite downfaulted against Black Rock Limestone on the north side of the Cleeve Fault. Good exposures of the Oolite are seen in the crags ¼ mile south of Warren House and in the sides of the track winding upwards to Wrington Hill, where the beds dip 10° west. The last-mentioned section begins at a point [4788 6472], 720 yd south of Warren House and extends for 250 yd in the sides of the track ascending westwards in a continuous exposure of white oolite, locally rather coarse and crinoidal and carrying *Palaeosmilia murchisoni, Athyris sp., productoids* and *Euomphalus*.

Goblin Combe Oolite is exposed in 30 ft cliffs ¼ mile south-west of Broadfield Farm, and is seen in a broad belt extending from east of the farm to a little north of Butcombe Court, dipping 20° south. Between the North Hill and Cleeve faults the area west of the Goblin Combe Oolite outcrop to Cleeve Hill is occupied by poorly exposed Clifton Down Limestone, apart from a small patch of Hotwells Limestone seen 500 yd north-east of Cleeve Court. The chert level occurs at the highest point of Cleeve Toot, and the Brockley Oolite occupies a band striking north-eastwards from Cleeve Court, the dips varying from 20°–30° north-west.

To the north and north-east of Woolmers the Brockley Oolite has been extensively mined for lead [461 649]. A quarry [4592 6538], 380 yd south-south-west of Cleeve Court affords a good section of calcite-mudstones and concretionary beds dipping 20° north-west. At the north-west corner of King's Wood there are two small outcrops of Hotwells Limestone. A quarry [4512 6510], 1220 yd south-west of Cleeve Court shows coarse oolitic and crinoidal limestone, with *Dibunophyllum, Palaeosmilia murchisoni* and *Gigantoproductus sp.* of *giganteus* type, dipping 20° north-north-west.

Area between North Hill and the Brockley Fault

A full succession from the Black Rock Limestone to the Hotwells Limestone dips uniformly 14°–20° north-west. Apart from Brockley Combe the ground is much wooded and exposures limited. The cliffs bordering Brockley Combe show splintery limestones and calcite-mudstones of the Clifton Down Limestone overlain by Brockley Oolite, seen in the crags in the zig-zag part of the combe. Extensive lead-mining has taken place in all the oolites.

North of the Brockley Fault

From the western margin of Broadfield Down at Brockley Hall, the Brockley Fault complex extends eastwards towards the head of Brockley Combe, where it affects the shelly and conglomeratic limestones of the Lower Lias and continues by way of a strong feature at Oatfield Batch [505 665] to Pottershill and the high ground at Upper Town, Felton. The faulted and folded Carboniferous Limestone rocks of Healls Scars and the marked ridge through Oatfield Batch and Oatfield Wood [510 663] may mark the main Variscan structure, but there is also evidence of strong post-Liassic movement. North of the structure the Dinantian rocks forming the northern part of Broadfield Down give rise to a roughly triangular upland plateau with its highest point (664 ft above OD) just north of Oatfield Batch. The limestone upland has a gentle northerly slope and terminates quarter of a mile south of Flax Bourton village where the limestones pass beneath Triassic rocks at about 210 ft above OD. The exposed succession belongs mainly to the Black Rock and Clifton Down formations, but on the north-western margins of the outcrop, between Brockley and Flax Bourton, the Hotwells Formation is seen at several localities, including Brockley Cottage [477 672] and Backwell village [493 682], where the basal beds of the Hotwells Limestone rest on Clifton Down Limestone. Two further small areas of Hotwells Limestone are seen near the northern limits of Broadfield Down, one [512 683] 300 yd west-south-west of Barrow Court and the other [514 681] about 300 yd to the south-east.

The oldest rocks are seen in the central area near Oatfield Batch where the Black Rock Limestone is strongly folded and faulted. The top of the Black Rock Limestone is exposed in a quarry at Hyatt's Farm [502 668], together with the lower beds of the Black Rock Dolomite, dipping northwards at about 35°. The thickness of the Black Rock Dolomite here is about 100 ft.

East of Hyatt's Farm the Carboniferous rocks are concealed beneath partly decalcified and silicified Mesozoic rocks (Harptree Beds). The Black Rock Limestone and Dolomite again crop out in a small strip extending from about 200 yd north-east of Oatfield Pool [5085 6680] to a point [5156 6672] about midway between Yewtree Farm and Freeman's Farm. Two small inliers of Gully Oolite occur north-north-east of the Black Rock Limestone at Pottershill. In an old quarry [5165 6605] north of Old Limekiln Lane the Black Rock Limestone dips 25° east-north-east.

Both the Gully Oolite and the Clifton Down Mudstone crop out in the angle between Freeman's Lane and the A38 road. The Gully Oolite forms much of the ground west of Freeman's Farm, extending to Hyattswood Farm and Healls Scars; dips, where seen, are gentle (10°–15°) but the width of outcrop shows great variability, suggesting that, like the Clifton Down Mudstone to the north, the detailed structure is complex, notwithstanding the continuity of the crop.

Exposures in the Clifton Down Mudstone are very poor. The base of the Goblin Combe Oolite is marked by a slight feature and the presence of small old workings for stone and lime. The Oolite consists mainly of low-dipping (12°–17°) hard grey crinoidal limestone and oolite occasionally yielding *Bellerophon* and *Palaeosmilia murchisoni*. A slight change of slope marks the top of the formation, which is succeeded by fine-grained splintery limestones typical of the base of the Clifton Down Limestone. Some of the best exposures of the Goblin Combe Oolite were formerly to be seen on the northern side of Freeman's Lane, south-east of Freeman's Farm and on either side of the A38 road near its junction with Freeman's Lane. These showed hard grey crinoidal limestone and coarse oolite with *Palaeosmilia murchisoni* and other fossils.

With the exception of the small areas of Hotwells Limestone mentioned above, the northern part of Broadfield Down is formed of massive Clifton Down Limestone, which has been extensively worked from large quarries. Among these is Backwell Quarry [490 680], in the upper part of the succession, and a more recently opened quarry west of Bourton Combe [503 683]. Working of Dial Quarry [528 666], situated east of the A38 road 1000 yd north-north-east of Upper Town, has long since ceased; it was described by Wallis (1922). Quarrying was active at Hartcliff rocks at the time of survey (1952). The Clifton Down Limestone, dipping about 35° north-east, is overlain by highly ferruginous Dolomitic Conglomerate, worked in the past for iron ore. Small's Quarry [5290 6635] at Hartcliff Rocks shows some oolitic limestone, as does Dial Quarry; it is also developed in the upper part of the formation in the northern part of Bourton Combe, but it is difficult to map over the rest of the northern part of Broadfield Down since exposures are so poor; the limits of the oolitic facies are clearly less well defined than in the southern and western parts of Broadfield Down.

In general the succession resembles that of the Avon Gorge. Wallis (1922, p.220) recorded *Davidsonina carbonaria* in Dial Quarry (Figure 13). This fossil is not recorded from the Avon Gorge though it has been found in the Mendips and in the northern fringes of the Bristol Coalfield. This distribution suggests that *D. carbonaria* is essentially basinal, being found within or at the margins of the Mendip basin and the structural depression following the Malvern Fault Zone. It is very rare or absent over the greater part of the shelf area. At Backwell Hill the lower part of the Clifton Down Limestone yields *Lithostrotion martini*, and *L. aranum* and *Composita*, some silicified, can also be found among the field brash 300 yd south-west of the Jubilee Stone [496 677].

South-east of Bourton Combe through Barrow Hill, Batches Wood [520 677] and Naish Lane to the A38 road, the Clifton Down Limestone is much disturbed and the stratigraphical relationships of the Dinantian and Namurian strata along the north-eastern margins of the uplands cannot be made out satisfactorily. According to Wallis (1922, pp.212,214) a small patch of 'D$_1$' limestones (basal Hotwells Limestone) was present at Rocks Wood [532 662] on the Hartcliff Rocks ridge, but this was not confirmed during the survey in 1949. Much of the limestone worked at Hartcliff Rocks is highly ferruginous. The associated fault appears to have been a locus for mineralisation and the Clifton Down Limestone on the northern side of the fault extending from Hartcliff Rocks to Barrow Hill and Bourton Combe has been extensively worked for lead, zinc and iron ores.

South of Hartcliff Rocks the Carboniferous Limestone is concealed beneath Triassic and Lower Jurassic strata. A borehole for water at Winford Orthopaedic Hospital [5340 6563] proved Carboniferous Limestone at a depth of 11½ ft and continued to 326 ft. The top 95 ft of limestones appear to be basal Clifton Down Limestone and the remainder Goblin Combe Oolite and Clifton Down Mudstone.

Lulsgate – Winford

Much of the central area of Broadfield Down is now occupied by Bristol's Lulsgate Airport, a plateau area underlain mainly by Black Rock Limestone. Only a very few exposures of gently dipping crinoidal limestone have been recorded. It is evident from the present disposition of the Mesozoic strata that the Viséan rocks which once overlay the Black Rock Group had been largely removed by late Triassic and early Jurassic times. Pre-Jurassic erosion had also removed most of the Black Rock Dolomite; some still exists in the peripheral areas of Broadfield Down, though none is seen at surface east of a line from North Hill to Pottershill, due partly to faulting and partly to Mesozoic cover in the area of Butcombe, Felton and Winford.

In a borehole at Stanshall's Lane, Felton [5216 6570] about 12ft of Lower Lias was proved resting on 418 ft of Black Rock Dolomite and Black Rock Limestone and 92 ft of Lower Limestone Shale. These beds were in turn underlain by 7 ft of highly faulted and fissured crinoidal limestone and dolomite, the exact relations of which are unknown.

Black Rock Limestone is overlain by Lower Lias limestone and conglomerate in the northern part of Lulsgate Quarry [517 658]. Here the Carboniferous rocks dip to the north-east, though they dip to the east and south-east in the southern parts of the quarry. The Black Rock Limestone is cut by numerous fissures striking N 20° W and infilled with red marl in the central and southern part of the face. A large fissure up to 30 ft wide, with an east-south-east strike is seen in the northern part of the quarry; this is filled with Lower Lias limestone and clay with some large limestone blocks and much calcite.

West of Lulsgate Quarry, the Black Rock Limestone crops out in a shallow valley north of Lulsgate Airfield before passing into a concealed area at Downside where the Carboniferous rocks have a cover of shelly and conglomeratic Lower Lias. The whole of this district has been strongly mineralised, both in the Carboniferous and Jurassic strata. Numerous old lead mines were worked hereabouts, some near Stone Farm [504 658] before 1746. Most of the workings were shallow and the strongest mineralisation appears to have been at or near the unconformity in the neighbourhood of post-Liassic faulting.

Farther south in the area of Winford Manor [5310 6415] and Crown Hill, Winford, there are a number of inliers of faulted Carboniferous Limestone and Millstone Grit rocks. The contact of the Clifton Down Limestone and the Hotwells Limestone was well exposed in an old quarry [5315 6440], 250 yd north-east of the crossroads at Winford Rectory. For the most part, however, the exposures of the higher Dinantian strata formerly seen in the old ironstone workings at Winford have deteriorated and are now of little value. A number of small inliers are shown on the published one-inch map. They formerly exhibited quartzitic sandstones and sandy or oolitic limestones characteristic of the Upper Cromhall Sandstone. The largest of these inliers is situated south of Winford Manor [535 634], but a more important section was formerly seen in shallow diggings at the top of Crown Hill [542 640]. Here the Winford Chert (basal Millstone Grit) could formerly be seen resting on thin limestones, sandstones and mudstones at the top of the Upper Cromhall Sandstone (Figure 13). This formation was proved beneath Triassic rocks in the Crown Inn Borehole [540 637], but the important feature of the Crown Hill section was the occurrence of *Tumulites (Eumorphoceras) sp.* in the chert beds: these have a south-east dip of 20° and underlie the quartzites formerly worked in Crown Hill Quarry [541 636]. Calculations based on the dip of the partly concealed Hotwells Limestone between Winford Manor and Crown Hill suggest that the Hotwells Limestone is not more than 250 ft in thickness and that the Upper Cromhall Sandstone is about 750 ft. This combined thickness of 1000 ft for the Hotwells Formation should be compared with the thickness of 500 ft calculated at Scar Farm east of Redhill, which indicates a westerly attenuation along the southern margin of Broadfield Down of about 120 ft per mile.

Cornwell Farm – Butcombe Court

A broad tract of Goblin Combe Oolite extends eastwards on the north side of the shallow dry valley followed by the road westwards from Kingdown to the A38. South of the valley lies a plateau-like area at Butcombe Court formed of Clifton Down Limestone diping south-south-east. The band of oolite occupying a median position within this tract north of Red Hill ends abruptly south-west of Butcombe Court, possibly due to faulting.

North of the crossroads [513 640], a minor road or trackway known as Old Barn Lane leads to Cornwell Farm [514 643]. In the floor of the track the Mesozoic cover has been eroded to expose strata beneath the Goblin Combe Oolite. These include Clifton Down Mudstone in which a band of tuff and ashy limestone can be seen. The volcanic rocks may not exceed 30 ft in thickness, but the exposure establishes their presence some ¾ mile east of their previously known limits.

CHAPTER 4

Millstone Grit (Namurian)

The name Quartzitic Sandstone Group has been given to the largely arenaceous sequence of rocks lying between the Hotwells Group, at the top of the Carboniferous Limestone and the Ashton Vale (*Gastrioceras subcrenatum*) Marine Band (see p.67) which marks the internationally accepted base of the Coal Measures (Kellaway and Welch, 1955). The earliest detailed account of these strata was given by H. Bolton (1907) who described 250 ft of strata, mainly sandstones, underlying the 'Chief Shell Bed' or Ashton Vale Marine Band at Ashton Vale Colliery. A boring for Bristol United Breweries at Lewins Mead, Bristol [5880 7337] gave a more extended record (Moore, 1941). Though the precise stratigraphical position of the rocks penetrated could not be determined, this borehole provided the first unequivocal palaeontological evidence for the existence of Namurian strata at Bristol. The most complete succession is that established in the Geological Survey's Ashton Park Borehole [5633 7146], which proved a full sequence of 606 ft 8 in (measured vertically through dips of 18° to 20°) between the base of the Ashton Vale Marine Band at 708 ft 6 in and the top of the Carboniferous Limestone Series at 1315 ft 2 in. In addition to the log of the borehole, Kellaway (1967, fig. 4, p.63) included a correlation with other localised sections in the Bristol Coalfield. A revised version of this correlation is given in Figure 14 which takes into account changes in the classification of the Carboniferous rocks as mapped in the Yate – Cromhall area, consequent upon the interpretation of more recent borehole evidence at Yate (Kellaway, 1970, pp.1042 – 1044; Wyatt *in* Cave, 1977, pp.307 – 314).

In general the strata which form the Quartzitic Sandstone

Plate 8 Clifton Down Limestone on the right is thrust over inverted Quartzitic Sandstone shales and cherts of the Millstone Grit. The fault is the Wicks Rocks Thrust. Quarry [7101 7333] at Wick (A10655)

Group are not unlike those encountered in the lower part of the Coal Measures: they consist of grey mudstones and seatearths with thin carbonaceous or coaly beds, and in places a marked preponderance of hard pale grey quartzitic sandstones throughout the middle part of the sequence. North of Bristol quartzitic sandstones are strongly developed at the top of the Quartzitic Sandstone Group and in the basal Coal Measures. In the absence of the Ashton Vale Marine Band definition of the boundary between the Namurian and Lower Westphalian is extremely difficult. The sandstones contain many lenticular bands and seams of pebbles, and massive conglomerates are not uncommon, the pebbles consisting for the most part of vein quartz, possibly derived from the quartz-conglomerates of the Upper Old Red Sandstone. Many of the sandstones are strongly stylolitic. The stylolites are lined with black carbonaceous mud, and some pass laterally into streaks of mudstone or muddy sandstone.

The development of pebbly sandstones, beds of seatearth and carbonaceous layers, and the paucity of marine shale shows that the Namurian rocks were deposited in an inland basin or depression which was only penetrated at infrequent intervals by shallow marine incursions. The deepest part of the depression seems to have extended from just north of Yate to Kingswood and the Chew valley, then turning gradually south-eastwards towards Radstock. Recycled Tremadoc acritarchs are said to be present in the mudstones at Yate suggesting that Cambrian as well as Old Red Sandstone rocks were undergoing erosion in the vicinity in Namurian times.

At the base of the group at Bristol and Winford, bands of hard dark grey and black splintery cherts and cherty mudstones are developed over about the bottom 50 ft, similar to those recorded at the base of the Millstone Grit to the south-east of Aberkenfig in South Wales (Woodland and Evans, 1964, p.9). These contain sponge spicules and other marine fossils including *Tumulites (Eumorphoceras) sp.* (Kellaway and Welch, 1955, p.19), as well as brachiopods and molluscs consistent with an E_1 (Pendleian) age[1]. For the most part the beds above the cherts are nonmarine and appear to have been laid down in hollows formed during the early phases of the Variscan movements. Sporadic thin marine layers occur among the quartzitic sandstones, but they have yielded only *Lingula mytilloides* and a few other brachiopods and molluscs. No other goniatites have been found beneath the Ashton Vale Marine Band[2], and so the exact age relations of the Quartzitic Sandstone Group as a whole remain obscure.

Plants are abundant and well preserved at some levels, notably in the mudstones between the base of the Coal Measures and the main development of sandstone, and again in isolated thin wisps of mudstone throughout the quartzitic sandstones. Two major palaeobotanical divisions can be recognised, the boundary being placed at a thin marine mudstone with *Lingula mytilloides* and *Ptyestia* cf. *acuta* found at 972 ft 10 in in the Ashton Park Borehole (Kellaway, 1967, p.102, pl.3), which lies above grey mudstones with plants. This marks the position of the so-called 'plant break' (Kidston, 1894), which was taken in former days to separate 'Upper Carboniferous' from 'Lower Carboniferous' floras. The plants in the uppermost mudstones, including *Alethopteris lonchitica*, *Mariopteris acuta*, *Neuropteris gigantea*, *N. heterophylla* and *N. scheuchzeri*, are similar in many respects to those identified from the $R_1 - R_2$ stages of the Namurian in South Wales (Dix, 1933), while those from the lower half of the quartzitic sandstone group, including *Sphenophyllum tenerrinum*, cf. *Neuropteris antecedens*, cf. *Diplotmema adiantoides* and *Rhodea feistmanteli* are more closely linked with the early Namurian or late Viséan.

The rocks of the Quartzitic Sandstone Group are largely concealed by Triassic strata and in those areas where they do crop out they are very poorly exposed. The thickest sequence preserved is found in the north-eastern part of the Bristol Coalfield between Yate and Cromhall (Figure 14), where it may reach about 1000 ft. South of Yate, at Wick, the thickness is about 600 ft and on the western margins of the main basin the Ashton Park Borehole proved a true thickness of about 570 ft. On the southern flanks of the Mendips between Wells and Cheddar, the Quartzitic Sandstone Group is about 175 ft thick according to Green and Welch (1965, p.52). Here the lower division of the Quartzitic Sandstone Group is absent, having been overstepped by the younger Namurian strata.

DETAILS OF STRATIGRAPHY

On the eastern margin of the Bristol Coalfield near Yate and Tanhouse, cherts and cementstones have been recognised, but these are of late Viséan (P_1) age (Smith, 1942). They are overlain by mudstones and sandstones of the basal Namurian, and the E_1 cherts are either absent or have passed laterally into mudstones. Between Yate and Bath, Namurian strata crop out in the Wick Inlier, where the upper boundary is defined by the Wick Marine Band with *Donaldina ashtonensis*, thought to represent the Ashton Vale Marine Band in the terrain west of the Wick Fault. East of the fault the higher beds of the formation are not exposed, but the base was seen in 1947 in an approach road to an old sandstone quarry near Cleeve Bridge [713 737]. Although the strata are contorted, it was possible to make out the following succession:

	Thickness	
	ft	in
QUARTZITIC SANDSTONE FORMATION		
Grey laminated mudstone	4	5
Black micaceous sandy mudstone with bands of pyritic mudstone; *Lingula sp.*, *Orbiculoidea sp.*	5	0
Soft blue-grey silty mudstone with a thin carbonaceous layer 5 ft below top	15	0
Hard splintery dark blue-grey cherty limestone and calcareous mudstone; *Lingula sp.*, *Pterinopecten sp.*, Eurypterid fragments	1	6
Silty and sandy mudstone	6	6
Ochreous limestone and calcareous mudstone on sandy shaly limestone	2	3

---

[1] For a full description of the zones and stages of the Namurian (Millstone Grit), the reader is referred to Ramsbottom et al., 1978.

[2] It was formerly supposed that *Gastrioceras cancellatum* (Lower G Zone) occurs at Vobster, but this was later identified by Dr C. J. Stubblefield as *G. subcrenatum*, indicating the presence of the Ashton Vale Marine Band (Kellaway and Welch, 1955).

64 CHAPTER FOUR MILLSTONE GRIT (NAMURIAN)

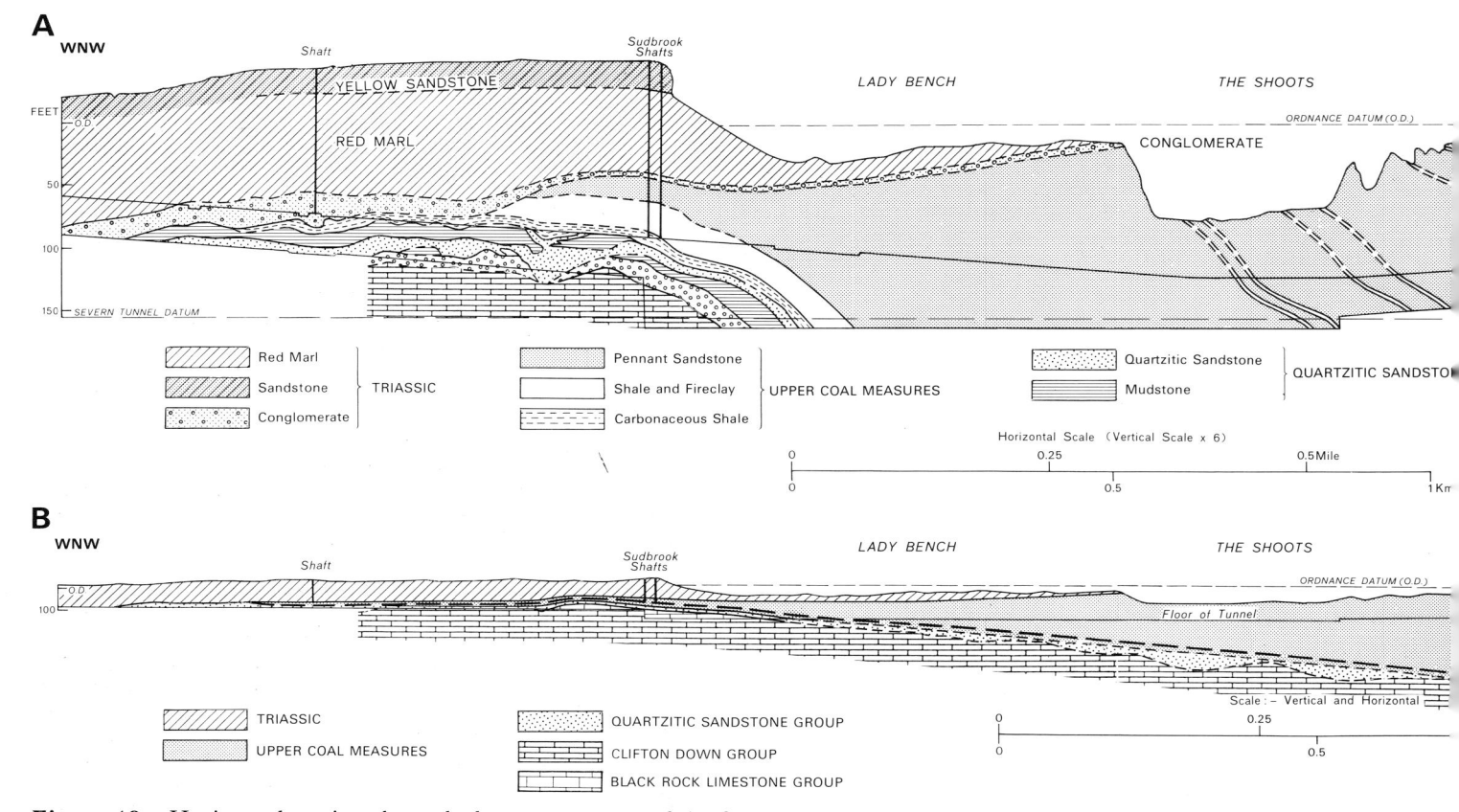

Figure 19 Horizontal section through the western part of the Severn Tunnel:
(A) Detailed section based on C. Richardson (1887): Vertical exaggeration ×6
(B) Generalised section showing the relationship of the Coal Measures to the underlying formations: natural scale

UPPER CROMHALL SANDSTONE (CARBONIFEROUS LIMESTONE SERIES)

Thinly bedded calcareous mudstone and decalcified sandy limestone, *Buxtonia scabricula*	12	0
Current-bedded quartzitic sandstone with lenticular grey limestone	5	0

The E_1 cherts are again absent and the succession as a whole seems less strongly marine than at Bristol and Winford.

North of Wick the position of the boundary between the Millstone Grit and the Coal Measures has long been a source of difficulty. The basal Coal Measures are largely concealed and the Ashton Vale Marine Band has not been found at crop; nor was it recognised in a borehole made at Limekiln Lane, Yate [7066 8589] and recorded by Wyatt (*in* Cave, 1977, pp.307–314). The reason for this is not clear, for subsequent palynological investigations carried out by Dr B. Owens have suggested that the uppermost 200 ft of interbedded mudstones and sandstones proved in the borehole may be of basal Westphalian age. The remaining 500 ft are almost certainly Namurian, with quartzitic sandstones developed intermittently throughout.

On the western margins of the main coal basin between Filton and Winford the Quartzitic Sandstone Group is fully and typically developed with E_1 cherts at the base and the Ashton Vale Marine Band at the top. Massive quartzitic sandstones crop out in west Bristol along a line from Kingsdown to Long Ashton. Here the succession is typified by the Ashton Park Borehole (Kellaway, 1967). The prominent feature of Brandon Hill, a mile and a half north-east of the borehole, is formed by pink-stained quartzites upwards of 600 ft thick, which make up 75 per cent of the total thickness of the Group. In the Harry Stoke area, five miles to the north-east, borings for coal, notably Harry Stoke A (p.85), proved massive quartzites both above and below the Ashton Vale Marine Band: this thick arenaceous facies appears to characterise a belt extending southwards from the Cromhall–Tytherington area to Bristol.

South of the Kingswood Anticline the Millstone Grit of the Somerset Coalfield is almost everywhere concealed beneath Coal Measures and Mesozoic strata, and it is not seen until the southern margin of the Radstock Basin is reached. Here it is poorly exposed in the northern limb of the Beacon Hill Pericline, as a narrow band marked mainly by isolated exposures of pebbly quartzite. It seems that the lower part of the formation is either very thin or missing altogether locally, since no trace of the basal cherts has yet been found (see section at Cookswood [6685 4795] in Figure 14). Quartzites appear to rest nonsequentially on the Hotwells Limestone and the entire thickness of the formation does not exceed 100 ft near Mells and 150 ft at Ashwick and Gurney Slade.

The close similarity between the rocks seen in the crop between Mells and Ashwick and those formerly exposed in Vobster Quarry is compelling evidence for the local origin of the rocks forming the structurally isolated masses seen at Soho, Vobster and Luckington (Welch, 1931a; 1931b).

Few coals of any note occur within the Quartzitic Sandstone Group, and fewer still have been worked, the seams for the most part being too thin and dirty. Two coals were formerly worked from shallow pits in the Tytherington–Cromhall area, but little is known about them. The lower seam, worked near Tapwell Bridge [6855 8940], was said to consist of up to 7½ ft of dirty coal and shale,

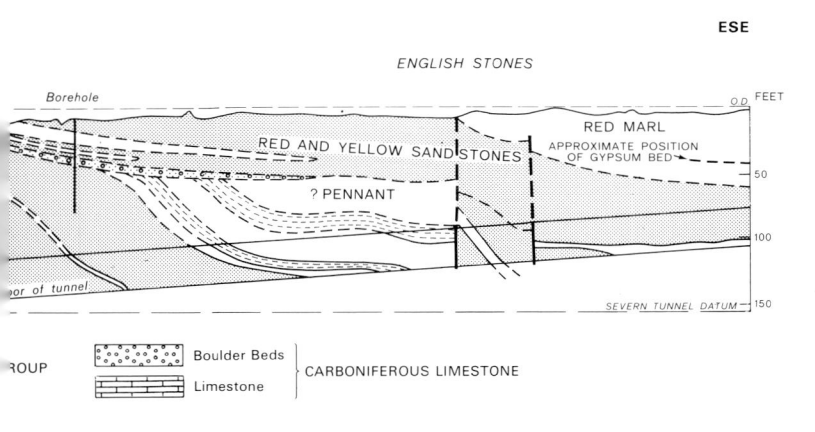

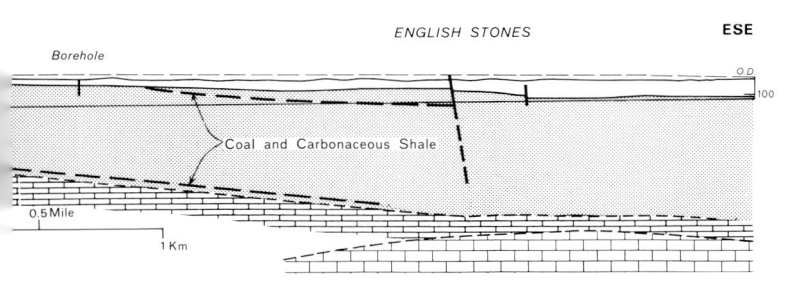

DETAILS OF STRATIGRAPHY 65

deposited in solution cavities in the limestone at a fairly early stage in the Variscan structural evolution. They are probably of Namurian age.

Most of the information concerning the Namurian rocks of the Severn Coal Basin has been obtained from tunnel sections and boreholes. In the Severn Railway Tunnel (Figure 19) coarse conglomerates and boulder beds, thought to be late Viséan, are overlain by mudstones and quartzitic sandstones presumed to belong to the Millstone Grit; they are in turn overlain by the Upper Coal Measures of the Severn Basin. At Portskewett, near Caldecote on the north side of the Severn Estuary, several boreholes (p.165) have proved Upper Carboniferous mudstones and sandstones lying beneath Triassic marls and dolomitic conglomerate, and resting on Carboniferous Limestone. A mudstone band overlying a coal streak and containing nondiagnostic fossils, including productids, was originally thought to be equivalent to the Crofts End or Winterbourne Marine Band, but Dr B. Owens reports that the well-preserved miospore population recovered from 53.6 m depth in the Portskewett Borehole 106 is diverse in composition and dominated by representatives of *Lycospora pusilla*. The presence of an association consisting of *Apiculatisporis varoscorneus, Raistrickia fulva, Ibrahimispores brevispinosus, Secarisporites remotus, Crassispora kosankei, Knoxisporites dissidius, Spelaeotriletes arenaceus* and *Pteroretis sp.*, suggests a Yeadonian (G_1) age to be probable. A lenticular development of interbedded mudstones and siltstones, 15 to 50 ft thick occurs beneath the marine band, some of the sandstones being pale grey and quartzitic with scattered quartz and sandstone pebbles, reminiscent of the Quartzitic Sandstone Group in the main basin to the east. Welch and Trotter (1961, p.84) describe deposits similar to these, filling steep-sided channels and solution cavities in the Carboniferous Limestone between Portskewett and Ifton, and they consider that these might represent part of the Millstone Grit. Not all the sandstones and mudstones filling cavities or solution pipes in the Carboniferous Limestone are of Namurian age, however. Some, as in the Clapton Basin, are of Westphalian age. No evidence of the presence of Namurian strata has been found in the Clapton Basin, where Upper Coal Measures rest unconformably on Lower Carboniferous and Upper Old Red Sandstone rocks (p.103).

The history of the Namurian movements and the true nature of the hiatus between Lower and Upper Carboniferous rocks is only imperfectly known in the west of the region. It is likely that the region around Clapton-in-Gordano was subject to repeated uplift from Viséan times onwards. A tectonic hinge line may be traced from Henbury south-westwards towards the Nailsea Basin. West of this line Millstone Grit strata rest unconformably on Clifton Down or older formations, while to the east the Quartzitic Sandstone Group is underlain by the Hotwells Limestone. At Winford, on the eastern margin of Broadfield Down, occurs the nearest approach to a conformable passage from late Viséan to early Namurian.

though only a 1 ft bed of white-ash coal appears to have been fit for working (Anstie, 1873, p.27). Some distance above, the Cromhall Vein was worked from bell-pits on the south and south-east sides of Cromhall Common. Some inferior coals were also worked on a small scale at Wick and near Barrow Gurney.

The Tapwell Bridge Seam appears to lie fairly low down in what is now regarded as the Millstone Grit, though there is no precise information concerning its age and correlation. It may equate with a thick white-ash coal formerly exposed at Cattybrook (see also p.95), but of this there is considerable doubt.

Westwards of Bristol there are clear indications that rapid attenuation, possibly accompanied by overlap or overstep of the Quartzitic Sandstone Group takes place along the western margin of the main coal basin, roughly along a line extending through Almondsbury and Henbury, and a comparable attenuation can be seen on the northern margin of the Nailsea Basin. At Tyntesfield [507 716] irregularly shaped, lenticular and piped masses of hard quartzitic sandstone are present in the Clifton Down Limestone. These were

CHAPTER 5

Coal Measures (Westphalian)

In the central area of the Bristol and Somerset Coalfield south of the Kingswood Anticline about 9000 ft of Coal Measures rocks overlie the Millstone Grit. The sequence is less complete where the Coal Measures rest unconformably on Old Red Sandstone and Dinantian rocks. Westphalian strata probably attain their maximum thickness in the Pensford Basin, north of Publow and south of Whitchurch. Elsewhere the sequence is less complete due to nondeposition or subsequent erosion of the higher strata, or to the attenuation, or absence of the lower beds due to unconformity and the attendant changes produced by overlap and overstep. The main area of exposed Coal Measures occurs in the Coalpit Heath Basin, extending some 4 miles east and 10 miles north-north-east of the city of Bristol. Several smaller inliers rise from beneath younger rocks to the west near Nailsea and Clapton in Gordano, and to the south around Corston, Pensford, Clutton and between Nettlebridge and Mells. Much larger areas of concealed Coal Measures exist beneath a cover of Mesozoic and later sediments. Some coal basins, such as the one beneath the Severn Estuary, are

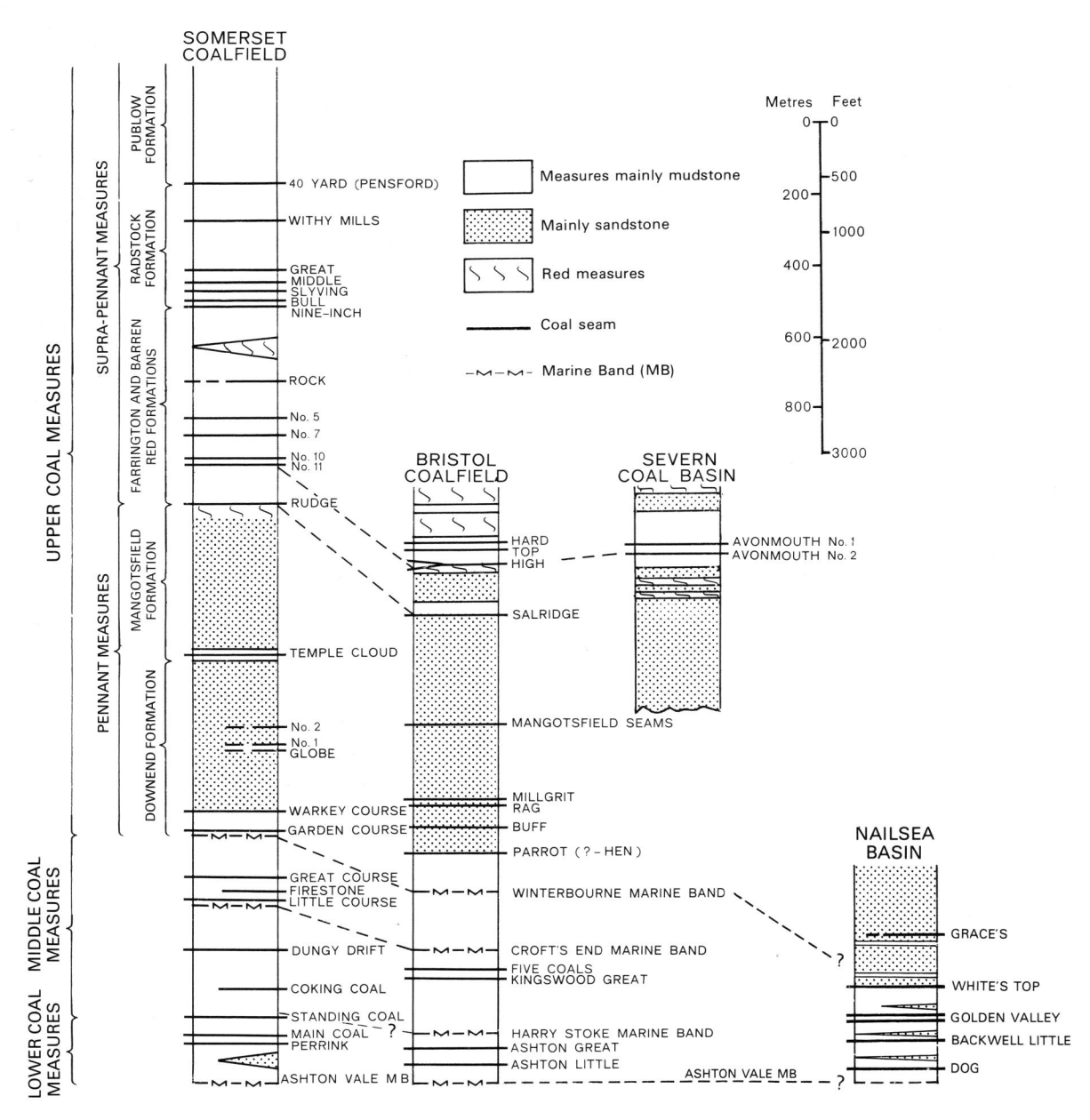

Figure 20 Generalised vertical sections of the Coal Measures of the Bristol and Somerset Coalfield

almost entirely concealed. In this instance the only surface evidence for the presence of this large tract of Coal Measures is provided by the small inliers at Cattybrook, Kingsweston and Portishead. All the remaining information has come from borings and underwater tunnels.

The Coal Measures sequence, like that elsewhere in north-west Europe consists of rhythmic alternations of mudstones, silty mudstones, siltstones and sandstones interspersed with numerous coals of varying thickness and their underlying seatearths. Generalised sections for the main areas of occurrence are shown in Figure 20, which also shows the stratigraphical classifications in current and recent use. Overall, three main lithological divisions are readily recognised: a lower largely argillaceous division, 2000–2500 ft thick, being separated from an upper similar division, about 4000 ft thick, by 2500 ft of predominantly arenaceous rocks, characterised by massive developments of feldspathic sandstones or subgreywackes, long known to miners as the Pennant Sandstone. In this respect the sequence is comparable with that found in South Wales and Kent. In the quality and development of its coal seams however, the Bristol and Somerset Coalfield resembles some of the minor coalfields in the Severn valley and the Massif Central, both in variability of thicknesses and rank of the coal and the presence of great thickness of fireclay and brown sandy sideritic mudstone. Coals occur at irregular intervals, but the principal seams worked in the past occur mainly in the upper and lower argillaceous divisions, those of the lower divisions being generally correlatable with groups of seams worked in other British coalfields.

This natural tripartite subdivision formed the basis of all the classification of the Coal Measures in the Bristol district in the past. However, modern formal stratigraphy throughout north-west Europe is based on the recognition of individual bands of marine mudstone with a characteristic fauna. These are widely distributed and therefore form a sound basis for correlation. Most Coal Measures sediments are of terrestrial origin, laid down in fresh or brackish water, though the principal marine bands appear to represent comparatively rare inundations of this environment by the open sea, possibly due to worldwide eustatic changes of ocean level.

These marine bands were first used in north-west Europe to define the lowest three subdivisions of the Westphalian, which is now regarded as broadly equivalent to a redefined Coal Measures as developed in most British coalfields, though the subdivisions of the Westphalian do not all coincide precisely with the now accepted three-fold grouping of the Coal Measures (see Table 2, and Stubblefield and Trotter, 1957). At Bristol the base of the Coal Measures is marked by the Ashton Vale Marine Band, a horizon characterised by the goniatite *Gastrioceras subcrenatum*, and equivalent to the Pot Clay Marine Band in Yorkshire and the East Midlands. The Lower Coal Measures are separated from the Middle Coal Measures at the base of the Harry Stoke Marine Band, characterised internationally by the goniatite *Anthraceratites vanderbeckei* and which has such equivalents as the Amman Marine Band in South Wales, the Clay Cross Marine Band in Yorkshire and the Queenslie Marine Band in Scotland. The Croft's End Marine Band is another important horizon for correlation purposes, being equivalent to the Cefn Coed Marine Band in South Wales and the Mansfield Marine Band in Yorkshire and the East Midlands. This horizon, marked by the presence of *Donetzoceras ('Anthracoceras') aegiranum*, is used on the continent to divide Westphalian B and C. The Upper Coal Measures is defined by the base of the Winterbourne Marine Band characterised by *Donetzoceras ('Anthracoceras') cambriense* and correlated with the Upper Cwmgorse Marine Band in South Wales and the Top Marine Band in Yorkshire; this is the highest of the true marine horizons in north-west Europe. Important though they are in many major coalfields, marine

Table 2 Stratigraphical classification of the Coal Measures

Series	Stages	Nonmarine bivalve zones	Marker goniatite bands	Lithostratigraphy		
Westphalian	Westphalian D	A. tenuis		Publow Formation Radstock Formation Barren Red Formation Farrington Formation	Supra-Pennant Measures	Upper Coal Measures
	Westphalian C	A. phillipsii	Winterborne (Cambriense) M.B.	Mangotsfield Formation Downend Formation	Pennant Measures	
		Upper similis-pulchra	Crofts End (Aegiranum) M.B.	Middle Coal Measures		
	Westphalian B	Lower similis-pulchra	Harry Stoke (Vanderbecki) M.B.			
		A. modiolaris				
	Westphalian A	C. communis	Ashton Vale (Subcrenatum) M.B.	Lower Coal Measures		
		C. lenisulcata				

bands do not occur above the base of the Upper Coal Measures. Thus the upper part of the Coal Measures in the Bristol and Somerset Coalfield cannot be classified in this way. Nonmarine lamellibranchs and plants are almost the only fossils available for this purpose and these are so erratically distributed in the Upper Coal Measures and have such long vertical ranges that great care is required in using them.

A zonal classification of the Coal Measures, based on variations in the nonmarine bivalve (or 'mussel') faunas, was first proposed by Davies and Trueman (1927). This was adapted to the Bristol and Somerset Coalfield by Moore and Trueman (1937). Following the widespread recognition of the marine bands and their use both in correlation and classification, the bivalve zones have tended to lose their importance; the faunas are nonetheless very useful for broad classification and even detailed correlation on a local scale.

The Upper Coal Measures of the Bristol district embraces the whole of the median arenaceous group, now called Pennant Measures, as well as the upper argillaceous group, here called the Supra-Pennant Measures. These terms replace the 'Pennant Series' and 'Upper Coal Series' of the Bristol district special sheet. The base of the 'Pennant Series' is diachronous, and does not correspond exactly with that of the Pennant Measures as defined above. In South Wales the entire Upper Coal Measures has been called Pennant Measures (Woodland et al., 1957) because the upper argillaceous group is only poorly developed in, and restricted to, parts of Gwent and West Glamorgan.

The Pennant Measures of the Bristol district are subdivided into the Downend and Mangotsfield formations, although these are not coloured separately on the published one-inch and 1:50 000 Geological Survey maps. The boundary between the two formations is taken at the base of the lowest of the Mangotsfield seams in the southern part of the Coalpit Heath Basin (Kellaway, 1970). The top of the Mangotsfield Formation and the base of the Supra-Pennant Measures is regarded by Dr M. A. Calver (in Ramsbottom et al., 1978, p.47) as the local boundary between Westphalian C and D, although the massive nature of the Pennant sandstones above and below the Mangotsfield seams makes it virtually impossible to apply any palaeontological or palaeobotanical criteria.

The base of the Supra-Pennant Measures is taken by Kellaway (1970) at the Rudge Seam in the Radstock Basin and the High Seam of the Coalpit Heath Basin, though these are not always well developed. The Supra-Pennant Measures are subdivided into four formations — Farrington, Barren Red, Radstock and Publow. The first three differ only in detail from those established under the same names by earlier workers, but the additional Publow Formation has been proposed (Kellaway, 1970) to cover the barren strata above the productive Radstock Formation. The Publow Formation is best developed in the centre of the Pensford Basin (p.176) but has been largely destroyed by erosion in the areas to the north and south.

Correlation problems of the Upper Coal Measures differ markedly from those of the Middle and Lower Coal Measures. It is only possible to use broadly based lithological comparisons after establishing lengthy stratigraphical sequences which have frequently to be determined in structurally deformed ground.

The definition of an upper limit for the Westphalian has also proved to be troublesome. The Radstock Formation and, by inference, the overlying Publow Formation, were attributed by Moore and Trueman (1937) to a Stephanian *prolifera* Zone. It has since been shown, however, that the flora of the Publow Formation is of Westphalian D age (Wagner, in Ramsbottom et al., 1978, p.47). The Publow Formation is therefore older than the Grovesend Beds and Woor Green seams of the South Wales and Forest of Dean coalfields in which Cantabrian floras have been found (Cleal, in Ramsbottom et al., 1978, p.47). It would appear therefore that the entire sequence of Coal Measures rocks in the Bristol district lies within the Westphalian of north-west Europe.

DISTRIBUTION AND THICKNESS

The distribution of Westphalian rocks, both at outcrop and concealed beneath younger strata, within the area covered by the Geological Survey's one-inch special sheet of the Bristol district is shown on Figure 21, together with the principal localities, mines and borings mentioned in the following text. The greater part of the combined Bristol and Somerset coalfields is present within the confines of the sheet, although the concealed eastern margin of the main field probably extends into areas covered by the one-inch Bath (265) and Frome (281) sheets. Coal Measures were also proved at Westbury (Wilts) in a boring at the old ironworks (Pringle 1922), where steeply dipping strata are probably separated from the main coalfield by a belt of folded and eroded Dinantian and older formations constituting a north-easterly extension of the rocks exposed in the eastern Mendips (Kellaway and Hancock, 1983, fig.5). The presence of another isolated mass of Lower Coal Measures on the southern flank of the Mendips at Ebbor (Green and Welch, 1965, p.56) also shows that the separation of the several Coal Measures basins of the region is due mainly to post-Carboniferous earth movements and that deposition of at least the Lower and Middle Coal Measures extended over the site of the Mendips.

The main basin at present recognisable is that generally referred to as the Bristol and Somerset Coalfield which occupies a broad tract of country to the north-east, east and south-east of Bristol, extending from Cromhall Common in the north to Nettlebridge and Mells in the south, a distance of some 26 miles. The principal coal basins are those of Coalpit Heath in the north and Pensford and Radstock in the south. The separation of the Coalpit Heath and Pensford basins is essentially structural and is due to the presence of the Kingswood Anticline while the boundary between the Pensford and Radstock basin is coincident with the 'Farmborough Fault'. Separate or largely separate basins occupy extensive tracts north of Avonmouth (the Severn Basin), in the Gordano valley, and between Nailsea and Wrington (Figure 21).

North of the Mendips and west of Wrington the extent of the Coal Measures is uncertain. A boring at Banwell (Green and Welch, 1965, pp.198–199) proved Upper Coal Measures beneath the Mesozoic cover, but there is no evidence as to the distribution of Coal Measures rocks in the Drift-covered area around Puxton and Kingston Seymour.

DISTRIBUTION AND THICKNESS 69

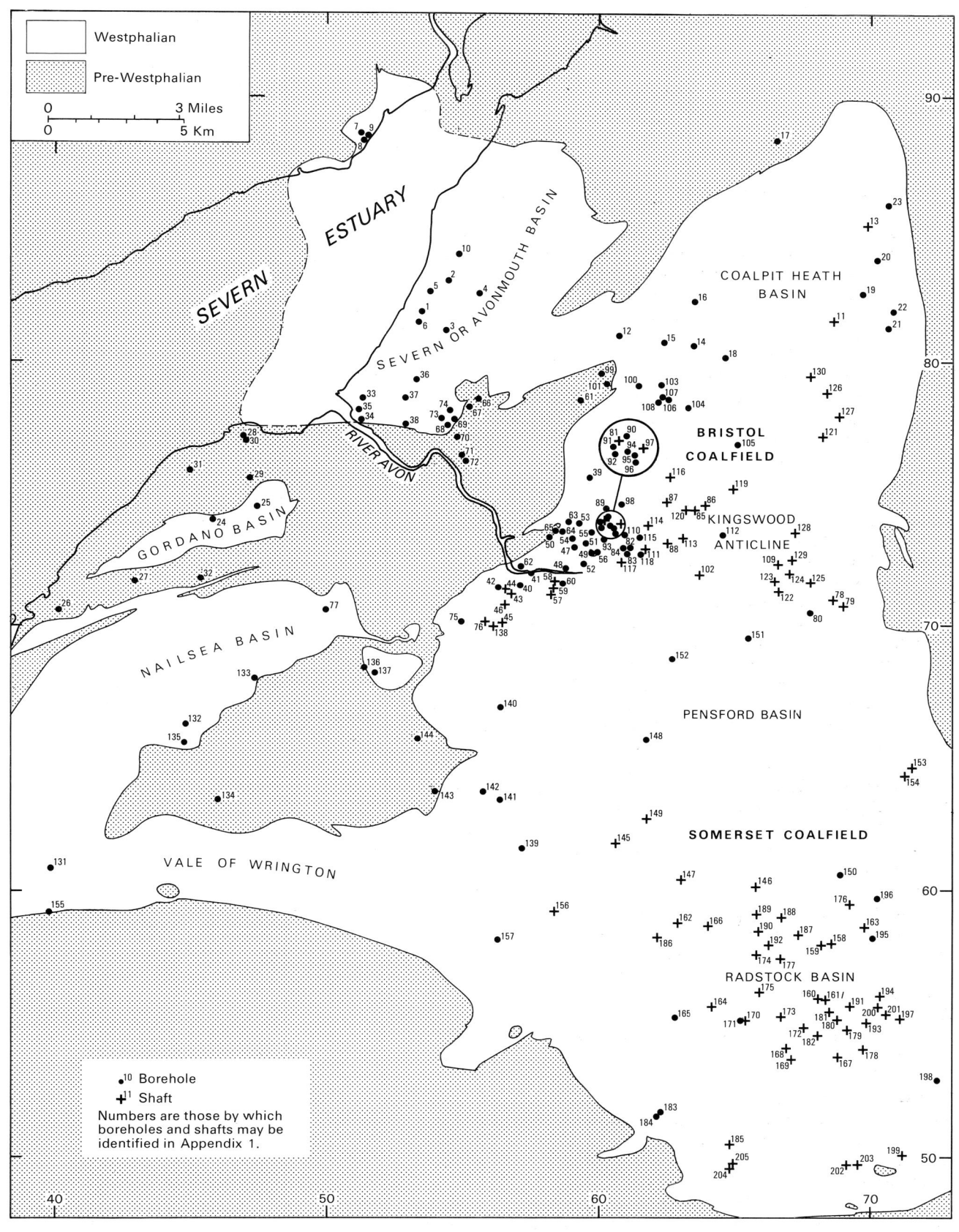

Figure 21 Map showing limits of area underlain by Westphalian rocks, and sites of principal shafts and boreholes included in Appendix 1

It is possible that some Coal Measures may have escaped pre-Triassic erosion, particularly between Kenn and Clevedon, but the depth to which denudation has been carried out is not known. It is, however, certain that a ridge of folded Carboniferous Limestone, overlain unconformably by Triassic rocks, extends from Broadfield Down to Yatton, thus bringing about a partial separation of the residual mass of Coal Measures in the Nailsea Basin in the north from those of the Banwell–Wrington area to the south.

In the Nailsea Basin (Figure 21) the northern and southern limits of the Coal Measures are fairly well known, though they and the underlying rocks are mostly concealed. It is probable that the Coal Measures of Nailsea were separated from those of Barrow Gurney and the main coalfield at Ashton and Bedminster by faulting and pre-Triassic erosion.

North of Nailsea, the small Clapton or Gordano Basin appears to be isolated from the much larger area of Coal Measures forming the Severn or Avonmouth Basin (Figure 21). The southern limit of this very extensive area of largely concealed Coal Measures is seen in the folded and faulted rocks on the foreshore between Portishead Point and Portishead Pier. Smaller areas of exposed Coal Measures, at Kingsweston on the north side of the Avon, and at Cattybrook near Over, mark the eastern margin of the Severn coal basin. Its northern margin lies south of Olveston and Aust and is everywhere concealed by Mesozoic and Drift deposits. Thence the boundary passes beneath the Severn Estuary to the vicinity of Portskewett in Gwent where Upper Coal Measures rest on thin Namurian rocks. Coal Measures sandstones (Pennant Measures) crop out on Lady Bench and other reefs in the bed of the Severn near Sudbrook Point (Figure 19), and Coal Measures resting on Namurian and older rocks were proved in the western part of the Severn Tunnel (Richardson, 1887).

The Coal Measures of the Severn Estuary are separated from those of the northern part of the Coal Heath Basin by a line of intense structural disturbance between Cattybrook and Alveston. East of this line the base of the Coal Measures is concealed by Mesozoic rocks, though the lowest Coal Measures are seen at intervals in the horse-shoe shaped crop extending from Tytherington to Cromhall and Yate. South of Yate the Coal Measures are again concealed by Mesozoic formations and they are not seen at surface until the Wick inlier is reached at the eastern end of the Kingswood Anticline. South of Wick the margin of the Coal Measures is concealed though its position can be inferred from borehole and other evidence. At Bath the precise position of the base of the concealed Coal Measures is not known, though it probably lies not far south of the city centre. South of Bath, however, the thick Mesozoic cover effectively conceals all the Palaeozoic rocks, and little is known of the eastern boundary of the main coalfield between Bath and the Mendips[1].

By far the greater part of the total area of Coal Measures thus defined is concealed by a blanket of Mesozoic rocks or younger sediment. Many of the exposed areas are formed by strongly disturbed and deeply weathered clay rocks. There are also large areas as in east Bristol, Kingswood and the Nettlebridge Valley where the original surface has been destroyed by mining activities, buried by waste material or concealed by urban development. Some useful information has been obtained from quarries, clay pits and cuttings but to all intents and purposes the investigation of the Coal Measures rocks hereabouts has been characteristic of a concealed rather than an exposed coalfield. Borehole data and information derived from mine shafts and underground sources have therefore played a dominant part in these investigations (Figure 21).

In the central part of the coalfield the greatest thickness of preserved Coal Measures, is found south of the Kingswood Anticline. Here the thickness of the Lower and Middle Coal Measures is about 2250–2500 ft and that of the Upper Coal Measures about 6500 ft (Figure 20). These residual thicknesses do not, however, have direct depositional significance. Thus the Lower and Middle Coal Measures were probably originally thicker in the south than they are in the north and west. A slight reduction in thickness is noticeable as the Upper Coal Measures are followed northwards from Radstock to the Avon valley at Brislington (Kellaway, 1970, pl.1) and further attenuation probably occurs north of Kingswood.

Between Soundwell and Hanham in the central region of the Kingswood Anticline, the combined thickness of the Lower and Middle Coal Measures is thought to be about 2200 ft, only slightly less than in the southern part of the coalfield between Stratton-on-the-Fosse and Mells. The deepest part of the sedimentary basin in which these measures were laid down, probably runs from Cromhall in the north towards Coalpit Heath and Downend, thence to Kingswood, Compton Dando, Midsomer Norton and Coleford in the south. The basin therefore had an orientation roughly parallel to that of the great Malvern Fault zone which traverses the coalfield from north to south (Kellaway and Hancock, 1983). In the Coalpit Heath Basin these faults have a north-north-easterly trend, but this changes in the vicinity of the Kingswood Anticline, first to a northerly or north-north-westerly direction and then to north-westerly as the southern margin of the coalfield is approached. It seems likely that the broad tract of exposed Lower and Middle Coal Measures north-east of Rangeworthy, the structurally contorted central region of the Kingswood Anticline and the violently contorted measures of the Nettlebridge Valley south of Radstock lie in the central region of an early Coal Measures downfold.

The associated positive fold belts marking the margins of this tract show attenuation of the Lower and Middle Coal Measures. On the west side the attenuation extends along a belt which is roughly parallel to the Ridgeway Fault. On the eastern side comparable attenuation and change of facies extends through the Codrington, Wick and the Grandmothers Rock inliers. The boundaries of these 'uplifts' are ill defined but the presence of positive structures can be deduced from changes in thickness of the Coal Measures. Thus, on the eastern flank of the main basin at Wapley and Pucklechurch, the reduction in thickness of the Lower and Middle Coal Measures was commented on by Anstie (1873) who attributed it to 'squeezing' of the measures. Not all the thinning is attributable to bed-by-bed attenuation, how-

---

1 Boreholes made in connection with the thermal water at Bath have proved Carboniferous Limestone (Kellaway, 1991, pp.97–125). Carboniferous Limestone has also been proved at the concealed eastern margin of the Radstock Basin at Tuckingmill near Midford (ST 7636 6163) at a depth of 224 m.

ever. Evidence of localised erosion of the upper surface of the Middle Coal Measures is present at Westerleigh (Kellaway, 1970). Here, coarse conglomerates largely composed of pebbles of Precambrian rocks collected on the eroded surface of the Middle Coal Measures, and the Winterbourne Marine Band may have been removed locally south of Westerleigh by erosion prior to the deposition of the Downend Formation (Upper Coal Measures).

Conglomerates containing felsites and other igneous rocks (Moore and Trueman, 1937, p.222) are found sparingly in the Middle Coal Measures and much more abundantly in the lower part of the Upper Coal Measures (Downend Formation) in the Coalpit Heath Basin. They occur on a more limited scale in the northern limb of the Kingswood Anticline and, with the possible exception of the roof of the Garden Course Vein, are almost unknown in the Radstock Basin. They are of north-eastern or eastern orgin and may have been recycled from Old Red Sandstone.

The term Pennant or Pennant Grit first appeared in Strachey's account of the Upper Coal Measures of Brislington (1725, p.972) where he describes the sandstone above the Rock Vein as a 'Rock of Paving-Stone called Pennant'. The main mass of the Pennant Sandstone (or Grit) forms the Pennant Measures of this account and is divided into the Downend Formation below and Mangotsfield Formation above. The Pennant sandstone (or subgreywacke) facies appears in a fine-grained form in the upper half of the Middle Coal Measures at the southern end of the Radstock Basin. Followed northwards the first appearance of the Pennant Sandstone facies lies near the Crofts End Marine Band in the south and rises northwards being at or near the Winterbourne Marine Band in the southern limit of the Kingswood Anticline. As the rocks are followed northwards to Rangeworthy the sandstones at the base of the Downend Formation are largely replaced by mudstone and fireclay. The boundary of the Mangotsfield Formation and the Farrington Formation remains largely unchanged but Pennant-type sandstone appears south of Iron Acton above the High Vein of Coalpit Heath i.e. in the Supra-Pennant Measures. The High Vein of Coalpit Heath correlates with Avonmouth No. 2 and the Coleford High Delf seam so the sandstones which first appear at the northern end of the Coalpit Heath Basin above the High Vein become the Pennant Sandstone of the Forest of Dean (Welsh and Trotter, 1961). This shift in the base of the subgreywacke facies is an important feature of the Westphalian succession.

Kelling (1974) has suggested that the source of the Pennant Sandstone of South Wales may lie to the south of the coalfield and the evidence pointing to the possible existence of an uplifted mass of Devonian rocks south of the Mendips is given below (p.24). If we exclude the Culm areas of central and east Devon, the general distribution of the coalfields in which Pennant Sandstones are present is limited by a line drawn from St Brides Bay to Shrewsbury, Kettering and Margate. South of this line Devonian rocks would have contributed detrital matter to the Westphalian rocks. North of the line Devonian rocks are missing and there are no typical Pennant Sandstones in the Coal Measures.

West of Bristol evidence for attenuation of the Lower and Middle Coal Measures is linked with overstep and overlap of these rocks by younger strata, notably of the Mangotsfield and Farrington formations. The western basins at Nailsea and beneath the Severn estuary are unlikely to provide much information of the detailed stratigraphy of the rocks below the Mangotsfield Formation, and it is the central area of the main basin of the coalfield that holds the key to the stratigraphical succession. North of the Kingswood Anticline the higher beds of the Coal Measures have mostly been destroyed by Permo-Triassic erosion leaving small isolated basins of productive Upper Coal Measures (mainly Farrington Formation) at Coalpit Heath and Avonmouth. The Coalpit Heath Basin is exposed, but the existence of the concealed Avonmouth Coal Basin was established only by drilling.

HISTORY OF RESEARCH AND DEVELOPMENT OF CLASSIFICATION

Among other pioneer geologists, John Strachey (1719, 1725) and William Smith (1815) were primarily concerned with coal mining, and the first systematic description of the rocks of the Coal Measures of Somerset was that of Buckland and Conybeare (1824). This work, like that of Greenwell (1854) and James McMurtrie (with whom Greenwell produced an important paper in 1864), was primarily concerned with establishing a stratigraphical succession based on the presence of groups of workable coal seams. This approach is also reflected in the report of the Royal Commission (Prestwich, 1871). Much information was collected for the Coal Commission by a mining engineer, John Anstie, whose monograph entitled 'The Coal Fields of Gloucestershire and Somersetshire and their resources' (1873) became the authoritative work on which local mining engineers relied. Somewhat later a small pamphlet by Cruttwell (1881) on the geology of Frome gave details of the sinking of a coal shaft at Buckland and of the adjacent Murtrey Borehole.

In the meantime, other investigators working in the northern area near Bristol had not been idle. Cossham (1865) had indicated that the so-called 'Millstone Grit' shown on the Old Series Geological Survey map of the Kingswood Anticline was, in fact, part of the Coal Measures. These and other siliceous Lower and Middle Westphalian rocks, which include the Holmes Rock and the Hard Venture Stone of Kingswood and Soundwell, closely resemble the Millstone Grit quartzitic sandstones. Not until the National Coal Board's Harry Stoke 'A' borehole was sunk in 1949, was it possible to show that the position of the Hard Venture Stone is immediately above the Ashton Vale Marine Band, now accepted as the base of the Coal Measures.

The natural grouping of the Coal Measures into Upper and Lower productive divisions separated by a great thickness of coarse gritty sandstone was recognised from early times, as may be seen from Buckland and Conybeare's classification (1824, p.252). They divided the Coal Measures into Upper coal shale, Pennant grit and Lower coal shale. Greenwell and McMurtrie (1864, p.19) proposed: Upper or Radstock Series; Second or Farrington Series; Pennant Rock; Lower Series. Subsequently the 'Lower Series' was divided by McMurtrie (1869a, p.48) into New Rock and Vobster Series. The other four series were then defined on the basis of the coal seams which they include. Thus the 1869 classification was as follows:

UPPER DIVISION:
First or Radstock Series: Nine Inch Vein to Withy Mills Vein
Unproductive strata with red shales
Second or Farrington Series: 17-inch Vein to Cathead[1] Vein

PENNANT ROCK

LOWER DIVISION
Third or New Rock Series: Standing Coal to Globe
Fourth or Vobster Series: Wilmot's[2] Vein to Fern Rag

McMurtrie's classification of 1901 is essentially a restatement of his 1869 scheme. In the Report to the Royal Commission, Prestwich (1871) adopted McMurtrie's classification, though he altered the limits of the New Rock Series to include coals from the Small Coal to the Hard Vein, while the Vobster Series included seams from the Red Axen to the Perkins. These divisions are based on coal seams rather than the general characteristics of the strata and little provision was made for the inclusion of other seams which might be discovered at a later date. This defect in the earlier classifications was partly remedied by Moore and Trueman (1937) who proposed the following scheme:

UPPER COAL SERIES
Radstock Group: Nine Inch Vein to Withy Mills Vein
Barren Red Group: Rock Vein to Nine Inch Vein
Farrington Group: No. 9 Vein (or its equivalents) to Rock Vein

PENNANT SERIES

LOWER COAL SERIES
New Rock Group: Coking Coal to Newbury No. 3 Vein
Vobster Group: Ashton Vale Marine Bed to Coking Coal

In a later work, Trueman (1947, p.lxix) discarded the New Rock and Vobster groups. This, in effect, returned the classification of the Lower Coal Series to a condition differing very little from that in which it was left in 1902. It therefore became abundantly clear that no further progress would be made unless the positions of the principal marine marker bands could be ascertained (Table 2).

The first major step had already been taken in 1907, when H. Bolton published a detailed description of a marine fauna including *Gastrioceras subcrenatum* from the base of the Coal Measures in Ashton Vale Colliery in the mining area of south Bristol. Next, a mining engineer, Mr E. H. Staples in about 1909 or 1910 discovered marine strata in the Toad Vein Branch at Deep Pit, Kingswood. The record of the exact position at which this discovery was made has been lost, but it is known to lie between the Kingswood Great Vein and the Kingswood Toad Vein. This marine band is, therefore, the Crofts End Marine Band (p.67), first recorded at the surface in Crofts End Brickpit near the top of Deep Pit (Kingswood) shaft by Moore and Trueman (1937, p.212). The specimens collected underground and preserved by Staples were subsequently presented by him to the Geological Survey and were identified by Dr M. A. Calver.

Between 1912 and 1920 four deep boreholes were sunk in the Coalpit Heath Basin, two at Westerleigh (1912–13) and one each at Winterbourne (1915–17) and Yate (1920). The Westerleigh borings were not examined by the Geological Survey and it is thought that no material from them has survived. The Winterbourne and Yate boreholes were logged by officers of the Geological Survey and the collected material was re-examined, first by Moore and Trueman (1939) and subsequently by Dr M. A. Calver. As a result it has been possible to define the position of the Crofts End and Winterbourne marine bands in these holes and to compare them with new material from the Harry Stoke and Stoke Gifford borings (p.166, 170–171) drilled between 1949 and 1954. On the basis of this work, the position of the principal marine bands within the Lower and Middle Coal Measures was established (Kellaway and Welch, 1955). The four principal marker bands, the Ashton Vale, Harry Stoke, Crofts End and Winterbourne marine bands now provide a firm basis for correlation, while three of them are used for defining the base limits of the Lower and Middle and Upper Coal Measures (Table 2).

Application of these discoveries to the Coalpit Heath Basin north of Chipping Sodbury, and the discovery of the crop of the Crofts End Marine Band at Broad Lane, Yate, has revealed that the sequence in these north-eastern areas is similar to that of Kingswood (Kellaway, 1970). This has resulted in substantial adjustments to the position of the boundaries of the Millstone Grit and Coal Measures rocks as originally shown on the Bristol district special sheet (1962), the corrected positions being given on the Malmesbury (251) sheet published in 1970.

Until recently, lack of evidence has prevented the application of this classification to the Radstock Basin. However, the positions of the Winterbourne and Crofts End marine bands were found in 1968 when demolition of New Rock Colliery enabled Dr B. Kelk and Mr R. J. Wyatt to examine the measures between the Garden Course and Dungy Drift seams (p.93). The Harry Stoke Marine Band is considered likely to lie above the Standing Coal or possibly the Coking Coal (Figure 20), but the position of the marker band has not yet been confirmed underground or at crop. Evidence of the presence of marine shale in structurally disturbed measures situated near the crop of the Coking Coal was found near Newbury Colliery during the course of the primary six-inch survey. At the time of its discovery it was thought that this shale might be the Crofts End Marine Band, but the latter is now established to lie below the Little Course Seam of New Rock Colliery, so the shale is correlated with the Harry Stoke Marine Band of Bristol.

The Winterbourne Marine Band lies at no great distance below the Garden Course seam of New Rock Colliery (Figure 20) so the boundary between the Middle and Upper Coal Measures can be defined. Unfortunately there is no comparable information available for the southern flank of the Kingswood Anticline and here the boundary is still indeterminate.

---

1 Cathead Vein, see p.107
2 Wilmot's Vein, a small thin, little known seam, probably lying very close to the Ashton Vale Marine Bed at the base of the Coal Measures.

PRINCIPAL COAL SEAMS

The coals of the Bristol and Somerset Coalfield vary greatly from one area to another and there are no individual seams which can be traced systematically throughout. Thus the thick coals of the Lower Coal Measures are known to occur only at Bristol south of Stoke Gifford and west of the Whitefaced Fault, and in the southern and south-western quadrant of the Radstock Basin. They include the Ashton Great Vein of Bristol and the Main Coal and Perrink of the Nettlebridge Valley. These coals are thin or absent in the eastern part of the Coalpit Heath Basin and the Kingswood Anticline where they are replaced by quartzitic sandstone, 'bastard ganister' and fireclay.

In the vicinity of Bristol and the central and western portion of the Kingswood Anticline, the principal coals of the Middle Coal Measures include the Kingswood Little and Kingswood Great Veins and the Lower Five Coals. These are most strongly developed between east Bristol and Warmley. They appear again at Yate though they die out or deteriorate farther north at Rangeworthy and Cromhall and to the south at Wapley. They reappear, however, south of the Avon valley and were worked in Twerton Colliery near Bath. The Kingswood seams become thinner as they are followed from Bristol towards Nailsea and, with the possible exception of the Dungy Drift, there do not appear to be any important correlatives in the southern part of the Radstock Basin.

In the south-western part of the Radstock Basin the most important group of coals above the Lower Coal Measures is the one including the Great Course, Firestone and Garden Course formerly worked in the Mells–New Rock area. These seams lie in the uppermost part of the Middle Coal Measures and in the basal part of the Upper Coal Measures. An analogous group of coals, including the Parrot, Buff, Rag and Millgrit seams of Oldland Common and Bitton, exists in the south-east quadrant of the Kingswood Anticline and extends to Newton St Loe, west of Bath. At the western end of the Kingswood Anticline, however, they have deteriorated. In the northern part of the Kingswood Anticline the only seam of any importance at or near the base of the Upper Coal Measures is the Hen Vein of Staple Hill and Fishponds. North and north-east of Stapleton no workable seams have survived in this position.

Although the Mangotsfield seams are important stratigraphically, none of the other coals in the Mangotsfield Formation is sufficiently continuous to merit reference here.

Above the Pennant Measures the only seams which can be traced from the Radstock Basin to the Coalpit Heath and Severn Coal basins are the Farrington seams. In the Radstock Basin these are mainly thin coals, seldom exceeding 2 to 3 ft in thickness. North of the Farmborough Fault Belt they are represented by the Bromley and Brislington seams. Two thick coal seams, Avonmouth No. 1 and No. 2 seams totalling about 10 to 12 ft of coal, are attributed to the Farrington Formation in the Severn Coal Basin (Figure 20). In the Coalpit Heath Basin these are represented by three (locally four) coals which have been worked over almost the whole extent of the Supra-Pennant Measures in the basin. Of these coals the High Vein of Coalpit Heath has a '*Leaia*'-rich roof and is regarded as being the correlative of the Coleford High Delf of the Forest of Dean (Welch and Trotter, 1961, fig.5). The Farrington seams are therefore unique in that they are the only group of coals which can be traced with any degree of certainty throughout the Bristol and Somerset Coalfield and which also extend into adjacent coalfields. Nevertheless, there are strong changes of character and facies as between the Farrington Formation of the Radstock and Severn basins and this has in the past given rise to strongly differing opinions as to the classification of these measures.

The Radstock Formation contains a number of thin seams seldom exceeding 2 to 3 ft in thickness. These are confined to the basins south of the Kingswood Anticline. The highest seam is a thin coal known as the Withy Mills (Figure 20), above which the measures are classified as belonging to the unproductive Publow Formation.

DETAILS OF STRATIGRAPHY

Lower and Middle Coal Measures

Ashton and Bedminster area (Figure 22)

The succession proved in the Lower Coal Measures at Ashton Vale Colliery [5656 7137] was described by Bolton (1907) in some detail and has been correlated with the sequence in the Ashton Park Borehole [5633 7146] (Kellaway, 1967). The **Ashton Vale Marine Band**, marking the base of the Lower Coal Measures, is about 20 ft thick, the name being restricted to the bed described by Bolton (1907) as the 'Chief Shell Bed'. Owing to their variability the overlying *Lingula* bands have not been named individually, though they have a bearing on the correlation of the rocks proved in the Winford boreholes (p.76–77, 172).

The lowest recognisable coal seam in Ashton Vale Colliery is recorded in some of the shaft sections as '**Smith's Coal**'. About 1 ft thick, it is situated about 60 ft above the top of the Ashton Vale Marine Band. In the Ashton Park Borehole [5633 7146] this seam is represented by a 6 in coal, with a fireclay floor and a marine roof, which is tentatively correlated with the Four Feet Vein of Easton Colliery [6065 7389] (Figure 22). About 45 ft above the Ashton Smith's Coal lies the **Ashton Gays Vein** which had a thickness of 20 to 30 in in the Exploratory Branch, but is apparently represented by only 6 in of coal at a depth of 615 ft 8 in in the Ashton Park Borehole. Bolton (1907) recorded marine strata about 11 ft above this seam though the mudstone roof appears to have included plant-bearing strata. In the Ashton Park Borehole the roof was particularly rich in *Lingula*. The probable correlative of the Gays Vein of Ashton Vale is thought to be the Seven Feet Vein of Easton Colliery (Figure 20). An interval of about 100 ft separates the Gays Vein from the **Ashton Little Vein**, which is the lowest coal to have been worked to any appreciable extent in the Ashton area. This coal, up to 24 in thick, was worked at Ashton Vale Colliery and probably at Starveall Pit [5652 7083] Bedminster, where it was said to have been proved in the shafts and workings.

Between the Ashton Little Vein and the **Ashton Great Vein**, the sequence in Ashton Vale Colliery is not known in detail. The distance between the two seams is of the order of 200 ft, whereas in the Starveall Shaft the separation is recorded as being only 165 ft (Anstie, 1873, p.48). The log of the Ashton Park Borehole (Kellaway 1967, p.93) records only one thin band of *Lingula*-bearing shale, at a depth of 461 ft, in this group of measures. This band, which may correspond roughly with the strata at 1315 ft in Winford No. 2 Borehole [5636 6343] (p.76–77), marks the top of the partly marine, partly nonmarine sequence which commences about 20 ft below the base of the Ashton Vale Marine Band and ex-

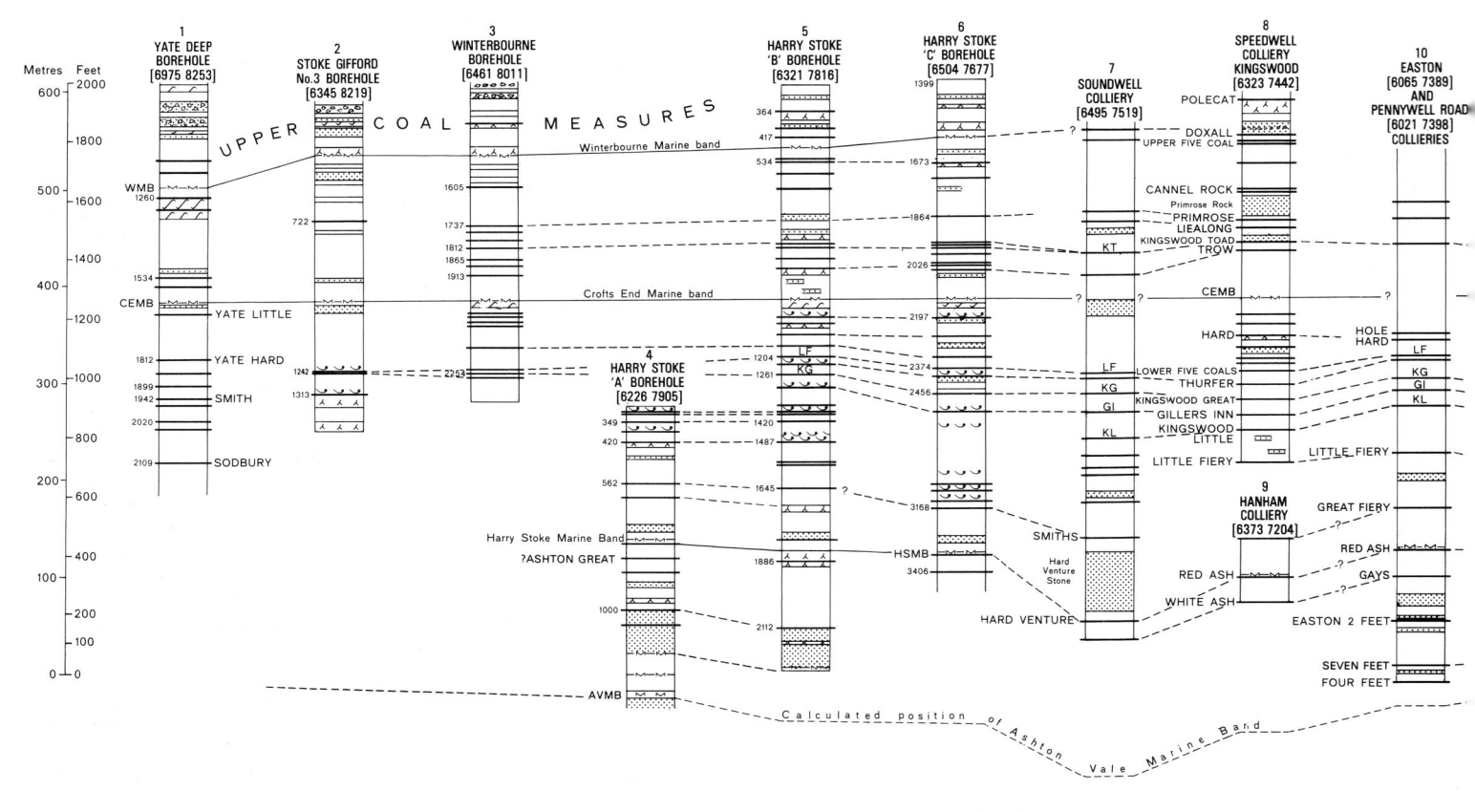

Figure 22 Comparative vertical sections of the Coal Measures between Yate and Winford

through about 265 ft of the overlying strata. Most of these intervening measures consist of mudstones rich in clay ironstone, interbedded with fireclay and with three or four thick beds of hard fine-grained sandstone and sandy mudstone.

The Ashton Great Vein is generally about 36 in thick, although about 48 in was proved in the Exploratory Branch at Ashton Vale Colliery and it is said to have been up to 68 in at Starveall (Anstie, 1873, p.48). Like many of the soft coals it shows rapid changes of thickness due to the squeezing of soft incompetent shale, coal and fireclay between more competent measures. 'Pods' of up to 30 ft of coal have been found in some squeezed masses of coal in the Ashton Vale workings. A section in the Ashton Great Vein of South Liberty Colliery [5648 7013] given on the abandoned mine plans shows the following: sandstone roof 3 ft 3 in, dark bind intermixed with bands of sandstone 4 ft 6 in, clod 16 in, coal 60 in, blacks 16 in, cockle bed 3 ft, on fireclay. This would appear to be representative of the less disturbed areas. A band of hard siliceous grit about 6 ft thick was interbedded with the mudstones separating the Ashton Great and Ashton Top veins in the Exploratory Branch at Ashton Vale. In the Ashton Park Borehole, however, this hard sandstone is nearly 60 ft thick and occupies about half the interval between the Ashton Great and Ashton Top veins.

According to the abandoned mine plans, the **Ashton Top Vein** appears to be a poor coal and in South Liberty Pit it was absent in many places. The strata above and below the seam in the Ashton Park Borehole were crushed or contorted. It was therefore impossible to confirm the presence of the Harry Stoke Marine Band in the roof of the coal. Bolton (1911) was unable to find marine strata in the roof of the Ashton Top Vein but it appears from the colliery records that the condition of the roof offered little hope of finding any fossils. In a section recorded by R. M. Dillwyn (MS) at South Liberty Colliery the roof of the seam consists of 'strong duns (rock in places)'. The coal, which is soft and gave 'all small coal', is about 36 in thick in two beds with a dirt parting in the middle. Detailed comparison of shaft records and borehole logs suggests the correlation of the Ashton Top Vein with the Hard Venture of Soundwell Colliery [6495 7519] and the Red Ash veins of Hanham [6373 7204] and Easton pits in the east Bristol–Kingswood area. However, it was believed by some that the Red Ash Vein of Easton is the correlative of a coal, probably the **Bedminster Toad Vein**, which was proved in Ashton Vale Shaft at a depth of about 348 ft. This would imply that the position of the Red Ash Vein of Easton and Hanham, and of the Harry Stoke Marine Band, is about 250 ft above the Ashton Top Vein. The correlation is possible, but a comparison of the sequence at Ashton Vale with that of Easton and Harry Stoke leads the writer to prefer that in Figure 20. In particular it will be observed that, while the Bedminster Toad Vein may not correlate precisely with the Little Fiery Vein of Easton, the Great Fiery may well be represented in the Ashton Vale shaft by one of the two seams underlying the Toad Vein.

If the Ashton Top Vein is correlated with the Red Ash Vein of Hanham and Easton, the latter may also be equated with the Gas Coal seam proved in the deep workings at Dean Lane Colliery, Bedminster [5835 7168]. So far as can be gathered from inadequate records, the Gas Coal of Dean Lane is a seam about 32 in thick (including a parting) situated about 650 ft below the Bedminster Great Vein, as is the Ashton Top Vein at Fraynes [5693 7121], Starveall and South Liberty collieries. The correlation of the Gas Coal of Dean Lane with the Red Ash Vein of Hanham Colliery (Figure 22) is also supported by the fact that the Hanham Red Ash

DETAILS OF STRATIGRAPHY

(e.g. Argus Pit [5818 7101]), but detailed correlation of these is at present impossible. It is fairly clear, however, that there is a general deterioration of these coals in the direction of Bedminster and Nailsea and the same process is observable in the overlying Bedminster Great Vein. It may well be that the Bedminster Little Vein occupies approximately the same position in the Middle Coal Measures as the Kingswood Little Vein.

Although thinner than the Kingswood Great Vein, which is its presumed correlative, the **Bedminster Great Vein** is said to possess much the same character and quality. It formed the mainstay of all the larger pits in the Bedminster area and was worked in South Liberty Colliery until its abandonment in 1926. The most northerly workings on the Bedminster Great Vein were those on the north side of Dean Lane Pit. These extend about 1¼ miles north-east of Dean Lane shafts, while the most southerly workings from South Liberty extend under Highridge Common to a point marked by an isolated hill known as the Peart, situated north of Dundry Church. In the intervening area it may safely be assumed that, with the exception of certain limited areas, the seam has been mostly worked from the sub-Triassic crop down to depths of 2000 to 3000 ft below the surface. Such continuity of working is remarkable in the Bristol Coalfield and is even more surprising when the highly folded nature of the Bedminster Great Vein in the southern workings of South Liberty Colliery is taken into account. However, the same feature is apparent in the Speedwell and Deep pit workings of the folded Kingswood Great Vein where the coal has been followed from flat-lying measures into folds with vertical limbs. At Dean Lane Colliery, the most northerly of the Bedminster pits, and in Starveall Pit to the south, the thicknesses given for the Bedminster Great Vein average about 42 in. A section on one of the old plans of Argus and Malago Vale pits gives: roof of duns, coal 6 in, clod 2 in, coal 16 in, fireclay and coal 24 in, rock 10 in, on duns. At South Liberty pit a typical section is: 'strong duns' resting on roof of duns 3 ft 9 in, clod 6 in, coal 42 in, fireclay 18 in, cockle bed [probably fireclay with small ironstone nodules] 10 in, stone 18 in; these thicknesses vary considerably 'in the vicinity of disturbed or faulty ground'. In the southern or Highridge part of the South Liberty workings the coal has been proved to be about 36 in thick. On the assumption that the Bedminster Great Vein is the correlative of the Kingswood Great Vein the seam shows an appreciable deterioration in thickness as it is traced from east Bristol (where it may be up to 60 in thick in one bed) to the southern part of the Bedminster area where it may not be more than 36 in.

Although the Ashton Great Vein and the Bedminster Great Vein appear to be distinct coal seams separated by over 600 ft of strata, the identification of the so-called Bedminster Great Vein in part of South Liberty workings has been questioned by Moore and Trueman (1937). Much turns on the source of the fossils which Bolton (1911) considered to have come from near the Bedminster Toad Vein, but which include forms also known from the Ashton Vale Marine Band. This led Moore and Trueman (1937, p.218) to suggest that repetition of the Ashton Great by thrusting may have taken place. However the extent of the Bedminster Great Vein workings was probably much greater than they assumed and this would militate against their proposal. It is possible, nonetheless, that a fault (or faults) placed the Ashton Vale Marine Band within reach of the Ashton Great Vein workings in South Liberty pit. Above all it will be recalled that Bolton (1911) was unable to locate the source of his material in situ above or near the Bedminster Toad Vein. On the available information there is no alternative to the rejection of Bolton's palaeontological evidence since the provenance of the material is unknown.

About 110 ft of measures, mainly mudstone, shale and fireclay separates the Bedminster Great Vein from the **Bedminster Top Vein**. At Dean Lane Colliery (Prestwich, 1871, p.56) the shaft may be in faulty ground, for the distance between the Bedminster Great Vein and Bedminster Top Vein is about 140 ft. Faulting or squeez-

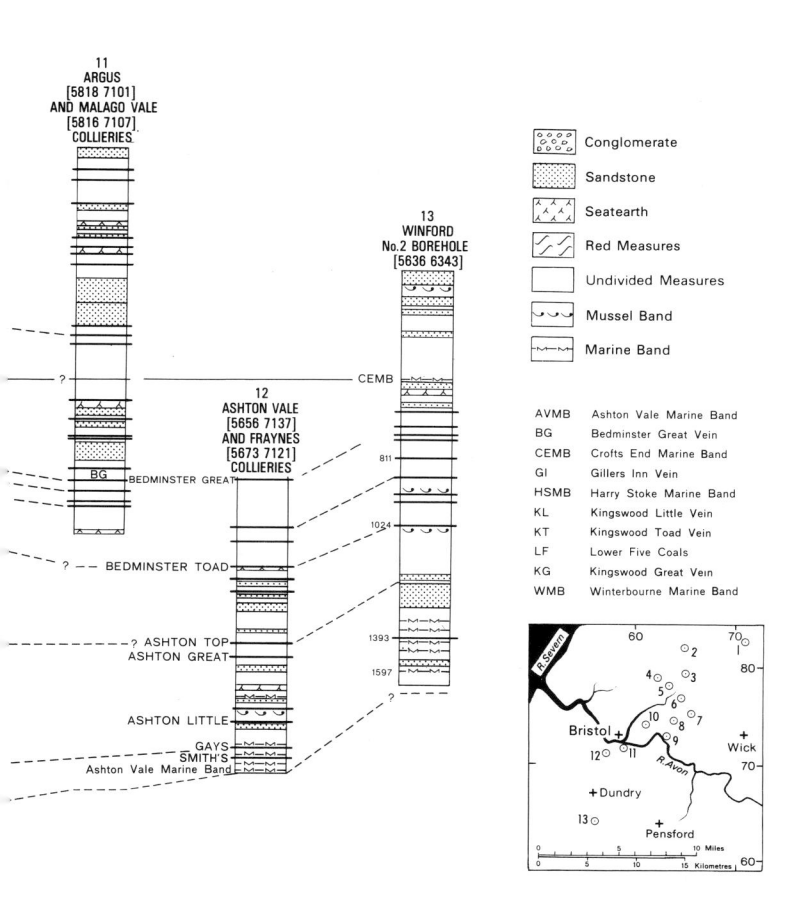

was well known as a good gas coal. Although a number of other seams have been recorded in the measures between the Ashton Top Vein and the Bedminster Great Vein, very little is known about them. At least three seams, Bedminster Smith's Coal, Bedminster Toad and Bedminster Little Veins were worked. Others have been recorded only in shaft sections.

Above the Ashton Top Vein there are several coals described in the shaft section of Ashton Vale Colliery [5656 7137]. One of these is likely to be the **Bedminster Toad Vein**, and it is possible that the coal, 24 in thick at 348 ft, is correctly identified as such in the original log. Unfortunately, the Starveall section, which might have been expected to supply the solution, is not altogether reliable and the Bedminster Toad Vein is not mentioned in Anstie's log of this shaft (Anstie, 1873, p.48). According to the various shaft sections and working plans of the Bedminster area, the distances from the Ashton Great Vein to the Bedminster Toad Vein are: Starveall 288 ft; Ashton Vale 297 ft; South Liberty 240 ft; Fraynes 300 ft. It is uncertain whether all the coals referred to as the 'Toad Vein' can be referred to the Bedminster Toad Vein. In Anstie's section of the Ashton Vale Colliery (Anstie, 1873, p.48) the coal (33 in) at 119 yd is probably the Bedminster Toad Vein. In Starveall Pit (Prestwich, 1871, p.56), where the Toad Vein appears to be between 280 and 290 ft above the Ashton Great Vein, the coal is said to be only 12 in thick, and in Anstie's section of the same pit (Anstie, 1873, p.48) this may be the unnamed coal proved in the shaft at 240 yd.

A thin coal known as the **Bedminster Little Vein**, said by Anstie (1873, p.48) and Prestwich (1871, p.56) to be 18 in thick, lies about 80 ft below the Bedminster Great Vein at Starveall Colliery and 114 ft below the Bedminster Great Vein at Malago Vale Pit [7816 7107]. This seam is usually regarded as the equivalent of the Kingswood Little (or Two Feet) Vein. Several of thin coals were proved beneath the Bedminster Great Vein in the Bedminster pits

ing of the measures may also be responsible for the recorded thickness of the Bedminster Top Vein being only 12 in in Dean Lane shaft, as distinct from its average thickness of 39 in both in the Dean Lane workings and in Starveall Shaft (Prestwich, 1871, p.56). At Malago Vale Colliery a typical section of the Bedminster Top Vein is given as: roof of duns 9 ft resting on shale 18 in, seam consisting of coal 6 in, clod 3 in, coal 12 in, clod 3 in, coal 12 in; floor of fireclay 2 ft 6 in, black clod with ironstone nodules 2 ft, coal 9 in, blacks. Complex seams of this kind where coal and shale alternate with harder measures show great local variability of thickness and internal structure due to shearing, crumpling and faulting.

Attempts were made to work the Bedminster Top Vein at almost all the Bedminster Collieries; some coal was won but in many pits the coal was too variable to be wrought profitably. At South Liberty Pit the plans show that the seam was worked over a very small area south-west of the shafts. The seam is not recorded in the shaft section, having perhaps been squeezed or faulted out locally, but the workings shows a general section as follows: roof of strong duns resting on coaly shale 18 in, clod 12 in, seam consisting of Top Coal 12 in, thin benching (parting), Bottom Coal 12 in, floor of fireclay 3 ft, resting on 'hard stone'. Other sections recorded from Malago Pit and from South Liberty Colliery are somewhat similar, though with rather more coal. At South Liberty the Top Coal is recorded locally as being 24 in. Generalised sections of the seam in the central part of the Bedminster area, around Starveall, Malago Vale and Deep Pit, usually give an average thickness of 32 to 39 in of coal.

Above the Bedminster Top Vein the measures consist predominantly of mudstone with some thin sandstone beds, mainly of hard fine-grained quartzitic type. A group of about 20 ft of hard sandstone beds, including the 'Top Vein Stone' of Malago Shaft, lies a short distance above the shale and mudstone which compose the immediate roof of the Bedminster Top Vein in both the Argus and Malago Vale shafts. These hard strata are succeeded by about 100 ft of mudstones and sandstones with three or four thin coal seams. The succeeding beds are crossed at a low angle by a belt of disturbance which may be due in part to the presence of incompetent mudstone and shale (Figure 22), which may well include the Crofts End Marine Band or rocks of equivalent age. Barren and contorted mudstones extend upwards to a position about 400 to 450 ft above the Bedminster Top Vein, above which appears a group of measures carrying four thin coals which are overlain by massive 'Pennant Stone' 100 to 150 ft thick with a 4 in coal in the middle. The base of this 'Pennant Stone' is taken as the base of the Pennant Measures in the Bedminster area, even though some of the basal strata may lie below the Winterbourne Marine Band.

Of the coals proved in the measures above the Bedminster Top Vein, none is workable, the thickest seam being about 20 in. Prestwich (1871, p.56) correlated the higher seams at Malago Vale pit with the Devil's (or Black), Rag and Millgrit veins of the Golden Valley and Oldland Common (p.98). The latter are now grouped within the Pennant Measures. Anstie (1873, pp.48–49) was probably referring to these seams when he said that 'Above the Bedminster 'Top' seam the only coals yet proved are four which occurred very near to the top of Deep Pit'. Discussing the correlation of these seams Anstie states that though the names are those of the Golden Valley coals there is 'no special reason to suppose the two groups to be identical, except the existence of certain sandstones presumed to represent the incoming of the Pennant Sandstone'. Although the higher strata at Argus and Malago Vale collieries are now classified as Pennant Measures this in no way clashes with Anstie's views on the correlation of these seams. The oncoming of the Pennant Sandstone is diachronous and there is no evidence to support Prestwich's correlation with the seams of Oldland and Golden Valley. Having rejected Prestwich's classification, Anstie was obliged to account for the apparent absence of a number of seams in the Bedminster area, and in this manner he was led to suggest that the higher coals of the old 'Lower Coal Series' of the Kingswood Anticline and also the basal Pennant seams of Oldland Common and the Golden Valley had not been proved in the Argus and Malago Vale shafts but existed to the deep in the concealed, and as yet unproved ground to the east.

From a consideration of the available sections it seems likely that Anstie was correct in his contention that the higher seams of the Middle Coal Measures of the Kingswood Anticline were not proved at Bedminster, but whether these seams are represented by recognisable coals in the concealed areas extending eastwards towards Bishopsworth is more doubtful. Deterioration of the coals in the upper part of the Middle Coal Measures seems in some degree to be linked with the south-westerly passage of productive measures into non-productive Pennant-type rocks. This impoverishment and change of facies of the 'Lower Coal Series' is accompanied by some attenuation, ultimately giving rise to the conditions seen in the Nailsea Basin.

Dundry and Winford

The workings of South Liberty Colliery extend southwards beneath the northern slopes of Dundry Hill at Highridge [563 678]. Here, although highly folded, the Bedminster Great Vein has a general north-north-easterly strike. The workings end about 2200 yd north of the Dundry (Elton Farm) Borehole [7636 6689] in which the Crofts End Marine Band was proved at a depth of 1434 to 1460 ft, i.e. 767 ft below OD. The level of the Bedminster Great Vein at Elton Farm would therefore be expected to be about 900 ft below OD

Two boreholes sunk by the National Coal Board in the Chew valley near Winford (Winford No. 1 and Winford No. 2 boreholes) fit fairly well into this picture. Winford No. 1 [5573 6375] proved the lower part of the Coal Measures, Winford No. 2 [5636 6343] penetrated the upper and middle part of the Middle Coal Measures including the Crofts End Marine Band (Figure 22).

The fossils recovered from Winford No. 1, including cf. *Anthraconsia regularis*, *Carbonicola venusta* and *Naiadites sp*. at 405 and 409 ft, suggest a horizon comparable with that of the lower *Anthraconaia modiolaris* Zone of South Wales. Winford No. 1 is therefore likely to have commenced below the position of the Harry Stoke Marine Band and penetrated Lower Coal Measures only. This interpretation is confirmed by the presence of *Carbonicola pseudorobusta* at 546 ft 10 in and *Curvirimula subovata*, *Naiadites sp*. and *Geisina arcuata* at 576 to 577 ft. These forms imply the presence of the mid to upper *Carbonicola communis* Zone.

At a lower level (about 690 to 691 ft) a marine fauna with *Caneyella sp. nov*. aff. *multirugata*, *Dunbarella* cf. *papyracea*, *Anthracoceratites arcuatilobus* and *Schartymites cornubiensis* is generally comparable with that of the Margam Marine Band (Woodland and others, 1957) of South Wales. Underlying the goniatite-bearing layer at Winford No. 1, *Lingula sp. (squamiformis)* occurs at 693 ft 9 in, showing that a *Lingula* phase precedes the goniatite/pectinoid phase of the main marine incursion. Winford No. 1 terminated at a depth of 787 ft but the strata beneath the marine horizon consisted of nonmarine mudstone, shale, coal and sandstone. If present, the Ashton Vale Marine Band must therefore lie below the bottom of the hole.

Winford No. 2 (Figure 22) proved higher beds in the Coal Measures succession, including the Crofts End Marine Band with its characteristic fauna between 532 ft 6 in and 547 ft 11 in. Dr M. A. Calver comments 'The band is roughly divided into two parts—the upper 11 ft ... lacking the richness of calcareous brachiopods seen elsewhere at this horizon and a lower 4 ft 3 in band with a more pectinoid/ goniatite aspect'. A feature of the lower band 'is the abundance of small, pyritised, coiled fossils, about 0.1 mm in size, which are possibly foraminifera belonging to *Ammodiscus* or goniatite spat. Beneath the Crofts End Marine Band, mudstone with nonmarine bivalves occurs at several horizons down to 1047 ft. These shells indicate mid to basal Lower *Anthracosia similis – Anthraconaia pulchra* Zone. The lowest horizon, i.e. between 1043 ft 9 in and 1047 ft, yields *Anthraconaia pulchella, A.* cf. *williamsoni, Naiadites productus* and rare specimens of *Carbonita humilis*. A similar assemblage is recorded from 380 ft 3 in to 385 ft 6 in in Harry Stoke A Borehole and at about 1482 ft in Harry Stoke B Borehole (p.172). Comparison of the sequences at Harry Stoke and Winford (Figure 22) suggests that the fauna proved in the latter hole lies below the Bedminster Toad and possibly below the Little Fiery Vein of Speedwell Pit and well above the Harry Stoke Marine Band.

Below the bivalve-bearing mudstone at 1047 ft lie nearly 300 ft of strata in which only plant remains occur. At 1319 ft dark silty mudstone with *Lingula mytilloides* marks the top of a group of mudstones and shales in which further marine bands are present at 1369 ft 6 in to 1377 ft 6 in, 1493 ft 7 in to 1494 ft 11 in and 1526 ft 3 in. Dr M. A. Calver suggests that all these marine horizons lie in the *Anthraconaia lenisulcata* Zone and that the *Anthracoceras/Dunbarella* fauna at 1369 ft 6 in is either the Margam Marine Band or the slightly lower Cefn Cribbwr (*G. listeri*) Marine Band. It is most unlikely that the whole of the *Anthraconaia modiolaris* and *Carbonicola communis* zones are missing in the Winford No. 2 Borehole but they are present in No. 1. There is some evidence of tectonic disturbance in the core and Mr D. R. A. Ponsford, who logged the No. 2 hole, drew attention to evidence of structural distortion at depths of 320 ft, 400 ft, 840 to 940 ft, and 1120 ft. The steeply dipping strata (up to 60°) proved at about 1120 to 1140 ft may mark the presence of a considerable fault which has led to the omission of the strata proved in the upper part of Winford No. 1 Borehole. This may also account in part for the (apparent) rapid reduction in the thickness of the Lower and Middle Coal Measures at Winford as compared with Ashton and Bedminster (Figure 22). It is possible that the coals proved at 735 ft 6 in, 928 ft 6 in, 950 and 1025 ft 2 in correspond to the principal Kingswood coals below the Crofts End Marine Band and that the sideritic coal at 928 ft 6 in may be the Gillers Inn but recovery was poor and proof is lacking.

East Bristol and Kingswood (the Kingswood Anticline)

STRUCTURAL PROBLEMS Among the earliest records of the sequence in the Lower and Middle Coal Measures of the Bristol area are the sections of Kingswood Collieries given by Cossham (1862, fig. opp. p.100) and of the Speedwell (Starveall) Shaft [6323 7442] drawn up by D. H. Williams in 1840 and subsequently published on Sheet 11 of the Vertical Sections of the Geological Survey (revised edition, 1873). On both these records the coal seams, or 'veins' as they were usually called, bear traditional names which date back in most cases to before the beginning of the 18th century. In order to understand their application it is necessary to know something of their origin and history.

Owing to the structural peculiarities of the Kingswood Anticline (Figure 23) and the partial concealment of the Lower and Middle Coal Measures beneath Triassic rocks, the lowest seams do not crop out at the surface and they were not worked on a large scale until the second half of the 19th century. The higher coals, however, have been worked more or less continuously over a period of at least 700

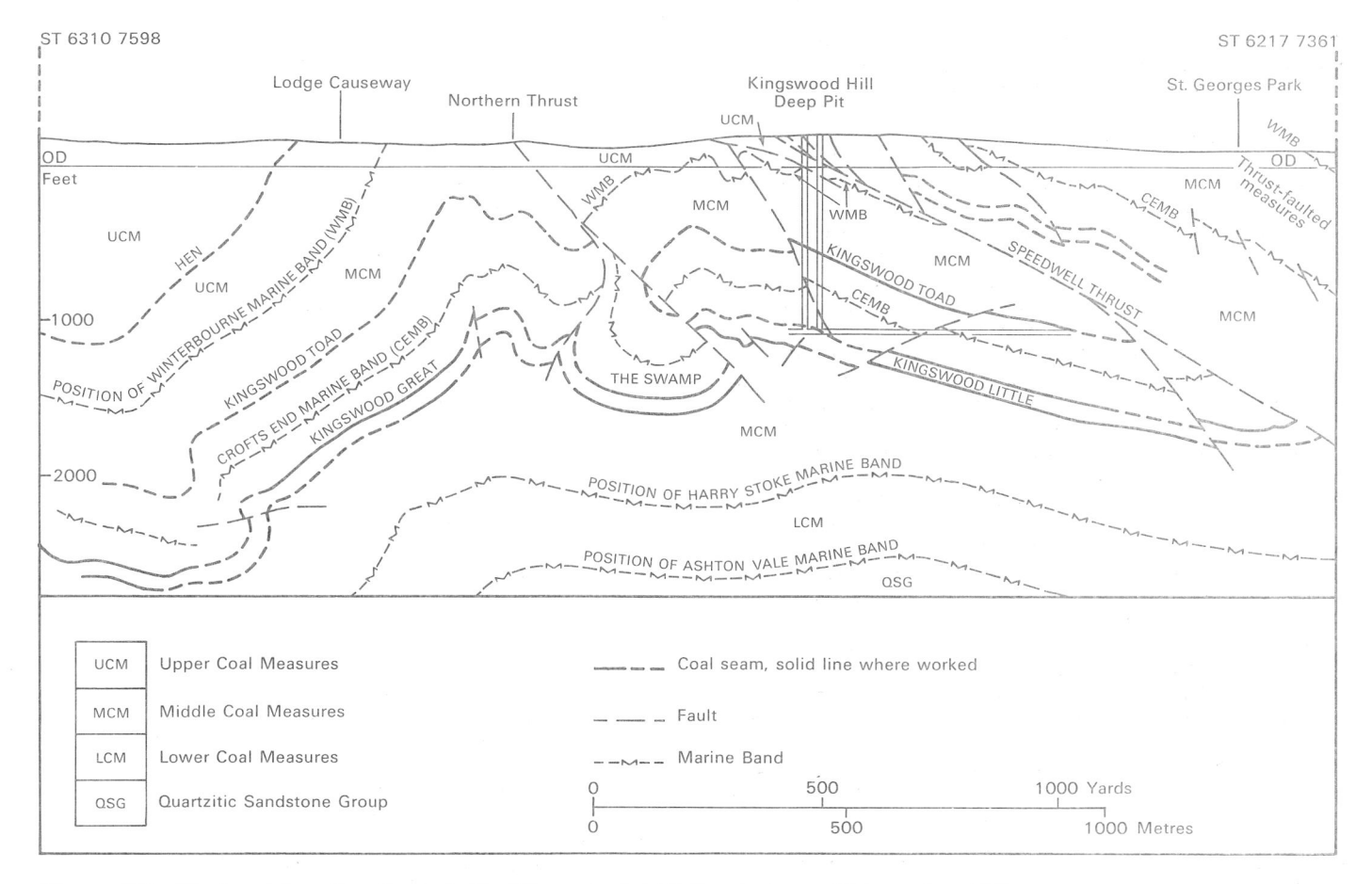

Figure 23 Horizontal section through the Kingswood Anticline between Fishponds and St George's Park, Bristol. The rocks above the Hen Vein are mostly massive Pennant Sandstone

years. During that time duplication of names has taken place, largely due to unsuspected repetition of the strata by folding and faulting.

The earliest works on the stratigraphy of the 'Lower Coal Series' of east Bristol and Kingswood were written before some of the most important mining discoveries were made. Thus the papers by Cossham (1865), Prestwich (1871), Anstie (1873), and Cossham, Wethered and Saise (1875) are useful records, but they predate the discovery of the Speedwell Thrust in 1884 (Figure 23) and the working of the thick coals in the basal Coal Measures at Easton and Pennywell Road [6021 7398]. In 1885 Cossham described the Speedwell Thrust, but beyond giving an indication of the existence of a reversed fault with a throw of 900 to 1050 ft, he did not describe the structure in detail; nor did he discuss all its stratigraphical implications. In 1907, Bolton described the Ashton Vale Marine Band of Ashton Vale Colliery (p.73), but between 1907 and 1937 relatively little progress was made in correlating the Coal Measures of south and east Bristol. Speedwell Colliery closed in 1936, and thenceforth there were few opportunities for checking the stratigraphy and structure of the Middle Coal Measures of the Kingswood Anticline. Thus Moore and Trueman (1937, figs. 7 and 10) were unaware that the type locality of the Crofts End Marine Band is situated above the Speedwell Thrust (Figure 23). The marine shale seen at Crofts End Clay pit [626 745] was thought to be in normal sequence with the worked coals below the thrust and the distance between the Crofts End Marine Band and the Kingswood Great Vein was therefore given as 1200 ft. Not until 1950 was it possible to show that the Crofts End Marine Band is only about 250 ft above the Kingswood Great Vein in the Harry Stoke borings (Figure 22).

When Prestwich (1871) and Anstie (1873) produced their descriptions of east Bristol and Kingswood, Soundwell Colliery [6495 7519] on the north-eastern part of the Kingswood Anticline was probably the only pit to have worked seams below the Kingswood Little Vein on an extensive scale in the Kingswood area. The lowest of these seams, the **Hard Venture Vein of Soundwell** (Figure 22), was tentatively correlated by Anstie with the Top Vein of Ashton Vale Colliery in south Bristol (1873, pp.40 and 48, seam No.xxv). Other seams, notably the Ashton Great Vein, had already been proved beneath the Ashton Top Vein in south Bristol, but these lower seams had not been worked in east Bristol when Anstie's account was written.

By the mid-19th century, the higher seams were exhausted and most collieries in the Kingswood Anticline were dependent on a group of seams extending from the Kingswood Little Vein up to the Lower Five Coals (Figure 22). Informally these seams are sometimes called the 'Kingswood Group', and are now known to lie in the Middle Coal Measures between the Harry Stoke and Crofts End Marine Bands. They are deeply buried over most of the western part of the Kingswood Anticline, but east of the Whitefaced Fault they crop out and have been known and worked from very early times. The most important seam in the group is the **Kingswood Great Vein** which is up to 60 in thick and is generally regarded as the correlative of the Bedminster Great Vein.

It will, therefore, be seen that in 1945 when development of the coalfield became an urgent necessity, the accurate identification of the 'Kingswood Group' of seams assumed prime importance. Writing in 1875 Cossham, Wethered and Saise stated: 'The terminology of these seams has been accepted over the whole field', and they go on to say that the miners have long regarded them as being the most valuable seams in the Kingswood Anticline. The seams were identified by peculiarities in their roof measures. In particular, the Gillers Inn Vein had characteristic features which enabled it to be used as a marker band.

Little working took place at depth on the western (downthrow) side of the Whitefaced Fault before the beginning of the 19th century. The older workings in this area extended down to the **Kingswood Toad Vein** which is separated by over 400 ft of barren ground from the **Lower Five Coals** beneath. To miners relying mainly on natural drainage to dewater their workings, only limited reserves of coal below the Kingswood Toad Vein were available in the low-lying areas of east Bristol, and it was not until about 1830 that the Kingswood Great Vein was worked at depth and in quantity at the Deep [6256 7458] and Speedwell pits of old Kingswood Colliery (Cossham, 1885, p.248). Eventually, by sinking and deepening the Deep and Speedwell No. 1 (or Starveall) shafts, all the seams between the Doxall and Kingswood Little veins were proved west of the Whitefaced Fault. Many of these seams were correlated with coals bearing similar names at the Old Lodge [6392 7452] and Soundwell collieries east of the Whitefaced Fault (Anstie, 1873).

However, in correlating the coals below the **Kingswood Little Vein** on the east and west sides of the Whitefaced Fault some problems were experienced by the old miners. These concerned the relations of the lowest seams proved at Soundwell (east of the fault) with those proved in the Speedwell and Deep pits on the west and with the so-called 'Parkers Veins' which were supposed to underlie the Deep and Speedwell coals. The resulting confusion makes it impossible for anyone who is not familiar with the surface and underground evidence to understand the accounts of Anstie (1873) and other older workers or to relate their results to those obtained in more recent times. Discussion of this problem is, therefore, essential if the contemporary records are to be made use of.[1]

According to the record of the lower part of the shaft section at Speedwell Pit the lowest coal, 30 in thick, known as the 'Parkers Ground Seam' was proved in steeply inclined measures in the sump, apparently lying in sequence beneath the Kingswood Little Vein. This seam was generally supposed to correlate with the highest member of a triad known collectively as the Parkers (or Parkers Ground) veins and individually as Parkers Little, Middle and Top veins. These coals were formerly worked along the crest of an anticline in the Royal Oak, Parkers Ground and New Lodge pits[2], situated north of Deep and Speedwell pits in the vicinity of what is now Forest Road, and were thought to dip southwards and pass beneath the Kingswood Little Vein workings of Speedwell and Deep pits (see Anstie, 1873, fig. 5). This erroneous conclusion was for many years a major obstacle to progress and no further advance was made until Cossham recognised the effect of the repetition of the Speedwell seams by the thrust fault and fold proved at Speedwell Pit in 1884 (Cossham, 1885). By this discovery he showed that the so-called Parkers Veins of the Lodge Causeway area are on the underside of the thrust and stratigraphically above, not below the Kingswood Great Vein (Figure 23).

This opened the way to exploitation of the Kingswood seams north of Speedwell and Deep pits. It also threw a fresh light on the question of equating the Soundwell, Speedwell and Easton successions, and of relating these to Ashton and Bedminster. Once the equivalence of the Little (or Little Toad) Vein of Soundwell and the Kingswood Little Vein of Speedwell is accepted (and there seems to be no valid reason for objecting to this correlation), then it follows that the correlatives of the Slate, Stony, Smith Coal and Hard Venture veins of Soundwell may be expected to underlie the Kingswood Little Vein in the area west of the Whitefaced Fault. Several versions of the section at Soundwell High Pit are in existence, including those given by Prestwich (1871) and Anstie (1873) and these show fairly close agreement with other (unpublished) mining records. From these accounts the following details relating to the seams below the Kingswood Little Vein have been abstracted:

---

1 In reading this account the reader is advised to consult the published six-inch-to-one-mile Geological Survey of sheets ST 67 NW and 67 SW.
2 Anstie (1873, p.32) refers to these as the 'Lodge Pits', a most unfortunate description which has led to them being mistaken for the Old Lodge Pit, Kingswood. Old Lodge Pit was situated on Lodge Hill, on the opposite or eastern side of the Whitefaced Fault.

Name of coal seam	Distance below Little Vein
Slate Vein, 24 in	58–78 ft
Stony Vein, 18 in	186–223 ft
Smith Coal, 24 in	334–366 ft
Hard Venture Vein, 36 in	616 ft
Coal, unnamed, 18 in, not in shaft	673 ft

These figures give the approximate depths beneath the Kingswood Little Vein at which the correlatives of these Soundwell seams may be expected to occur west of the Whitefaced Fault. They also serve to explain an apparently anomalous statement by Anstie (1983, p.32) to the effect that 'Parkers Top Seam' is absent at Speedwell, since it would appear that he regarded 'Parker's Top Seam' as being the correlative of the Smith Coal of Soundwell (1873, p.41) and not, as might have been expected, of some higher seam such as the Soundwell Slate or Stony. From the Soundwell section it will be seen that the Slate Vein, lying up to 78 ft below the Kingswood Little Vein is much more likely to correlate with the so-called 'Parker's Ground Vein' of Speedwell Sump than is the Smith Coal of Soundwell. The latter is situated about 350 ft below the Little Vein and is, therefore, unlikely to have been found in the sump of Speedwell shaft. Anstie caused some confusion by applying the term 'Parker's Top' to a seam proved beneath the Kingswood Little Vein at Easton Colliery. In the working plans and sections of Easton, however, the first two major seams below the Kingswood Little Vein are described in descending order as the Little Fiery and Big Fiery veins. The Little Fiery of Easton may correlate with the so-called 'Parker's Ground Vein' of Speedwell, but owing to structural complications there is some uncertainty about this.

Exploration of the basal Coal Measures at Pennywell Road and Easton collieries commenced soon after the publication of Anstie's book in 1873. Anstie (p.38) observed that the old colliery at Jeffries Hill, Hanham [6372 7206], was being redeveloped at the time of writing and this led eventually to the discovery and working of the Kingswood Great Vein and the **Hanham Red Ash and White Ash veins**, though the basal Coal Measures were not proved. At Easton and Pennywell Road collieries, over 600 ft of strata were proved beneath the Kingswood Little Vein. These measures included seven seams of coal, of which at least five were in work at various times between 1874 and 1885.

In addition to abandonment plans the available records include some documents dating from the beginning of the 19th century and supplementary notes and sections made by E. H. Staples, H. E. Monks and A. H. Bennet. From these accounts it is clear that there are a number of folds and faults, including low-angle slides, in the Whitehall–Easton area. The Easton sequence of the coals beneath the Kingswood Little Vein is shown on the sections accompanying the abandonment plans of the colliery. According to notes by H. E. Monks (*in* Staples MS) the distance at Easton from the Kingswood Great Vein to the **Easton Red Ash Vein** is 212 yd compared with 245 yd separating the Great Vein from the Hard Venture Vein at Soundwell. Monks regarded the Slate Vein of Soundwell as the correlative of a 'coal-stain' under the Little Toad or Kingswood Little Vein at Easton, and the Stony Vein of Soundwell as the equivalent of the Little Fiery Vein of Easton. From this it would follow that the Hard Venture Vein and the lowest coal proved at Soundwell are equivalent to the Easton Red Ash Vein and Gays Vein respectively. This correlation serves to link the measures below the Kingswood Little Vein on the west side of the Whitefaced Fault with those on the east side at Soundwell.

The old Soundwell Colliery workings are situated on the eastern side of the Whitefaced Fault on the northern limb of the Kingswood Anticline, but there have been more recent workings on equivalent seams at Hanham Colliery on the southern limb. The principal seams worked here were the Red Ash (or Jubilee) Vein and the White Ash Vein, the latter being some 90 ft below the Red Ash Vein.

Higher seams including the Gillers Inn and Kingswood Great Vein were worked from various level branches at Hanham, but the measures between the Red Ash Vein and the Pennant Sandstone were too highly disturbed to encourage extensive development. The Hanham Red Ash Vein has a characteristic roof of dark shale from which *Lingula* is said to have been obtained, and this may indicate the presence of the Harry Stoke Marine Band. The Hanham workings terminated in the west against the Whitefaced Fault which has a westerly downthrow of about 160 yd. From this and other evidence it can be deduced that the Kingswood Great Vein lies about 600 to 650 ft above the Hanham Red Ash Vein, a distance that is comparable with the interval between the Great Vein and Hard Venture Vein at Soundwell and between the Kingswood Great Vein and Red Ash Vein at Easton (Figure 22). The similarity of thickness, working properties and relative position of the Red Ash veins of Hanham and Easton indicate that they are the same coal, and as is shown below, there is some evidence suggesting that the Easton Red Ash, like the Hanham Red Ash Vein, may have a marine roof.

East of the Whitefaced Fault no workable coals are known beneath the Hanham White Ash or its presumed equivalent at Soundwell though, according to Cossham, Wethered and Saise (1875) a branch was driven from the bottom of Soundwell Colliery (probably from the High Pit) .. '200 yards past the Hard Venture Series, and is probably the only branch which has approached so near to the Millstone Grit.' No details of this branch have survived, but from the absence of any workings or developments it appears unlikely that coals of workable thickness are present in the basal Lower Coal Measures in this area. Thus there is no record of coals having been worked beneath the Hard Venture Vein at New Cheltenham Pit [6528 7447], nor below the White Ash Vein of Hanham Colliery. The lowest workable coals known in the Kingswood Anticline are, therefore, the **Easton Two Feet, Seven Feet and Four Feet seams** proved at the western end of the anticline at Easton and Pennywell Road collieries. On the assumption that Anstie (1873) was correct in correlating the Hard Venture Vein of Soundwell with the Ashton Top Vein, it follows that the Easton Two Feet, Seven Feet and Four Feet seams may be the Little Vein, Gay's and Smith's Vein respectively of Ashton Vale Colliery (Figure 22).

GENERAL SUCCESSION (ASHTON VALE MARINE BAND TO HEN VEIN)
South of Easton and Whitehall [6181 7380] collieries a gap of about ½ mile separates the workings on the Kingswood Great Vein from the most northerly workings on the Bedminster Great Vein of Dean Lane pit (p.75). The ground between the two areas is almost certainly faulted, but an attempt was made by Alfred Bennett, a local mine owner, to link the workings by deepening the old Great Western Shaft [6091 7249] in St Phillip's Marsh, about 1500 yd west-north-west of St Anne's Park Station. This plan was frustrated by legal difficulties and the shaft sinking was eventually abandoned at a depth of 901 ft. The presumed equivalence of the Bedminster Great Vein and the Kingswood Great Vein cannot therefore be proved conclusively. According to information supplied to Prestwich (1871, p.61) and Anstie (1873, p.49) there is 'a trace' of the Gillers Inn Vein beneath the Bedminster Great Vein at Bedminster, and if the Gillers Inn was identified by means of the Worm Bed and Jingleboys (p.81) the correlation is likely to be sound. The Bedminster Little, Great and Top Veins therefore appear to correlate with the Kingswood Little, Great and Thurfer-with-Lower Five Coals respectively. Although the Ashton Vale Marine Band has not been proved at Easton, it is possible to correlate the basal Coal Measures at Harry Stoke with those of Ashton and thus to make effective comparison of the measures between the Ashton Vale Marine Band and the Kingswood Great Vein in the area extending from Ashton and Bedminster through Speedwell and Easton to Harry Stoke (Figure 22).

CHAPTER FIVE COAL MEASURES (WESTPHALIAN)

The Ashton Vale Marine Band marking the base of the Lower Coal Measures in the Bristol and Somerset Coalfield has nowhere been proved in the area of the Kingswood Anticline. The lowest seam, the **Easton Four Feet**, believed to lie about 100 ft above the base of the formation, is known only from its section at Easton Colliery given on AMP 2962B. Its section is as follows: Stone, lower earth 6 in, soft fireclay 6 in, coal 51 in, fireclay 18 in, stone 3 ft. The seam has a good roof, the measures between the Four Feet and Seven Feet consisting largely of sandstone and hard duns. The Four Feet Vein has been worked to some extent at Easton and at Pennywell Road, but the nature and quality of the coal has not been recorded. With the exception of the Smith's Coal of Ashton Vale Colliery (p.73) no correlatives are known elsewhere in the Bristol Coalfield.

Named on some plans as the Big Vein, Nine Feet, Ten Feet or Twelve Feet Vein, the **Easton Seven Feet** is a thick composite seam situated about 75 ft above the Four Feet. The seam appears to be very variable in thickness and section, possibly due in part to deformation of the coal and associated shale and fireclay. The section given on AMP 2962B (Easton Colliery) is as follows: Lower earth 5 ft, soft fireclay 4 in, coal 30 in, dirt 9 in, coal 9 in, coal 33 in, fireclay 3 in. On the North Incline of Easton Colliery the seam was apparently represented by three coals whose thicknesses, given in ascending order, are 30 in, 84 in, and 36 in. In the lower part of the deep shaft at Easton, however, the same coal band is described as 'coal 12 ft in two laps'. On the incline section the upper part of the thick coal may be the Easton Three Feet Vein which was worked as a separate seam over a small area.

According to the mine plans the **Easton Three Feet** was cut in a branch near the top of the Seven Feet Vein workings. The seam section is given as lower earth 3 in, fireclay 36 in, coal 36 in, lower earth 3 in. Between the Easton Seven Feet and Easton Gay's Vein the distance is about 300 ft. The section through this group of measures is uncertain though there is at least one coal, the Easton Two Feet, proved in the shaft and inclines of Easton Colliery. The thickness of this seam appears to vary from 18 to 26 in, but it may be impersistent or owe its position to low-angle faulting.

The **Gay's Vein** at Easton is said to have had a good roof, but being only 16 in thick was very little worked either here or at Pennywell Road Colliery. Its presumed equivalent at Hanham, the **White Ash Vein**, is said to have been a good coal and was worked with a thickness of 24 in. At Soundwell the Hanham White Ash may be represented by 18 in of coal lying beneath the Hard Venture Vein, but there is no evidence that the coal was workable. It will be seen that the Gay's Vein of Easton is in no way related to the Gay's Vein of Ashton Vale Colliery (Figure 22).

Lying about 90 ft above the White Ash or Gay's Vein is the **Red Ash Vein of Hanham and Easton** collieries. This seam was worked extensively at Hanham and to a smaller extent at Easton and Pennywell Road. At Easton the coal is about 30 to 32 in thick and at Hanham it averages 36 in. The Hard Venture Vein of Soundwell, 36 in thick, probably represents the Red Ash Vein on the northern side of the Kingswood Anticline in the area east of the Whitefaced Fault. At Hanham where the coal was used for gas-making, the Red Ash normally comprises a single coal, but locally it consists of two coals each 16 in thick with a very thin parting. Large quantities of marine shale have been found on Hanham tip (Bolton, 1911; Moore and Trueman, 1937) and it is now considered probable that this material came from a position on or near the roof of the Red Ash Vein, which presumably equates with the **Harry Stoke Marine Band**.

Several coal seams have been worked and named in the belt of measures extending upwards from the Harry Stoke Marine Band to the Gillers Inn Vein. Detailed correlation of these coals, which include the **Kingswood Little Vein**, is generally difficult and in some cases impossible. The measures are structurally incompetent and there are many coaly and carbonaceous bands of varying thickness and persistence. Some of them, including the Kingswood Little Vein, appear to attain workable thickness over a limited areas, but elsewhere are thin or pass into layers of carbonaceous 'muck' or 'blacks'.

The Big Fiery seam appears to lie some 140 ft above the Harry Stoke Marine Band and is thought to equate with the Smith's Coal of Soundwell and Hanham. At Pennywell Road Colliery it is said to be 36 in thick with a good roof (probably of duns) and a floor of blacks and fireclay. The seam was said to yield large quantities of small coal and to be too soft and shaly to be considered of good quality. The **Little Fiery Vein** was said to have been 10 in thick at Pennywell Road Pit where it had a sandstone roof and was about 180 ft above the Big Fiery Vein. It was proved, but not worked to any appreciable extent, at Easton Colliery and may be represented at Soundwell by the Stony Vein, said to be 18 in thick. The distance between the Big and Little Fiery veins at Easton is about 240 ft according to the colliery section. However, it is known that there is some repetition by thrusting in the measures where the Little Fiery Vein was found in the colliery workings and this may account for the difference. According to the Pennywell Road section, the Little Fiery Vein is overlain by a bed of sandstone 8 ft thick. Above this the measures extending to the base of the Little or Two Feet Vein consist mainly of argillaceous strata. The seam proved in the Lower Branch and Return Airway of Hanham Colliery under the name of 'Two Foot Vein' is probably the Little Fiery Vein of Easton and not the Kingswood Little or Two Feet Vein. At Hanham the Little Fiery coal was very variable in section owing to squeezing and distortion but generally about 24 in thick with a floor of shale and fireclay and a roof consisting of 2 ft of shale overlain by at least 5 ft of sandstone.

At Soundwell the Slate Vein is said to have been about 24 in thick. It may very well correlate with a thin seam known to exist locally in the Speedwell and Deep pits (Cossham, 1879, p.413) at a depth of about 100 ft below the Kingswood Little or Two Feet Vein, and also with a coal 'stain' or thin coaly band found in a comparable position at Easton Colliery. The position of the so-called Parkers Ground Vein of Speedwell Shaft relative to the Slate Vein of Soundwell is obscure. If it is assumed that it equates with the Little Fiery Vein of Easton, then the Stony Vein of Soundwell would be represented by the thin coal beneath the Kingswood Little Vein as suggested above.

The **Kingswood Little or Two Feet Vein** is one of the best known and most extensively worked coals in the Kingswood Anticline (Figure 23). Its original name of Little Toad Vein points to some resemblance to the Old or Great Toad Vein situated some 600 to 650 ft above. The Old or Great Toad Vein is now described as the Kingswood Toad Vein (p.84) in order to distinguish it both from the Kingswood Little or Two Feet Vein and from the Bedminster Toad Vein (p.75). By some it was thought that the word 'Toad' is derived from the German *tot* meaning 'dead', and that the Great and Little Toad Veins of Kingswood and the Toad Vein of Bedminster resemble one another by virtue of their position above or within thick belts of 'dead ground' where there is little or no other workable coal. Large areas of the Little Seam have been worked in the Kingswood Anticline, especially at Speedwell, Deep, Easton and Whitehall pits. It is generally described as of good quality with a firm mudstone roof. At Easton the coal is in one place 18 to 24 in thick but it thins southwards towards Bedminster. In the northeastern workings near Eastville the thickness is about 24 in and the section is comparable with that proved in the Speedwell and Deep pits where the average thickness is about 27 in.

At the Speedwell and Deep pits the seam has a shale roof and a fireclay floor. There is some evidence to suggest that the maximum thickness is found in the vicinity of Fishponds and Soundwell. At Fishponds the workings from Speedwell Pit proved 27 in of coal, while south of Deep Pit near St George's Park the coal was only 24 in thick. At Soundwell the seam was said to be 24 in and at

Siston Common Pit [6694 7390] 26 in thick (Anstie, 1873, pp.34,40), but as is shown below it deteriorates northwards towards Harry Stoke and to the south-west of Easton (Prestwich 1871, p.61). Some genuine Kingswood Little Vein may have been worked at Hanham, but the indications are that some of the faulted pieces of coal called 'Two Feet Vein' at Hanham are actually the Little Fiery Vein of Easton (see above).

The strata above the Kingswood Little Vein up to the Lower Five Coals was proved in the AB Branch on the north-side workings at Deep Pit, Kingswood at about 1750 ft below OD. The section recorded by E. H. Staples in about 1905 reads:

	Thickness		Depth below Five Coals	
	ft	in	ft	in
Lower Five Coals: coal	4	3	—	—
Fireclay and cockles	6	0	6	0
Shale	18	0	24	0
Thurfer Vein	1	6	25	6
Fireclay and ironstone	3	0	28	6
Duns	1	0	29	6
Stone	5	6	35	0
Greys	6	4	41	4
Stone	3	0	44	4
Greys	3	0	47	4
Stone	10	0	57	4
Very hard duns	7	0	64	4
fault				
Coal	0	6	64	10
Duns	10	0	74	10
Coal	0	4	75	2
Clod	3	6	78	8
Kingswood Great Vein: coal	5	0	83	8
Fireclay and ironstone balls	4	0	87	8
Stone	1	8	89	4
Duns	1	9	91	1
Fireclay	0	7	91	8
Very hard duns	14	0	105	8
Greys	8	0	113	8
Thick-bedded shale	20	0	133	8
Worm band	0	6	134	2
Shale	0	8	134	10
Jingles (or Jingleboys)	0	6	135	4
Very dark shale	0	6	135	10
Gillers Inn Vein: coal and dirt	4	3	140	1
Fireclay and stains of coal	2	9	142	10
Thick-bedded duns	14	0	156	10
Black chalk	0	2	157	0
Coal and dirt	4	2	161	2
Shale	5	0	166	2
fault				
Coal and coal blacks, contorted	4	0	170	2
Duns, contorted	20	0	190	2
Thin-bedded shale	4	0	194	2
fault				
Measures (unspecified)	3	0	197	2
Black chalk	0	1½	197	3½
Fireclay, ironstone balls and coal stains	5	0	202	3½
Stone	7	0	209	3½
Thin-bedded shale	5	0	214	3½
Kingswood Little or Two Foot Vein	2	0	216	3½

About 45 to 50 ft above the Kingswood Little Vein lies the **Gillers Inn Vein**. Small quantities of coal have been recovered from this seam at Kingswood, Soundwell, Hanham, Pennywell Road and Easton collieries (Prestwich, 1871, p.61). The Gillers Inn Vein is, however, of considerable historical interest by virtue of its association with a band of blackband ironstone formerly worked in mines at Lodge Hill and Hopewell Hill, Kingswood. It has long been used as a stratigraphical 'marker band' by the old miners, principally as a guide to the position of the Kingswood Little Vein below and the Kingswood Great Vein above. The presence of workable ironstone in the roof of the Gillers Inn Vein clearly drew the attention of the old miners to the impure ironstone band known as the 'Worm Bed', which was described by Cossham, Wethered and Saise (1875) as 'a good landmark in the Bristol coalfield ... The surface of this remarkable bed is covered with what appears to be remains of some form of vegetation of a small and succulent character (probably some species of algae), and really looks like a number of fossil worms — hence the name'. These trace fossils have now been identified as *Planolites montanus* and the Worm Band can be regarded as a good nonmarine marker band.

Sections of the Gillers Inn Vein and its roof and floor show considerable variation but generally indicate that the coal band rests on a fireclay floor and that it consists of a number of layers of coal separated by shale or 'dirt'. The roof shale with ironstone has at its base a characteristic hard black shale known as 'Jingleboys' or 'Jingles', which was cut when the roof of Gillers Inn was ripped to obtain ironstone or to make headroom above the coal. Fragments falling from the roof of the working made a clinking or jingling sound, which gave rise to the name of the shale. The best recorded sections of the Gillers Inn Vein in the Kingswood Anticline were made by E.H. Staples at Deep Pit and Hanham Colliery. In the AB cross-measures branch at Deep Pit, the detailed section is as follows: coal 4 in, dirt 5 in, coal 15 in, dirt 6 in, coal 3 in, dirt 2 in, coal 16 in, on fireclay. The best section recorded from Hanham Colliery was taken from the Long Branch: roof of shale, Worm Band, shale about 15 in, Jingleboys 12 in, resting on coal 10 in, parting, coal 12 in, fireclay 8 in, coal 4 in, benching 1 in, resting on fireclay. According to Cossham, Wethered and Saise (1875, p.418), the Gillers Inn Vein section (probably in the southern workings from Deep and Speedwell pits) was shale, (locally termed 'Jingleboys') 8 in, coal 8 in, blacks 4 in, coal 6 in. The thicknesses given by Anstie (1873) and Prestwich (1871) for the collieries between Easton and Warmley are all less than 24 in and this suggests that the coal was worked mainly in association with the overlying ironstone. It was seldom mined at depth in those places where the overlying ironstone is missing or of poor quality.

The important **Kingswood Great Vein** lies about 30 ft above the Gillers Inn Vein in the type area of the Deep and Speedwell pits where the average section is: roof of duns resting on blacks 5 in, clod 6 in, coal 54 in, overlying fireclay 18 in. According to Cossham, Wethered and Saise (1875) the average thickness of the Great Vein in the area is 54 to 60 in, the roof being a tender one needing heavy timbering though the coal was easily wrought and free from dirt. At Easton Colliery the section in the Easton Great Vein (which is correlated with the Kingswood Great Vein) was recorded by R. M. Dillwyn as clift, resting on a rough clod roof 4 ft, coal in one bed 57 in. According to Anstie (1873, p.31–34) the thickness of the seam was 54 in at Easton, 48 in at Speedwell and 36 to 42 in at Soundwell and Siston. The Kingswood Great Vein, like the underlying Kingswood Little Vein, attains its maximum thickness in east Bristol, being up to 60 in thick in some of the northern workings of Easton, Deep and Speedwell pits. No fossils other than poorly preserved plant remains have yet been recorded from the roof of the Kingswood Great Vein either in the Kingswood Anticline or in the Harry Stoke area. Bolton's references (1911) to the presence of *Lingula* and various nonmarine bivalves in shale above the Easton Great Vein are highly questionable. He did not collect in situ and may have been misinformed as to the source of his material. The writer was informed by E. H. Staples that the Easton and

Kingswood Great Vein are one and the same seam, the Easton workings having holed through to the Kingswood workings at one point.

Correlation of various seams which have been described as the 'Great Vein' in the Kingswood Anticline presents less difficulty than might be imagined. The 'Great Veins' of Easton, Deep, Speedwell, Lodge, Soundwell and Siston collieries can all be referred to the Kingswood Great Vein. At Whitehall Colliery the so-called 'Doxall Vein', structurally isolated from the main workings in the Kingswood Great Vein and the 'Doxall' of Hanham Colliery, subsequently described as 'Great Vein', can also be referred to the Kingswood Great Vein. At Warmley at the eastern end of the Kingswood Anticline the correlation of the Warmley Great Vein with the Kingswood Great Vein was questioned by Anstie (1873, p.34) and by Cossham, Wethered and Saise (1875) but Moore and Trueman (1937, pp.214–215) considered that the palaeontological evidence supports this correlation.

About 40 to 50 ft of strata separate the Kingswood Great Vein from the overlying **Thurfer Vein** in the Kingswood area and the intervening strata appear to be mainly mudstones with nodular ironstones. The Thurfer Vein is regarded as distinct from the Lower Five Coals in the western part of the Kingswood Anticline but at Siston Colliery the 'Five Coal' Seam, which is 84 in thick, includes the Thurfer Coal, 27 in thick, at the base. The thickness of the Thurfer Vein and its separation from the Lower Five Coals in the Kingswood Anticline between Easton and Siston is shown below:

	Average thickness of Thurfer Vein	Separation of Thurfer Vein and Lower Five Coals
Easton	14 in	42 ft
Speedwell and Lodge Hill	12 to 18 in	30 ft
Kingswood and Hanham	16 in	38 ft
Upper Soundwell	24 in	20 ft
Siston	27 in	Nil

At Warmley it is generally supposed that the Thurfer Vein has become a part of the so-called Ragged Seam (Prestwich, 1871, p.54), situated about 70 ft above the Great Vein and presumed to be the correlative of the Five Coals Vein of Siston (see above). At Soundwell the Thurfer Vein was generally workable but at Speedwell and Deep pits, workings were usually restricted to small areas of coal proved in branches connecting workings on the Kingswood Great Vein and Lower Five Coals.

Like the Thurfer Vein, the **Lower Five Coals** is best developed near the eastern end of the Kingswood Anticline, notably between Soundwell and Siston. It is not known with certainty at outcrop but the higher of two contorted thick coals formerly exposed in the south-eastern corner of the now-abandoned northern quarry of the Bristol Brick Co. at Crofts End [625 746] may have been the Lower Five Coals. The thicknesses given for the seam in the older records show some local variation. This may be due partly to structural deformation of the kind which causes local thickening of the Lower Five Coals in the Harry Stoke Drift Mine (p.86), but the regional differences are genuine.

At Siston, Prestwich (1871, p.54) and Anstie (1873, p.34) recorded the 'Five Coal' seam as being 84 in thick, the full section being 'Top' coal 8 in, 'Stool's Legs' 8 in, 'Benching' coal 14 in, 'Smith' coal 27 in, 'Thurfer' coal 27 in. Siston is the only district where the typical five-coal structure has been found. According to Cossham, Wethered and Saise (1875, p.418) the prefix 'Lower' probably originated in the Kingswood area where the Upper Five Coals Vein has a structure that is somewhat similar to that of the Lower Five Coals of Deep and Soundwell pits. In general the coal is about 18 in thick at Easton Colliery where it has been worked only to a limited extent. Passing eastwards to Deep and Speedwell pits the section of the seam is given on one old section as duns 4 in, stone 6 in, coal 10 in, shale 3 in, coal 6 in, soft shale 20 in, coal 18 in, duns and fireclay 4 ft. Another section shows the seam as having increased to about 24 to 36 in, while east of the Whitefaced Fault at Lodge Pit it is about 42 in thick with a four-coal structure. According to Prestwich (1871, p.55) the four component coals at Lodge Pit were known as the 'Top Coal', 'Ten-inch', 'Profit' and 'Bottom coal' and, though individual thicknesses are not given, the four coals above the Thurfer Vein appear to be thinner than the equivalent group at Siston where the thickness was 57 in.

At Soundwell the coal is about 54 to 60 in thick exclusive of the Thurfer coal. The 'Five Coals' Seam of Siston Colliery is generally believed to pass southwards into the 'Ragged Seam' of Crown Colliery, Warmley [6725 7350], described by Prestwich (1871, p.54) and Anstie (1873, p.34) as being 48 in thick, the coal being of 'middling' quality with 'bands of shale'. This suggests a rather rapid southerly deterioration and is consonant with the apparent absence of any thick workable coal in a comparable position in Hanham and Whitehall collieries or in south Bristol. It is possible that the Thurfer Vein and Lower Five Coals are both present at Hanham, the former being represented by a coal 20 in thick, the latter by two coals of which the lower one is 32 in thick with a 3 in parting and the upper one is two 7 in coals separated by an 8 in parting. There is much structural disturbance at Hanham and it is impossible to reconstruct the detailed sequence, but the colliery section suggests that the Thurfer Vein is separated from the four-coal seam by at least 35 to 40 ft of measures and that the component coals have become attenuated and have split into two groups.

At Deep and Speedwell pits the thickness of the seam is 24 to 36 in with a compound structure as indicated by the following sections from the Speedwell workings: duns and black shale 3 ft 6 in, hard duns 4 ft, stone 6 in, coal 10 in, shale 3 in, coal 6 in, soft shale 20 in, coal 18 in, on heavy duns and fireclay 4 ft. Another record from Speedwell Colliery shows a four-fold structure: duns, coal 6 in, clod 4 in, coal 14 in, blacks 5 in, coal 6 in, clod 4 in, coal 21 in, on fireclay 18 in. Another section located in the northern Speedwell workings (A. Savage, MS) situated approximately beneath the Parish Hall, St Mary's Church, Fishponds [633 760] showed coal 9 in, soft shale 4 in, coal 6 in, soft shale (sometimes coal) 8 in, coal 20 in. The local passage of the 8 in shale band into coal suggests that in this and similar cases the top two coals and the bottom coal are persistent and may be correlated with the Siston 'Top coal', 'Stools Legs' and 'Smith Coal' respectively. The 'Binching coal' of Siston is apparently represented mainly by shale at this locality.

Between the Lower Five Coals and Kingswood Toad Vein only the **Hard Vein** of Speedwell and Deep pits and the Hole Vein of Easton are worthy of mention. By some authors, notably Prestwich (1871, p.61), these are regarded as the same seam, though the distance between the Lower Five Coals and the Hard Vein appears to be about 110 ft at Speedwell compared with about 70 ft between the Five Coals and Hole at Easton. At Easton the **Hole Vein** is about 18 in thick and lies about 70 ft above the Lower Five Coals. It has been worked to some extent at Easton, Whitehall and Pennywell Road collieries and the working plans show that it was not unduly affected by small-scale faulting. The measures between the Lower Five Coals and the Hard Vein at Kingswood consist mainly of hard mudstone and sandstone and the roof of the Hard coal appears to be equally firm.

The section of the Hard Vein in Speedwell Shaft reads: roof of duns resting on 5 ft 5 in of 'hard duns', good coal 17 in, on soapy fireclay 8 ft. Cossham, Wethered and Saise (1875) describe the Hard Vein as 'about 18 in thick ... coal of excellent quality and firm texture'. The Kingswood Hard Vein is quite distinct from the Hard

Plate 9 Coal seam in the Middle Coal Measures repeated by an overlap fault or thrust, on the south limb of the Kingswood Anticline. Bristol Brick Company's Pit [624 746] at Crofts End (A8298)

Vein of Soundwell which lies at a much higher level (Figure 22).

Several coals including Norman's Vein, the Pigs Cheek Vein and the Trow or Trough Vein lie between the Kingswood Hard and the Kingswood Toad Veins, but all appear to be thin or dirty and are of little importance. These measures include also the **Crofts End Marine Band**. Although it is now known to lie about 250 to 350 ft above the Kingswood Great Vein, and roughly halfway between the Lower Five Coals and the Kingswood Toad Vein of Speedwell Colliery (Figure 22) determination of the precise position of the marine bed in the Kingswood and east Bristol sequence is partly dependent on external evidence. The exposure at the type locality in the clay pit at the now defunct Crofts End Brickworks [626 745] was described by Moore and Trueman (1937, p.212) who measured the section on the east side of the main incline near the bottom of the pit. This working (Plate 9), like the two large clay workings of the Bristol Brick Company some 300 yards to the west, is now disused and the sections obliterated. A second exposure of the Crofts End Marine Band situated at the top of the southern working face of the clay pit was found during the official survey. This exposure is separated from the incline section by a thrust fault with a throw of about 50 ft and adds nothing to the stratigraphical information obtained from the exposures near the incline. The crushed and contorted strata below the marine band appear to be in sequence; they include at least one bed of carbonaceous shale with nonmarine bivalves which may correlate with a similar band proved at a depth of 1045 ft and 2196 ft in Harry Stoke B and C borings respectively (Figure 22).

There is little to add to the description of the marine band given by Moore and Trueman (1937), but it may be observed that the 'sandstone' layer recorded within the band is a carbonate rock, though somewhat arenaceous. A thin coal usually underlies the marine strata, while the measures above consist of soft grey shale, becoming silty and sandy upwards. These show some 'ghostly' markings which may have been bivalves.

The mainly shale and mudstone sequence between the Crofts End Marine Band and the **Kingswood Toad Vein** contains some sandstone and sandy mudstone.

Being fairly accessible over much of the Kingswood Anticline the Kingswood (Great) Toad Vein was worked extensively long before mining records were properly kept. According to Cossham, Wethered and Saise (1875, p.418) it is 'an important seam of good steam coal, about 3 ft thick'. Above the roof shale with nodular ironstone, the Speedwell colliery section shows a bed of 'stone', presumably sandstone, about 20 ft thick.

A coal lying about 40 ft above the Toad Vein and generally 12 to 18 in thick is the Liealong Vein. It was said by Cossham, Wethered and Saise (1875) to have yielded coal of good quality and to have been worked to a considerable extent, though at shallow depth, by the old miners of Kingswood. The reference to 'Liealong Blacks' in the Speedwell shaft section suggests that it deteriorates northwards at depth and passes into 'blacks' or coaly shale. Though thin, averaging only about 16 in, the coal was of sufficient value to encourage the miners to work it even though it could only be dug in a recumbent position. The roof of the Liealong Vein is said to be mudstone or shale containing much ironstone, mostly in nodular form.

Above the Liealong and separated from it by some 30 ft of mudstone is the **Primrose Vein** which has a thickness of about 36 in. The coal is almost certainly of poor quality, probably with a high ash and sulphur content. It was said by Cossham, Wethered and Saise (1875) to be divided into two by a band of dirt and their description strongly suggests that oxidising solutions reaching the coal by way of a thick sandstone in the roof, caused the formation of iron sulphates and the eventual deposition of hydrated iron oxides in the joints and bedding planes of the coal. This process has been observed to take place in a coal seam (which may be the Primrose Vein) exposed in the Hollybrook Brick Co.'s pit, Chester Park [637 746]. Above the Primrose Vein and separated from it by a thin bed of shale is the sandstone band called the Primrose Rock, which is about 50 ft thick in the Speedwell section.

The Rock Vein immediately above is said to have yielded good coal, but being only 12 in thick it was little worked. A seam was worked under this name at Belgium Pit [6353 7435] but it does not appear to have been valuable. It may also be the seam known as Netham Vein at Hanham. The name may reflect the position of the coal which lies within a group of massive sandstones including the Primrose Rock. Above the coal lies about 10 or 11 ft of mudstone and fireclay on which lies a 'Bog Head Coal or Cannel', 19 in thick and overlain by a bed of sandstone 4 ft 4 in thick.

The **Upper Five Coals Vein** is said to take its name from a resemblance to the Lower Five Coals of Kingswood. In measures cut by thrust faults, the similarity of the Doxall Vein to the Kingswood Great Vein and of the Upper Five coals to the Lower Five Coals has been a source of confusion, notably in the Whitehall Colliery area. Cossham, Wethered and Saise (1875, p.418) gave a detailed section of the Upper Five Coals: duns 2 ft 5 in on coal 20 in, dirt 8 in, coal 16 in, dirt 25 in, coal 34 in; the floor is said to be similar in character to the roof, both being firm compact mudstone. The Upper Five Coals was worked at Belgium Pit, and was said by Anstie (1873, pp.41–42) to correlate with the Stubbs Seam of Walter's Pit, situated north of the Soundwell shafts, though the precise site is unknown. Prestwich (1871, p.61) described the Upper Five Coals as 24 in thick, and suggested that the Rock Vein and Upper Five Coals Vein were synonymous. This, however, appears unlikely, since the Rock Vein is distinct both in character, thickness and position. A seam of comparable section and position is known at Hanham where the measures appear to consist largely of sandstone.

Only by comparison with the sequences proved at Harry Stoke, Winterbourne and Yate can the approximate position of the **Winterbourne Marine Band** be determined throughout Kingswood and east Bristol, since there is no direct evidence of its presence. It is not impossible that the mudstone strata immediately above the Upper Five Coals may be in part marine, and this has been assumed to be the case in the present account. In the southern limb of the Kingswood Anticline the Winterbourne Marine Band is thought to lie about 300 ft below the Parrot Vein, which may be a correlative of the Hen Vein of the northern limb (Figures 20 and 23).

This uncertainty about the position of the Winterbourne Marine Band, which normally marks the top of the Middle Coal Measures, makes it convenient to describe strata which probably belong to the Downend Formation (Upper Coal Measures, p.95) in the Kingswood Anticline together with the undoubted Lower and Middle Coal Measures section, thus retaining in effect the old 'Lower Coal Series' division in this problematic area. In the northern limb of the Kingswood Anticline the Hen Vein (see below) and overlying strata are known to be of *Anthraconauta phillipsii* Zone age at Fishponds, but whether this fossil occurs in underlying strata is uncertain.

Some 20 to 30 ft above the Lower Five Coals lies the **Doxall Vein**, the name being said to be a corruption of 'Dogs Hole' (Cossham, Wethered and Saise, 1875, p.418). This is generally supposed to correlate with Brittain's Seam, 15 in thick, of Walter's Pit[1] and Soundwell (Prestwich, 1871, p.61; Anstie, 1873, p.41). Few sections of the Doxall Vein are known, but the coal may average 22 to 24 in thick in the Speedwell–Soundwell area. In the Speedwell area the roof consists of about 2 ft of shale underlying the Doxall Stone, which is a sandstone of 'Pennant-like' type with one or more beds of conglomerate, 25 to 30 ft thick, reminiscent of the Garden Course of New Rock Colliery (p.100). At Hanham Colliery there does not appear to be any coal which can be correlated with the Doxall Vein. Thick Pennant-like sandstones are developed at about this position and may have replaced the seam. At Whitehall Colliery the measures are so disturbed that correlation is virtually impossible. The thickness and distribution of the Doxall Vein in the southern part of the Kingswood Anticline are therefore unknown. North of Speedwell and at Soundwell it has been tentatively identified in some of the Harry Stoke borings. However, all the possible correlatives are thin and of no economic value.

The distance from the Doxall Vein to the Hen Vein can be only roughly determined and little is known about the seams which occur in the intervening strata. In the view of Anstie (1873, p.42) and other early geologists the apparent absence of workable coals between the Doxall and Hen veins in the northern limb of the Kingswood Anticline was due to faulting. In this way Anstie accounted for the statement of the miners that the seams on reaching the anticlinal from the south 'roll over and die out'. However, it seems more probable that the change is due to the original deterioration of the seams rather than to structural causes.

On the southern limb of the Kingswood Anticline the seams between the Doxall Vein and the **New Smith's Coal** are little known, although some may have been worked quite extensively in the distant past. Among them are the so-called 'Dolly' and 'Plox' veins, but the nomenclature of these coals was described by Anstie in 1873 as 'local', the seams being 'not always in the order in which they are given by local miners'. Attempts have been made to resolve these difficulties and to relate the Kingswood coals to those of Warmley and the Golden Valley at Bitton, but no success has been achieved.

---

1 The position of Walter's Pit is uncertain but it may be an old shaft about 200 yd north of Soundwell High Pit.

Even the correlation of the 'New Smith's Coal' marking the base of the local Pennant Sandstone division is open to question and the position of the upper boundary of the old 'Lower Coal Series' is only a rough approximation. Pennant-like sandstones first appear as thin layers below the Upper Five Coals, though these have been neglected for purposes of classification. In order to avoid further confusion the so-called 'Plox' and New Smith's Coal of Warmley are described with the coals of the Downend Formation (Upper Coal Measures).

Owing to the difficulties resulting from intense structural deformation and the discontinuous nature of the exposures, there are several other localities where the Lower and Middle Coal Measures have been seen, but where the evidence is insufficient to enable the stratigraphical position of the exposed measures to be accurately determined. These localities include two large excavations made for brickclay, one at the former Hollybrook Brick Works at Gipsy Lane, Chester Park [637 746], the other at Mount Hill Brickworks, Mount Hill, Kingswood [649 730]. The Hollybrook Brickworks section showed southward-dipping sandstones and shales in the main workings, but at the eastern end of the quarry face at the foot of Lodge Hill, there was exposed for a time a remarkable section showing the Whitefaced Fault, with the adhering white clay minerals that give it its name. East of the fault the beds are vertical and have a northerly strike. They include a coal seam, apparently about 48 in thick, with a strong quartzitic sandstone in the roof. It is possible that this is the Kingswood Toad Vein. Another coal seam, seen in the main workings, may be the Primrose Vein. Unfortunately there are no marine strata or mussel-bands known from this locality and the sequence in the main workings is broken by low-angled thrust faults.

At Mount Hill the structural problems are equally severe and palaeontological evidence is lacking. Fossil plants found at an adjacent locality at Cockroad are of no great value (Moore and Trueman, 1937, pp.212 – 213). Farther east, at Warmley, it has been suggested by Moore and Trueman (1937, pp.214 – 215) that the highly contorted measures seen at Warmley Brickworks [672 738] are adjacent to, and perhaps include, the Kingswood Warmley (or Kingswood) Great Vein. Other measures overlain by massive Pennant-like sandstones and conglomerates in Haskin's Pit [6735 7275] may lie near the position of the Doxall Vein and may be of early Upper Coal Measures age. Here, as at other localities, no animal fossils have yet been found and palaeontological evidence is provided only by plants. Among the localities where nonmarine bivalves are known are Crofts End Brickworks, the cemetery of Kingswood Parish Church and the waste tip of Siston Colliery. All three localities have yielded *Naiadites*, which has only a very limited value as a stratigraphical index fossil.

North-east Bristol (Harry Stoke and Downend)

The Lower and Middle Coal Measures are concealed beneath Mesozoic rocks in the Harry Stoke area and hidden by higher Pennant sandstones (Upper Coal Measures) around Downend. Their stratigraphy is in consequence known only from boreholes and mine workings (Figure 22). The Harry Stoke series of boreholes was drilled by the National Coal Board to prove the principal seams of the Lower and Middle Coal Measures (Lower Coal Series) and together they provide the substance of the following account. Harry Stoke A [6226 7905], sunk in 1949, proved the measures extending upwards from just below the Ashton Vale Marine Band to about 200 ft below the Kingswood Great Vein. Harry Stoke B [6321 7816], drilled near Hambrook in 1950, proved almost the entire thickness of Lower and Middle Coal Measures with the exception of the basal portion. Boreholes D, E and F [6249 7849, 6234 7857 and 6220 7837] followed in 1950 – 51 in the neighbourhood of the Harry Stoke Drift Mine and provided added evidence of the local stratigraphy. Finally, Harry Stoke C [6504 7677] was sunk in 1955 at Downend to prove the existence of the workable coals lying at depth in the area between Harry Stoke and the northern limit of the Speedwell workings on the Kingswood Anticline. This borehole also provided additional information regarding the sequence at the top of the Middle Coal Measures and in the Downend Formation (Upper Coal Measures.). The Harry Stoke Drift Mine provided further evidence concerning the strata adjacent to the Kingswood Great Vein and the Lower Five Coals.

The lowest measures proved in Harry Stoke A Borehole consist of about 50 ft of intensely hard quartzitic sandstone. These beds closely resemble other quartzitic sandstones proved at higher stratigraphical levels, but being beneath the Ashton Vale Marine Band they are now classified as a part of the Quartzitic Sandstone Group or Millstone Grit.

The **Ashton Vale Marine Band** consists of grey laminated mudstone with some ironstone nodules and sideritised mudstone, 17 ft 9 in thick at 1291 ft 9 in, in Harry Stoke A Borehole. The base rests on hard grey quartzitic sandstone with carbonaceous patches and some mudstone pellets. A few thin sandy or silty beds occur but are not conspicuous. Much of the mudstone is highly fossiliferous, goniatites including *Gastrioceras subcrenatum* being abundant. Above the Ashton Vale Marine Band a group of strata extending upwards for about 135 ft is composed of sandy mudstone and fireclay with beds of hard quartzitic sandstone containing one pebbly or conglomeratic band and two bands of marine mudstone at 1219 ft 1 in and 1158 ft. Similar marine strata occur above the Ashton Vale Marine Band at Ashton Vale Colliery and in the Ashton Park Borehole (Kellaway, 1967); one such band proved in Harry Stoke B Borehole at 2246 ft 9 in is taken as marking the position of the uppermost of these marine layers in Harry Stoke A. The band yields *Dunbarella* and may be of approximately the same age as the marine mudstones proved at 1378 ft in Winford No. 2 Borehole (Figure 22).

Succeeding the marine strata in Harry Stoke A and B boreholes is a group of hard quartzitic sandstones 120 to 140 ft in thickness, more massive and thicker in A and with some conglomeratic bands. The 30 in coal proved in Harry Stoke B at 2111 ft 8 in may well represent the Ashton Little (p.73).

At 819 ft 2 in in Harry Stoke A lies a carbonaceous band consisting of two thin coals with a dirt parting. This is presumed to correlate with the Ashton Great Vein (p.73). The 36 in coal at 764 ft 1 in lying beneath the Harry Stoke Marine Band is the correlative of the 30 in coal at 3347 ft 3 in in Harry Stoke C, and this coal appears to be the Red Ash Vein of Hanham and Easton (p.79) which is now equated with the Top Vein of Ashton Colliery (see p.75). At Harry Stoke A the coal is high in sulphur (about 4 per cent). In south Wales, the Amman Rider which underlies the Amman Marine Band in the same position in the sequence, is also sulphur-rich.

The **Harry Stoke Marine Band** which forms the roof of the 36 in coal at 764 ft 1 in in Harry Stoke A is composed of about 1 ft 7 in of black pyritic mudstone with circular, somewhat flattened pyritic concretions having a diameter of 1 to 1½ in. The pyritic bed is overlain by 7 ft 9 in of dark grey laminated and partly fissile mudstone with lenticular and nodular ironstone, some of the nodules being veined with pyrite. By far the most abundant fossil in this bed is *Lingula mytilloides* but the band as a whole yields only a poor fauna, lacking both numbers and variety when compared with the other marine marker bands of the Bristol Coalfield. The top of the Harry Stoke Marine Band is marked in Harry Stoke A by barren silty mudstone, slightly micaceous and containing poorly preserved plant remains. In Harry Stoke C the marine strata yield *L. mytilloides* but the beds are only 2 ft 6 in thick. Neither the marine band nor its underlying coal were recorded in Harry Stoke B and F boreholes. It is not clear whether they have been lost in a 'washout' or cut out by faulting, but their position appears to be marked by much contorted mudstone and fireclay at 1856 ft in Borehole B.

Above the Harry Stoke Marine Band (or its inferred position) lies a group of barren strata averaging about 175 ft in thickness and composed predominantly of sandstone, sandy mudstone, fireclay and clunch. Neither coals nor mussel-bands have been found in this group though there are a few plant-bearing beds.

These barren strata are succeeded by a pair of coal seams which were proved at 562 ft 6 in and 606 ft 9 in in Harry Stoke A and which can be recognised in Harry Stoke B, C and F boreholes where they lie at about the same distance above the Harry Stoke Marine Band. One of these seams may correlate with the Smith's Coal of Soundwell (Figure 22).

Above these coals and upwards to the Crofts End Marine Band there appear at intervals thin bands carrying nonmarine bivalves. The individual beds are irregular in their distribution and the associated faunas usually lack variety. The lowest was proved in Harry Stoke F at a depth of 682 ft and yielded *Anthraconaia* cf. *pulchella*. Between the depths of 318 ft 3 in and 420 ft in Harry Stoke A lies a group of strata which can be matched with the strata proved in Harry Stoke B at a depth of 1387 ft 3 in to 1487 ft 4 in. The same group can be recognised in Harry Stoke F at a depth of about 420 ft to 535 ft. At each locality the group commences with a thin coal seam 22 in thick in A, 18 in in B and 14 in in F. In Harry Stoke C the group is less easily recognised, though it appears to lie between 2855 ft 3 in and 2936 ft 7 in. The 18 in seam at 1487 ft 4 in in Harry Stoke B carries a mudstone roof yielding nonmarine bivalves including *Anthraconaia sp. nov.* cf. *williamsoni* and *Naiadites* cf. *productus*. This was not located in the other three borings, though a somewhat similar fauna including *A.* cf. *williamsoni* occurs in adjacent bands, notably about 30 ft above the last-named coal in A, B and F boreholes and at a depth of 2898 ft 11 in to 2907 ft 6 in in Harry Stoke C. At all four localities the mussel-bands overlie thin coal smuts or layers of carbonaceous shale resting on fireclay. The uppermost of this group of mussel-bands overlies a thin coal at 319 ft 1 in in Harry Stoke A and at 1390 ft 9 in in Borehole B.

Although the Kingswood Little Vein cannot be identified with certainty in the Harry Stoke area, no such doubt attaches to the recognition of the **Gillers Inn Vein**, which, with its characteristic roof including the Jingleboys and Worm Band (see p.81), has been proved in all the borings. The seam shows considerable variation in both thickness and physical composition, and similar variations have also been noted underground in the Harry Stoke Drift Mine, where the coal was worked to a small extent. These may be partly structural in origin and, as with so many coal seams in the area, it is difficult to give an 'average' section. A thin band of sideritic mudstone showing traces of the 'worm' structure was proved 57 ft 7 in below the Gillers Inn in Harry Stoke F Borehole (p.171). There is, however, little resemblance between this part of the succession and the roof of the Gillers Inn Vein, since neither the 'Jingleboys' nor the coal seam are present at the lower level where the measures include shale with nonmarine bivalves. Nonmarine bivalves are found above the Worm Band in Harry Stoke B, C and F borings, and one or more bands of ironstone are usually present in the roof shale.

The measures above the Gillers Inn Vein include some thin slumped beds, but otherwise there are few features of interest and the upper beds, which contain some rootlets, are succeeded by the fireclay floor of the Kingswood Great Vein. With the exception of Harry Stoke A, every borehole sunk in the Harry Stoke area has proved this coal and it was worked in the Harry Stoke Drift Mine. The seam hereabouts shows considerable variation in thickness, in part at least due to structural deformation. In general it averages 30 to 36 in, being thinner than in the neighbouring parts of the Kingswood Anticline (p.81).

The roof of the **Kingswood Great Vein** is composed of weak shaly fireclay or friable mudstone which proved troublesome in the underground workings at Harry Stoke Drift. Cosham, Wethered and Saise (1875) said that this was also a feature of the seam in the Kingswood Anticline. As at Speedwell Pit, the Kingswood Great Vein consists of a single coal without partings at Harry Stoke, and it contains sporadic lumps of sideritic plant-bearing carbonaceous mudstone known to the miners as 'coal-stones'. The floor consists of a medium to dark grey fireclay with some carbonaceous matter and pyrite films on joint surfaces.

About 40 ft of strata separates the Kingswood Great Vein from the **Thurfer Vein** in Harry Stoke B, these measures consisting mainly of mudstone and fireclay, some of it sandy. The comparable measures in Harry Stoke C and D boreholes contain rather more sandstone. The presumed Thurfer Vein attains a thickness of 20 in in Harry Stoke C, and appears to be about 18 in thick over much of the area. It is characterised by a mudstone roof which usually carries *Lioestheria* and *Naiadites* in shale resting on the coal or separated from it by a thin band of barren shale. In Harry Stoke C the distance from the Thurfer Vein to the **Lower Five Coals** is about 26 ft 9 in, the intervening strata being mainly silty mudstone. Farther to the west, notably in Harry Stoke D, the measures pass up into sandy mudstone, sandstone and fireclay forming the floor of the Lower Five Coals.

It has already been pointed out that only three, or at the most four, of the five component coals of the so-called Lower Five Coals are present in the western part of the Kingswood Anticline; conditions in the Harry Stoke area are similar. The composite nature of the seam is not noticeable in the logs of the Harry Stoke borings, though the coal is of variable thickness. In the Harry Stoke D Borehole, where the seam was recorded as being only 19 in thick, subsequent working of the coal in the immediate vicinity of the boring shows it to have been partly replaced by a sandstone 'washout'. In places, sandstone-filled channels extend to the floor of the seam, the coal having completely disappeared. Such large-scale replacement of coal by sandstone is rare. The roof of the seam, though composed of relatively soft mudstone in Harry Stoke B Borehole, has been proved to be sandstone or hard sandy mudstone in Harry Stoke C, D and E boreholes and it is mainly hard and sandy in the workings of the Harry Stoke Drift Mine. In working the coal at this mine some large rolls or 'bladders' of coal up to 15 ft in thickness were encountered. These may owe their origin to bedding plane slip and turbulent movement due to variations in lithology and competence of the roof measures.

About 40 ft above the Lower Five Coals in Harry Stoke B and D boreholes and about 65 ft above in Harry Stoke C lies a coal seam some 16 in thick. The intervening measures consist almost entirely of sandstone in Harry Stoke D but are largely mudstone or sandy mudstone in boreholes B and C. Above the 16 in coal the strata consist of sandy mudstone, fireclay and clunch, with a band of hard grey sandstone in Harry Stoke B. These measures are succeeded in turn by a coal which in B, C and D boreholes, appears to maintain its thickness of 16 in This coal is probably the Hard Vein of Speedwell Colliery (p.82) and lies about 80 ft below the base of the Crofts End Marine Band. No coals of value are found in these strata though there are some important stratigraphical features, notably a band of mudstone with *Naidites* cf. *obliquus* and *Carbonita humilis* overlying dirty coal and shale at 2194 ft 4 in in Harry Stoke C. This appears to correlate with a shale containing *Naiadites* cf. *productus* resting on dirty coal and shale in Harry Stoke B at 1045 ft 1 in. Further assistance in correlating these measures is provided by a thin bed of carbonaceous shale with a roof of plant-bearing mudstone at 2177 ft 4 in in Harry Stoke C, which is represented by a similar development in Harry Stoke B at 1034 ft. Between this bed and the base of the Crofts End Marine Band in Harry Stoke B and C boreholes there are thin layers of red or brownish mottled mudstones.

Both at Downend (Harry Stoke C) and Hambrook (Harry Stoke B) the **Crofts End Marine Band** is over 10 ft thick, while farther west at Harry Stoke D Borehole it was only slightly thinner. In each case marine strata rest on thin coal, generally about 1½ in thick. The marine band consists largely of dark grey shale or laminated mudstone with some nodular and pyritic ironstone. In Harry Stoke C a band of calcareous silty mudstone comparable with the carbonate band recognised at Crofts End Brickworks (p.83) was proved. Even where the matrix is not calcareous the marine band contains many chonetoids, productoids and other calcareous brachiopods, and the shells seen in section can form chalky-looking bands which contrast with the darker and less shelly layers. At the top of the band the dark or medium-dark grey marine mudstone passes almost imperceptibly into paler grey, laminated, slightly silty mudstones about 20 ft thick. These beds are in turn succeeded by 50 to 60 ft of grey slightly sandy and micaceous mudstone with thin bands of slumped or streaky sandstone. This group of barren mudstones is recognisable above the Crofts End Marine Band in the Stoke Gifford and Winterbourne borings and possibly in the shaft sections of the Kingswood and Bedminster areas to the south.

In the Harry Stoke area, only Harry Stoke B and C borings have proved the measures above the Crofts End Marine Band, but the correlation between the two borings is good, the most notable difference being the absence in Harry Stoke B of a coal proved at 1864 ft 1 in in Harry Stoke C. This which appears to correlate with a similar coal in the Winterbourne Borehole at 1737 ft 2 in and probably with the Primrose Vein of Speedwell Colliery (p.84). Comparison of the measures lying between the Crofts End and Winterbourne marine bands at Winterbourne and Harry Stoke with those believed to occupy a similar position in the Kingswood Anticline, show that many of the coals are highly irregular in their development. In general, however, they fall into two groups, the lower one including the Trow, Kingswood Toad and Primrose Vein being separated by a group of barren strata (including the Primrose Rock of Speedwell Colliery) from a higher group including the Rock Coal or Rock Vein of Kingswood. The **Kingswood Toad Vein**, which is about 36 in thick at Speedwell, appears to be represented in Harry Stoke B by 14 in of coal at 810 ft 2 in, and doubtfully in Harry Stoke C by some 10 in of coal and shale at 1940 ft 1 in. The Primrose Vein provides an even better illustration of the local vagaries of coal seam development. This coal is well developed at Speedwell and appears to lie at 1864 ft 1 in in Harry Stoke C; it is, however, apparently missing in Harry Stoke B, where the equivalent position is marked by a band of quartzitic sandstone with coaly wisps at 729 ft 10 in. The roof of the seam in Harry Stoke C is carbonaceous mudstone overlain by pebbly sandstone or conglomerate, with mudstone pellets, pebbles of ironstone and quartz, tiny coal pebbles and coaly streaks. The upper surface of the mudstone shows irregularities and hollows due to erosion prior to the deposition of the sandstone. These and other features suggest that the Primrose Vein is prone to replacement by sandstone 'washouts' and probably provide an explanation of its absence in Harry Stoke B. Were it not for the obvious variability of the coals it would appear reasonable to correlate the seams at 1672 ft 7 in in C and 533 ft 9 in in Harry Stoke B with the Doxall Vein.

With regard to the lithology of this uppermost group as a whole, there are some features of general interest including the development of red and mottled measures in Harry Stoke B Borehole at 640 to 647 ft 6 in. The commonest rock type in the group is, however, hard grey sandy or slightly micaceous mudstone. Beds of fireclay and clunch also occur, and some of these beds contain a considerable quantity of nodular or lenticular ironstone. Sandstones are mostly of the rather fine-grained muddy or quartzitic types so common in the measures below. At 540 ft 3 in to 544 ft 1 in in Harry Stoke B there appears the lowest bed of the coarse subgreywacke or Pennant-like sandstone which predominates in the Upper Coal Measures.

The **Winterbourne Marine Band**, marking the top of the Middle Coal Measures, is known from both the Harry Stoke B and C boreholes. It consists of dark grey mudstone with paler patches, with sporadic ironstone bands and nodules and some scattered pyrite. It is 6 ft thick at 481 ft 6 in in Borehole B and 3 ft thick at 1594 ft 2 in in Borehole C, underlain by a coal 6 to 8 in thick. Marine fossils include brachiopods, both horny and calcareous, bivalves and crinoid columnals.

Stoke Gifford (Figure 22)

Very little was known of the Lower and Middle Coal Measures on the western side of the Coalpit Heath Basin prior to the sinking of three boreholes by the National Coal Board in 1953–54. Exploration of this north-western margin of the coalfield followed the discovery of productive seams in the Middle Coal Measures at Harry Stoke and Downend (see preceding section). Of the three borings sunk in the Stoke Gifford area, No. 1 [6346 8053], situated at the now disused Hambrook Brickyard, proved measures believed to lie in the Downend Formation (Upper Coal Measures) and is discussed later.

Stoke Gifford No. 3 [6345 8219], at Leyland Court, is situated within the area covered by the one-inch Chepstow (250) sheet and has been described in the relevant Memoir (Welch and Trotter, 1961, pp.149–153). Stoke Gifford No. 2 [6233 8967] at Bailey's Farm proved Middle Coal Measures, from 164 ft 9 in above to 577 ft 8 in below the Crofts End Marine Band. It did not penetrate as far as the Harry Stoke Marine Band and the total thickness of the Middle Coal Measures is thus not known. The results of the Stoke Gifford boreholes were economically disappointing in that a two-coal seam in Stoke Gifford No. 3 at 1241 ft 9 in was the only coal of any consequence proved (Figure 20). Welch and Trotter (1961, p.109) suggest that this seam may represent the combined Kingswood Great and Thurfer veins. Stoke Gifford No. 2 Borehole should have assisted with the correlation of the seams proved in Stoke Gifford No. 3. Unfortunately, there is a belt of disturbed strata with dips of up to 70° and some loss of core between 580 ft and 730 ft in No.2. This is almost certainly due to a thrust fault or very sharp fold which produces an apparent thickening of the measures between the Crofts End Marine Band and the principal seams (Two Foot and Lower Five Coals). It is possible that carbonaceous mudstone and fireclay at about 580 ft may be at approximately the same stratigraphical horizon as similar carbonaceous mudstone and fireclay between 730 and 740 ft. If this conclusion is correct, the thicknesses of strata proved in Harry Stoke B, Stoke Gifford 2 and Stoke Gifford 3 is as follows:

Boreholes	Gillers Inn Vein to base of Crofts End Marine Band	Base of Crofts End Marine Band to top of Winterbourne Marine Band
Harry Stoke B	296 ft 1 in	532 ft 11 in
Stoke Gifford No. 2	not proved	460 ft 0 in
Stoke Gifford No. 3	318 ft 7 in	501 ft 0 in

(The thicknesses are as penetrated in the boreholes and are not corrected for dip, which is normally about 20°.)

The Gillers Inn Vein appears to be the seam at 969 ft 4 in in the Stoke Gifford No. 2 Borehole, ironstones with worm-like structures being recovered from the roof-mudstones of this 12 in coal. It may equate with a coal (not recovered) at 1313 ft 1 in in No. 3 Borehole. The Kingswood Great Vein and the Thurfer were presumed by Welch and Trotter (1961, p.152) to have come together to form a compound seam with the section: top coal 45 in, dirt 19 in, coal 12 in, but it is less certain where these coals lie in No. 2 Borehole. The only other seam of merit proved (top coal 26 in, dirt 5 in, coal 6 in) was at 722 ft 4 in in Stoke Gifford No. 3. This was tentatively equated with the Doxall, but this correlation is very doubtful.

The **Croft's End Marine Band** attained a thickness of 16 ft 2 in between depths of 978 ft 4 in and 994 ft 6 in in Stoke Gifford No. 3 Borehole, and 15 ft 1 in between 494 ft 3 in and 509 ft 4 in in Stoke

Gifford No. 2. It contained an abundant and varied fauna in both boreholes including foraminifera, brachiopods, both horny and calcareous, and bivalves. The **Winterbourne Marine Band** was proved to be 6 ft 4 in at a depth of 499 ft 10 in in Gifford No. 3 with foraminifera, brachiopods, crinoids and bivalves. The upper group at Stoke Gifford No. 3 contains numerous bands of red or grey mottled mudstone which are not present in the corresponding group at Harry Stoke.

Winterbourne, Yate and Rangeworthy (Figures 22, 24)

This area comprises most of the Coalpit Heath Basin, the north–south elongated basinal syncline lying north of the Kingswood Anticline. It is bounded on the west by the Whitefaced Fault, on the south by a line drawn from Winterbourne to Westerleigh Common roughly along the line of the Kidney Hill Fault, and on the north-west, north and east by the outcrop of the base of the Coal Measures. Throughout this large area the Lower and Middle Coal Measures are very imperfectly known, and even the coals that have been worked locally can be only doubtfully correlated with those known elsewhere in the coalfield. The only continuous sections of any note are those proved in the Winterbourne [6461 8010] and Yate Deep [6975 8253] boreholes (Figure 20). The former, drilled in 1915–17, penetrated 11 ft of Triassic red marls and 1474 ft of Upper Coal Measures before entering Middle Coal Measures; it ended at 2338 ft 6 in, probably without entering the Lower Coal Measures. The Yate Deep Borehole, drilled in 1918–20, commenced in Pennant Measures, and entered the Middle Coal Measures at 1211 ft, ending at 2203 ft 4 in, again without reaching Lower Coal Measures. None of the coals recorded can be correlated with certainty.

Few exposures of the measures exist. Over the extensive central portion of the Coalpit Heath Basin the Lower and Middle Coal Measures lie beneath considerable thicknesses of Pennant and Supra-Pennant Measures, while over much of the north-western and south-eastern margins they are concealed by Triassic and Jurassic rocks. Only on the extreme northern and north-eastern flanks of the basin, between Cromhall and Yate, do the Lower and Middle Coal Measures crop out, and here the relief is low and exposures few and far between. There is much evidence of past shallow mining, but most of the workings were in the 19th century or earlier and the records that have been preserved are scanty indeed. There is therefore little underground information concerning the stratigraphy.

On the western margins of the area the Ashton Vale Marine Band has not been found north of Harry Stoke, and on the east it is unknown north of Wick. The Yate (Limekilns Lane) Borehole (p.64) was sunk with the purpose of defining locally the base of the Coal Measures, but no marine strata were encountered. Palynological investigations suggested to Dr B. Owens (*in* Cave, 1977, p.61) that perhaps it should be drawn at a depth of about 200 ft, but it seems more likely that the borehole was drilled just east of the outcrop of the Ashton Vale Marine Band (Cave, 1977, p.59) into strata near the top of the Millstone Grit.

Both Weaver (1824) and Anstie (1873) gave descriptions of the Cromhall Veins, the lowest coal seams known in the northern part of the Coalpit Heath Basin. These appear to be interbedded with shale and sandstone, and are probably underlain by more massive quartzitic sandstone. These coals are shown as lying within the Quartzitic Sandstone Group of the Millstone Grit on the published six-inch sheet ST 68 NE, but it is now thought that they lie within the Lower Coal Measures. The hard sandstone that underlies them crops out fairly continuously, and it may be the yellowish quartzitic sandstone exposed [7050 8823] about 400 yd west-north-west of Barbers Court Farm, west of Wickwar. This sandstone is seen intermittently to Hall End and beyond. Evidence from the Limekilns Lane Borehole suggests that the base of the Coal Measures lies 115 ft or more above the top of this sandstone.

Between Cromhall and Rangeworthy only two coals of any consequence are known between the Cromhall veins and the Upper Coal Measures. These are the **Yate Hard Vein** and the **Smith Coal**, the distance between them being about 90 ft. The Yate Hard is said to be about 28 in thick, while the Smith is very variable between 2 and 7 ft. At Yate Lower Common, the Yate Hard Vein is about 24 in thick and the Smith Coal 24 to 30 in. The Sodbury Seam, 24 in thick, makes its appearance in this area about 60 ft below the Smith Coal, and the measures below appear to be unproductive. Still farther south the Yate Little Vein, the highest of the Yate Seams, makes its appearance north of Engine Common. This coal is thin, seldom exceeding 18 in; it lies about 240 ft above the Yate Hard Vein. Only one coal is known to exist below the Sodbury Seam at Hall End: this may be the coal proved immediately below the Harry Stoke Marine Band in the Harry Stoke A and C boreholes (Figure 22); if so, the thickness of the Lower Coal Measures may be assessed as 350 to 400 ft.

In the tract of singularly flat and featureless ground which marks the crop of the Lower and Middle Coal Measures at Engine Common and Yate, there is little evidence apart from that provided by old shafts and workings. A fairly complete section through the Yate measures is provided by Dogtrap Pit [7014 8363] on the north side of Broad Lane. This is a particularly useful record in that the Chipping Sodbury RDC boreholes (p.167) at Broad Lane are only distant some 215 yds north-east of the old mine shaft.

Dogtrap Pit (Surface level 235 ft AOD)

Yate Little Seam	at 315 ft
Black Vein	at 590 ft
Yate Hard Vein	at 614 ft
Smith Coal	at 730 ft
Sodbury Seam	at 850 ft

The **Crofts End Marine Band** was encountered immediately below the surface in the Broad Lane boreholes, and with a dip of about 30° to the west-south-west, the thickness of measures lying between the Crofts End horizon and the Yate Hard Vein can be calculated as approximately 260 ft, which is almost identical with the distance between the Crofts End Marine Band and the Kingswood Great Vein of Kingswood and east Bristol.

The most northerly of the collieries which worked the Yate Hard Vein is Old Wood Colliery at Rangeworthy [6997 8515]. Anstie (1893, pp.27–28) saw this pit being sunk on or before 1871. The Yate Hard Vein was proved at a depth of 253 to 255 ft, and so it would appear that the Old Wood shaft must be on or near the crop of the Crofts End Marine Band. Among a number of shafts situated along this well-marked clay 'slack' are Staley's Pit [7007 8425], Kedge Pit [7014 8393] and Long's Pit [7040 8317].

The most informative of the old sections at Yate is that of Eggshill Colliery [7078 8237] where the section was:

Black Seam, 26 in	at 41 ft
Yate Hard Vein, 17 in	at 137 ft
Smith Coal, 12 in	at 165 ft
Fireclay Seam, 18 in	at 190 ft
Sodbury Seam, 22 in	at 242 ft

South of Eggshill Colliery the crops of the Yate Hard Vein, Smith Coal and Sodbury Seam can be traced to the northern margin of the Mesozoic rocks, which obscure the Lower and Middle Coal Measures from east of Westerleigh Common to beyond Siston.

The position of the crop of the Crofts End Marine Band, is therefore known on the eastern flank of the basin at Yate. On the western side it is concealed by Mesozoic rocks but it was proved in the Winterbourne Borehole at a depth of 2008 ft 3 in. It consisted of 18½ ft of shale at 2008 ft 3 in, carrying a rich and varied fauna, of crinoid columnals, brachiopods and bivalves, and goniatites in-

DETAILS OF STRATIGRAPHY 89

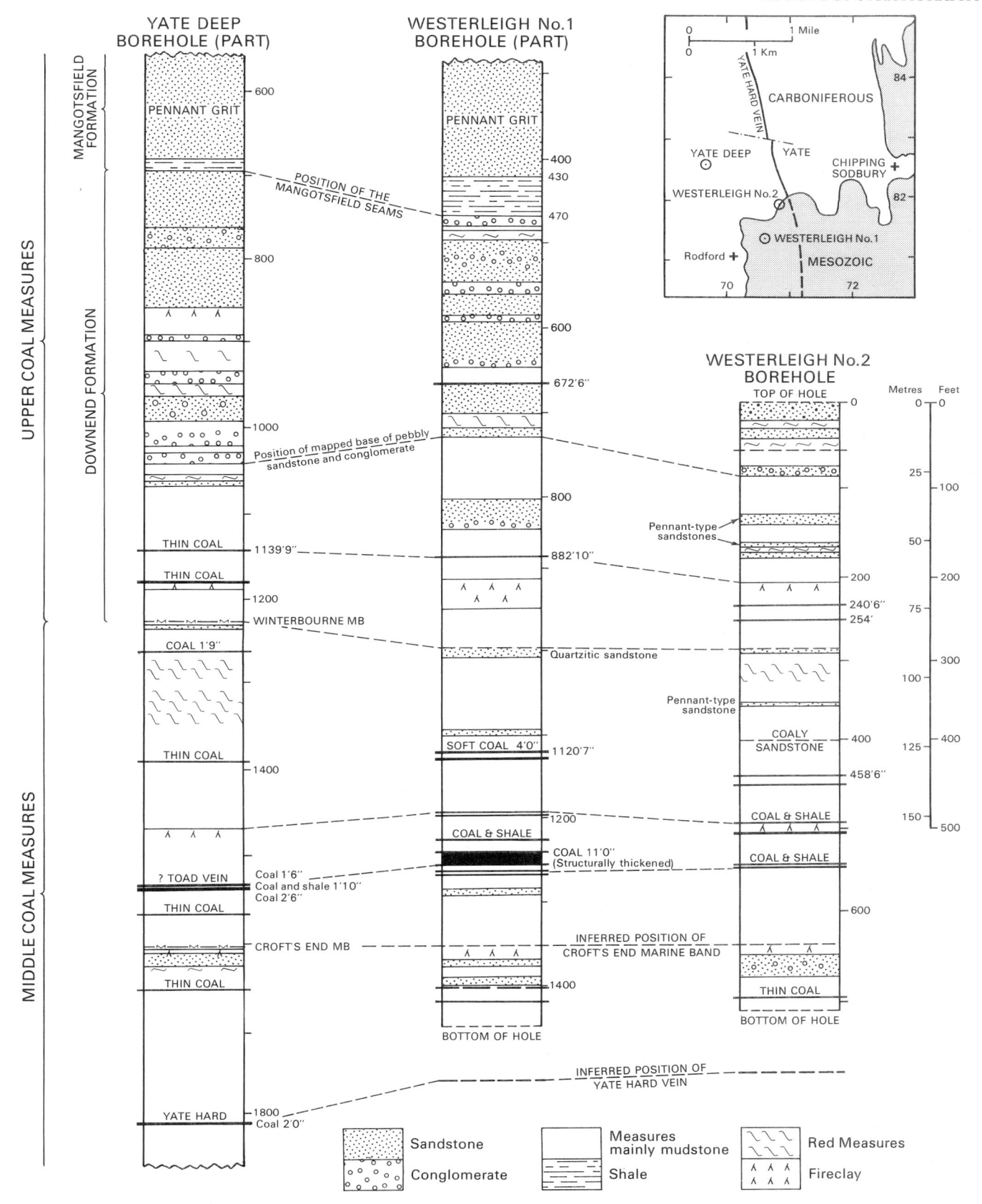

Figure 24 Comparative vertical sections of the Middle and Upper Coal Measures in the Yate Deep and Westerleigh boreholes

cluding *Anthracoceras* cf. *hindi* as well as foraminifera. In the Yate Deep Borehole (Figures 22, 24) 6 ft of shale with ironstone at 1616 ft contained a similar though less abundant fauna. The band was encountered between 6 ft 3 in and 21 ft 5 in in the Chipping Sodbury RDC's Broad Lane No. 1 Borehole [7024 8382], and between 30 ft 6 in and 40 ft 10 in in No. 2 Borehole [7024 8382], the fauna again being rich and varied.

The **Winterbourne Marine Band**, marking the top of the Middle Coal Measures has not been found at crop and is known only from the Winterbourne and Yate Deep boreholes. In the former 9 ft of sandy shale at 1494 ft yielded brachiopods and bivalves, and in the latter the marker horizon was equated with a layer containing *Lingula* and *Orbiculoidea* at 1227 ft.

Westerleigh and Pucklechurch (Figure 24)

South of Yate the Lower and Middle Coal Measures are concealed beneath Triassic and Jurassic strata. Beds at or near the base of the Upper Coal Measures were formerly exposed in a clay-pit at Westerleigh Road, Yate, described by Moore and Trueman (1937). The eastern end of this clay pit is thought to lie near the position of the Winterbourne Marine Band though no marine strata have yet been seen there. Westerleigh No. 2 Borehole [7077 8185] was sunk in 1913 near the southern margin of the old clay workings. No. 1 Borehole [7061 8127] (p.167), sunk in 1912, was situated about 650 yd south-south-west of No. 2 and commenced in higher strata, the Mangotsfield seams (or the equivalent shale band) being proved between 430 and 470 ft (Figure 24). Although no fossils have been preserved, a detailed comparison with the succession in the Yate Deep Borehole suggests that the base of the Upper Coal Measures in Westerleigh No. 1 is at about 905 ft and in Westerleigh No. 2 at about 201 ft 6 in (Figure 24). This correlation fits all the available data and suggests that the position of the Crofts End Marine Band may be at about 1355 ft in Borehole No. 1 and about 634 ft 6 in in No. 2. It therefore appears that the Westerleigh boreholes, which were intended to prove the Yate Hard Vein and adjacent seams, did not do so. If, as is suggested, the coals proved at about 1242 ft in No. 1 and 547 ft in No. 2 are situated above the Crofts End Marine Band, they may well correlate with the Kingswood Toad Vein (Figure 24). Deeper borings would therefore have been required to prove the Yate Hard Vein and other adjacent coals formerly worked at Yate.

South of Westerleigh the crops of the Yate seams are concealed and the only pit of any consequence is at Wapley [713 797] two miles south of Yate. Anstie (1873, p.29) gives a description of this old colliery in which only one seam, the 'Smith-coal', is said to have been regular, the other two seams found in the pit 'proving very irregular in thickness and sometimes disappearing.' A section taken from D.H. Williams MS Section book (1840) gives the following account of Wapley Pit [713 797]:

	Thickness		Depth	
	ft	in	ft	in
LOWER LIAS AND RHAETIC				
Lias	8	0	8	0
TRIAS				
New Red Marl	150	0	158	0
COAL MEASURES				
Dunns (mudstone)	30	0	188	0
Coal	1	8	189	8
Pan (fireclay)	2	0	191	8
Dunns	60	0	251	8
Coal (Hard or Wapley Vein)	3	0	254	8
Pan	5	0	259	8
Dunns and Rock	240	0	499	8

Correlation of the two seams is doubtful. They may equate with Yate Hard and Yate Smith Coal, but the section given by Anstie (1873, p.29) does not agree with William's account.

Wapley Pit was the most southerly colliery working the Yate seams and the terrain between this colliery and the eastern end of the Kingswood Anticline is virtually unknown. Carboniferous Limestone crops out at Codrington and Doynton, enabling an approximate eastern limit to the coalfield to be drawn. Beneath the extensive tract of Lower Lias at Pucklechurch the base of the Upper Coal Measures probably assumes a gently curving course from near Siston to Westerleigh. Anstie (1873) and others have suggested that the pit at Wapley was badly sited and that considerable reserves of coal remain buried in this concealed area. The probability that the seams have deteriorated or passed into fireclay must, however, be borne in mind.

Bridge Yate, Wick and Bath

At the eastern end of the Kingswood Anticline there is a marked falling off in the density of coal workings. Evidence of the presence of old coal pits could formerly be seen on Bridge Yate Common and at Chesley Hill Farm [6868 7343]. It is probable that the seams cropping out in the area between Webbs Heath and the Naishcombe Hill Fault, on the east side of Bridge Yate, belong to the group of coals (including the Kingswood Great Vein) which were worked at Warmley and Siston Common. Between Chesley Hill and Holbrook Common the measures are concealed by Triassic marls, but south of the Highfield Fault, in the valley of the River Boyd west of Wick, evidence of old coal pits has been found near New Mills [6945 7259] and also at The Green [6959 7235]. The situation of these very old workings, relative to the concealed crops of the seams worked in the Golden Valley to the south and the Chesley Hill–Bridge Yate area to the north-west, suggests that these coals may belong to the lower part of the Middle Coal Measures. They may well be a continuation of the coal or coals worked at Webbs Heath and Chesley Hill and are likely to be older than the lowest seams (New Smith's Coal and Kenn Moor Seam) formerly worked in the Golden Valley New Pit (Anstie, 1873, fig. 8).

A small area of basal Coal Measures is exposed in the Boyd Valley at the western end of the Wick Inlier [7025 7315]. It includes an impersistent band of fossiliferous black shale up to 1¼ ft thick which is believed to be the **Ashton Vale Marine Band** (Kellaway, 1967, p.57). The strata above this shale include hard sandstones like those at the base of the formation at Soundwell and Harry Stoke. There is no evidence of coal working below the level of the Kingswood Seams at Holbrook Common or Wick and it may be assumed that, as at Yate and Warmley, the Lower Coal Measures and the lower part of the Middle Coal Measures are devoid of workable coals.

South of Wick the position of the base of the concealed Coal Measures can be roughly estimated from the presence of Upper Cromhall Sandstone (topmost Carboniferous Limestone) in the bed of the brook west of Grandmother's Rock [706 731]. Nothing further is seen of the formations below the Upper Coal Measures. Pennant Measures crop out in the valley of the Bristol Avon at Corston but the Middle Coal Measures are concealed. South-east of Corston, coal in the Middle Coal Measures has been worked at Twerton Colliery. The old shaft is situated on the crop of the Lower Lias in Pennyquick Bottom [715 646] where the waste heap (now partly levelled) can still be seen. Anstie (1873, p.46) gave an account of the seams here, which are strongly folded and faulted according to the abandoned mine plans. Morris (*in* Woodward, 1876) is said to have found 'bivalve shells' at Twerton in the 'coal shale'. The great variation in thickness of the so-called 'Great Seam' of Twerton Colliery recalls the Lower Five Coals rather than the Kingswood Great Vein.

South of Twerton the position and structure of the Lower and Middle Coal Measures is unknown and the rocks are not seen again at the surface until they reappear at the eastern end of the Radstock basin near Mells. The concealed Coal Measures of Westbury (Wiltshire) lie on the southern flanks of the concealed extension of the eastern Mendips.

Southern part of the Radstock Basin: Nettlebridge Valley

In the Nettlebridge Valley the Lower and Middle Coal Measures are continuously exposed for a distance of about 5 miles between Mells Park in the east and Downside in the west. Over the greater part of the area, however, the rocks are so highly contorted that it is difficult, if not impossible, to identify the individual coal seams. Information from mining is very incomplete, while the confused state of the ground resulting from old shallow workings, dating in some cases from the 14th century, makes it impossible to elucidate much from surface evidence (Figure 28). One of the collieries where the measures were said to be the least affected by structural complexities was Moorewood [6421 4847], but this pit was closed before the primary six-inch survey commenced.

West of Downside the Lower and Middle Coal Measures, concealed beneath Triassic marls, have been proved in borings near Emborough, as well as farther west, near Chewton Mendip and West Harptree. Here also the measures are greatly folded and contorted.

The occurrence of *Gastrioceras subcrenatum* in contorted strata at Vobster (Kellaway and Welch, 1955, p.19), and its presence together with other marine fossils at Harridge Wood [6530 4804], 1½ miles south-west of Holcombe, confirms that the Ashton Vale Marine Band must lie just above the the Quartzitic Sandstone Group on the northern flanks of the Mendips. South of the Mendips at Ebbor [519 490], east of Westbury, the marine band has again been detected in a similar position. It is not known anywhere throughout the south-western margins of the concealed Radstock Basin to the west of the Nettlebridge Valley, though its presence at Gratwicke Hall, north of Broadfield Down near Barrow Gurney suggests that it is probably continuous throughout this area.

In the general succession of coals in the Lower and Middle Coal Measures of the Nettlebridge Valley, the lowest seam appears to be that known as **Wilmot's Vein**. Said to be 18 in thick, little is known about it, although Greenwell (1854, p.261) described it as occurring 120 ft below the Red Axen Vein (or Firestone). This seam, 24 in thick, was worked in places in the Nettlebridge Valley and probably by small adits on the north side of Lechmore Water (Emborough Pond) near Old Down. The next highest seam, the **White Axen**, appears to lie about 200 ft above the top of the Quartzitic Sandstone between Ashwick and Gurney Slade Bottom. Its thickness is given as 30 in on Vertical Section No. 52, 48 in by Prestwich (1871, p.62) and 36 in by Buckland and Conybeare (1824, p.271). It was worked along its crop south of Moorewood, and at the eastern end of the valley in Melcombe Wood, Mells. Some of the old pits along Crock's Bottom, north of Moorewood, may have been sunk to this vein, and the old pits close to Emborough Church probably worked it. Anstie (1873, p.56) states that the seams wrought here were the Fern Rag and Stone Rag, but from the position of the pits in relation to the outcrop of the Quartzitic Sandstone this statement would appear to be incorrect. Support for the view that the seam occurs low in the Lower Coal Measures is given by Moore and Trueman (1937, p.202). Prestwich states that the White Axen is commonly accompanied by water-bearing sandstones, which are probably the rocks exposed near a line of old workings on the south bank of Emborough Pond.

A considerable thickness of pale yellowish, partly calcareous sandstone lies between the White Axen seam and the **Perrink** in the Lower Coal Measures at Ham and Edford. Along the Nettlebridge valley this sandstone gives rise to a feature near Ham, and when traced eastwards its outcrop widens rapidly towards Coleford, though part of this apparent increase in thickness may be due to folding and faulting. The sandstone is well exposed in the old canal cutting at Bennett's Hill Farm, near Edford.

The coals succeeding the White Axen form an important group formerly worked along the whole length of the Nettlebridge Valley, and used in smelting Corallian ironstone at the Westbury (Wilts) ironworks. The Perrink (or Blackstone) is the most important coal in the Lower and Middle Coal Measures: it was worked along its crop throughout the Nettlebridge Valley. It was excellent coking coal, and was in such demand that it was extensively worked in the fantastically contorted ground of Vobster Breach [6976 4890] and Edford [6723 4891] collieries (Figure 25). Less disturbed conditions prevailed at Goodeaves, Moons and Ringing pits, Highbury where the coal was nearly horizontal but inverted (Prestwich, 1871, sect.4). Another gently inclined patch of inverted coal was worked until about 1850 in the Coal Barton pits, (Prestwich, 1871, pl.xiii, fig.3). The name Blackstone, as a synonym for Perrink, may be derived from the dark grey clift roof with ironstone nodules.

The only available detailed sections of the Perrink are from the abandonment plans of Moorewood and Edford collieries. The Moorewood section shows: black shale roof, coal 24 in, parting 2 in, coal 24 in, clod 10 in, hard pan. At Edford the section was: black clift roof, coal 30 in, clift floor. From the sinking of Mendip (formerly Strap) Pit [6482 4957] (p.176), Mr H. E. Hippisley records (in MS) coal 42 in, clod 9 in, on pan at a depth of 1714 ft: this seam may be the Perrink. From the black shales with associated ironstone nodules supposed to come from the roof of the Perrink, Moore and Trueman collected *Carbonicola communis*, *C.* cf. *pseudorobusta*, and *Naiadites flexuosus* (1934, p.204) on the tips of Moorewood and Breach collieries.

About 80 ft of mudstone separate the Perrink from the **Main Coal**, with an 18 in seam known as the Whing or Strap about 25 ft below the Main; in the Vobster district a 20 in seam known as the Shelley lies 6 ft below the Main. The Main has been extensively worked from Moorewood to Vobster and, like the Perrink, was sent to the Westbury ironworks. Its soft nature and bad roof, as well as the frequent development of large 'bladders' of coal in disturbed ground, made working difficult. At Moorewood the section is black shale roof, top coal 24 in, clod 12 in, bottom coal 30 in, soft shale; at Edford, where the measures are inverted, it is fireclay roof, coal 48 to 72 in, clay floor. In the Strap Pit sinking Hippisley records, at a depth of 1599 ft, 'Coal 20 ft 6 in, coal 8 in, pan 12 in' but notes on the back of his manuscript section 'vein about 7 ft to 9 ft thick, a roll in it made it 28 ft 6 in where the shaft cut it'.

The **Stone Rag** lies some 60 ft above the Main Coal, and has been worked in the same areas as the latter, though to a considerably lesser extent. According to Prestwich it is a good coking coal, but spoilt for smith coal by containing thin bands of 'mother coal' (fusain). The section at Edford is: strong clift roof, coal 36 to 48 in, soft clay-clift floor. In places the coal was reported to reach a thickness of 72 in.

A further 60 ft of mild clift with ironstone nodules separates the Stone Rag from the **Fern Rag**. The localities in which the seam has been worked are much the same as for the Stone Rag. It is a good smith's coal of rather variable thickness but has a bad roof. At Old Moorewood the coal was reported to be 48 in thick, whilst the Edford section is: clift roof, shab, coal 30 in, soft clift floor. A thin seam 18 in thick, called the Blue Pot, said to be of good quality, occurs 25 ft above the Fern Rag. Buckland and Conybeare (1824, p.271) recorded it from the Ashwick estate, and a small area of the coal was worked at Edford Colliery about 100 yd north-east of the shaft.

The **Standing Coal** or New Hit Seam has been recognised both at Moorewood and Vobster where it lies respectively 70 ft and 50 ft above the Fern Rag. Edford Colliery plans show workings in the seam to the north-east of the shaft. The level branch connecting

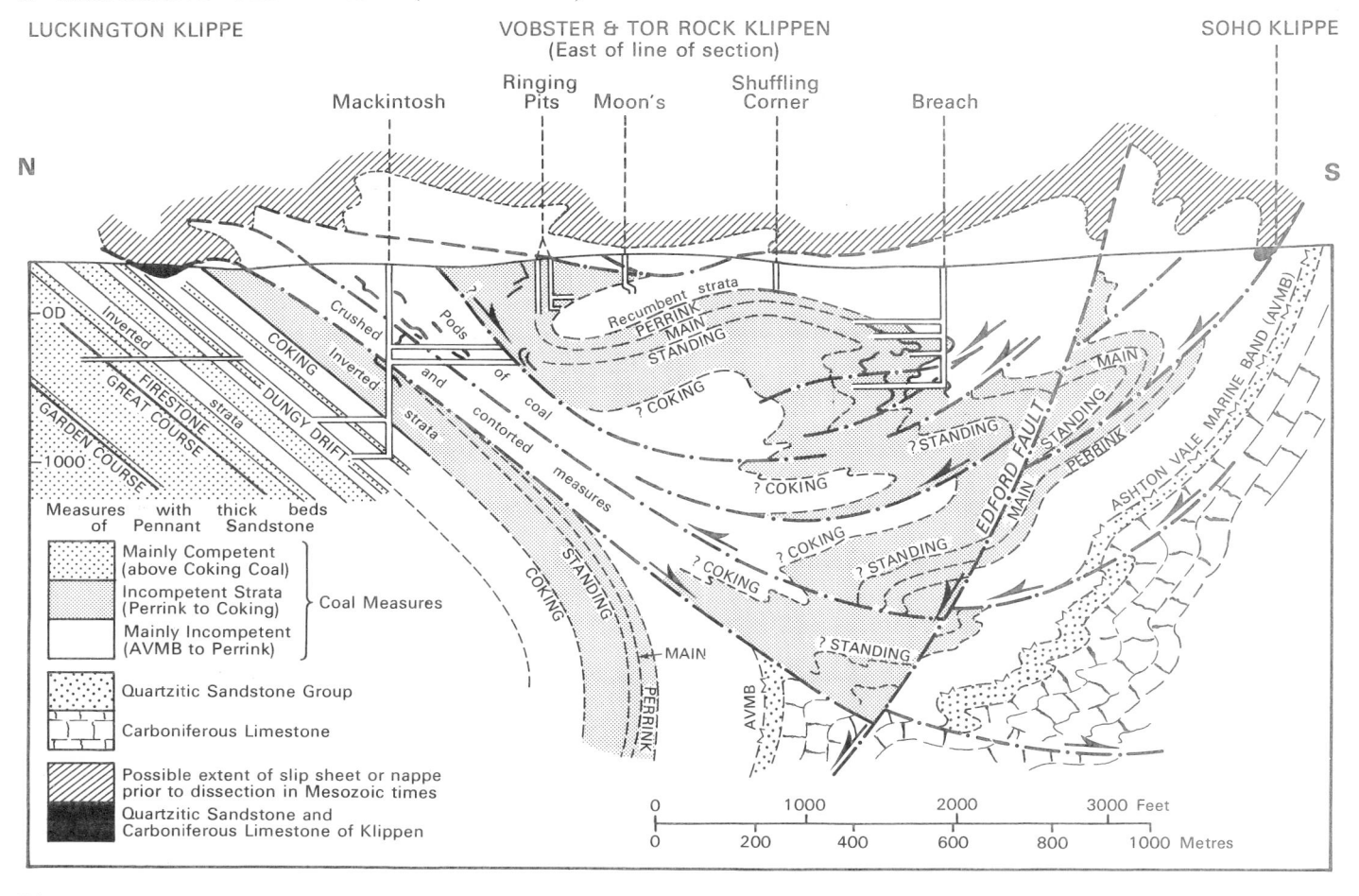

Figure 25 Horizontal section through the southern part of the Radstock Basin between Luckington and Soho

New Rock and Moorewood pits intersected the seam 46 yd from the Fern Rag drift road, but no coal appears to have been worked here. Prestwich states the seam is of irregular thickness, reaching up to 14 ft at Edford. The coal was brittle and gave out gas quickly when first opened. Partings of cannel and pyrite occur in the coal, which altogether seems to have been of rather poor quality.

The classification of the next 600 ft or so of measures presents considerable difficulty owing to structural distortion and faulting (Figure 25). In the numerous small old pits, fragments of the same seam have been worked under different names, which may account for the great number of seams listed by Buckland and Conybeare (1824, pp.270–271). The most complete succession is at Mackintosh Colliery, where the measures above the **Coking Coal** appear to be unfaulted. Although the faulting south of the pit is violent (Figure 23), it is assumed that the Coking Coal lies about 200 ft above the Standing Coal. In Buckland and Conybeare (1824), and also in Prestwich's (1871) account, the Coking Coal has not been recognised, and the succession of seams occurring in the 440 ft of measures between the Standing Coal and the Perkins Course, taken in ascending order, was said to be the South and North Shoots, Golden Candlestick, Branch and Foot, all coals which were said to have been worked south of the Nettlebridge river, to the south and south-west of Holcombe; but even in 1871 Prestwich could give little information about them. In the Moorewood–New Rock level branch a number of inferior seams of dirty coal, the thickest 42 in, were cut. These may be the deteriorating representatives of seams worked to the east.

The Coking Coal was a thick seam worked from the Mackintosh Pit westwards for over 1000 yd. The section given by Mr C. Heal of Midsomer Norton was: strong clift, rotten shab roof with brasses 2 ft 9 in, coal 60 in, shab partings, coal 5 in, shab 12 in, clift floor. Some 44 ft above, the Privy Coal was worked near Mackintosh Pit [6914 4972] bottom in 1881–98. The section given on the abandonment plan reads: working stone 5 in, shab 6 in, coal 18 in, binching 8 in, coal 18 in. Somewhere near the position of the Coking Coal or the Privy Coal a marine shale was found on the surface 110 yd 68°W of Newbury Colliery shaft [6957 4977]: since the roof of the Coking Coal is described as being pyritic it may well have come from this seam. It seems very likely that this marine shale is the **Harry Stoke Marine Band**, and thus that the boundary between the Lower and Middle Coal Measures should be taken at the Coking Coal.

The next 80 ft of measures above the Privy Coal contain several seams about 12 in thick. These measures, as well as those immediately below the Privy Coal, contain abundant nodular ironstone. Between these clift beds with ironstone nodules and the next workable seam, the **Argyle Drift**, lie 160 ft of ironstone-free clift with two 24 in coal seams. The Argyle Drift at Mackintosh Pit had the following section: shab roof, coal 33 in, clift floor. This seam may be the 'New Seam' at Mells Colliery [7114 5008] reached at the end of the south level branch, and it may also be the Perkins Course or the Hard Coal worked farther west in the Holcombe–Pitcot area.

The **Hard Coal**, according to Prestwich (1871, p.26), was 'a pretty fair household coal 30 in thick.' The section of Barlake Colliery [6605 4925] shows workings in both the Perkins Course and Hard Coal, and the Pitcot shaft [6529 4709] was actually sunk for 480 ft in one of these seams standing vertically. Only a short distance west of Barlake and Pitcot, the Strap shaft shows (if Hippisley's correlation is correct) no workable coals for a distance of 300 ft above the Main Coal, though there is a great development of fireclay. This westerly thinning and deterioration is further seen in the New Rock section in which no workable seam is found in the 640 ft of measures between the Standing Coal and the base of the Pennant. In the Strap sinking, Hippisley records a coal seam 72 in thick, some 300 ft above the Main Coal, but its identity is not known.

The **Dungy Drift** appears to be developed over most of the Nettlebridge area about 60 to 70 ft above the Argyle Drift. It has been worked in the Mells, Newbury and Mackintosh pits, but elsewhere the thickening of the dirt partings rendered it unworkable. At New Rock Colliery [6479 5057], 48 in of coal and rubbish cut in the New Rock–Moorewood level branch 150 yd beyond the Standing Coal may represent the Dungy Drift. The seam, according to Prestwich, has been worked from Newbury as far west as Barlake. It is a fiery seam containing a good deal of stone and rubbish and the sections vary considerably from place to place. At Newbury (according to Mr C. Heal) the section is: clift roof, brass 7 in, shab 18 in, coal 34 in, clod 15 in, coal 11 in, shale parting 3 in, coal 11 in on boardy clift floor. At Mells Colliery the section (given on the Old Mells Plan No. 11) shows: clift, brasses 6 in, shab 32 in, coal 34 in, clod 3 in, coal 42 in, shale 10 in, on clift. In the lower level workings of both Mells and Newbury, dirt partings split the coal to form a three-coal seam, e.g. top coal 33 to 39 in, middle coal 10 to 20 in, bottom coal 4 to 6 in.

In the succeeding 300 ft of strata between the Dungy Drift and the base of the Pennant Measures, no coal of workable thickness is known. At New Rock Colliery the position of the **Crofts End Marine Band** was detected in 1969, about 230 ft above the Dungy Drift.

The proportion of sandstone to shale gradually increases above the Dungy Drift and the upper part of the Middle Coal Measures contains a high proportion of hard quartzose sandstone and subgreywacke of Pennant type. There is, therefore, a great improvement in structural competence in the upper part of the Middle Coal Measures as well as in the overlying Pennant Measures. Even when the rocks are inverted, as they are east of Pitcot, the coals are relatively free from faulting and crumpling; from the mining standpoint the only disadvantage of inversion is that the fireclay normally underlying the coal appears in the roof of the workings. Four named coal seams have been worked in the measures above the Dungy Drift. The **Little Course** which was formerly taken as the base of the old Pennant Series, has been recognised from Mells to Pitcot (see Figure 28). At Strap Pit and New Rock it has not been detected[1]. In the Holcombe–Mackintosh area its usual thickness is 18 in, but towards Mells it expands to 32 in, though the section is very variable. The seam was worked at Mells Colliery and a section near the Newbury take showed: coal 12 in, clod 16 in, coal 10 in. The **Firestone** was a good household coal extensively worked at Mells. Like the Little Course it thins westwards being only 22 in thick at Newbury and according to Prestwich (1871, p.62) does not prove at Sweetleaze (adjoining Strap Pit). A section at Mells colliery shows: fireclay 5 ft, black shale 3 in, coal 30 in, black shale 3 in, on grey sandstone.

Some 100 ft of sandstone separates the Firestone from the **Great Course**, one of the more extensively worked seams, though rather variable in quality and section. In the most southerly workings at Mells (Figure 28) the section was: fireclay 5 ft, dark friable shale 3 in, top coal 10 to 24 in, binching 2 to 9 in, middle coal 9 to 17 in, binching 7 to 10 in, bottom coal 22 to 30 in on sandstone. Towards Newbury the binching partings become thinner and the coal improves in section to: greys, fireclay 5 ft 4 in, coal 48 in, on greys. At Holcombe and Barlake a thickness of 49 in is recorded (Buckland and Conybeare, 1824, p.273). At Pitcot pit the section of the seam (here not inverted) was: solid roof, over coal 9 in, mid coal 27 in, smith's coal 24 in; the coal was described as good but rather soft. Another section at the same colliery records: solid roof, rubbish 1 ft to 3 ft 9 in, over coal 12 in, rubbish 0 to 6 in, mid coal 12 to 21 in, binching 1 to 3 in, smith's coal 27 to 34 in. Still farther west dirt partings are prevalent and the coal is thinner, as may be seen from the following section at Moorewood Colliery (taken from the abandonment plan): greys roof, clift 4 ft, coal 9 in, binching 2 in, coal 8 in, binching 2 in, coal 27 in on black shale 10 ft. In the recently abandoned workings at New Rock colliery grey mudstones usually form the roof but clift often feathers in between.

The Strap is a small seam 18 to 20 in thick lying 140 to 180 ft above the Great Course. Prestwich stated it was 'a hard coal, pretty good for household purposes, but injured by streaks of mother-coal fusain'. The seam was formerly worked at the Strap Pit.

Ston Easton and Harptree

Except for a small area of the lowest beds exposed at Emborough in the area between Downside Abbey and West Harptree, the Lower and Middle Coal Measures are concealed beneath Triassic and Jurassic rocks, and the stratigraphical succession and structure are known only from borings. Old workings for coal seen in the inlier at Emborough are thought to have been based on the White Axen Vein, a seam 30 in thick occupying a position low in the succession (Green and Welch, 1965, p.57). Between Emborough and Compton Martin, several small inliers give an indication of the position of the Quartzitic Sandstone Formation and these act as a guide to the approximate position of the concealed base of the Coal Measures.

In an attempt to elucidate the structure and to find coal three boreholes were sunk between 1873 and 1875 near Chewton Mendip. None of these was deep enough to supply significant information about the Coal Measures. A better result was obtained in 1911 at Gurney Farm, West Harptree [5630 5810] (p.173), near the southern end of the Chew Valley Lake. Four coal seams were proved, one with a recorded thickness of 67 in. Unfortunately, the log gives no indication of the dip. A considerable mass of sandstone and sandy shale associated with red mudstone was proved between the base of the Dolomitic Conglomerate at 209 ft 6 in and the shale above the highest of the coal seams at 519 ft 4 in. One interpretation of these records is that the boring proved the basal part of the Pennant Measures or the top of the Middle Coal Measures, i.e. part of the sequence worked at New Rock Colliery.

An alternative view is that the seams are the Farrington (Bromley) coals, overlain by the red measures and sandstones of the Barren Red Formation, formerly worked in Bishop Sutton Colliery. The incrops of the Bromley seams have been proved from Stanton Drew to Bishop Sutton, except beneath the grounds of Sutton Court where, because the ground was not worked, evidence is lacking. If the crop is projected in a south-westerly then southerly direction to pass west of Lower Gurney Farm then the immediate structural problems are greatly eased. This implies that the major thrusts associated with the 'Farmborough Fault' and which are supposed to cross Fry's Bottom north of Fry's Bottom Pit [631 604] and to pass through the deepside workings of Bishop Sutton Colliery [583 594] are curving southwards to join the Stock Hill–Biddle wrench fault system (Green and Welsh, 1965, pl. 5). In this event the conjectural line of the Farmborough Fault, as shown, will require some modification.

Of much more recent date are Ston Easton No. 1 [6225 5174] and No. 2 [6211 5158] boreholes (p.174), sunk near Old Down in

---

[1] Prestwich's section of Strap Pit (1871, p.48, no. 81) appears to contradict this statement. Hippisley's detailed log of the sinking of Strap Pit shows neither Little Course nor Firestone.

1952–53. In both holes highly contorted rocks were proved with evidence of very strong disharmonic movement. Investigations carried out elsewhere, notably in the attempted development of Strap (or Mendip) Pit in the Nettlebridge Valley, have shown conclusively that the whole of the argillaceous Lower and Middle Coal Measures has been affected by these disharmonic movements. Two of the major seams of the Nettlebridge Valley, possibly the Main and Perrink, were encountered at 937 ft 6 in and 1065 ft 3 in respectively in No. 1 Borehole. The coals, 55 ft 3 in and 4 ft 7 in thick, were so highly folded and distorted that no further investigations were made.

South of the Mendips

Speculation as to the possible existence of Coal Measures south of the Mendips dates from the early days of geology. The presence of Lower Coal Measures, including the Ashton Vale Marine Band has been established on the southern slopes of the Mendips at Ebbor (Green and Welch, 1965, p.57, pl.v) where the workings were described by Greenwell (1892). Small areas of Lower Coal Measures may be preserved in other downfolded or faulted areas at the southern margin of the Mendips. These include the rocks proved in the Rodney Stoke Borehole [4874 5012] (Green and Welch, 1965, p.141), on the south side of the Rodney Stoke Fault, a structure with a southerly downthrow of about 2000 ft. Other areas where tracts of Upper Carboniferous strata may have been downfaulted lie between Cheddar and Axbridge and on the north side of the Clift Woods Thrust near Wells. Farther east the Coal Measures proved in the Westbury (Wilts.) Borehole [861 520] (Pringle, 1922) may also occupy a faulted or downfolded tract on the southern margin of the buried continuation of the eastern Mendips fold belt. The age of the strata proved at Westbury is uncertain, but the lithology suggests Lower or lower Middle Coal Measures. Like the Somerset occurrences they are probably confined to a limited tract of downfolded or faulted rocks.

Nailsea (Figure 20)

In the low-lying area between the Carboniferous Limestone uplands of Broadfield Down and the Clevedon–Failand ridge, the Nailsea Coalfield occupies gently rising ground reaching a little over 100 ft above OD. The Coal Measures are preserved in a shallow synclinal fold with a westerly plunge. Much of the outcrop is occupied by Pennant Sandstone, but the Lower and Middle Coal Measures occupy a crescent-shaped area at the northern and eastern margins. Most of the coal workings are very old and those records still extant are incomplete and lacking in detail. Very little is known about the character of the coals, their succession or their correlation. Excluding Grace's Seam, which lies within massive Pennant Sandstone, all the seams lie within the argillaceous sequence formerly called Lower Coal Series. These measures are about 900 ft thick in the Backwell and Nailsea area, resting upon the hard quartzitic sandstones of the Millstone Grit.

On the northern margin of the coalfield at Court Hill, Tickenham, sections in the M5 motorway and adjacent excavations have shown pockets of hard quartzose sandstones with some sandy mudstones preserved locally beneath Pennant-type sandstone resting unconformably on Lower Carboniferous limestone and dolomite. It therefore seems that drastic changes in the thickness and succession of the Coal Measures rocks take place between the Backwell area in the south and south-east, where the succession is relatively complete, and Tickenham and Court Hill in the north, where the greater part of the Lower and Middle Coal Measures are missing.

So far as is known the largest disturbance affecting the productive measures is a north-west–south-east fault between Nailsea and Backwell, which, in its south-eastern part, is said to have a throw of 14 fathoms (84 ft), to the south-west. Practically all the workable coals have been exhausted on the upthrow side of the fault, but on the downthrow side only limited amounts of White's Top Vein have been worked from White Oak [4716 6952] and Young Wood [4728 6926] pits, in both of which the seam was subject to a good deal of minor folding and faulting.

The Ashton Vale Marine Band is unknown and it would appear that the base of the Coal Measures is nowhere exposed. The Winterbourne Marine Band has not yet been proved and so the boundary between the Middle and Upper Coal Measures remains undetermined. In general it would appear that, as in the Bedminster area of Bristol, Pennant-like conditions set in at a stratigraphically low level, not very far above the only good and extensively worked seam, known as the White's Top, and that the Middle Coal Measures extends some way up into the Pennant sandstones. The generalised sequence shown in Figure 18 has been taken from Geological Six-inch Sheet ST47SE, compiled from various sources. Buckland and Conybeare (1824) gave a list of seams most of which appear on the Geological Survey Vertical Section No. 49 (published 1873); in this the distance given between the Golden Valley and Backwell seams is conjectural, since no colliery shaft or workings penetrated the intervening measures. Little is known about the sequence as a whole, and the following account of certain of the worked seams must be read in conjunction with Figure 20.

The lowest known seam is the **Crow Vein**, 14 in thick and lying not far above the base of the sequence. About 29 ft above lie the **Spider Delf and Dog veins**, about 25 in and 34 in thick respectively. About 200 acres of the Dog Vein are shown on the 1871 Royal Commission map as having been worked from the Backwell Common pits ('Teague's Colliery' of Buckland and Conybeare, 1824). The old workings on the north side of the coalfield 1¼ miles E by N of Tickenham church may be on the crop of this seam, which probably continues eastwards beneath the Trias to a point some 1¼ miles south-east of Wraxall church whence it swings around the nose of the synclinal fold and assumes a south-westerly strike. Watercress Farm Borehole [5000 7050], near Tyntesfield (Richardson, 1928, pp.99–100), proved a 36 in coal seam at a depth of 126 ft. This may be the Dog Vein. From the tips of Backwell Common pits, Seavill (1936) collected *Anthraconauta minima* (renamed by Dr M. A. Calver *Curvirimula minima*), suggesting that the coals worked in the colliery belong to the *C. communis* Zone of the Lower Coal Measures. According to Buckland and Conybeare (1824) both the Dog Vein and the underlying Spider Delf were regarded as 'good coals', although the latter was said to possess thin clay partings and to swell into large irregular masses.

Between the Dog and the Golden Valley seams, lenticular sandstones are developed at three horizons, about 100 ft, 200 ft and 300 ft above the Dog Vein. The lowest band, a hard, fine-grained, pebbly grit, having a maximum thickness of about 40 ft, forms a fairly strong surface feature extending eastwards from Nailsea Heath on which Wraxall House is built. The middle band of sandstone is lenticular, up to 25 ft in thickness, and consists of hard white grit and sandstone, pebbly in places. The uppermost band, up to 50 ft thick, consists of brown sandstone.

The **Golden Valley seam**, 33 in thick, was worked from the Golden Valley Pit [4844 7060] and from Nailsea Heath Colliery old pit [4765 7111]. The **Golden Valley Top seam**, 20 in thick, was also worked to a limited extent from the latter colliery. In 1950 a trial shaft [4826 7124], 400 yd south-south-west of Wraxall House proved the crop of a seam which, from its position, may be the Golden Valley. The section here is: clay and shale 14 to 21 in, coal 10 to 12 in, clay (impersistent) 1½ in, coal 18½ in, fireclay floor. About 40 to 50 ft above the Golden Valley Top seam there is a fairly persistent brown sandstone in the form of two bands separated by shale; together these reach a maximum thicknesss of 100 ft and give rise to a low feature on the northern side of the basin. Probably the same sandstone produces the higher ground of the small 'island' on which the Backwell Common shafts were sited.

The **Dungy seam**, about 16 in thick, was worked to the deep of Golden Valley pit. Nothing is known of the quality of the seam, but the name suggests it was probably a dirty coal.

The **White's Top** was the most extensively wrought vein of the Nailsea Basin and is said to average 42 in. According to Anstie (1873, p.53) the coal thickens southwards from 48 in to 54 in. At Heath Pit [4668 7078] the depth to the seam is 86 ft, at Glasshouse Pit [4800 7087] 110 ft, at Golden Valley Pit 108 ft, at Doublescreen Pit [4772 7047] 360 ft, and at Farler's Pit [4777 6980] 315 ft. On the downthrow side of the main fault, Young Wood Colliery reached the seam at 387 ft, whilst to the deep White Oak Pit cut the coal at 492 ft.

As already remarked, correlation of the coals of the Nailsea Coalfield with those elsewhere in Bristol and Somerset is difficult. It would seem, however, that the Crow, Spider Delf and Dog veins represent the Ashton seams (including possibly the Ashton Little and Ashton Great veins. The best quality seam worked was White's Top Vein, and for this reason it has been correlated with the Bedminster Great Vein. However, this involves the assumption that massive Pennant sandstones occur well below the position of the Crofts End Marine Band and that, in the Nailsea Basin, a large part of the Middle Coal Measures is in a Pennant Sandstone facies.

An alternative explanation is that the Nailsea Basin presents an attenuated sucession which is partly transitional from that of Ashton and Bedminster but which also has features in common with the New Rock area in the south-western part of the Somerset Coalfield. The Backwell Little or the Smith Coal may equate with the Bedminster Toad Vein, and the Golden Valley seams may then, as proposed by Anstie (1873, tab.i), represent the Bedminster Great Vein. In this case the group of coals that includes the Under Little, Dungy and White's Top of Nailsea could correlate with the worked seams above the Crofts End Marine Band at New Rock Colliery, i.e. with the Little, Firestone and Great Course. This correlation has the merit of suggesting that deposition of the massive Pennant Sandstone and subgreywacke facies at Nailsea was roughly contemporaneous with that of the New Rock area.

Barrow Gurney

Productive Coal Measures were said to have been proved in an old borehole at Barrow Gurney, the site of which is not known. At Gratwicke Hall Home Farm [5818 6820] a borehole drilled in 1943 at the bottom of a 190 ft well penetrated Lower Coal Measures at 272 ft. A 4 ft layer of grey shale from 287 ft to 291 ft yielded fossils characteristic of the Ashton Vale Marine Band, including *Gastrioceras subcrenatum, G.* cf. *coronatum, Anthracoceras sp.* and *Lingula mytilloides*. The extent of the concealed Coal Measures in the Barrow Gurney area is not known.

Chew Valley and Vale of Wrington

West of Bishop Sutton and the Chew Valley Lake an extensive tract of concealed Coal Measures separates the Carboniferous Limestone uplands of Broadfield Down to the north from the Blackdown Pericline to the south. The upper part of the Chew valley, now partly flooded by the lake, includes Upper Coal Measures (Pennant Measures) exposed at Moorledge, south-east of Chew Magna; the workings of Bishop Sutton Colliery (Farrington Formation) extend to the eastern margin of the flooded area. Middle Coal Measures are thought to be present in the vicinity of the dam cut-off trench of the Chew Valley Lake near Chew Stoke (Moore, 1938). From the results of the Winford No. 2 Borehole it would therefore appear that Middle Coal Measures are likely to form the sub-Triassic surface immediately west of Chew Magna and to extend southwards to the northern end of the lake. West of the Chew Valley Lake and over the greater part of the Vale of Wrington, the structure and stratigraphy of the Coal Measures is unknown.

At the western end of the Vale, Upper Coal Measures were proved in a borehole at Banwell Moor [3995 6087] (Green and Welch, 1965, pp.198 – 199). Farther east, a boring at Langford Lodge [459 611] is said to have proved 72 ft of Keuper Marl resting on 68 ft of Coal Measures (Richardson, 1928, p.44). It is considered that the mass of Carboniferous Limestone which forms an isolated hillock protruding through the cover of Keuper rocks at Churchill [453 602] probably rests on contorted Lower and Middle Coal Measures (Green and Welch, 1965, fig.11).

Cattybrook and Over

Across most of the Severn Coal Basin (Woodward, 1876), Lower and Middle Coal Measures rocks are missing. Thus at Severnside, Portskewett, Portishead and Kingsweston, Pennant sandstones rest unconformably on Carboniferous Limestone or thin Millstone Grit. Only in the eastern part of the basin, at Cattybrook and Over, is there any evidence of the presence of sub-Pennant measures. Sections in Lower Coal Measures can be seen in Cattybrook Clay pit (Hancock and Williams, 1977), but the exposures formerly observed in old workings alongside the railway have long since been obliterated. These showed folded, vertical and steeply dipping mudstones, coal seams and quartzitic sandstones, abutting against the structurally contorted Upper Cromhall Sandstone which was exposed in deep railway cuttings farther east (Smith and Reynolds, 1929). Evidence of coal working by the Birmingham Mining Company dating from 1756 was found at the time that the railway cuttings were made.

Accounts of older sections at Cattybrook have been given by Richardson (in Prestwich, 1871, p.65) and Woodward (1876, pp.42 – 43). Woodward records that one seam or carbonaceous band about 4 ft thick yields a white ash on combustion. It has generally been assumed that the thickness of this coal seam implies that it is of Coal Measures age, but the description recalls the dirty Millstone Grit coal formerly worked at Tapwell Bridge near Cromhall, 7 miles north-east of Cattybrook.

In the clay pit at Cattybrook brickworks the exposed Coal Measures consist mainly of barren mudstone and sandstone. No nonmarine bivalves or marine fossils have been found, but the plants are said to indicate a *Modiolaris* Zone age (Moore and Trueman, 1942, p.38). The succession between the Patchway Tunnel and Cattybrook brickworks, therefore, shows progressively younger rocks from east to west. The close proximity of the barren mudstones to the top of the Carboniferous Limestone may be due largely to attenuation of the Millstone Grit and Lower and Middle Coal Measures strata rather than to large-scale translatory movement. The Ridgeway Thrust at Cattybrook and Over is a belt of strongly faulted and overfolded strata with inverted Hotwells Limestone resting on Upper Cromhall Sandstone at Knoll Park.

Pennant Measures

Downend Formation

The Downend Formation, lying between the Winterbourne Marine Band and the Mangotsfield Seams, comprises a variable assemblage of grey with red beds, conglomerates and thin coal seams in the lower part, passing upwards into massive Pennant sandstones in which poor coals are developed locally. Small forms of the nonmarine bivalve *Anthraconauta phillipsii* constitute the principle element of the coal-roof faunas though they are neither abundant nor well preserved. Plants occur sporadically through the grey argillaceous beds.

96 CHAPTER FIVE COAL MEASURES (WESTPHALIAN)

Northern limb of Kingswood Anticline and the Coalpit Heath Basin
(Figures 23, 26, 32)

In the Downend, Mangotsfield and Staple Hill area the Downend Formation is probably at its thickest, about 2000 to 2200 ft of measures being present. Farther west it has been destroyed by post-Coal-Measures erosion. Eastwards it can be traced without substantial change of thickness to the vicinity of the fault between Syston Farm [6683 7504] and Oakleigh House [6725 7593]. Still farther east the measures are largely concealed beneath Mesozoic rocks and the Downend Formation is not seen again until it appears on the surface on the eastern margin of the coalfield at Yate. Between Syston Farm and Yate its thickness is reduced from over 2000 ft at Downend to 400 to 500 ft. Calculations based on the probable position of the incrop of the Yate Hard and other adjacent coal seams at Wapley suggest that further rapid attenuation takes place in the vicinity of Shortwood Hill and that the formation is strongly attenuated in the area between Pucklechurch and Yate.

From Yate the Downend Formation continues round the eastern margin of the Coalpit Heath Basin to the Iron Acton Fault at Rangeworthy. On the western side of the basin it is concealed beneath Mesozoic rocks, though its boundaries can be traced locally using borehole data. There is, however, a considerable area between Patchway and Tytherington where very little is known about the stratigraphy and structure of the Pennant Measures generally, and the concealed boundaries cannot be drawn with any degree of confidence north of Stoke Gifford.

Between Stoke Gifford and Stapleton borings made in search of coal provide a substantial amount of detailed information. The Harry Stoke B, Harry Stoke C, Winterbourne, and Stoke Gifford No. 3 boreholes all proved the **Winterbourne Marine Band** and the overlying Downend Formation strata (Figure 26). At Winterbourne the borehole commenced in measures which are thought to be close to and just below the Mangotsfield seams, and here the total thickness of the Downend Formation is about 1500 ft. Of these strata, the lowest 600 ft consist of grey and red mudstone with numerous thin bands of conglomerate and pebbly sandstone. The overlying 900 ft consist of Pennant Sandstone with some conglomerate or pebbly layers and also sporadic shale bands. No workable coals were proved, but a thin seam at 1093 ft 6 in has a roof rich in *Euestheria simoni* and *Anthraconauta phillipsii*, suggesting the possible presence of the Hen Vein (see below) about 400 ft above the Winterbourne Marine Band. In the Harry Stoke C Borehole a coal, thought to be about 36 in thick by gamma logging, was proved at 1191 ft, that is about 400 ft above the Winterbourne Marine Band. The equivalent coal in Harry Stoke B Borehole was proved at 40 ft 6 in, about 440 ft above the Winterbourne Marine Band.

The most northerly borehole proving the Downend Formation on the western side of the Coalpit Heath basin is Stoke Gifford No.3 at Leyland Court (Welch and Trotter, 1961, pp.149–150). This proved the Winterbourne Marine Band at a depth of 493 ft 6 in to 499 ft 10 in. Only 183 ft 6 in of beds belonging to the Downend Formation were proved in this borehole, though a somewhat greater thickness was encountered in Stoke Gifford No.1 Borehole (Figure 24) where the succession is comparable with that of the Winterbourne Borehole. The 15 in dirty seam proved at 428 ft 10 in in the No.1 Borehole may correlate with the coal at 40 ft 6 in in Harry Stoke B. These boreholes suggest a progressive westerly and north-westerly lowering of the level at which the Pennant Sandstone facies enters the succession, a change which can also be observed on the surface at the north-western margin of the Kingswood Anticline.

On the eastern side of the Bristol Coalfield between Westerleigh and Cromhall there are no workable coals in the Downend Formation. Between Pucklechurch and Stapleton on the northern limb of the Kingswood Anticline, however, four seams, known in ascending order as the Hen, Chick, Cock and Stinking veins, occur in the upper part of the formation, which consists mainly of Pennant Sandstone interbedded with thin layers of shale. These arenaceous strata rest on a lower division in which red and grey mudstones, fireclays and conglomerates, with thin coals, are presumed to overlie the Winterbourne Marine Band, the position of which has nowhere been located in this area. All four coals were extensively worked between Eastville and Staple Hill, probably at an early stage in the development of the coalfield, though their identification and correlation throughout the area remains in a confused state, the principal reason being the loss of accurate mining data relating to ground which has since become heavily built upon. The generally accepted sequence of the measures between the Hen Vein and the Stinking Vein is that given by Anstie (1873, p.43). He gives the distance between the Hen and the Chick at Castle's Ridgeway Pit [6263 7563] as about 50 yd, and the interval between the Chick and Cock is about the same. The distance between the Cock and Stinking veins according to Anstie is 160 yd, though a note by E. H. Staples amends this to 260 yd. The former measurement of 160 yd agrees with the interval shown on the Geological Survey's 1:10 560 map ST 67 NE.

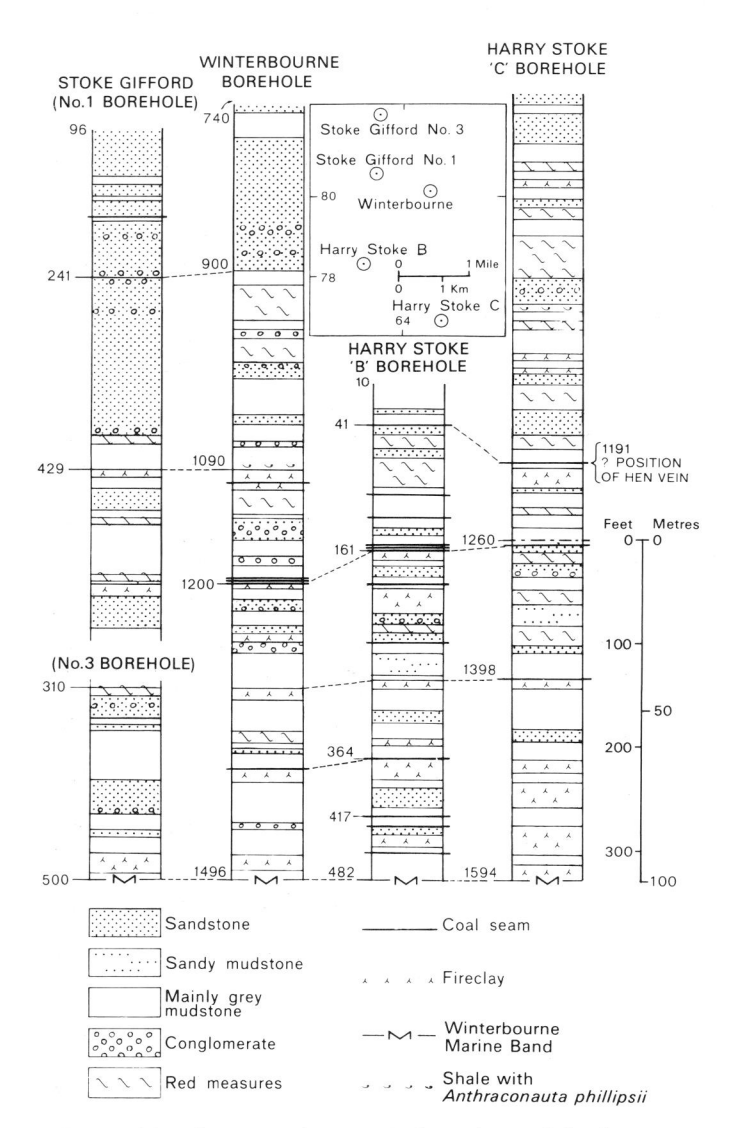

Figure 26 Comparative vertical sections of the lower part of the Downend Formation in the Winterbourne area

The **Hen Vein**, which is generally about 36 in thick, has been worked almost continuously from the Whitefaced Fault at Thicket Avenue [6423 7567] to the Frome Valley north of Eastville Park [6165 7548]. Excavations made on the south side of the now abandoned railway line 300 yd east of the bridge at New Station Road, Fishponds [6364 7558] proved black shale with abundant *Anthraconauta phillipsii* in the roof of the coal. The position of the crop can only be roughly determined and very little information has survived regarding the old workings. However, the Hen Vein is said to have been proved in Castle's Ridgeway Pit at a depth of 600 ft. To the north the crops of the Chick, Cock and Stinking veins have been calculated, there being little surface evidence to act as a check on position. Sandstone and conglomerate were recorded by Moore and Trueman (1937) in the measures above the Cock Vein in Forest Road [6383 7575]. Other conglomeratic sandstones can be seen in the river banks [6183 7563] on the east side of the Frome valley above the supposed crop of the Stinking Vein. North-north-westerly or northerly dips of about 60° were recorded on the Hen Vein near the disused Fishponds Railway Station, diminishing northwards to about 15° in the vicinity of Delabere Road [6393 7636]. Farther west, in the vicinity of the Frome valley, the dip of the measures between the Hen Vein and the Stinking Vein increases to 75–80°. West of the Frome the Downend Formation is concealed by red Triassic sand and sandstone. North of the crop of the Stinking Vein a belt of coarse Pennant sandstone and conglomerate extends from the River Frome [6188 7570] to Fishponds [6300 7591], overlain by fairly massive Pennant Sandstone, which was formerly worked to a depth of about 20 ft at Fishponds Fire Station [6270 7590].

Roughly following an east-north-easterly course from Stapleton Bridge [6117 7560] to Fishponds Training College [6336 7625] is the axis of a fold, or fractured fold, along which the dip of the massive Pennant Sandstone changes from a generally northern direction at Fishponds and Eastville to south-easterly or south-south-easterly at Stapleton and Broomhill. Situated on this line and immediately to the south of Manor Road is an old sandstone quarry [6250 7600] in which the Pennant Sandstone dips very gently to the east. The strike of the massive Pennant Sandstone turns through a right angle immediately south of the old sandstone workings and it seems likely that the structure includes a major fault.

Massive Pennant sandstones are again seen in the Frome gorge above Stapleton. Evidence of disturbance resulting in strong changes in the strike of the rocks is seen near the bend in the river above Snuff Mills Park: below the bend the rocks have a southerly dip, above it an easterly one. Another sharp change is seen near Frenchay Bridge where the dip swings from south-easterly to east-north-easterly. One of the last working Pennant Stone quarries is situated on the east bank of the river near Frenchay Bridge [639 771]. In this quarry (closed after the 1939–45 war), which was about 100 ft in depth, very massive Pennant sandstone was worked for curbstones and flags as well as for general building purposes. The worked sandstone appears to underlie the Stinking Vein and is relatively free of shale intercalations and conglomeratic bands. It is noticeable that at Fishponds and Frenchay the higher parts of the Downend Formation above the Stinking Vein mainly consist of massive sandstones with some lenticular bands of sandy mudstone. Pebbly sandstones and conglomerates are mainly concentrated in the lower beds. About 40 ft of current-bedded sandstones with some shale bands are seen in an old quarry on the south bank of the river, east of Frenchay Bridge [641 772]. Associated with the current-bedded sandstones are massive sandstones with a curious type of bedding which, seen in cross section, presents the appearance of a mass of interlocking lobate or rhomboidal units. These beds are exposed at several localities along the Frome, notably near Oldbury Court [632 768].

Above Frenchay the Frome is joined by a valley eroded along the crop of the Stinking Vein shale at Downend. The shale slack continues along the west side of the valley, the river having cut a deep gorge in the overlying Pennant Sandstone. The shale slack marking the position of the Stinking Vein (here reduced to a thin coaly band) is parallel with the river gorge and distant some 200 yd to the west of it. At the Cleeve Road Bridge [644 777] current-bedded sandstone with shale bands is present in old quarries on the east bank and between Bromley Heath and White Hill, the Frome and its tributary, the Bradley Brook are mainly entrenched in a sandstone gorge. A thin coal is seen at the termination of the Stinking Vein slack near Hambrook [6441 7855]. North of this locality it has not been possible to trace the seam and it may have died out. An easterly dip of about 10° is maintained however at least as far north as Hambrook. Traces of old coal workings occur in a tributary gully [6490 7885] south of the Whiteshill Common, which may represent the Mangotsfield Seams. If this should prove to be the case then it would follow that the distance between the Stinking Vein and the Mangotsfield Seams has been reduced to about 300 ft, that is roughly half the distance between the Stinking Vein and the Mangotsfield Seams at Downend and Mangotsfield.

At Mangotsfield the Hen Vein was formerly exposed at the now disused railway station [6651 7532] and is said to have been seen also in a subway beneath the track. North of the station the crops of the Chick, Cock, Stinking and Mangotsfield seams can be identified on the surface at Rodway Hill. These north-easterly dipping coals are displaced by a transcurrent fault which passes just east of Stockwell House [6595 7712] and which shows a dextral slip. West of the fault the position of the Stinking Vein of Rodway Hill can be identified by a broad slack which is roughly coincident with the northern part of Burley Grove [657 767] and swings into a westerly direction before passing through the centre of Downend. Harry Stoke 'C' Borehole was sunk at Downend [6507 7677] on the position of the crop of the Stinking Vein and penetrated 1557 ft of Downend Formation strata before passing through the Winterbourne Marine Band at 1591 ft 2 in. From Downend the shale slack can be followed to the Whitefaced Fault near Overndale Road [6470 7700] and thence in a north-westerly direction to the Frome valley [6430 7731] where the shale assumes a northerly strike, roughly parallel with the Whitefaced Fault.

In the Downend area an additional shale band is developed above the Stinking Vein: it is seen as a well-marked linear depression extending from near Stanbridge House [657 768] and extending in a north-westerly direction to the Whitefaced Fault [650 779].

It is probable that nearly all the seams of the Downend Formation have been worked to exhaustion in the accessible areas of Rodway Hill and Mangotsfield. Evidence of the presence of crop workings and bell pits have been found, but there is little or no documentary evidence. In the old railway cutting at Charn Hill [6592 7552] and also in quarries on Rodway Hill Pennant sandstones with pebbly and conglomeratic beds occur both below and above the Hen Vein and also above the Chick Vein. Thick sandstones of Pennant type form most of the ground extending up to the base of the Mangotsfield Seams.

Between the eastern end of the Kingswood Anticline and Westerleigh Common the Downend Formation is concealed by Mesozoic rocks. Two boreholes, Westerleigh No. 1 and No. 2 were sunk near the northern margin of the concealed area and both of these borings are thought to have penetrated the Downend Formation (Figure 24). At the eastern margin of Westerleigh Common a large excavation for brick clay [708 819] was described by Moore and Trueman (1937). This has now been filled in but at the time of the survey the following section was measured:

	Thickness ft
Deep red-brown shale	60
Conglomerate	10
Shale	10
Gritty sandstone	24
Shale with thin sandstone	230

98 CHAPTER FIVE COAL MEASURES (WESTPHALIAN)

This section has been carefully searched in the past for evidence of marine bands, though without success. A comparison of the details of the Yate Deep and Westerleigh boreholes (Figure 24) suggests, however, that the base of the Downend Formation may not have been proved in the brickworks pit at Westerleigh, the crop of the Winterbourne Marine Band being on or very near the eastern margin of the clay pit.

No exposures of any consequence are seen between Westerleigh and Yate, but in the railway cutting about 500 yd north-west of Yate railway station pebbly sandstone underlies a band of sandy clay. Conglomeratic sandstone is also seen in a bank of the River Frome 1350 yd west-north-west of Yate Church. In the Yate Borehole [6975 8253] conglomerate bands in the Downend Formation total about 280 ft (Figure 24). North of Yate the position of the base of the Downend Formation can be calculated with reference to the lowest conglomerate bed which gives rise to a slight feature in an otherwise almost featureless terrain. North of Engine Common the conglomerate, which is up to about 180 ft thick at Rangeworthy becomes lenticular and is intermittently developed or absent farther north.

From Yate southwards the conglomerate of the Downend Formation is directly overlain by shale marking the base of the Mangotsfield Formation, but between New Road, Rangeworthy [6932 8584] and Engine Common [6960 8437] a mass of Pennant-type sandstone is developed beneath the basal shale of the Mangotsfield Formation. This extends as far south as Engine Common.

Southern limb of Kingswood Anticline (Figure 27)

Between Great Western Colliery [6091 7249] at St Phillips March, Bristol and Newton St Loe, west of Bath, a fairly compact group of coals occupying the lower part of the Downend Formation and very probably the upper part of the Middle Coal Measures has been worked almost continuously (Figure 27). The thickness of measures between the highest of these coals, known as the **Millgrit Vein**, and

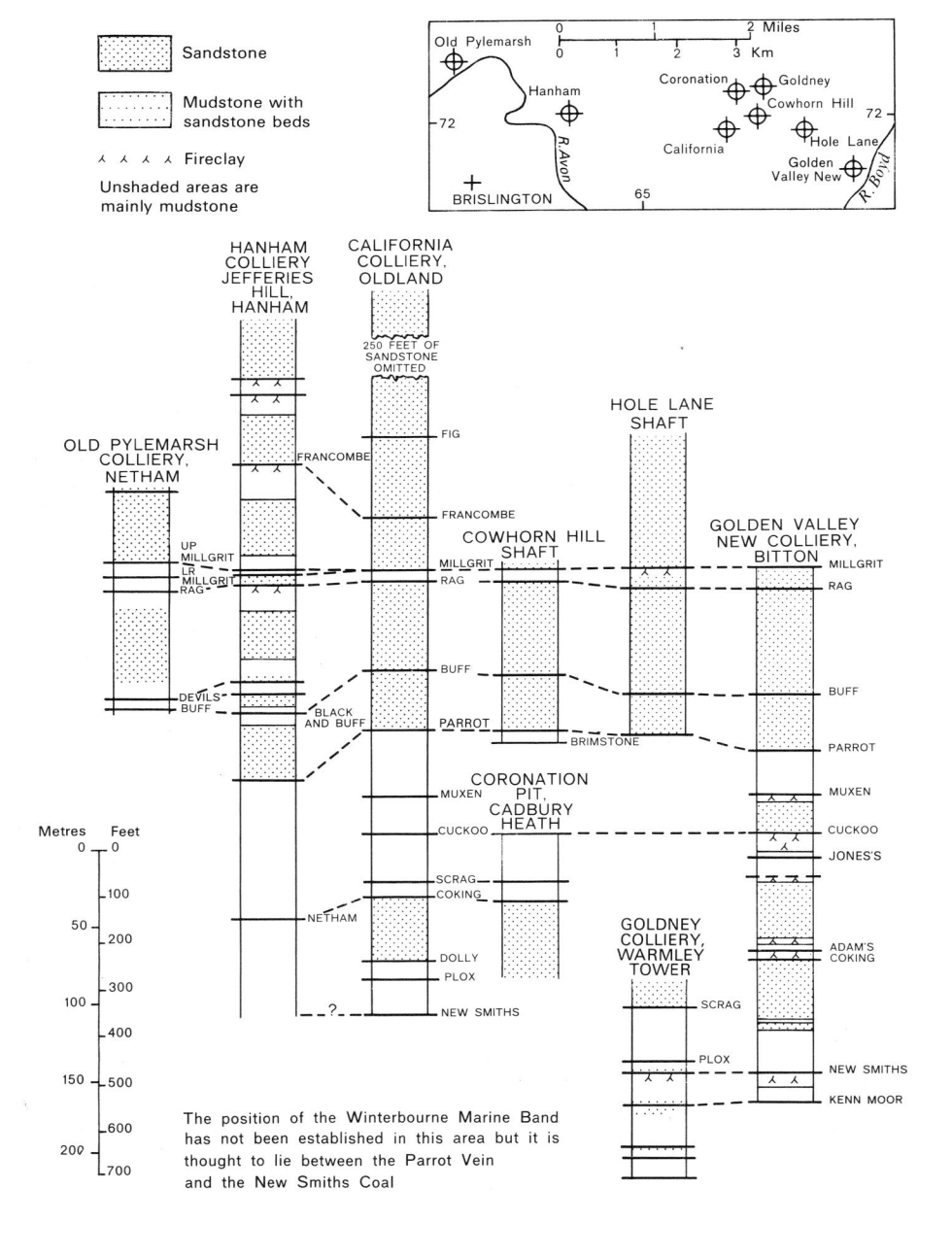

Figure 27 Correlation of the coal seams in the lower part of the Downend Formation and the upper part of the Middle Coal Measures on the southern side of the Kingswood Anticline

the **New Smith's Coal** varies from about 500 ft near Newton St Loe to over 1100 ft at Netham, in east Bristol, where the measures pass beneath Triassic sandstones. The Winterbourne Marine Band has nowhere been located in this area and the base of the Downend Formation cannot be accurately determined: it is almost certainly situated between the **Parrot Vein** and the New Smith's coal, which are separated by about 700 ft at Bitton (Figure 27).

The seams achieved their maximum development in the central part of the area around Oldland Common and Bitton and most were still workable at Hanham. Deterioration sets in westwards and at Pylemarsh and Netham the Parrot Vein was unworkable. Still farther west, at Bedminster, very little working has taken place and the seams appear to be very thin or missing altogether. This change is comparable with a similar deterioration of the coals of the Downend Formation west of New Rock Colliery at the south-western extremity of the Radstock Basin.

At California Colliery [6654 7140] in the Oldland Common area, the generalised succession through the productive part of the Downend Formation is:

	Thickness		Depth	
	ft	in	ft	in
Millgrit Vein (1 to 10 ft) average	5	0	5	0
Shale	30	0	35	0
Pennant Sandstone	30	0	65	0
Rag Vein (1 to 13 ft) average	3	0	68	0
Shale	132	0	200	0
Pennant Sandstone	18	0	218	0
Dibble Vein average	1	0	219	0
Shale	36	0	255	0
Buff Vein (1 to 30 ft) average	3	0	258	0
Pennant Sandstone	120	0	378	0
Parrot Vein (1 to 10 ft) average	2	0	380	0

Hereabouts the interval from the Parrot Vein to the New Smith's Coal varies from 380 to 480 ft, the changes being due to rapid variations in thickness, both of the coals and intervening measures.

The best section of the beds below the Parrot Vein is the one given by Cossham, Wethered and Saise (1875). This relates to the 'Pound Branch', a cross-measures level heading driven S85°E from the bottom of Golden Valley Old Pit [6901 7080] at about 340 ft below OD.

	Thickness		Depth	
	ft	in	ft	in
Parrot Seam	1	6	1	6
Measures (unspecified)	65	0	66	6
Thin bedded duns	0	6	67	0
Duns	18	0	85	0
Muxen Seam (blacks)	1	0	86	0
Fireclay	10	0	96	0
Duns with blacks at base	8	0	104	0
Cuckoo stone	68	0	172	0
Cuckoo Seam	1	6	173	6
Fireclay	33	0	206	6
Duns with ironbands	20	0	226	6
Jones's Seam	1	6	228	0
Fireclay	20	0	248	0
Duns with Ragged Seam (6 in) at base	20	0	268	0
Blacks and fireclay	15	0	283	0
Pennant Sandstone with trace of blacks at base	122	0	405	0
Fireclay	15	0	420	0
Duns	14	0	434	0
Adam's Seam (? = Scrag Vein of Oldland)	1	0	435	0
Fireclay	25	0	460	0
Coking Coal Seam	2	0	462	0
Pennant Sandstone	130	0	592	0
Duns	10	0	602	0
Pennant Sandstone	15	0	617	0
Duns, blacks at base	60	0	677	0
Fireclay, blacks	10	0	687	0
Duns and blacks	12	0	699	0
Fireclay and duns	15	0	714	0
New Smith's Coal	2	0	716	0
Fireclay with median band of ironstone	30	0	746	0
Duns with ironbands	32	0	778	0
Kenn Moor Seam	2	0	780	0

Nearly all the seams from the Parrot Vein upwards show rapid local variations in thickness. The intervening measures contain much lenticular sandstone and this has also given rise to changes in overall thickness. There is also evidence of seam splitting. Thus the Millgrit Vein is said to split into an Upper and Lower Millgrit at Hanham [6373 7204] and Old Pylemarsh [6174 7298] collieries.

The identities of the crops shown on the published maps (six-inch ST 67 SW, one-inch sheet 264) are doubtful, especially at Whites Hill and Hanham where the rocks may be structurally disturbed. The base of the group of coals is marked by a very strong quartzose sandstone which emerges from beneath the Triassic rocks at Redfield [6175 7355] and continues thence along the southern margin of St George's Park where it forms a ridge or feature extending by way of Summerhill to the western end of Air Balloon Road, near the old shaft of Air Balloon Pit [6309 7332]. Here sandstones and quartz conglomerates crop out on the ridge overlooking the valley leading to Crews Hole.

At Stibbs Hill [6339 7337] the coal seams formerly worked in the Old Air Balloon Pit (Anstie 1873, p.39) are supposed to crop out. These include the 'Smith Coal' and the Scrag Vein. The latter is reputed to overlie the New Smiths Coal of Warmley. No evidence of the presence of these coal crops could be found, but this may be due to the great age of the workings and the heavily built-up ground.

The pebbly sandstones appear to be faulted between Stibbs Hill and Potterswood, but reappear as a folded mass of sandstone and conglomerate on the eastern side of the Whitefaced Fault at Fry's Pit [6410 7293]. The tortuous crop of a coal seam which may be the New Smiths Coal lies on the underside of the folded sandstone and conglomerate. The sandstone is coarse grained with tiny greenish-coloured specks, contrasting strongly with the fine-grained quartzitic sandstone in the underlying measures. It is displaced by a fault but is recognisable at Luggers Hill [6510 7256] and Lower Barrscourt Farm [6595 7245], where it appears to lie between the crops of the New Smiths Coal and the Coking Coal of Warmley. Coarse sandstones and conglomerates formerly exposed at Haskins' claypit at Warmley Tower [672 728] appear to mark the continuation of the same beds but they pass beneath Mercia Mudstone west-north-west of Cann Farm [683 725] and are not seen again.

From Hanham to Oldland Common the productive part of the Downend Formation can be traced by belts of old workings to the margin of the concealed area at Redfield Hill and Bitton. The principal coals worked in this belt were the Parrot, Buff, Rag and Millgrit veins, which, with the New Smiths Coal, were extensively worked. Though commonly less than 24 in thick the Parrot Vein was especially valued because of its superior quality. Anstie (1873, p.37, fig. 8) noted that by 1871 the lowest level on the workings of the Parrot Vein (there 18 in thick) in the New Golden Valley Pit [6860 7102] was already 640 yd below surface.

Above the Millgrit Vein two or three thin coals including the Fig and Francombe veins (see California Colliery, Figure 27) have been proved in massive Pennant Sandstone but none was of workable thickness. About 540 ft of Pennant Sandstone overlies the Millgrit Vein at California No.1 Pit and a further 500 ft of sandstone is present south of the shaft. The crop of the Mangotsfield Seams is

100 CHAPTER FIVE COAL MEASURES (WESTPHALIAN)

thought to be present in the old tramway south of Oldland [6705 7105] so the total thickness of the Downend Formation above the Millgrit Vein is about 1050 ft. Taking the base of the Downend Formation at the New Smiths Coal gives a total thickness of about 1200 ft, though this is subject to considerable local variation due both to sedimentary and tectonic effects.

South of the Bitton fault the only known workings in coals of the Downend Formation were at Globe Pit, Newton St Loe [7022 6538]. According to Anstie (1873), the succession at Globe Pit is as follows: Millgrit Seam (24 to 30 in) at 100 yd; Rag Seam (24 in) at 108 yd; Coke Seam (24 to 28 in) at 160 yd; Little Seam (26 in) at 176 yd; Black Seam (36 to 96 in) at 188 yd. The coals are said to dip at 1:4 to the west-north-west; they were reputed to be of good quality and were chiefly used for coking.

The Millgrit and Rag seams of Globe Colliery probably correlate with their namesakes at Bitton and Hanham; less certainty attaches to the correlation of the Coke, Little and Black seams. The position of the Winterbourne Marine Band is unknown but is thought to be below the Black (Kellaway, 1970, p.1).

North side of Nettlebridge Valley (Figure 28)

The details of the Downend Formation succession in this old mining area are not very well known. The best section appears to be that of New Rock Colliery [6479 5057]. The lowest seam of significance is the **Garden Course** with a thickness of about 42 in. It lies about 150 above the Strap Seam, the intervening measures consisting mainly of clift (silty mudstones) within which the **Winterbourne Marine Band** is probably developed. The Garden Course is probably the best and most constant seam in the whole of the district, being

Figure 28 Sections through the productive Coal Measures of the southern Radstock Basin. The gentle dip of the beds in the west at New Rock Colliery steepens eastwards, until at Barlake the strata are inverted and are even more overturned in the easternmost colliery at Mells

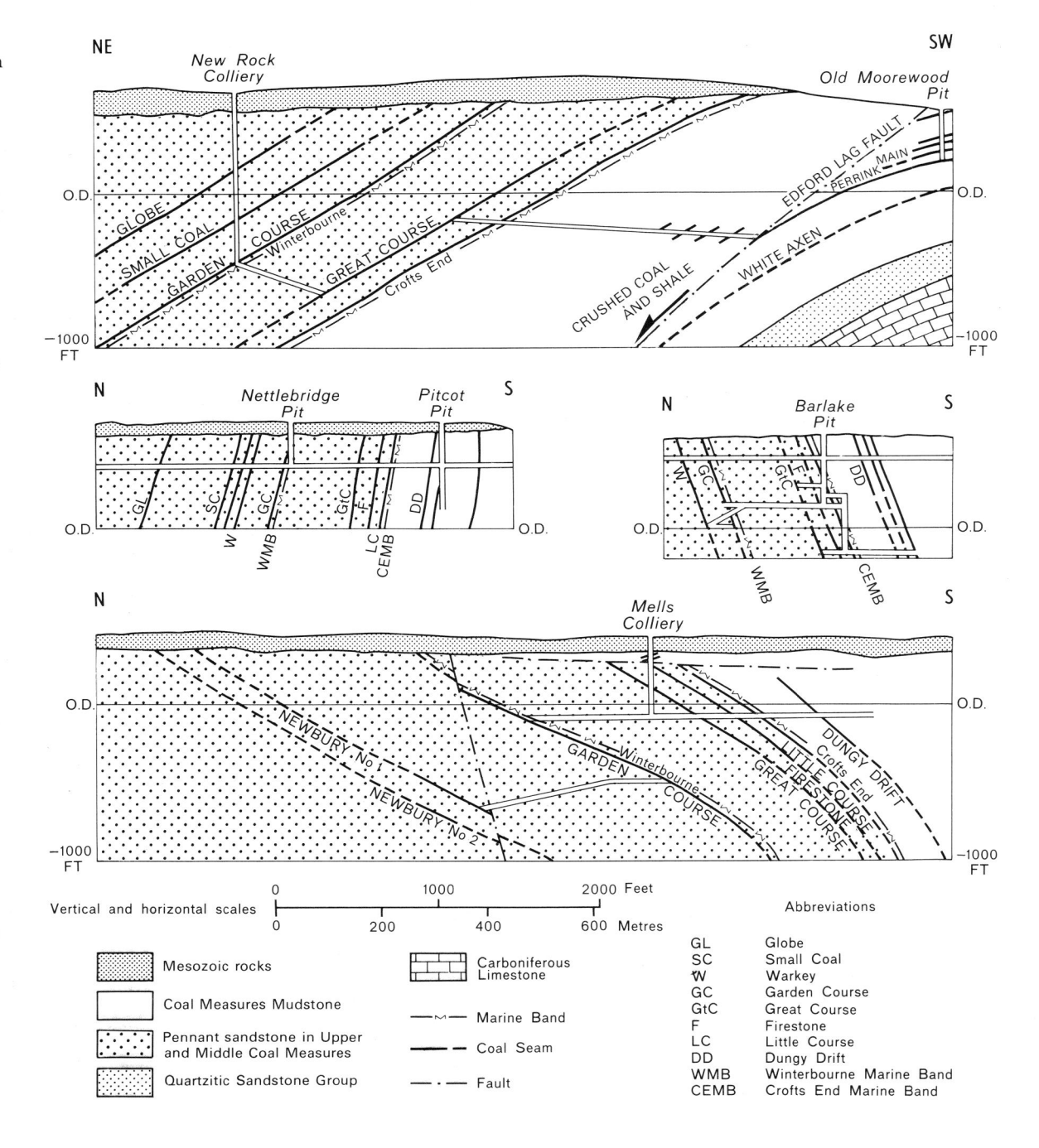

Abbreviations
GL   Globe
SC   Small Coal
W   Warkey
GC   Garden Course
GtC Great Course
F   Firestone
LC   Little Course
DD   Dungy Drift
WMB Winterbourne Marine Band
CEMB Crofts End Marine Band

worked formerly from Mells to New Rock (Figure 28). At Mells Colliery [7114 5008] the average section (in inverted strata) was: hard fireclay roof, coal 30 in, sandstone. However the section was very variable due to compressional structures. One section in the lowest working level showed: clift, coaly shale 6 in, stone 14 in, binching 2 in, coal 66 in, clod and clift 36 in; however, only 50 yd away the coal was only 6 in thick. At Newbury Colliery [6957 4977] the section was: clift roof, working stone 8 in, coal 34 in, clod 8 in, clift floor. At New Rock Colliery the section of the seam given on Plan No.2 was: clift roof, stony coal 2 in, top coal mixed with streaks of shale 16 to 24 in, clod 12 to 30 in, bottom coal 24 in, pan.

Locally a conglomeratic band is developed above the Garden Course which contains large quartz pebbles together with pebbles of pink and green igneous rocks.

Approximately 140 ft above the Garden Course lies the thick two-coal seam known as the **Warkey Course**. It has a thick dirt parting, and contains streaks of fusain which render it of inferior quality. The section in the New Rock Colliery shaft was: coal 11 in, shale 57 in, coal 15 in; at Newbury: coal 2 in, dirt 18 in, coal 20 in; whilst at Mells the seam identified as the Warkey Course contained only 12 in of coal.

The Warkey Course marks the highest of the coals that can be recognised throughout the whole length of the Nettlebridge Valley. For some distance above the measures consist largely of mudstones with coals that appear to be local in their distribution. West of Holcombe there appears to be one group of worked coals — the Two Coal, Small Coal and Globe veins — whilst to the east another — the Newbury Nos. 1, 2 and 3 — was mined; so far there has been no direct correlation between these two groups (Figure 28).

In the western area between Holcombe and New Rock some 80 ft of sandstone with partings of mudstone and several false seams separate the Warkey Course from the Two Coal Vein. At New Rock Colliery it consists of top coal 10 in, fireclay 36 in, bottom coal 12 in, and carries in its mudstone roof a few nonmarine bivalves originally identified as '*Anthracomya sp.* juv', but presumably *Anthraconauta phillipsii*. Some 50 ft above, the **Small Coal Vein** has been proved over much the same area. Prestwich (1871, p.62) stated that it was 'a red ash coal and makes very hard coke, but rather sulphury'. The seam was formerly worked at Pitcot [6535 4930] and New Rock collieries and was reopened at the latter for a short time during the Second World War with a section: clift roof, binching 4 to 6 in, top coal 18 to 30 in, bottom coal 24 to 48 in, pan floor. At New Rock Colliery 330 ft of measures separate the Small Coal Vein from the Globe Vein. These measures consist in ascending order of mudstone and sandy mudstone 115 ft, thin coal, hard sandstone 180 ft, mudstone 35 ft. At Pitcot the separation is 400 ft and at Holcombe 250 ft. The **Globe Vein** was reputed to be one of the best house coals in the district and at New Rock Colliery showed the following section: grey mudstone roof, coal 12 in, occasionally pinching out, rubbish 36 in, coal 26 in, pan floor. The Globe is the highest worked coal in this western area and what lies above it is a matter of some conjecture. Some evidence may be forthcoming from the record of the 'Chilcompton New Pit', in which Buckland and Conybeare (1824, p.272) regarded the lowest coal at a depth of 533 ft 7 in as the 'Warkey Course' and the seam at 438 ft 7 in as the 'Small Coal'. This 'Chilcompton New Pit' is probably the one now known as New Rock, in which case the so-called 'Warkey Course' would appear to be the Globe Vein. In the New Rock shaft at a distance of 91 ft above the Old Globe Coal Hole (at 541 ft depth) 'Daniel's Holing', was located at precisely the level at which the seam called 'Small Coal' was cut in the 'Chilcompton New Pit'. It would seem therefore from the Buckland and Conybeare section that two coals of 36 in thickness and one of 25 in lie above the Globe.

East of Holcombe a different nomenclature is used for the three seams proved and worked at the Charmborough [6786 5110], Newbury [6958 4999] and Mells [7114 5008] collieries. Named the **Newbury Nos. 1, 2 and 3 veins** they appear to have been first proved when the long level branch was driven north from the Garden Course at Newbury Colliery in the hope of finding the Globe Vein, the accepted view having been that No. 1 and No. 2 veins were the equivalents of the Two Coal and Small Coal veins and that the Globe Vein lay ahead. When it is realised that only a little over one mile separates Holcombe and Charmborough, it is hard to believe that the measures would expand rapidly enough to place the Globe Vein above the No. 3.

At Charmborough Colliery the No. 1 Vein lies about 450 ft above the Warkey Course the coal being 20 to 24 in thick with a shale roof containing *Anthraconauta phillipsii*. At Newbury Colliery the separation is recorded as being about 600 ft and the section of the seam: clift 24 in, branching clift 30 in, binching 13 in, coal 15½ in, binching 7 in, coal 18 in, blacks 7 in, on greys. At Mells Colliery the No. 1 Vein is some 500 ft above the Warkey Course with a section: variable roof, soft coal 6 in, dirt 9 in, coal 18 in, clod 12 in, coal 24 to 30 in, clod 3 in, on greys.

The No. 2 Vein lies about 160 ft above the No. 1. At Mells Colliery its section was given as: clift roof, blacks 2½ in, working stone 18 in, coal 39½ in, clift; at Newbury: branchy clift 3 ft, coal 46 in, clift; and at Charmborough: planty clift, clod, coal 33 in, clod (locally replacing much of the coal), clift.

The No. 3 Vein was worked only for a short period during the 1914 – 18 war. The section was: good roof, coal 12 in, shaly binching 12 in, coal 22 to 24 in, blacks 12 in, soft clift. At Charmborough the seam was proved to be 20 in thick but was never worked.

Little is known of the succession above the No. 3 Vein. In Holcombe Wood and in the ground to the north-east a seam has been worked from a number of shallow pits [670 510], on the tips of which *Anthraconauta phillipsii* and *A. tenuis* occur in abundance. Calculating the distance from the crops of lower known seams, the Holcombe Wood seam is not far removed from Charmborough No. 3 Vein: on six-inch sheet ST 65 SE it is tentatively referred to the Temple Cloud Seam, which is taken to mark the base of the Mangotsfield Formation.

Mangotsfield Formation

Pennant-type lithologies continue to dominate in the Mangotsfield Formation which overlies the Downend Formation without major change (Figure 24). Despite minor variations in the thickness of the individual sandstones and shales, the Mangotsfield Formation is relatively uniform in character throughout. The base is marked by the Mangotsfield Seams, the name given to a compound coal horizon associated with fairly thick shales, usually giving rise to a well-marked slack or hollow on the surface. In general, the formation comprises a massive arenaceous lower division and an upper part consisting of thick beds of shale with intercalatious of Pennant sandstones and one or more thin coals. *Anthraconauta phillipsii* occurs sporadically in the roof of the Mangotsfield Seams (Moore and Trueman 1937, p.225) though much of the shale is barren or carries scattered plants. The upper limit of the Mangotsfield Formation is taken at the base of the High Vein of Coalpit Heath.

Coalpit Heath Basin

On the eastern and south-western side of the Coalpit Heath Basin the base of the Mangotsfield Formation is well defined by a belt of shale containing the **Mangotsfield Seams**, here clearly separated into two seams: these are the only coals of any consequence that have been worked in the formation (Figure 32). Another thin coal, the so-called Made-for-Ever Seam, is reputed to have been worked in the Frampton Cotterill area, though this has not been confirmed. It lies in the thick shaly upper part of the Mangotsfield Formation which is confined to the western and south-western parts of the basin, west of a line from Frampton Cotterell to Pucklechurch.

In the area between Iron Acton, Engine Common (Yate), Winterbourne and Pucklechurch the general succession and thickness of the Mangotsfield Formation can be summarised as follows:

	Iron Acton – Engine Common ft	Frampton Cotterell ft	Winterbourne – Mangotsfield ft
Upper Division			
Shale with some sandstone	150	130	100
Pennant Sandstone	250	200	350
Shale with some sandstone	—	80	140
Made-for-Ever Seam	200	—	—
Shale with some sandstone	—	50	400
Pennant Sandstone	1280	c.1000	480
Mangotsfield Seams (coal and shales)	c.20	?	10–20
Total	1900	1460+	1485

Much of the upper shaly portion of the Mangotsfield Formation has a distinctive plum-red colouration. Some of this tinting may be due to Triassic red-staining as much of the present land surface is at or near the level of the ancient Triassic peneplain. At Parkfield Colliery, Pucklechurch [6885 7778], however, a cross-measures drivage, made in 1893, proved 150 ft of 'red dunns' at a depth of 600 ft below OD, which suggests that some, at least, of the red colouration may be primary. In this drivage the Frampton Cotterell Sandstone proved to be 200 ft thick, and two coaly bands 45 ft and 70 ft below the sandstone may lie in the approximate position of the **Made-for-Ever Seam**.

In the Mangotsfield area the shale belt containing the Mangotsfield Seams can be traced from just south of Shortwood across a well-defined north–south zone of faulting and continuing in a north-westerly direction through Mangotsfield to a point ½ mile north-west of the church, where it is intersected by the Stockwell Hill Fault. The two seams have been worked from Church Farm and Wallsend collieries at Mangotsfield. Church Farm Colliery consisted of two pits, Deep (Buller's) Pit and Land Pit. The former [6678 7643], situated 450 yd E37°N of Mangotsfield Church, was some 230 ft deep to the Lower Seam, 22 to 24 in thick, the Upper Seam, 33 to 36 in thick, lying 8 ft higher. The measures dip about 22° to the north-east. At Wallsend Colliery, the deepest pit, known as Common Pit [6659 7675] 470 yds N2°E of Mangotsfield Church, was only 100 ft deep. According to an account by J. Sherborne (1908) the coal (? the Upper Seam) was 28 in thick, in three beds overlain by 4 to 7 ft of black dunns under Pennant Sandstone. The floor was said to be a soft fireclay 4 ft thick resting on black dunns. The figures given for the thickness of measures separating the two seams vary considerably, 8 to 15 ft being shown on the abandonment plans of the Mangotsfield collieries, whereas Prestwich (1871, p.54) and Anstie (1873, p.65) both gave 30 ft.

The Mangotsfield Seam appears on the west side of the Fault near Stockwell House [6602 7724] and the crop can be traced thence in a north-north-west direction to the eastern side of the Whitefaced Fault where the Coal Measures are concealed by Mercia Mudstone west of Bough Farm [6517 7835]. The abundance of spoil along the depression suggests that considerable outcrop coaling has taken place.

The massive sandstones of the Pennant have, in the past, been much quarried for building stone. A quarry [641 772] 340 yd south-east of Frenchay church shows 40 ft of massive current-bedded sandstone with irregular thin shale bands dipping at 10° in an east-north-easterly direction. Other exposures in similar gently dipping sandstones appear at intervals along the east side of the River Frome between the last-mentioned quarry and Hambrook. Quarries on the west side of Bury Hill show 40 ft of current-bedded sandstone dipping at 15–20° to the east-north-east.

The broad curving crop of the main sandstone division of the Mangotsfield Formation extends from Pye Corner to Iron Acton and consists chiefly of massive sandstone with impersistent bands of grey sandy shale. North of Bury Hill on the south side of the Frome river are several quarries in massive sandstone with an easterly dip of 8°–12°. In one of these quarries [651 794] 30 ft of massive sandstone is overlain by 20 ft of sandy shale which underlies a further 60 ft of massive sandstone. For a distance of some 750 yd west of the railway station at Winterbourne, the cutting shows massive current-bedded sandstone dipping at 8°–12° to the east. Details of the lower part of the Pennant succession, here largely concealed by Triassic rocks, are shown by the Winterbourne Borehole [6460 8022] 1100 yds S27°W of Winterbourne church (Cantrill and Smith, 1919). A quarry [649 815] 1040 yd north-east of Winterbourne church showed up to 40 ft of current-bedded sandstone with vertical joints and fissures filled with thin bands of limonitic iron ore.

A coal horizon at the position of the Made-for-Ever Seam has been proved (by augering) to extend from the railway line north of Winterbourne Down in a south-south-easterly direction past Kendleshire Farm [663 795] to the Buryhill Fault north-east of Wick Wick Farm. A coal bearing this name is shown on a horizontal section drawn through the Frampton Cotterell area by Hill and Fairley (1874, pl.i, p.704) where it was reputed to be 30 in thick; there is no record however, of any such seam having been worked in this area. Traces of coal at a level corresponding to that of the Made-for-Ever Seam at Frampton Cotterell have also been found south of Kendleshire towards Emmerson's Green.

At Winterbourne the Made-for-Ever Seam is situated about 140 ft below the base of the Frampton Cotterell Sandstone. This massive rock is about 245 ft thick at Frampton Cotterell but thickens northwards at the expense of the shales in the upper division of the Mangotsfield Formation. At Iron Acton where the total thickness of the Upper division is reduced to 500 ft the Frampton Cotterell Sandstone attains a thickness of 360 ft.

The upper shale division of the Mangotsfield Formation extends north-westwards from Emmerson's Green towards Frampton Cotterell. It consists of shale with subordinate sandstone bands. About 1000 ft above the base of the Mangotsfield Formation the massive Frampton Cotterell Sandstone forms a low feature stretching from a point east of Emmerson's Green to Folly Bridge [6689 7830]. North-west of Folly Bridge the whole of the upper shale division is concealed by Trias marl which extends northwards to the east–west-trending Buryhill Fault north of Wick Wick Farm. About 200 ft below the Frampton Cotterell Sandstone there are traces of a small coal seam which was said to be 2 ft thick where it was proved in a small trial pit 590 yds north-east of Blackhorse. This is thought to be the Made-for-Ever Seam.

South of Nibley on the east side of the basin the upper argillaceous division is about 550 ft thick, including 300 ft of Frampton Cotterell Sandstone. Near the base of the latter a shale parting gradually develops southwards and increases in thickness towards Westerleigh where it unites with the lower shale band of the group. A short distance north of Rodford two old quarries show massive Frampton Cotterell Sandstone dipping at 26° to the west-south-west.

Clutton and Temple Cloud

Rocks of Pennant lithologies occupy a triangular area in the centre of the Pensford–Radstock basin south of Clutton. They are bounded on the west by the Clutton Fault and on the south by the Timsbury Fault. The lowest known coal crops out 150 to 200 yd west of the main Bristol road in Temple Cloud village, and was worked from an adit [6200 5773] situated immediately north of the road to Cameley, 270 yd from its junction with the Bristol road. This **Temple Cloud Vein**, according to H. E. Hippisley's manuscript notes was 4 ft thick in three leaves. It is considered to represent the Mangotsfield Seams of the Coalpit Heath Basin and in consequence is taken to mark the base of the Mangotsfield Formation in this area.

Thin seams occur above Temple Cloud Vein but they have not been worked to any great extent. Some 200 ft above the Temple Cloud a small seam was struck at a depth of 419 ft in a shaft at Temple Cloud Brickworks [6211 5815], and according to Hippisley this seam cropped in the foundations of the Divisional Court nearby [6221 5803]. The coal, which was said to be soft and was used only for engine coal in the brickyard, had the following section: coal 16 in, binching 6 in, coal 6 in, pan. A little over 600 ft above this Brickyard Seam a thin coal occurs in the most easterly sandstone quarry of Cloud Hill [6313 5782]. This coal lies about 440 ft below the **Rudge Vein**, the uppermost 240 ft of measures being mainly mudstones. Wherever the measures immediately below the Rudge Vein have been seen in the Radstock Basin they include red or mottled red and grey mudstones, as they do in the Coalpit Heath Basin (see above).

PENNANT MEASURES, UNDIFFERENTIATED

To the west of the main Bristol and Somerset Coalfield the isolated Coal Measures basins of Nailsea, Clapton in the Gordano Valley and on the margins of the Severn estuary north of Avonmouth all contain thick successions of undoubted Pennant Measures lithologies. Knowledge of the full sequences is invariably incomplete and correlation between the individual basins and in particular into the main coalfields to the east still remains to be established and in consequence classification of the measures into constituent formations remains uncertain. The Lower and Middle Coal Measures are only in part developed and there is widespread overstep of successively higher Upper Coal Measures beds on to Lower Carboniferous and older rocks. Over much of the area the lower part of the Downend Formation may be lacking and although the Mangotsfield Formation must be present in the Clapton and Severn basins, the absence of any firm correlatives of the Mangotsfield Seams themselves makes definition of that formation very doubtful. In the following account of the three western basins, therefore, the Pennant Measures are treated as a single individual group.

Nailsea Basin

The Winterbourne Marine Band has not been proved in the Nailsea Basin and so the true base of the Pennant Measures and of the Downend Formation cannot be defined. Pennant Sandstone lithologies set in at a stratigraphically low level in the western reaches of the coalfield not far above the White's Top Seam (p.95), and it may well be that the Middle Coal Measures extends some way up into the Pennant sandstones. For practical reasons the White's Top was taken as the base of the previously defined Pennant Series. According to the Geological Survey's Vertical Section No. 49 (published 1873), the White's Top is overlain by some 550 ft of measures, mainly Pennant sandstones, in which one coal, **Grace's Seam**, has been worked to a limited extent. It is shown lying 450 ft above the White's Top, but this figure appears to be largely conjectural. The seam is 36 in thick and was worked north and north-west of Nailsea Church, the last pit to close being West End (Grace's) Colliery [4580 7000]. Here the seam lies at a depth of 330 ft below surface and dips 1 in 5 south-south-westwards. Anstie recorded (1873, pp.53–54) that the coal was rather sulphury, and that water in the sandstones made working difficult.

On the thrust-faulted Carboniferous Limestone ridge at Court Hill, which lies between the Nailsea and Clapton basins, small masses of Coal Measures sandstones and shales rest in solution cavities and pockets on the surface of the Black Rock Dolomite. Other masses were observed in the cuttings made during the construction of the M5 Motorway. The basal sandstones are mainly hard grey and red-stained quartzose rocks and these are overlain by current-bedded Pennant-type sandstones interbedded with sandy mudstones. Owing to the intense structural disturbance it is not possible to make out any consecutive succession.

About 1 mile north-east of Tickenham Church Pennant sandstones are visible immediately north of Hale's Farm [473 722]. The crop is about 700 yd long and 100 yd wide and is bounded on the south side by a thrust fault which brings up the Clifton Down Mudstone. At the eastern end of the outcrop the Pennant sandstones appear to rest upon Clifton Down Limestone. Exposures are few, but current-bedded sandstone was seen dipping at 40° to the south at a point 200 yd north-east of Hale's Farm.

Clapton Basin

The largest area of exposed Coal Measures is that seen around Clapton-in-Gordano, a triangular area of Pennant-type sandstones with subordinate mudstone and shale bands containing at least one coal. Most of the southern boundary is formed by the fault which thrusts Black Rock Limestone and Lower Limestone Shale over the Pennant sandstones, but at its eastern end the thrust dies out and the Pennant sandstones rest unconformably on Lower Limestone Shale east of Naish House [477 733]. Along their eastern margin about ç mile east of Clapton village, coarse pebbly sandstones and conglomerates with quartz pebbles rest unconformably upon Portishead Beds of the Upper Old Red Sandstone. Immediately north-west of Clapton the Pennant rocks pass beneath Triassic Dolomitic Conglomerate and Mercia Mudstone. In the east the conglomeratic sandstones show westerly or north-westerly dips varying from 18° to 40°, whilst in the centre they are to the south-west and of small amount suggesting the presence of a plunging anticline.

The whole of the Clapton Basin sequence can be referred to the Pennant Measures with some certainty, though their exact stratigraphical position is not known. Fossil plants recorded by Moore and Trueman (1937) suggest an upper Westphalian age though without any precision.

In a boring sunk for water near Sperrings Farm, Clapton-in-Gordano (p.167) [4748 7435] over 600 ft of Pennant-type sandstones with slumped beds and numerous bands of conglomerate were recorded. These rocks appear to underlie shales within which coal was formerly worked extensively between Brooks Farm [4735 7415] and Cockheap Wood [4680 7320]. West of this belt of workings a boring was recorded by Anstie (1873, p.55) [?c.4592 7408] about 450 yd north-west of the Inn at Clapton. This boring proved Pennant sandstones to a depth of 510 ft with a band of shale between 156 ft and 222 ft in which there was a 10 in coal band. It is not known whether this is the coal worked at Clapton, but none of the sandstones seen in the boring appears to have been conglomeratic.

From the Park at East Clevedon [415 719] a narrow outcrop of Pennant Sandstone extends in an east-north-easterly direction to Clapton Wick [455 726]. This outcrop is limited on its south side by the reversed fault which thrusts Carboniferous Limestone against the Pennant, continuity of the outcrop being broken by the East Clevedon gap floored by Quaternary sediments. The visible junction between the Pennant Sandstone and older rocks is usually a faulted one, but at The Park Pennant Sandstone appears to rest unconformably upon the lower beds of the Clifton Down Limestone and dips at 30° in a south-easterly direction. The best exposure of the formation is the large quarry (Conygar Quarry) [421 722] 600 yds north-north-west of Clevedon Court, where some 50 ft of massive sandstones are overlain by about 20 ft of sandy shale containing a thin seam of dirty coal, the rocks dipping 20° to the south-south-east. Higher beds occupy the steep northern slope of Court Hill but are rarely exposed.

West of Clapton all the sandstones appear to be even grained and no massive conglomerates have been observed. This suggests that the coarse conglomeratic and pebbly rocks which occur on the eastern margins of the basin have been overstepped or overlapped. It may well be that in the east beds of Downend Formation age rest on strata ranging from Old Red Sandstone to Lower Carboniferous and that west of Clapton beds extending high into the Mangotsfield Formation come to rest directly on Lower Carboniferous limestones. In pursuit of this model of Clapton Basin stratigraphy it may well be that the coal worked at Clapton (see above) represents the Mangotsfield Seams horizon, but of this there is no supporting palaeontological evidence.

Separated from the main Clapton Pennant Measures outcrop by about two miles of Quaternary, Triassic, Lower Carboniferous and Old Red Sandstone rocks, a small area of Pennant sandstones appears from beneath the Dolomitic Conglomerate on the foreshore west of Portishead Pier; farther west other very small outcrops are visible. The beds appear in the crest of an anticlinal fold the axis of which trends in an east-south-easterly direction. The sandstone mass is bounded on its south side by Black Rock Dolomite which forms most of the cliff and hillside inland. Whilst faulting and crushing has undoubtedly taken place along this junction the magnitude of the fault is probably much smaller than the juxtaposition of Carboniferous Limestone and Pennant Measures would suggest. Sections in the cooling tunnels driven seawards from the Portishead Power Station show that several seams of coal were intersected. The ground is very disturbed and it is probable that not more than two seams are represented, a lower one 3 ft thick separated by some 22 ft of shale from an upper coal 2 ft thick. These seams are possibly on the horizon of the Mangotsfield Seams.

Severn Basin

Upper Coal Measures form the greater part of the Severn Coal Basin, though Middle Coal Measures are known to exist at the eastern margin at Cattybrook (p.95). The northern part of the Severn Coal Basin has been described by Welch and Trotter (1961, pp.105–107). The southern limit extends from Henbury, by way of Penpole Point [530 776] to the foreshore at Portishead (Kellaway 1970, fig. 5). Between Portishead and Portskewett the western boundary of the basin lies beneath the Severn estuary, though its position is difficult to determine, since the area is traversed by the Denny Island Fault Belt. East of this structural line, the deepest part of the basin lies between Pilning and Avonmouth. Here the Farrington Formation is preserved along the axis of the basin and borings have proved two coal seams of workable thickness, Avonmouth No. 1 and Avonmouth No. 2, the latter being taken as the equivalent of the High Vein of Coalpit Heath and thus marking the base of the Farrington Formation.

The full extent of the Middle Coal Measures is uncertain but they are absent at Portishead and Kings Weston (Figures 16 and 18), and were not proved in the turbo-drill borehole at Severnside (p.166). The western limit of the Middle Coal Measures would therefore seem to lie along a line drawn from Olveston to Henbury. This is a line of major tectonic disturbance and it is probable that there has been some lateral shortening resulting from folding and thrusting. These features are illustrated in a generalised form on the horizontal section of the revised 1:50 000 edition of the Chepstow (250) Sheet.

Boreholes made in the Portskewett area (p.166) suggest that beneath a cover of Triassic rocks, Upper Coal Measures rest unconformably on Millstone Grit shales and sandstones. This evidence has clarified the interpretation of the Severn Tunnel sections (Figure 19). The Portskewett boreholes, Severn tunnel sections, and sections seen in tunnels at Kings Weston and Portishead, now supported by the results of the Severnside turbo-drill hole show conclusively that a major unconformity is present beneath the Upper Coal Measures. This develops over and to the west of the 'Lower Severn Axis' (Kellaway and Welch, 1948, pp.8–10). Along the axial region, in the area extending from Severnside towards Avonmouth Docks the Upper Westphalian rocks rest unconformably on Tournaisian or older formations. The Severn Tunnel section shows that Viséan rocks, themselves overlain unconformably by Namurian strata, are preserved beneath a major unconformity farther west (Figure 19).

The question therefore arises as to how much of the so-called Pennant Measures of the Severn Coal Basin should be attributed to the Downend Formation and how much to the Mangotsfield Formation. It has been suggested that there may be an unconformity beneath the Mynyddislwyn Seam of South Wales (Woodland and Evans, 1964, p.65). Since this seam may equate with the Avonmouth No. 2 Seam, the possibility that an unconformity exists at this horizon has to be considered. However, the measures underlying Avonmouth No. 2 Seam are in many ways similar to those of the Mangotsfield Formation of the Coalpit Heath Basin and if there is any such break no obvious evidence of its existence can be detected, either at Avonmouth or Coalpit Heath.

All these measures west of the Olveston–Henbury line were attributed by Kellaway (1970, fig. 4) to the Mangotsfield Formation, but this assumes that the thick coal proved in the Portskewett boreholes (p.166) correlates with one or both of the Mangotsfield Seams. The presence of *Anthraconauta phillipsii* above this coal in the Portskewett boreholes is consistent with this interpretation. However, it is possible that the Portskewett coals may represent seams in the Downend Formation in which *A. phillipsii* also occurs.

In the Severn Coal Basin the total thickness of Pennant Measures underlying Avonmouth No. 2 Seam can hardly be less than about 1400 ft (Welch and Trotter 1961, p.105), of which the lowest 500 ft was penetrated in the Severn Tunnel. Approximately 590 ft were proved at Severnside[1]. The sequence consists mainly of sandstones with subordinate sandy shales and a few thin coals. As in the main coalfield to the east the lower part is dominated by massive Pennant sandstones. In the Avonmouth (Chittening No. 2) Washingpool Farm Borehole [5347 8152] at least 337 ft of interbedded sandstones and mudstones with red measures, succeeded by grey mudstones with thin carbonaceous bands extend the sequence up to the Avonmouth No. 2 Seam. Near the top mudstones and shale associated with dirty coal carry a fauna of bivalves, ostracods and *Leaia*, similar to that above the No. 2 Seam itself. *Anthraconauta phillipsii, A. tenuis. Carbonita pungens, Hemicycloleaia boltoni, Leaia bristolensis* and *L. parallella*.

Supra-Pennant Measures

Farrington Formation

The Farrington Formation comprises the lowest group of essentially grey argillaceous strata with productive coals (Figure 20) overlying the Pennant Measures: it is separated from a higher group of grey measures also with productive coals by a development of red and

---

1 The use of the term 'Severnside' as applied to the celestine deposits has been discontinued.

mottled mudstones, seatearths and sandstones without coals — the Barren Red Formation. The Farrington Formation is best developed in the Radstock Basin of the Somerset Coalfield where a sequence of grey mudstones and sporadic subordinate sandstones, containing numerous coals, mostly thin but associated with comparatively thick fireclays, varies from about 1000 to 1400 ft in thickness. In the Pensford Basin, and to the north in the Bristol Coalfield, the presumed equivalents of the Farrington Formation of Radstock are much thinner, 200 ft or less, a feature which may be caused partly by thinning and partly by the oncoming of red measures lower in the succession.

The name 'Farrington Group' appears to have been initiated by Greenwell and McMurtrie (1864), being taken from the now abandoned Farrington Colliery. Apart from a small outcrop extending from Fry's Bottom Colliery [6300 6042] through Greyfield and extending intermittently from Hallatrow to Farrington Gurney, the formation is concealed south of the Farmborough Fault belt. North of the Fault belt the coals formerly worked at Bishop Sutton and at Bromley colliery near Stanton Drew fall within this formation, but exposed measures are confined to a very narrow belt crossing the Bromley Horst.

At the time of the primary six-inch survey about 1950 all the Somerset collieries, except Pensford Colliery, were producing coal from the Farrington Formation. These, like the coals of the Radstock Formation, are generally thinner than the seams of the Lower and Middle Coal Measures. None of the productive coal seams is free of faulting, but those of the Upper Coal Measures, and in particular the Farrington Formation, have not been subjected to such intense deformation by disharmonic folding as have the seams below the Pennant Measures. The lack of thickness of the coals of the Upper Coal Measures is therefore offset, to some extent, by shallower depth of burial and better structural conditions, and it is this which permitted thin-seam working to survive in the Radstock Basin long after mining of much thicker coals of the Lower and Middle Coal Measures had ceased.

Radstock Basin

The Radstock Basin comprises a synclinal area of Coal Measures lying south of the Farmborough Fault and mostly concealed beneath Mesozoic strata. The Farrington Formation, lying towards the central part of the basin between Radstock, Timsbury and Farrington Gurney, contains five worked coals only three of which exceed 24 in (Figure 29).

Moore and Trueman (1937, p.228) originally selected the base of the No. 9 Seam (No. 10 or Bright's Vein in the Kilmersdon–Farrington area) as the base of the Farrington Group (now Formation). This is some 400 ft higher than the position now advocated. Kellaway (1970) accepted the Rudge as the base of the Farrington Formation but indicated that where the Rudge is not developed, or cannot be identified, the lowest worked seam may be taken as a substitute.

The adoption of the position of the **Rudge Vein** for the boundary is supported by the evidence from the Old Mills Underground Borehole [6516 5516] which proved the strata from the Brights (No. 10) Vein to the Rudge Vein (Figure 29). Commenting on the fossil evidence from this boring, Dr M. A. Calver stated: 'The strata between the Brights and the Rudge veins are clearly referable to the *A. tenuis* Zone; the name-fossil is typically represented and *A. phillipsii* is also present. *Anthraconaia* aff. *pruvosti* at 279 ft is also consistent with a position in the *A. tenuis* Zone. The plants collected from the borehole were identified by Dr R. Crookall who considered that they showed general agreement with those from the Farrington Formation. The suggested lowering of the base of the Farrington Formation from the Brights Seam to the Rudge Vein, is therefore consistent with the floral evidence'.

The top of the Farrington Formation is more difficult to define. Where it is present in the south-eastern part of the Radstock Basin, the **Rock Vein** forms a convenient line, as suggested by Moore and Trueman (1937). Elsewhere, at locations where this vein cannot be recognised, the topmost seam, such as the Cathead or Stinking Vein of Old Mills, can be used. With such a poorly defined limit it is difficult to give precise thicknesses for the formation. In general a little over 1000 ft separates the Rudge from the Rock Vein (see Figure 29) and isopachs of the measures between Nos. 5 and 10 veins suggest a thickening of the beds in a south-westerly direction.

At Rudge (or Clutton Ham) Pit [6286 5858] where the Rudge Vein is typically developed, a characteristic section [6266 5856] shows shale roof 4 ft, coal 17 in, 'cockroach' holding 8 in, shells or blacks 30 in, hard pan. Below the seam was 'confused red gritty ground of dicey texture'. The coal was worked by means of a long drift-road extending almost to beneath the church, and to the south-east of the pit are traces of several small crop workings. At Grayfield Colliery [6400 5868] a small area of about 2 acres of Rudge Vein was worked in 1910 from a level branch 570 yd from the back of the pit. Here the section was: clift roof, coal 11 in, stone 3 in, coal 2 in, blacks 15 in, coal 3 in, pan.

In the Marsh Lane Colliery [6321 5526] at Farrington Gurney the Rudge Vein was worked with a section: coaly shale 5 in, coal 14 in, soft coal 1 in, shale 4 in, coal 4 in, fireclay. In the last face to be worked the section was: clift roof, hard shale 2 to 3 in, coal 16 in, binching 2 to 2½ in. Some 60 ft below the Rudge, 'red measures' were met in a cross-measures road and also in a borehole [6280 5526] to the west of the colliery at the foot of Rush Hill (Moore and Trueman, 1936). Anstie (1873, p.81) remarks that the Rudge Vein was said to have been wrought to the rise of Farrington Gurney Colliery, but no traces of any pit or workings were observed during the recent survey. The underground borehole [6516 5515] at Old Mills Colliery from the Brights to prove the Rudge Vein showed a section of only 14 in of coal with 'red measures' developed some 40 ft below. At Lower Writhlington Colliery [7051 5531] a back roadway proved a 13 in seam, with 'red measures' 60 ft below, which is tentatively correlated with the Rudge Vein.

The 400 ft of strata separating the Rudge from the No. 10 Vein consists of mudstone and sandstone, the latter being more strongly and thickly developed to the north-east of a line joining Writhlington and Greyfield, whereas to the south-west, at Old Mills and Farrington collieries, mudstone replaces much of this sandstone. Throughout the area there is a median group of fireclays, at which position thin coals are developed (Figure 29), one of which, in the Norton Hill–Old Mills district, was worked under the name of the No. 11 Vein. A section in this coal taken at a point one mile south-south-east of Norton Hill Shaft showed: shale roof, coal 10½ in, fireclay 24 in, coal 2 in, shale ½ in, coal 33 in, carbonaceous shale.

In the succeeding measures the principal coal seams of the Farrington Formation have been traditionally numbered in accordance with a classification set out at Ludlows Pit, Radstock [6914 5476], the highest numbers indicating the lowest seams. The National Coal Board adapted this system for its use after 1947 and the following account is based on their nomenclature.

The **No. 10 Vein** was probably the thickest and best quality of all the seams worked in the Supra-Pennant Measures. South-east of Kilmersdon and on the west side of the Clandown Fault, where it was known as the 'New Great Vein', it had the following section: massive blue shale roof, top coal 30 in, thin black shale parting, bottom coal 12 to 18 in. Traced westwards to Norton Hill Colliery [6690 5409] the seam, known there as the 'Big Vein', was worked as a single seam of two leaves separated by a thick dirt parting; in the north of the workings the section was: clift roof, stone 6 in, top coal 22 in, dirt 15 in, bottom coal 20 in; and in the south towards the Writhlington area: clift roof, clod 5 in, top coal 25 in, dirt 5 in, bottom coal 19 in. To the north-west, in Old Mills Colliery [6527 5516]

106 CHAPTER FIVE COAL MEASURES (WESTPHALIAN)

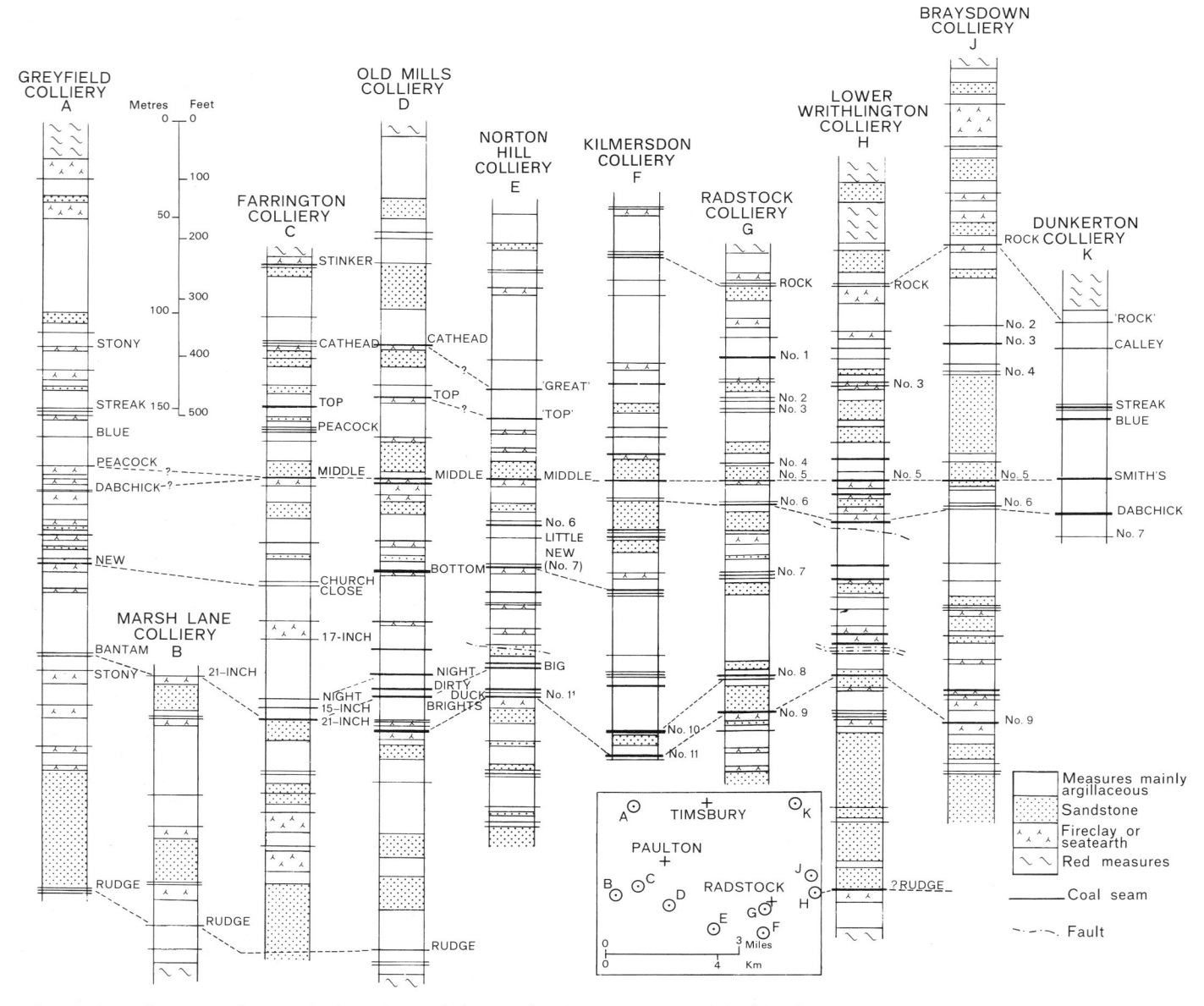

Figure 29 Comparative vertical sections of the productive measures of the Farrington Formation in the Radstock Basin

further splitting into three distinct and named seams takes place: Night Vein 18 in, measures 20 ft, Dirty Duck Vein 16 in, measures 4 ft 9 in, Bright's Vein 21 in; only the Bright's Vein was worked.

At Farrington and Marsh Lane collieries still farther west, the equivalent of the Bright's Vein, known as the 21 Inch Vein, was worked, as was the Night Vein at Farrington, its section being: clift roof, coal 11½ in, binching ½ in, coal 5½ in. At Greyfield Colliery, about 2 miles to the north, the 21 Inch Vein was known as the Bantam Vein, also a split seam: coal 12 in, dirt 14 in, coal 5 in.

Thus from virtually one seam of thick coal at Kilmersdon there is a gradual splitting to the west and north-west, but north and north-eastwards from Kilmersdon there appears to be an even more rapid splitting for it is considered from the evidence of the flora in the seam roofs (Moore, 1938, p. 274) that the Nos. 8 and 9 veins of Ludlows, Middle Pit [6874 5511] and Braysdown [7038 5600] collieries together represent the **Kilmersdon New Great Vein** (Figure 29).

The nearest point to the New Great Vein workings at which Nos. 8 and 9 veins are known to retain their identity as separate and worked seams is approximately half a mile away, in the south-eastern workings of Ludlows Pit. The section in the branch from No. 7 to No. 9 at Ludlows is:

	Thickness		Depth	
	ft	in	ft	in
No. 8 Vein: clift roof, dirt 7 in, coal 2 in dirt 2 in, coal 14 in, blacks 3 in	2	4	2	4
Pan	6	0	8	4
Inferior coal 27 in	2	3	10	7
Cockly clift	14	0	24	7
Coaly bind 4 in, coal 14 in, binching 2 in	1	8	26	3
Pan	7	0	33	3
Greys (sandstone)	36	0	69	3
No. 9 Vein: coal 22 in	1	10	71	1

The **No. 9 Vein** has been worked to the rise of Middle Pit and Ludlows and to the north of Braysdown. The seam usually consisted of 20 to 22 in of bright coal with a good sandstone roof. Locally however, grey shale lenticles with ironstone nodules were present between the coal and the sandstone. Moore (1938, p.271) records a varied flora from the shale intercalation at Braysdown including *Neuropteris* and *Sphenopteris*. In an exploration branch (Scott's Branch) at Lower Writhlington a seam, probably the No. 9, was proved at a distance of about 330 ft below the No. 5 Vein (see Figure 38); it had a sandstone roof, binching 4 in, coal 18 in, pan floor.

The **No. 8 Vein** was worked for a short time (1930–31) at Ludlows Colliery, where Moore (1938, p.272) gave the following section at the working face: roof hard blue shale, coal (inferior) 3 to 4 in, main coal 18 in, floor soft. The seam appears to have had an even poorer section at Braysdown Colliery where Moore (1938) records: roof hard blue shale, coal 4 in, black shale and thin coal 8 in, thin bedded dark shale 6 ft, coal 12 to 18 in, floor soft. The section recorded at the same colliery in Mitchard's Branch was: clift roof, coal 4 in, shale 2½ in, coal 7 in, shale and pan 7 ft, coal 12 in, pan 6 ft 6 in.

The **No. 7 Vein** was known as the **New Vein** at Norton Hill Colliery, the Bottom Vein at Old Mills, the Church Close Vein at Farrington and Greyfield and the 'No. 8' at Writhlington. It is separated from No. 8 Vein by about 200 ft of measures in which numerous 'coal horizons' occur but in which there are no seams of workable thickness or quality (Figure 29). Thus in the branch dipple off the West Level branch below No. 7 Vein at Norton Hill the following seams were cut: hard stony coal 24 in at 42 ft 6 in below No. 7, and coal, coal band and rubbish 36 in some 30 ft lower. At Farrington Colliery a seam called the 17 Inch Vein has been recorded in this position (Prestwich, 1871, p.59).

No. 7 vein is a fairly thick seam of brittle coal with dirt partings: it has its thickest development at Old Mills and Norton Hill, but it deteriorates to the south-east and becomes unworkable at Braysdown, Ludlows and Kilmersdon. The following sections have been recorded; at Norton Hill Colliery—clift roof, shale 0 to 3 in, coal 14 to 17 in, dirt 3 to 14 in, coal 14 to 17 in, shaly coal 0 to 3 in, on pan; Old Mills Colliery—clift roof, coal 8 in, binching 2 to 3 in, coal 20 in, shale 3 in, on pan; Farrington Colliery—clift roof, coal 6 in, binching 6 in, coal 14 in; Old Grove Colliery [6587 5842]—coal 9 in, stone 4 in, coal 15 in; Greyfield Colliery—coal 7 in, stone 5 in, coal 27 in; Kilmersdon Colliery—coal 17½ in, batt 9½ in, coal 9½ in, batt 10 in, on pan; Ludlows Colliery—planty shale roof, coal 9 in, shale 12 in, coal 12 in, shale 24 in, coal 8 to 24 in.

Some 120 to 160 ft of predominantly argillaceous strata separate the No. 7 Vein from the No. 6. A 14 in coal, called the Little Vein, is recorded 20 ft below the No. 6 Vein in Norton Hill shaft, but this seems to be local development. The **No. 6 Vein** is a rather poor coal with dirt partings. Sections are recorded as follows: at Norton Hill Colliery—clift roof, coal 13 in, dirt 11 in, coal 15 in; Welton Hill Colliery [6662 5524]—clift roof, coal 4 in, dirt 4 in, coal 7 in, coaly dirt, coal 17 in; Braysdown Colliery—clod roof, coal 6 in, shaly coal 8 in, binching 2 in, coal 14 in, on pan; Ludlows Colliery—coal 10½ in, stone 4 in, coal 15 in, hard floor.

About 30 ft of clift separates the No. 6 Vein from the **No. 5 Vein** (Figure 30), which was known as the **Middle Vein** at Norton Hill and Farrington collieries, the Deep Middle Vein at Kilmersdon Colliery, and Smith's Coal at Dunkerton Colliery [6985 5859] (Figures 29 and 30). At Greyfield Colliery the seam splits into two distinct coals, known as the **Dabchick** below and the **Peacock** above separated by about 39 ft of clift and fireclay. In thickness the seam ranges from about 18 in at Dunkerton in the north-east to 28 in at Farrington in the south-west, where a median band of stone appears, marking the onset of the split towards Greyfield. The seam usually had a hard roof formed of sandstone—the Middle Vein Greys—though locally a planty shale developed between the coal and the sandstone.

The Dabchick Vein at Greyfield Colliery comprised: black shale roof 4½ in, coal 8 in, coal and shale 4 in, coal 14 in, parting 1 in, coal 2 in, on pan; at Clutton (Rudge) Colliery—coal 8 in, dirt 4 in, coal 17 in, dirt 3 in, coal 4 in, pan. The Peacock Vein at Greyfield had a section: shale roof, coal and shale 3 in, coal 18 in, shale ½ in, coal 3½ in; and at Clutton—coal 20 in, dirt 1 in, coal 6 in.

The coals above the No. 5 Vein and its equivalents show great variation of both thickness and distribution over the Radstock Basin (see Figure 29); none of them can be recognised continuously over any great distance, and few have been worked to any extent. At Ludlows Colliery a complete succession was formerly visible from the No. 5 Vein to the Rock Vein, whilst at Braysdown Colliery a cross-measures drivage through the same measures was known as the 'Badger Gug' and recorded by Moore (1938, pp.280–281). It is not possible to correlate any of the seams here exposed with those of Old Mills, Farrington, and Greyfield or with the Dunkerton succession. The following account records what is known of individual seams in this part of the sequence, even though their full relationships remain obscure; their positions, where known, are shown in Figure 29.

The **Top Vein**, 18 in thick, was formerly worked in the south-west of the area at Farrington, Old Mills and Norton Hill collieries. It may be represented by a 10 in seam at Kilmersdon, but it cannot be recognised at Ludlows, though it may possibly equate with the so-called No. 4 Vein.

The Cathead Vein of Farrington and Old Mills collieries may possibly equate with the Great Vein at Norton Hill Colliery and the Stony Vein at Greyfield. It was an unworkable seam developed over the same area as the Top Vein, its name deriving from the occurrence of large stones or 'catheads' in the seam.

The Streak Vein was a three-leaved seam formerly worked at Clutton, Greyfield and Old Grove collieries as well as at Dunkerton, where it was known as No. 3 Vein. Farther south the seam cannot be recognised, though Prestwich (1871) considered it might be the equivalent of Top Vein of Farrington and Old Mills; from its section it might equate with the Peacock at Farrington. A seam of somewhat similar section (No. 4 Vein) was proved at Camerton but never developed.

At Old Welton Colliery [6754 5486] two coals, termed the No. 2 and No. 3 veins, occur close together about 150 ft above the No. 5 Vein. Their sections were as follows: No. 3 Vein, coal 18 in, clod 8 in, binching 6 in, blacks 32 in, blacks and clod 8 ft; No. 2 Vein, coal 18 in, blacks 19 in, coal 14 in, on pan 12 in.

The **Rock Vein** is the highest worked seam in the Farrington Formation. It is best developed in the south-east of the area and has been extensively worked from Lower Writhlington Colliery and to a less extent from Braysdown (see Figure 38). The synonym **Badger Vein** in the Radstock area is derived from the thick dark shale with coaly streaks which overlies the coal. A typical section shows: clift roof, 'badger' 16 in, thin black mudstone, coal 18 in, on pan. The mudstone roof contains nonmarine bivalves, notably *Anthraconauta tenuis, A. phillipsii* and *Anthraconaia pringlei* (Bolton, 1911, p.322; Dix and Trueman 1929). Sandy shale replaces the shelly mudstone westwards and in the same direction the coal thins and the 'badger' thickens. The Rock Vein was recorded as coal 3 in, shale 14 in, coal 17 in at 1306 ft 3 in in the Ludlows shaft and as coal 5 in, blacks 21 in, coal 14 in at 1377 ft 2 in in the Radstock Middle Pit. The seam was not recognised with any certainty at Norton Hill Colliery.

In the northern part of the area a small area of coal called 'Badger Vein' was worked some 370 yd south-south-east of Dunkerton shaft, and at Camerton a coal correlated with the Rock Vein was cut in the shaft at a depth of 1457 ft. A section showed strong dark clift with coal threads 15 in, coal 18 in, pan. Moore states (p.289) that a little coal from this seam was raised at the colliery.

At Greyfield Colliery the topmost seam to be worked was marked on the abandonment plan as the Rock Vein (on old sections it was called the Tommy Collier's Vein), but it is not known whether it

108 CHAPTER FIVE COAL MEASURES (WESTPHALIAN)

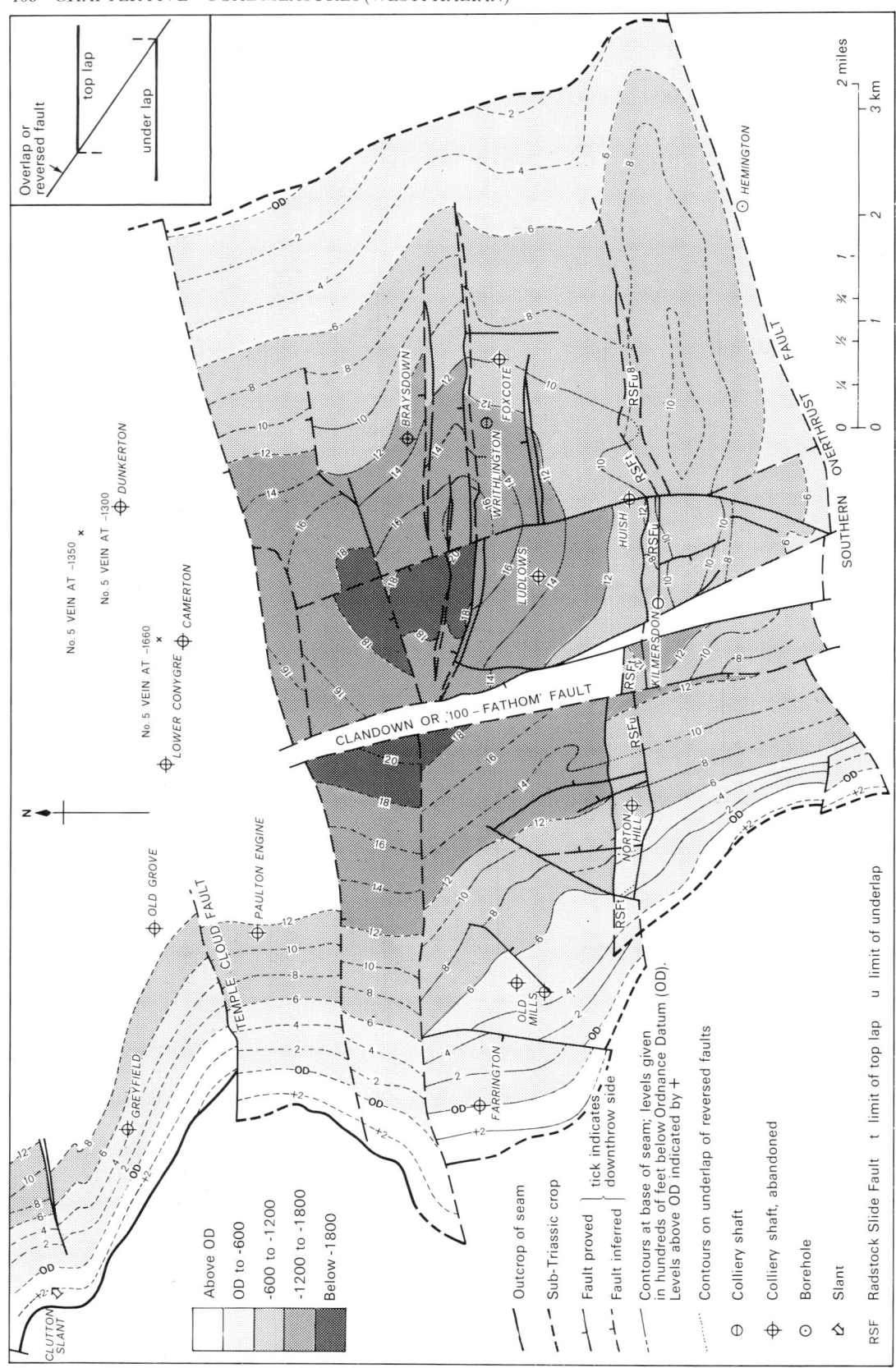

Figure 30 Structure contours on the No. 5 (Middle) Vein of the Farrington Formation south of the Farmborough Fault. The form of the Radstock Basin is shown. The Radstock Slide and other thrusts are displaced or terminated by the pre-Triassic Clandown or 100 Fathom Fault

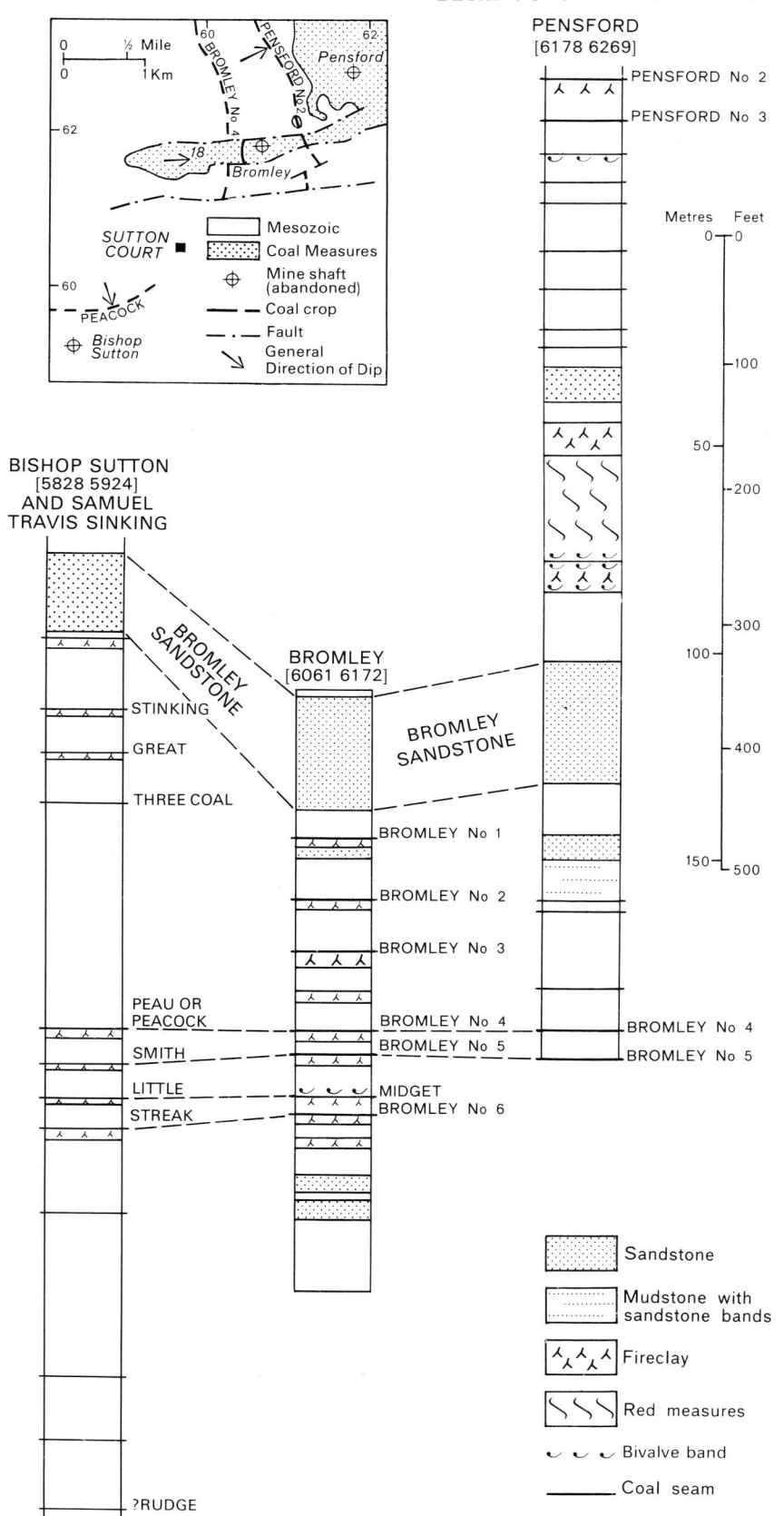

Figure 31 Comparative vertical sections showing the probable correlation of the coals formerly worked at Bishop Sutton, Bromley and Pensford collieries. The thickness of the measures between the Three Coal and Peacock veins at Bishop Sutton is probably due to repetition by thrusting

was the same seam as that worked under the same name in the Radstock area. At Greyfield the section was given as 17 to 21 in of coal overlain by 5 to 12 in of clod and shale: at Clutton nearby it was recorded as: clift roof, 'badger' shale 5 in, coal 21 in, dirt 1½ in, coal 1½ in, on pan.

Among the features shown by the Farrington Formation, the attenuation, splitting and deterioration of the coal seams are of considerable interest. Several of the seams, notably No. 10 Vein, show the development of intercalations of shale or sandstone in the coal, splitting the individual seams each into several thin coals which then become attenuated. This process tends to take place in a northerly direction. At the same time, deterioration of coals which pass into carbonaceous fireclay and sandy mudstone takes place in other seams (notably No. 6) near the south-eastern margin of the basin.

Pensford Basin

The Pensford Basin is separated from the Radstock Basin by the Farmborough Fault Belt and it passes northwards without break into the Kingswood Anticline where the beds underlying the Supra-Pennant Measures crop out. Both the Pennant Measures and lower Coal Measures strata are imperfectly known within the main basin, but they are presumed to be present everywhere both at depth and subcropping beneath Mesozoic cover around the edges.

As in the Radstock Basin two groups of strata with productive coals are separated by about 350 ft of barren measures including red and mottled beds. These are known as the Bromley seams below and the Pensford seams above (Figure 31), and they appear to correspond broadly with the Farrington and Radstock formations of the Radstock Basin though exact correlation of the seams is not possible. There is, therefore, no clear definition of the upper and lower limits of the Farrington Formation.

Most mining engineers since Anstie (1873) have taken the view that the Bromley seams belong to the Farrington Formation and the Pensford seams to the Radstock and this correlation is accepted here. Steart (1911) was of the opinion that both groups of seams were stratigraphically higher than those of the Radstock Formation, whereas Moore and Trueman (1937), following Prestwich (1871), expressed the view that the Pensford seams equate with the Farrington coals and the Bromley seams with the Pennant Measures. Their conclusion rested heavily on the discovery of *Anthraconaia prolifera* in plant-bearing shale in the Radstock Formation. This and other palaeontological evidence led Moore and Trueman (1937) to include the Radstock Formation with the so-called *prolifera* Zone and to regard it as Stephanian. This solution has always raised difficulties. The presence of the Pensford No. 2 seam was established in the Hursley Hill Borehole [6180 6565] (p.172) and showed that the flora of the associated measures was not younger than Westphalian D (Kellaway, 1970, p.1047). The flora of the Radstock Formation is certainly not younger than this and most likely is of *tenuis* Zone age. Abolition of the *prolifera* Zone has therefore been proposed (Calver, 1969; Ramsbottom et al., 1978, p.4).

The only surface exposures of the Bromley seams lie within the narrow band of the Bromley Horst, some 300 yd wide, west of Bromley Colliery [6061 6173]. From this area northwards to the Avon valley, the Farrington Formation is concealed by Mesozoic rocks save for a heavily built-up area at Brislington.

It is thought that the crop of a small seam about 930 yd west of Bromley shafts may be the equivalent of the Rudge Vein and thus mark the base of the Farrington Formation. Seven coal seams are developed in the Bromley group. They are numbered 1 to 6 in downward sequence with a named thin coal — the Midget — overlying No. 6, though only three, Nos 4, 5 and 6, have been worked from Bromley Colliery (Figure 31).

The **Bromley No. 6 Seam** is a thin seam with rather high ash and sulphur content. A typical section is: top coal 7 in, shale and coal 15 in. About 13 ft higher in the sequence lies the Midget Seam. A section measured in the branch from No. 5 to No. 6 seams was: clift and mudstone 6 in, clean coal 10 in, coaly shale 4 in, hard coal 3½ in, on pan and ganisterised mudstone 14 in. Although too thin to be worked the seam proved to be a useful marker, since the roof mudstone carried nonmarine bivalves, including *Anthraconauta* aff. *phillipsii* and *A.* aff. *tenuis*.

Some 31 ft above the No. 5 Seam, a coal of good quality but with a bad mudstone roof containing abundant plants, showed the following section: hard clift overlying a soft shaly clift 6 to 15 in, coal 24 in, hard floor. **The No. 4 Seam** was the most extensively worked seam in the colliery. It is separated from No. 5 Seam by 10 ft of sandy mudstone, and the immediate roof proved to be richly plant-bearing. The coal was 20 to 24 in thick.

Little is known of the Nos. 3, 2 and 1 seams which were all unworkable, nor were they identified at Pensford Colliery, where it appears they may have been cut out by sandstone. The Bromley shaft-section gave the following thicknesses: No. 3 Seam, 21 in; No. 2 Seam, 27 in; No. 1 Seam, 24 in.

The workable Bromley seams maintain their quality and thickness within the colliery area and are less subject to the formation of squeezed and distorted masses of coal than are the Pensford seams above. When traced northwards through the Bromley workings an attenuation of the measures occurs between the seams, and it is possible that still farther north accretion of some of the seams and attenuation of others may take place.

The top of the Bromley Group is placed at the base of the massive Bromley Sandstone, heavily waterlogged in the Pensford area, which lies some 150 to 190 ft above the No. 4 Seam in Bromley shaft.

To the south-west of Bromley Colliery, and situated almost within the Farmborough Fault Belt, is the old mining district of Bishop Sutton. The last working pit closed in 1929, before the primary six-inch survey took place, and little is known about the seams or their correlation. The measures generally appear to be considerably faulted. The three main seams closely resemble the Bromley Nos. 4, 5 and 6 and the Bishop Sutton group of seams is therefore tentatively correlated with the Bromley group. Steart (1911) considered the Bishop Sutton seams to be the equivalents of the coals at New Rock Colliery (now grouped as basal Pennant but formerly 'New Rock Group' at the top of the Lower Coal Series), but this view is now generally discounted.

The earliest account of mining here was given by John Strachey (1719, pp.969–970) in which seven seams are mentioned separated from one another by 35 to 40 ft of measures (Figures 3 and 4). The same general succession is reported by Buckland and Conybeare (1824) and in 1871 by the Royal Commission. Later sections (Hippisley MS and Abandonment Plan No. 9945) show considerable differences from the earlier accounts, suggesting that an extra 140 ft of measures without any workable coals have been introduced by thrust faulting between the Three Coal Vein and the Peacock Vein. H.E. Hippisley records a section entitled 'Mr Samuel Travis sinking 305 ft 6 in below the Little Vein' at the bottom of which is shown a 17 in seam of coal. This may be the Rudge Vein. The section as given in the Hippesley MS is shown in Figure 31. The principal seams are as follows.

The Streak Vein, about 10 in thick, was a highly bituminous coal. The Little Vein, 23 to 30 in thick, was worked over a small area during the last years of the colliery. According to the Royal Commission Report (Prestwich, 1871, p.60) the seam section was: coal 18 in, binching 16 to 18 in, coal 5 in. The Smith Vein, 26 to 36 in, was but little worked. The Peacock or Peau Vein, 20 to 24 in, was worked on a small scale in 1929. Strachey (1719, p.970) stated that the clift over the vein was 'variegated with cockle-shells and fern-branches' from which the seam could always be recognised. This description suggests that the Peau Vein is Bromley No. 4 which has a roof yielding plants and *Anthraconauta*. A section of the coal was given as: coal 18 in, pan 18 in, coal 6 in. According to

Buckland and Conybeare (1824) some 60 yd of clift separated the Peacock Vein from the Three Coal Vein, a seam but little worked: they gave an overall thickness of 54 in of coal in three leaves with stone partings. The Cathead or Great Vein although tender and irregular was worked to a considerable extent; its average thickness was 30 in. The Stinking Vein, a hard and sulphury coal, about 24 in thick with shale partings, was never worked. Some 63 ft above the Stinking Vein a mass of sandy shale and sandstone, 69½ ft thick, may be the equivalent of the Bromley Sandstone.

In the northern part of the Pensford Basin, workings in exposed Coal Measures above the Mangotsfield Formation around Brislington have long been abandoned. The sequence is imperfectly known, and there is difficulty in correlating the seams worked with those at Bromley or Bishop Sutton. The existence of a coal, the Rock Seam, near the top of a group of massive Pennant sandstones suggests an analogy with the seam identified with the Rudge at Bromley (see above) and so the Brislington seams may be regarded as representatives of the Bromley group, a view first expressed by Anstie (1873).

The Brislington workings which date from the 16th century or earlier were first described[1] by Strachey (1719, p.972), who gave the following (downwards) succession: Uppermost, Top or Trolley Vein, 36 to 60 in, measures 36 ft, Pot Vein 18 in, measures 42 ft, Trench Vein 24 to 30 in, measures, 22 ft, sandstone 20 ft, Rock Vein. The Trolley Vein is thought to crop out in Water Lane, Brislington [6160 7038], where 42 to 48 in of coal was seen in 1948.

Strachey states that the seams have been worked from Brislington through Queen Charlton to Burnett, but become thinner when traced in this south-easterly direction. This deterioration affects both Mangotsfield and Bromley seams as well as the Pensford coals. The line of deterioration is thought to run in a southerly direction, from Scotland Bottom a little east of Brislington, through Queen Charlton to about 1½ miles east of Pensford Village. From here it continues with diminishing effect into the Radstock Basin. This line of deterioration is believed to be associated with an ancient north – south line of basinal subsidence with 'drowning' and interbedding of mud and coal peat in Coal Measures times.

The investigation of the Farmborough Fault (1719–1951)

John Strachey's records of the coal seams worked between Brislington and Farrington Gurney encapsulate knowledge of the 17th and early 18th century miners. He exhibits close familiarity with the coals of the Farrington Formation while making little reference to the coals of the overlying Radstock Formation which, presumably had yet to be developed on appreciable scale. In his sections Strachey used generalised profiles and exaggerated vertical scales, making it difficult to fit these to the geology and physical relief (Fuller, 1969). Strachey's earliest section (Figure 3) is said to be four miles in length, drawn from Farrington Gurney to Stowey, but his third version (1727) shows Farrington replaced by High Littleton.

By correlating Bishop Sutton and Farrington seams and assuming a regular dip and strike Strachey estimated the vertical displacement due to faulting of the Farrington seams between Bishop Sutton and Farrington as being of the order of 1⅓ mile. This figure (Figure 3) makes no allowance for folding or change of strike and is excessive. Nevertheless he understood that there was a major structure south of Sutton Court, much longer than the local 'ridgs' (or faults) which form the northern and southern limits of the ancient Farrington seam workings of Stanton Drew. This huge disturbance and the waterbearing sandstones which are a feature of this area obliged the last working colliery at Bishop Sutton to close in 1929 (Williams, 1976). No single colliery has ever succeeded in working through the Farmborough Fault either from the north or the south and the difficulties which were encountered by the miners had become legendary by the middle of the 19th century.

The belt of disturbed ground is about ½ mile in width at Fry's Bottom between Clutton and Chelwood and it may become wider when followed eastwards beneath the Mesozoic cover at Farmborough and Marksbury Plain. North of Priston it trends towards Englishcombe and Bath where the fault zone is thought to have increased in width owing to the development of additional thrusts above and below the principal sole or thrust plane.

West of Fry's Bottom the fault zone can be traced to Bishop Sutton when the deepside colliery working's encountered seam repetition and a wide belt of disturbed ground. South west of the colliery the identification of the coals proved in the Lower Gurney Farm borehole (p.173) is crucial. If as supposed, the coals proved in the borehole are the Farrington seams, then the Farmborough Fault is most likely to curve southwards to join the Stock Hill – Biddle Fault zone of the central Mendips (Green and Welch, 1965, pl. 5). This implies that the Farmborough Fault is related to this dominantly transcurrent fault system which cuts the Dinantian and Old Red Sandstone rocks and any Silurian rocks which exist below. The Lamb Leer Fault may be one of its terminal fractures and folds, the Harptree Klippe being situated above the main thrust.

Exploration of the Farmborough Fault at depth east of Frys Bottom did not take place until the latter part of the 19th century. Radstock Old Pit was sunk to the Radstock Great Vein at 163 yds and the remaining Radstock seams proved to a depth of 358 yd in 1763. In later years, as the number of pits increased in the area north of Radstock, exploratory branches were driven northwards from Frys Bottom, Hayswood and Conygre collieries, but none proved workable Coal Measures in the fault zone (Anstie, 1873, p.91). In 1840 Charles Hollway sunk two shafts [6576 6011], one of which proved 1400 ft of measures near Farmborough village but only a few thin coals were proved even through the Radstock coals are of normal number and thickness within 400 yds of the trial shaft. Most mining engineers concluded that the thin dirty coals probed at Farmborough were the deteriorated Farrington and Radstock seams. Anstie (1873, pp.92–93) thought differently; he suggested that the strata proved in the Farmborough shafts were measures above the Radstock seams, implying that the Farmborough Fault has a northerly downthrow. The log of the shaft sinking strongly supports Anstie's contention since the rocks resemble those of the Publow Formation proved in the Hursley Hill Borehole (p.176). However, this was not the complete answer to the problem and it was not until 1917 that E. H. Staples was able to show that repetition by low angle reversed faulting or thrusting of the Supra-Pennant Measures has taken place on the northern side of the Lower Conygre Colliery and at Priston and Dunkerton. In the northerly workings of Upper Conygre Pit, Timsbury the principal (or uppermost) fault plane has a southerly dip of about 10° but this may not be true elsewhere. At Camerton the northside workings proved a net downthrow of 960 ft with a foreshortening (or total measured overlap) of 750 yds. Sections showing examples of this faulting are seen in Figures 39 and 40.

In addition to the compressional structures there are normal faults affecting the Coal Measures in the Farmborough Fault zone. Some of these are post-Jurassic age, probably Lower Cretaceous or Tertiary. In addition there are several belts of steeply inclined faults which may be of post-Upper Westphalian age, but which cut the Variscan folds and compressional structures and also show evidence of much younger Mesozoic reactivation. Among these is the Bromley Horst which postdates the formation of the depositional basin and the main Variscan folding movements at the southern end of the Pensford Basin. It may be in part of Permo-Triassic origin, modified by (?)Lower Cretaceous reactivation. The mar-

---

1 The Brislington succession may have been known to the Tudor coalminers. The seams are thick and the structure is relatively simple with a fairly large exposed area.

ginal faults have depressed the surface of the Coal Measures by about 100 ft on the north and 150 ft on the south side of the structural ridge. This structure can be traced eastwards to Stanton Wick. Hunstrete and Stanton Prior appears to converge with the Farmborough Fault. So far as can be judged it has no effect upon the thickness or facies of the Westphalian strata. It is probably one of the two 'ridgs' (ridges) referred to by Strachey (1719) as forming the northern and southern boundaries of the shallow Stanton Drew workings on the Farrington seams.

The northern 'ridge' is the boundary fault of Pensford and Bromley colliery workings. This has a northerly downthrow of at least 900 ft, and crosses the coalfield from Butcombe in the west to Newton St Loe and Twerton in the east and constitutes a major line of dislocation. It affects the whole thickness of the Coal Measures and transects the north–south basinal fold while showing evidence of strong post-Jurassic reactivation as, for example, in the highly faulted tract at Fairy Hill in the Chew valley north of Compton Dando. This fault system may act as a channel for rising thermal water at Bath where Quaternary reactivation has taken place. The Kingswood Anticline like the Farmborough Fault includes thrusts with southerly dipping fault planes (Figure 23) but their inclination in the south limb tends to be rather steep (Plate 8). The Kingswood Anticline shows strong evidence of structural evolution in Upper Westphalian times, with marked differences in the successions on the southern and northern limbs of the fold. North of Kingswood the Farrington seams of the Coalpit Heath basin are of a very different facies from those of the Radstock and Pensford basins, the coals being thicker and more regular in their distribution, with abundant nonmarine lamellibranch-bearing mudstones associated with shales rich in the phyllopod *Leaia*. This organism is exceedingly rare in the Pensford basin only a few examples having been found. This characteristic *Leaia*-rich facies is found in the Upper Coal Measures in Oxfordshire, the Coalpit Heath and Avonmouth basins and in the Forest of Dean.

Only in the south-east corner of the Coalpit Heath Basin is there anything suggesting a link with the Farrington Formation of the Pensford Basin. Here in the workings of Parkfield Colliery some splitting of the thick coals of Coalpit Heath takes place.

Where the Radstock Formation is cut by the Farmborough Fault there appears to be evidence of seam splitting, more especially at Camerton and Dunkerton and it is thought that this represents a continuation of the zone of deterioration of the Farrington and Radstock seams as seen in the Pensford Basin, distorted and foreshortened by tectonic movement. The absence of *Anthraconauta* and *Leaia* in the Radstock Formation of the Radstock Basin and difference between the coals north and south of the Farmborough Fault indicate that growth of the fault system led to the increasing isolation of the southern or Radstock basin in late Westphalian times. This is also supported by evidence for strong uplift of the region south of Radstock provided by the southwards thickening of the red measures in the Barren Red Formation.

Determination of the throw of the Farmborough Fault at the base of the Supra-Pennant measures is very difficult to estimate though it is unlikely to be less than 1000 ft north of Camerton. The major thrusts in the Bristol–Kingswood–Wick area show a distinct tendency to increase in throw from west to east.

Coalpit Heath Basin

The Farrington Formation in the Coalpit Heath Basin is defined as extending from the base of the High Vein to the top of the Hard Vein (Anstie, 1873) (see Figures 32 and 33). The measures embraced seldom exceed 200 ft and they contain three (locally four) seams which have been worked. The chief faunal characteristics are abundant '*Leaia*' associated with *Anthraconauta tenuis* and *A. phillipsii*, particularly in the roof of the High Vein, and this has led to the suggested correlation of that seam with the No. 2 Seam of Avonmouth (see below) and the Coleford High Delf in the Forest of Dean.

The Supra-Pennant Measures occupy an area of nearly 6 square miles and crop out in a roughly oval basin of which the longer axis extends from Iron Acton in the north to Shortwood in the south, and the shorter axis from Serridge in the west to Westerleigh in the east. The basin is traversed by two main faults or fault-belts which, intersecting at right angles, divide the area into unequal quadrants and formed natural boundaries to the colliery workings (Figure 33). The smallest quadrant is that in the south-west, the remainder increasing in size as they are followed in a clockwise direction. The west–east fault extending from Winterbourne Down to Westerleigh was known to the old miners as the Kidney Hill Fault. Its southerly throw is variable in amount, but is estimated to be of the order of 140 ft at a point about 1000 yd south-west of Westerleigh. The north–south fault, usually termed the Coalpit Heath Fault, can be traced from Iron Acton to Tubb's Bottom and thence to Ram Hill where it intersects the Kidney Hill Fault and continues southwards through Henfield to Shortwood. Near Lyde Farm several branches radiate from the main fracture. The throw of the Coalpit Heath Fault is easterly and variable in amount, being about 300 ft south of Ram Hill, but near Shortwood it appears to be of the order of at least 1000 ft.

The principal seams have been worked extensively in the past, and most of the pits were worked to exhaustion. Seam nomenclature at Parkfield Colliery [6885 7779] in the south-east quadrant (Figure 33) differs from that used in the rest of the basin, due to the splitting into two separate worked seams of the lowest coal (High Vein). Four seams were thus worked at Parkfield, while elsewhere there were only three:

Coalpit Heath Colliery	*Parkfield Colliery*
Hard Vein	Hard Vein
Hollybush Vein	Top Vein
High Vein	Hollybush Vein & Great Vein

The High Vein with an average thickness of 60 in was worked throughout the basin. In the northern and western areas the seam consists of two leaves of coal separated by 4 to 6 in of clod: the parting thickens south-eastwards and to the south of Parkfield, at Brandybottom [6823 7715] and Shortwood the Great and Hollybush Veins are separated by 30 to 50 ft of shale.

In the southernmost workings of the Coalpit Heath Colliery the High Vein is directly overlain by sandstone, but westwards and northwards mudstone feathers in between the coal and the sandstone and is up to 6 ft thick near Froglane Pit [6870 8155] and 12 ft at New Engine Pit [6871 7938]. The shale overlying the coal at Coalpit Heath contains *Anthraconauta tenuis* associated with '*Leaia*' and ostracods, but at Parkfield the Great Vein roof carried plants (Moore and Trueman, 1937, p.232).

Between the High Vein and the overlying Hollybush Vein (Top Vein) there is a series of mainly argillaceous measures varying in thickness from about 67 ft at Parkfield to 88 ft at Coalpit Heath.

The Hollybush (Top) Vein, usually 28 to 30 in thick, varies considerably in thickness over the basin. In the north-east and east it was too thin to be worked, but it appears to have been extensively wrought from New Engine Pit [6780 7938] northwards to Ram Hill Pit [6791 8025], and in the south-east from the Kidney Hill Fault southwards through Parkfield to Shortwood. Some 700 yd north of Parkfield Pit the coal was only 6 to 7 in and this thinning continued northwards into the workings of Coalpit Heath Colliery, apart from a small area immediately south of Froglane Pit where the seam is 31 to 33 in. Unfortunately here it was separated from the overlying Stinker Vein by only 1½ to 2 ft of mudstone which made working impracticable, especially as the floor was a heaving fireclay. From the roof of the Hollybush Vein at Parkfield, Moore and Trueman (1937, p.232) recorded *Anthraconaia* cf. *pruvosti* and *Euestheria simoni*: a rich flora has also been recorded (Crookall, 1955, pp.151–152).

Between the Hollybush and Hard veins a group of dominantly argillaceous measures ranges from about 60 ft in thickness at Coalpit Heath Colliery to 110 ft at Parkfield. In the west of the basin a thin (6 to 18 in) seam of very sulphury coal—the Stinking Vein or Stinker—occurs. As mentioned above, this rests almost directly on the Hollybush Vein in places, but in others as much as 34 ft of mudstone and sandstone intervenes.

The Hard Vein was a much-worked seam of fairly uniform distribution over the entire basin (Figure 33). It varies in thickness from 22 in in the south to 34 in in the north. A strong mudstone roof has yielded a rich flora including *Acitheca polymorpha*, *Eupecopteris fletti* and *Alethopteris radstockensis* (Moore and Trueman 1937, p. 233).

SOUTH-WEST QUADRANT   South of the road from Folly Bridge [6687 7830] to Henfield the Supra-Pennant Measures are largely concealed by Triassic rocks, but between the Coalpit Heath Fault and a subsidiary fault extending from Henfield towards Folly Bridge, Farrington and Barren Red measures are exposed in a small synclinal fold. A line of old crop-workings on what appears to be the Hard Vein trends in a north-north-westerly direction about ¼ mile east of Folly Bridge.

North of the Folly Bridge—Henfield road the measures strike in a north-north-westerly direction to the Kidney Hill Fault, which here has a southerly throw of 72 ft according to a plan dated 1850. In this region the High, Hollybush and Hard veins have all been extensively wrought, both from ancient crop workings and pits at Henfield. The deepest shaft, New Engine Pit, was sunk to a depth of 508 ft 10 in, the High Vein being cut at 502 ft 10 in, the Hollybush (30 in thick) at 430 ft 4 in and the Hard Vein (30 in thick) at 350 ft 10 in. The dip of the beds is about 1 in 9, in an east-north-easterly direction. The coal section of the High Vein was given by Buckland and Conybeare (1824, p.275) as: top coal 24 in, binching 6 in, middle coal 24 in, bottom coal 24 in. The Stinking Vein, 12 in thick, lies 34 ft above the Hollybush. Some 30 ft above the Hard Vein is the Rag Vein, 12 in thick, overlain by 5 ft of thin coals alternating with black soft mudstone.

NORTH-WEST QUADRANT   Apart from some local disturbance in the proximity of two westerly throwing faults which trend in a north-north-westerly direction from the Kidney Hill Fault near Serridge, the outcrop of the productive Farrington measures curves gently in a northerly direction to a point near Iron Acton. The coals have long been exhausted by crop-workings and shallow pits about which only meagre information is available. The chief pits showing depths to the High Vein are: Serridge Engine [6747 7964] 284 ft; Ram Hill 558 ft; Upper Whimsey [6781 8070] 210 ft; Half Moon [6796 8107] 240 ft, coal dipping east at 1 in 9; Oxbridge [6814 8151] 288 ft.

West of Mayshill the Coalpit Heath Fault, which has hitherto followed a more or less northerly course, turns north-westwards for ¼ mile before resuming the northerly course to Iron Acton and cuts out the outcrop of the Hard Vein north of Tubbs Bottom (Figure 33). About ¼ mile south-east of Algars Manor a sandstone, up to 60 ft thick, is developed above the High Vein and is exposed in the banks of the River Frome where it forms a small gorge.

NORTH-EAST QUADRANT   Except along the western edge near to the Coalpit Heath Fault where the beds show some disturbance, the unbroken synclinal nature of the structure is clearly shown by the curving outcrops of the shale and sandstone belts, the axis of the fold passing almost due south through Frog Lane Pit to the centre of an isolated patch of sandstone 1200 yd west of Westerleigh Church. Dips are relatively low, being from 1 in 8 to 1 in 10 in an easterly or south-easterly direction on the western limb of the syncline, and between 1 in 3 and 1 in 5 in a westerly direction on the eastern limb (Figure 33).

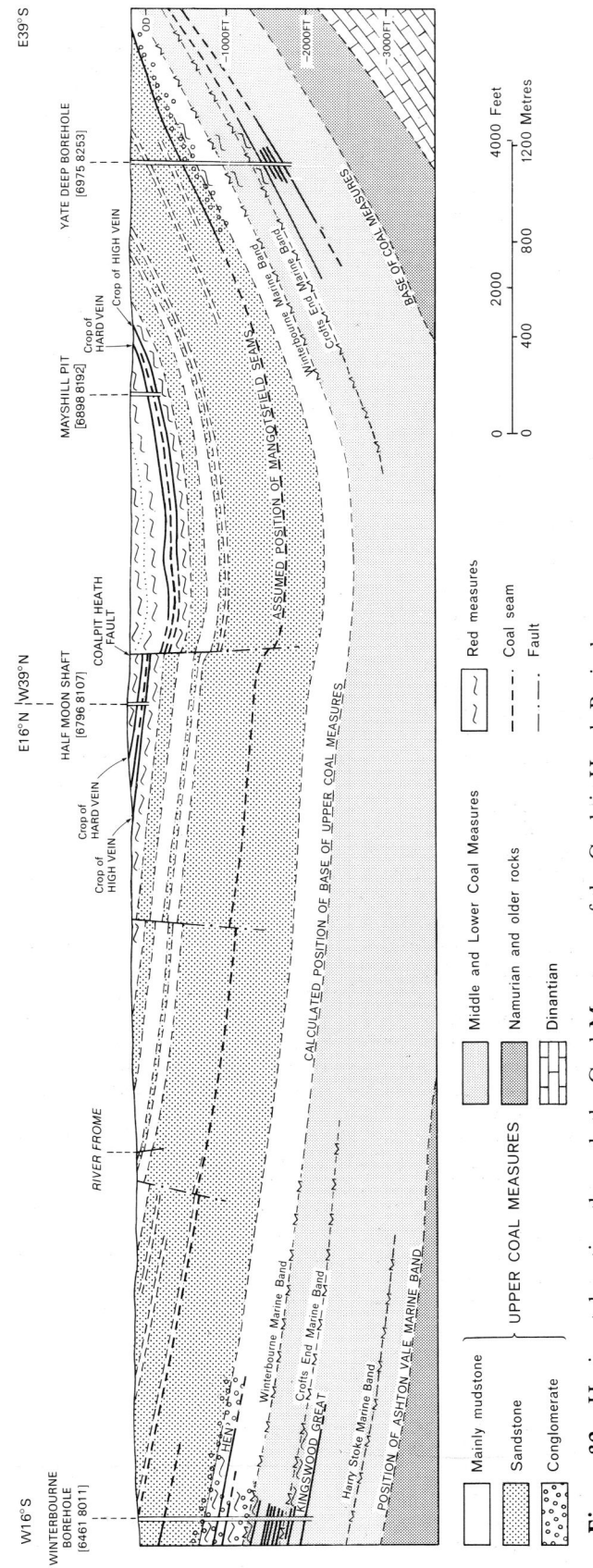

Figure 32 Horizontal section through the Coal Measures of the Coalpit Heath Basin between Winterbourne and Yate

Unlike the coal-workings of the north-west quadrant, most of the coal extraction has been confined to one large colliery. A small amount of crop working of the High and Hard veins has taken place, but this was restricted by the fact that seams crop out in the alluvial flat of the River Frome and water difficulties must always have been present. The Hollybush Vein appears to have been of workable thickness only between Iron Acton and Nibley, so that over most of the quadrant only the High and Hard veins have been wrought.

Above the High Vein a thick current-bedded sandstone is developed in the Iron Acton–Nibley area, and locally also near Westerleigh. In the former locality the rock is exposed dipping at 10° to the south-west in the railway cutting 750 yd south-east of Iron Acton Church. Over 14 ft of sandstone rests directly upon the Hard Vein in the railway cutting 400 yd south-west of Westerleigh Church.

At a distance of 780 yd S17°E of Iron Acton Church there is an old engine pit [6827 8276] which is said to have been sunk 276 ft to the High Vein. About ¾ mile to the south-east is the Old Nibley Colliery which consists of two shafts—the High Vein Pit [6918 8204] and the Hard Vein Pit [6911 8198]. The former shaft cut the Hollybush (Top) Vein at 49 ft 6 in, but the seam was never worked; the High Vein was proved at 137 ft and the seam worked pillar-and-stall along a level course extending some 300 yd on either side of the shaft. Dips varied from 14° to 22° in a south-westerly direction. The Hard Vein Pit cut the Hard Vein at a depth of 95 ft and the shaft was connected to the High Vein Pit by a level branch at 148 ft 6 in. The Hard Vein was worked pillar-and-stall from a level course extending from a point 450 yd north-west of the shaft to a point 1500 yd south, the workings being ventilated by a shaft 1180 yd south of the pit. Dips average 29° to the south-west. There are no known sections either of the High or Hard veins.

The chief colliery to work the coals of the north-east quadrant was the now-abandoned Coalpit Heath Colliery which consisted of Froglane Pit [6870 8155]. This was connected to the old Mayshill Pit [6897 8191] some 510 yd to the north-east, which was used for ventilation. Coalpit Heath Colliery closed during the 1950s due to almost complete exhaustion of all workable seams. At Mayshill Pit the Hard Vein was cut at 248 ft, the Hollybush at 300 ft 6 in and the High Vein at 392 ft. The Froglane Pit intersected the Hard Vein at 480 ft 6 in, the Hollybush at 538 ft 11 in, and the High Vein at 625 ft 3 in; shaft bottom was at 655 ft 3 in.

The Hard Vein workings were reached by a level branch extending from the High Vein in the shaft for a distance of 500 yd southwards. About 230 yd from the shaft the branch intersected the Hollybush Vein and from this point limited workings of the seam were driven on both sides. A section of the High Vein in the northern part of the colliery showed: roof, rashes 12 in to 6 ft, coal 16 in, rashes 4 in, coal 36 in, bastard fireclay floor. In the southern part of the workings the rock roof extended down to the coal and a section of the seam showed: sandstone, coal 26 in, parting, coal 36 in, bastard fireclay floor.

The Hollybush Vein showed great variation in section and the seam where worked showed: soft shale roof, rashes 2 in, top coal 5 in, black clod 4 in, coal 28 in, heaving fireclay floor. In an exploratory drivage 850 yd south-east of Froglane Pit the coal was seen to thin rapidly from 26 to 12 in, and at a point 1200 yd south of the pit it was absent.

The Hard Vein, some 150 ft above the High Vein, was 28 to 34 in thick and had a strong mudstone roof and bastard fireclay floor: it was worked extensively.

SOUTH-EAST QUADRANT In this, the largest quadrant, the greater part of the Farrington Formation outcrop is concealed by Mesozoic deposits and only in the area extending westwards from Oakleighgreen Farm [691 787] to the Coalpit Heath Fault, and in a triangular region stretching north from Shortwood to Lyde Green are the rocks exposed. It is therefore somewhat unusual to find that some of the earliest workings in the coalfield took place along the fringes of the basin in which the measures were so extensively concealed. From the Kidney Hill Fault to Shortwood there is a line of old pits which have been sunk through a considerable thickness of Mesozoic rocks before reaching the Coal Measures. From north to south these pits include: Dudley Pit [6947 7817] of which a section, made in 1855, shows Hard Vein, 30 in, at 185 ft 6 in, Hollybush (Top) Vein, 5½ in, at 289 ft, High (Hollybush and Great veins), 81 in, at 347 ft 3 in: Puffers Pit [6938 7800] of which a section, dated 1834, shows Top (Hollybush) Vein at 324 ft and High Vein at 396 ft. Cook's Pit [6924 7750], was sunk 285 ft to the High (Hollybush and Great) Vein and rise workings in this and the Hard Vein were shown on old plans, while a connection in the Hard Vein extended to Whimsey or Bryant's Pit [6933 7771]. In later years Cook's Pit acted as the upcast shaft to Parkfield Colliery.

About ¼ mile to the south-west of Cook's Pit is Great Cart Pit of which no details are known. Between Great Cart Pit and Shortwood pits are a number of old abandoned pits including Engine Bottom, Hangbeggar, Wood and Quarry pits, but no records relating to them exist.

Shortwood Colliery, which was closed in 1909, consisted of three interconnected pits — Lapwater [6798 7679], Chaffhouse [6790 7669] and Cook's (or Fryar's) [6776 7637] — but only the two last mentioned appear to have been worked to any extent. The colliery workings are intersected by north-north-westerly faults diverging from the Coal Pit Heath Fault and throwing westwards, the largest of which has a throw of 60 ft and passes immediately west of Lapwater Pit. Abandonment plans indicate both Chaffhouse and Cook's pits reaching the Hard Vein at 300 ft and Lapwater at 330 ft. Most of the Chaffhouse workings appear to have been in the Hard Vein which dips northwards at 1 in 3. A section of the seam was given as: sandstone top, clod 1 ft, coal 28 in, fireclay floor. From Cook's Pit a level branch extended south-westwards cutting the Top, Hollybush, and Great veins, all of which appear to have been worked to a limited extent.

The principal workings in the south-east quadrant were from Parkfield and Parkfield South collieries, the latter consisting of Brandybottom Pit [6283 7715] and New Pit [6819 7712]. A section of Brandybottom Pit is given by Prestwich (1871, p.54). From the Parkfield shafts workings have extended in a westerly or north-westerly direction to the centre of the basin and all coal has been virtually exhausted. The closure of the colliery in 1936 was, however, accelerated by the incursion of water into the workings along the line of the Coalpit Heath Fault where minor folding associated with the fault also occurs. At Dudley Pit the section of the High Vein was: top coal 40 in, clod 15 in, bottom coal 26 in; at Parkfield: top coal (Hollybush) 36 in, soft fireclay 18 in, bottom coal (Great Vein) 34 in; and at Brandybottom: Hollybush Vein 30 in, measures 33 ft 6 in, Great Vein 30 in. Finally at Shortwood, a section by Thos. Austin, dated 1853, gives 50 ft of measures separating the two seams.

Regarding the thickness of the Top and Hard veins in the quadrant there is little information apart from that shown in shaft sections. The Top Vein shows much variation in thickness in the same manner as in the north-east quadrant. A section recorded in 1853 at a point 220 yd north-west of Brandybottom Pit shows: sound shale roof, soft clod parting, hard coal 22 in, soft coal 11 in, 'kelves' or slaty coal 6 in, underclay.

During the widening of the railway cutting at Shortwood in October 1948 a section through the productive measures was examined by G. W. Green and G. A. Kellaway. The beds were seen to be steeply dipping or vertical and in places inverted so that accurate measurements were not possible. Some 540 ft north of the north parapet of the roadbridge [6734 7598] a vertical coal (?Great Vein)

Figure 33 Structure contour map of the Hard Vein, Coalpit Heath Basin

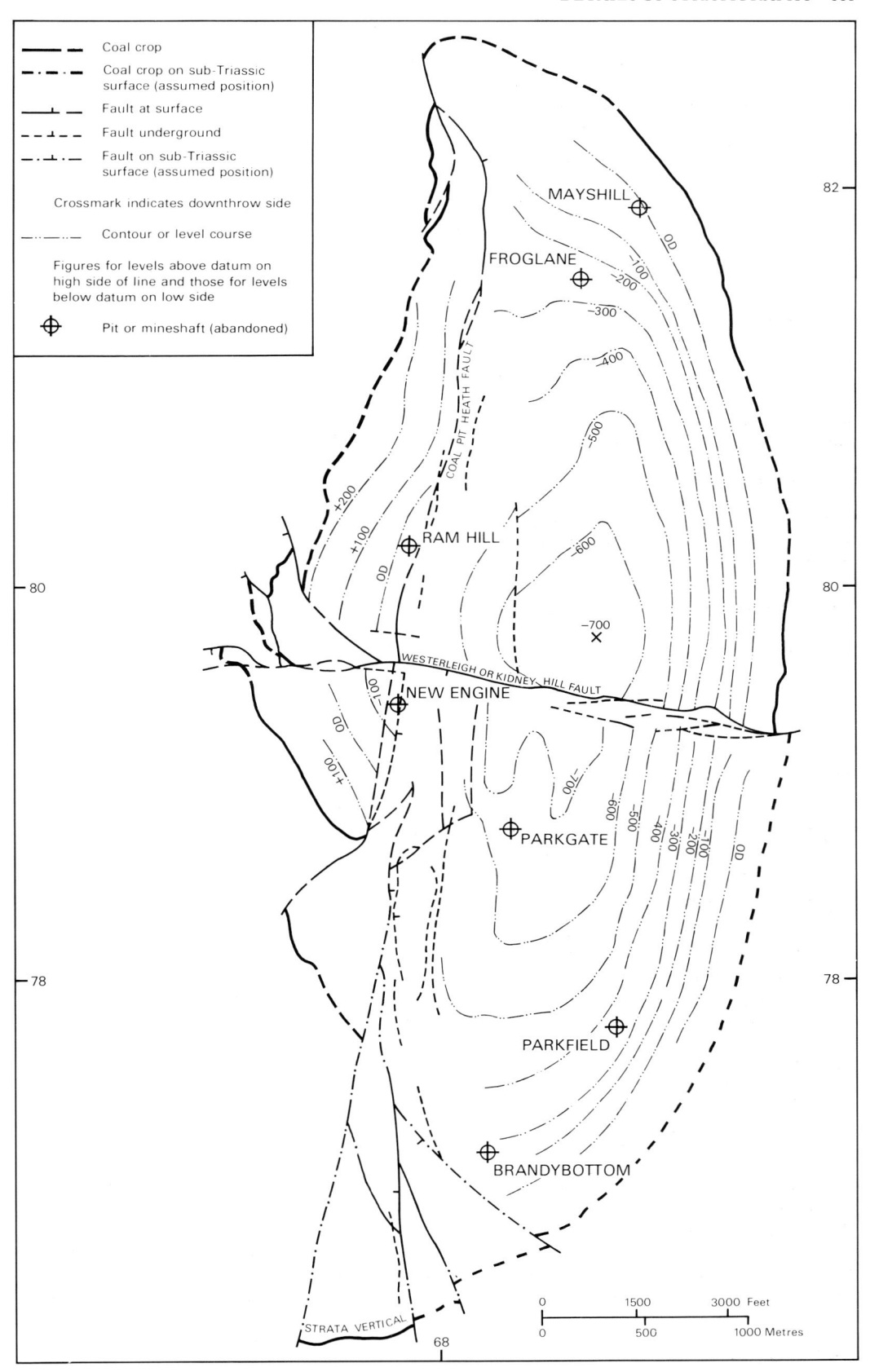

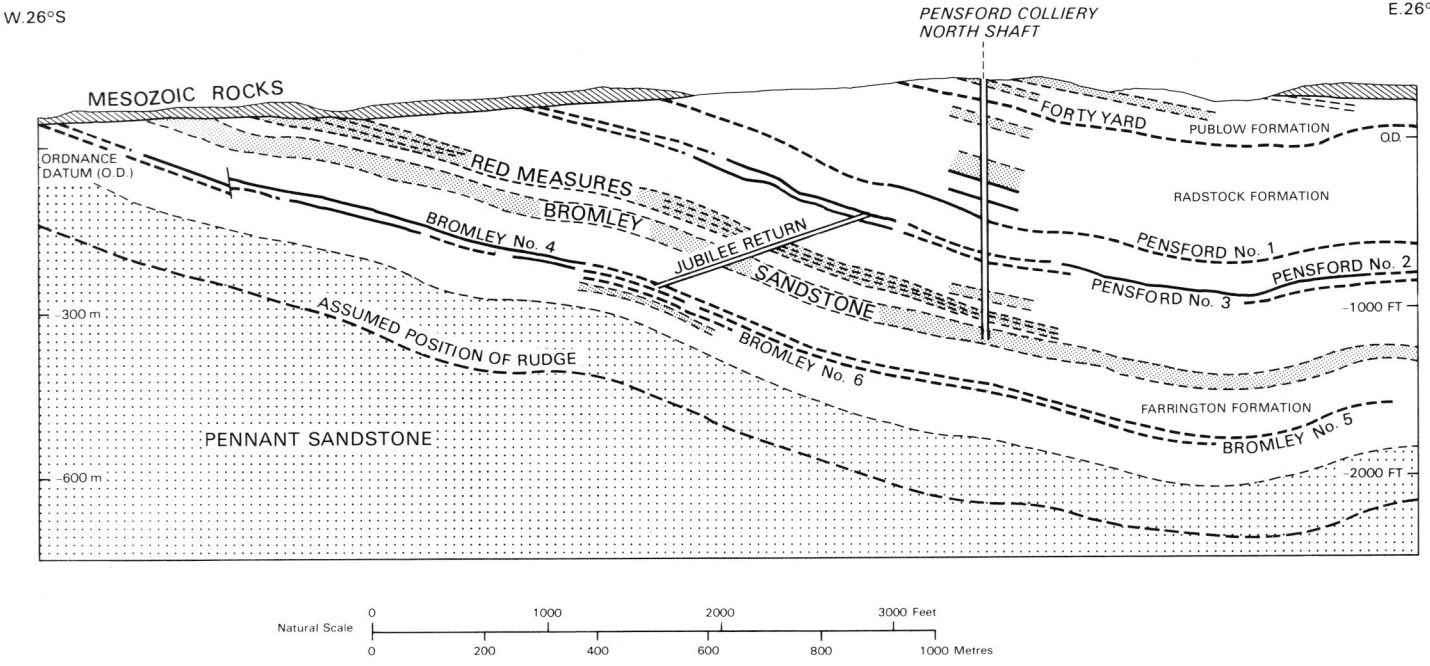

Figure 34 East–west section through Pensford Colliery

was observed; 50 ft farther north another vertical coal, which had apparently been worked, may represent the Hollybush Vein. The Top Vein may possibly be represented by a 12 in coal, vertical or inverted, which occurs about 30 ft above the Hollybush. Some 42 ft of inverted mudstone and sandstone separate the Top Vein from a further seam, 18 in thick, which may be the Stinking Vein. At 720 ft north of the bridge a 27 in coal dips 70° in a northerly direction. This appears to have been worked and may be the Hard Vein. Above this seam between 768 and 790 ft from the bridge is a belt of hard grey sandstone dipping at 65° to the north. Two thin coal seams 860 ft north of the bridge may represent the Rag Vein.

Severn Basin

Pennant sandstones with only thin coals, belonging to the Mangotsfield Formation, make up most of the Severn Basin (Figure 21), but in the deepest part of the elongated syncline extending north-eastwards from near Avonmouth towards Pilning two thick coals, the Avonmouth Nos. 1 and 2 Seams occupy an area of between 2½ and 3 sq. miles. They were first proved in 1928 in a boring [5347 8152] at Washingpool Farm, north-west of Hallen. In 1953–54 the National Coal Board drilled a further five holes which established the nature of the seams and the shape and depth of the basin of Supra-Pennant Measures. The two seams are contained within a sequence of about 300 ft of grey mudstones and seatearths with sporadic bands of sandstone lying between massive Pennant sandstones and a succession of barren sandstones and red and mottled mudstones probably reaching 800 ft in thickness. The comparison between the grey argillaceous group of measures with the two thick coals and the Farrington Formation of the main coalfields to the east is obvious, and is strongly supported by the occurrence of *Anthraconauta tenuis* and *A. phillipsii* associated with abundant '*Leaia*' in the beds above the Avonmouth No. 2 Seam, thus invoking direct correlation with the High Vein of the Coalpit Heath Basin. The following account is taken largely from that given in the Monmouth and Chepstow Memoir (Welch and Trotter, 1961, pp.106–107).

Some 60 ft of shales intervene between the No. 2 Seam and the underlying massive Pennant sandstone, again inviting comparison with the shales locally developed beneath the High Vein. The Avonmouth No. 2 Seam varies in thickness from 49 to 62 in and consists of two layers of bright coal each 22 to 25 in thick separated by a band of fireclay or carbonaceous mudstone about 15 in. The measures between the Avonmouth No. 2 and No. 1 seams vary from 90 ft in the west of the syncline to 110 ft on the east. On the western side a sandstone is developed in the lower half of the sequence, especially in Avonmouth No. 2 Boring [5444 8305] at Vimpennys Common where it extends almost down to the No. 2 Seam. About 20 to 30 ft below the Avonmouth No. 1 Seam there is a thick development of seatearths with thin bands of coal, and the remainder of the succession consists of grey silty mudstone with abundant *Anthraconauta* and '*Leaia*'. Avonmouth No. 1 Seam, ranging from 60 to 89 in, consists of layers of coal and inferior coal separated by dirt and rashings. The roof is mudstone; immediately above the coal this is laminated and full of plant remains.

BARREN RED FORMATION

McMurtrie (1869a, p.48) first defined the rocks lying between the Farrington seams and those of Radstock as 'unproductive strata with red shales'. These became known as the Barren Red Group (now Formation) and were redefined by Moore and Trueman (1936) as embracing the strata between the Rock Vein (top of the Farrington Formation) and the Nine-Inch Vein (base of the Radstock Formation). This definition related primarily to the Writhlington succession. However, neither the Cathead Vein nor the Rock Vein can be recognised throughout the Radstock Basin and in practice the term Barren Red Formation is loosely defined so as to include all the unproductive measures between the Farrington and Radstock formations (see Figure 20). Having no special faunal or floral characteristics which can be relied on as a means of determining its boundaries, it is purely lithostratigraphical in concept. The comparable rocks north of the Farmborough Fault Belt lie between the productive measures with the Bromley seams below, and Pensford seams above.

In general the Barren Red Formation is composed of sediments which do not differ greatly from those of the productive measures. The more massive sandstones, for example the waterlogged Bromley Sandstone, are of Pennant or subgreywacke type, but there is a preponderance of gritty fireclay and clunch. The red measures are brownish or maroon when fresh, becoming reddish purple or petunia-coloured where deeply weathered. Some of the rocks are grey when fresh but rapidly acquire a brownish tinge with exposure to the air.

Radstock Basin

The thickness of the Barren Red Formation is said to be as much as 750 ft in the south (McMurtrie, 1901), but both the barren strata and the associated red measures thin in a northerly direction. North of Braysdown Pit they appear to be about 500 ft thick and of this thickness only 200 ft are red. The beds crop out in the northern part of the basin from North End, Clutton [620 600] to High Littleton [650 575], and also in the banks of the River Somer [655 547] to the south-east of Old Mills Colliery. Exposures are rarely visible since the rocks are usually concealed by small hummocky landslips caused by the large proportion of fireclay within the group. The best exposure of the uppermost 290 ft of barren measures including 'red shales' formerly existed in the now grass-grown railway cutting 1000 yards north-north-east of Clutton Church. The section below the Nine-Inch Vein (see Figure 35), measured by H. E. Hippisley (Hippisley MS) is reproduced below. The thicknesses of the individual beds have been measured from the original horizontal section drawn at 30 ft to one inch.

	Thickness		Depth	
	ft	in	ft	in
Concealed, probably pan	2	11	2	11
Arenaceous shales	0	11	3	10
Sandstone	2	9	6	7
Argillaceous shales	6	6	13	1
Coal and shale	0	11	14	0
Pan	3	1	17	1
Arenaceous shales	0	9	17	10
Shales, grey, argillaceous	5	10	23	8
Arenaceous shales	1	7	25	3
Arenaceous shales and soft sandstone	4	8	29	11
Sandstone	1	5	31	4
Concealed strata	6	0	37	4
Shells of coal	3	1	40	5
Pan	2	7	43	0
Arenaceous shale	4	7	47	7
Dark-coloured pan	0	9	48	4
Argillaceous shale	1	7	49	11
Trace of coal	0	5	50	4
Pan	1	5	51	9
Clod	0	2	51	11
Argillaceous shale	3	2	55	1
Strata not described: from ornamentation probably dark carbonaceous shale	3	2	58	3
Pan	2	9	61	0
Arenaceous–argillaceous shale	1	7	62	7
Argillaceous shale	1	8	64	3
Sandstone	1	10	66	1
Argillaceous shale	2	10	68	11
Sandstone	0	9	69	8
Arenaceous shale	2	10	72	6
Argillaceous shale	14	2	86	8
Coal and shale	2	6	89	2
Pan	2	7	91	9
Sandstone	1	9	93	6
Hard arenaceous shales	3	6	97	0
Softer arenaceous shales	0	9	97	9
Coaly shale and pan	0	6	98	3
Ironstone	0	6	98	9
Pan	1	8	100	5
Sandstone	1	6	101	11
Arenaceous–argillaceous shale	2	5	104	4
Pan	1	5	105	9
Sandstone	0	4	106	1
Pan with ironstone nodules	3	11	110	0
Hard arenaceous shale	8	6	118	6
'Reeds' (probably argillaceous shale with plants)	0	10	119	4
Pan	1	3	120	7
Argillaceous–arenaceous shales with 3 in nodular bed	5	10	126	5
Dark pan	6	2	132	7
Brown shale	6	11	139	6
Red shale with brown streaks	11	5	150	11
Grey sandstone	1	10	152	9
Red shale	8	5	161	2
Grey arenaceous shale	3	6	164	8
Red shale	3	2	167	10
Grey argillaceous shale	0	7	168	5
Red shale	6	9	175	2
Hard grey-brown arenaceous shale or sandstone	0	5	175	7
Red shale	10	9	186	4
Red-grey shale	1	9	188	1
Hard grey sandstone	0	7	188	8
Red-grey shale	24	7	213	3
Grey argillaceous shale	2	2	215	5
Sandstone	1	8	217	1
Grey arenaceous shale	1	7	218	8
Red shale mixed with hard brown arenaceous shale	6	7	225	3
Red shale	4	10	230	1
Hard arenaceous shale	1	4	231	5
Red shale	3	2	234	7
Grey arenaceous and red shale mixed	3	2	237	9
Red shale	4	9	242	6
Brown arenaceous shale	6	9	249	3
Red shale	1	10	251	1
Brown shale	5	7	256	8
Red shale	0	4	257	0
Brown shale	17	10	274	10
Red shale	0	5	275	3
Brown shale	11	5	286	8
Dark blue-grey shale	10	0	296	8

Pensford Basin

On the assumption that the barren measures between the productive Bromley and Pensford seams are equivalent to the Barren Red Formation of the Radstock Basin, then the thickness of the formation has not diminished greatly north of the Farmborough Fault Belt, though the intercalated red measures are only about 80 ft thick at Pensford (Figure 35). The only exposed area of Barren Red Measures on which there is any reliable information is a narrow tract of ground east of Bromley Pit. Here, in the centre of the Bromley Horst [609 617] the very soft fireclays and red measures can be identified on the surface by means of small slips which have formed in the weathered clays. The Bromley Sandstone also crops out on the surface and was proved in the shaft of the colliery (Figure 31).

Between the Bromley Horst and the crop of the Farrington coals at Brislington in south-east Bristol the Supra-Pennant Measures are concealed by Mesozoic rocks. It is probable that the Barren Red Formation crops out at Brislington in the area immediately to the south of the crop of the worked coals, but the featureless ground is mainly built-up and there are no definite records of the Barren Red or Radstock formations having been proved. There is surface evidence of a coal, possibly the Pensford No. 2 seam, in the low ground betwen Flowers Hill Farm [6220 6985] and Flowers Hill. Some 600 ft of measures intervene between this coal and the Trolley Vein (p.111), the highest of the Brislington seams in the Farrington Formation: they appear to be barren and to include fireclay and a substantial thickness of Pennant-type sandstone. They are approximately the same thickness as the measures between the Pensford No. 2 seam and the base of the Barren Red Formation at Pensford and Bromley collieries, and the thick sandstone mapped above the Trolley Vein of Brislington may well be the Bromley Sandstone.

The best sections of the Barren Red Formation in the Pensford Basin are those which have been measured in the Jubilee Main and Jubilee Return branches of Pensford Colliery [6178 6269], and in the Hursley Hill Borehole [6180 6565] (p.176). Part of the Barren Red Formation was also proved in the shafts of Pensford and Bromley collieries (Figure 31). The sections at Pensford Colliery (Figure 34) were logged by G. A Kellaway and F. B. A. Welch with the assistance of the surveyors of the National Coal Board. The branches were parallel, cross-measures drivages, inclined to the west and linking the Pensford No. 2 and Pensford No. 3 workings near pit-bottom with the Bromley (Farrington) seams. They therefore penetrate the entire thickness of the Barren Red Formation, though the ground was not entirely free from faulting. The section in the Main Branch is given below:

	Thickness ft	in	Depth ft	in
RADSTOCK FORMATION (for section see p.124)				
BARREN RED FORMATION				
Hard sandy mudstone	20	0	20	0
Sandstone	2	6	22	6
Mudstone with plants	8	0	30	6
Soft coaly shale with mudstone and fireclay	0	6	31	0
Hard mudstone with ironstone nodules	7	0	38	0
Sandstone	7	6	45	6
Sandy mudstone, passing down into mudstone with ironstone nodules	18	0	63	6
Soft grey fireclay	10	0	73	6
Hard grey mudstone and bastard ganister	7	0	80	6
Red mudstone and red and grey mottled mudstone and fireclay	89	0	169	6
Mudstone, grey, partly ganisterised, with ironstone nodules	34	0	203	6
Soft, light grey mudstone and fireclay, passing down into mudstone with ironstone nodules	21	0	224	6
Shaly mudstone (faulty) with abundant nonmarine bivalves	4	0	228	6
Fault—base of Formation not seen				

The section in the Return Branch repeats the red measures seen in the Main Branch and continues downwards to the Bromley No. 4 Seam (Farrington Formation):

	Thickness ft	in	Depth ft	in
BARREN RED FORMATION				
Red and mottled mudstone and fireclay with a red-stained shell of *Anthraconauta sp.* in slumped red and grey mudstone 10 ft above the base	86	8	86	8
Dark fireclay and mudstone with a band of shale at the base with *Spirorbis sp.*, *Anthraconauta* cf. *phillipsii*, *A.* cf. *tenuis* and *Carbonita sp.*	23	0	109	8
Mudstone	17	0	126	8
Clod	2	0	128	8
Mudstone with band of shale at base yielding *Anthracomya pruvosti*, ostracods and *Eustheria simoni* (shale is cut by a small fault apparently with easterly downthrow)	25	0	153	8
Bromley Sandstone: hard grey Pennant-type sandstone and sandy shale	87	0	240	8
Coal	0	8	241	4
Mudstone	41	0	282	4
Sandstone and sandy shale	36	0	318	4
Coaly sandstone	7	0	325	4
FARRINGTON FORMATION				
Coal (on south side only)	—	—	—	—
Hard mudstone	11	0	336	4
?Bromley No. 1 Seam: Seam: coal	2	0	338	4
Mudstone and fireclay with 1 in coal at base	18	0	356	4
Mudstone and fireclay	9	0	365	4
Hard mudstone	20	0	385	4
Hard sandy mudstone	9	0	394	4
Mudstone (faulty): ?No. 3 Seam missing	28	0	422	4
Bromley No. 4 Seam: coal	1	10	424	2

The base of the Barren Red Formation is taken at the first of the Bromley coals. In practice this gives rise to difficulty when coal seams appear to be replaced locally by arenaceous rocks at the base of the Bromley Sandstone. The combined sections give a thickness of about 406 ft for the Barren Red Formation and about 100 ft for the measures between the Barren Red Formation and the floor of the Bromley No. 4 Seam.

In the Hursley Hill Borehole (Figure 35), the the top of the Barren Red Formation is thought to be at 1920 ft 10 in, beneath a dirty coal and carbonaceous shale. The colour of the rocks is mainly described in the log as 'grey' but many of the rocks changed colour on long-continued exposure to the air, and some, notably the friable fireclay recorded at 2152 ft 2 in to 2173 ft 4 in, acquired a brownish or red-brown hue. The strata below 2173 ft 4 in consist predominantly of very hard sandstone with mudstone, coal and ironstone fragments. The lowest sandstone proved in the borehole at 2395 ft to 2400 ft was a coarse conglomerate with quartz pebbles and fragments of coal and ironstone. Some of the sandstones have erosional surfaces at the base, the implication being that large sandstone washouts infilled hollows eroded in mudstones with thin coal seams and ironstone nodules. It had been intended to prove the Bromley seams in the borehole, but owing to technical difficulties it was impossible to drill below 2400 ft and the hole had to be abandoned still in the Barren Red Formation. The bottom 128 ft 8 in was in sandstone, presumed to be the Bromley Sandstone.

Coalpit Heath Basin

Above the Farrington Formation the Coal Measures surviving in the centre of the Coalpit Heath Basin are predominantly red and plum coloured mudstones with subordinate bands of sandstone which are only prominently developed in the region of Mayshill. A belt of grey silty mudstones, some 120 ft thick, lies about 600 ft above the base of the red measures, which may represent the Radstock Formation, though there is no evidence to support this, and further red mudstones occur above. About 900 ft of these barren, predominantly red, measures have survived Permo-Triassic erosion.

Radstock Formation

Strata above the Barren Red Formation are confined to the Radstock and Pensford basins of the Somerset Coalfield. In both these areas the red and grey strata devoid of production coals are succeeded by a group of measures with numerous thin coals, some of which have been worked both north and south of the Farmborough Fault Belt. Correlation in detail between these two areas has always been a problem, but there seems little doubt concerning the general equivalence of the measures now considered as the Radstock Formation in the two basins. The formation has not been proved at depth within the Farmborough Fault Belt and all attempts to work the Radstock seams through the disturbed zone to link with the Pensford seams ended in failure.

The base of the Radstock Formation in the Radstock Basin is generally taken at the Nine Inch Seam and the top at the Withy Mills Seam (McMurtrie, 1867), the main coals being confined to the lower half of the succession. In the Pensford Basin the formation may be taken to extend from the No. 8 Seam, or just below (see p.176) upwards to the Forty Yard Seam (so called from its position in the Pensford Colliery North Shaft, Figure 34), although the equivalence of these to the Nine Inch and Withy Mills seams has yet to be proved. Here also the main seams, collectively known as the Pensford coals, occur in the lower part of the sequence (Figures 34 and 35).

Because of the local difficulties of correlation as well as considerable structural problems, it is not easy to determine true thicknesses for the Radstock Formation. In the Radstock Basin the maximum thickness would appear to be about 820 ft, at Pensford about 1030 ft, and in the Hursley Hill Borehole 918 ft (Figure 35). There would appear thus to be an appreciable thickening of the measures as they are traced from Fry's Bottom and Timsbury into the Farmborough Fault Belt and beyond.

Optimum coal-bearing conditions occur in the southerly parts of the Radstock Basin where the measures as a whole are thinnest; six seams have been widely recognised, four of which have been worked extensively. In the Pensford Basin, where the measures are thicker, only the Pensford No. 3 and No. 2 seams proved workable.

The principal coal seams in the Radstock Basin are the Slyving, Middle and Great veins. At Pensford the Nos. 2, 3 and 4 seams may correlate with Radstock's Slyving Vein and the coals beneath. The Nos. 2 and 3 seams, recognisable from their chemical and other characteristics, were identified in the Hursley Hill Borehole.

Only a few of the principal coals can be traced continuously into the eastern and northern parts of the Radstock Basin and a similar behaviour is even more strongly marked in the Pensford Basin. This deterioration is due primarily to progressive intercalation of argillaceous and silty sediments accompanied by attenuation of the coal layers themselves. Locally in the Pensford Basin thin layers of coal and shale with nonmarine bivalves are found interbedded, indicating the temporary drowning of the coal forests by mud-bearing fresh water. There is relatively little evidence of seam deterioration due to passage of coal into grey fireclay. A more common cause of coals disappearing is the development of erosional surfaces or channels, which were then overlain or infilled by sandstone containing debris derived from the eroded beds.

The whole of the Radstock Formation is essentially of Westphalian D age (Wagner *in* Ramsbottom et al., 1978, p.47). Layers of dark grey laminated mudstone with *Anthraconauta* cf. *tenuis* amd *Anthracomya pruvosti* are found associated with the Pensford seams. These bivalves, and the accompanying phyllopod *Leaia*, have not been found in strata above the Barren Red Formation in the Radstock Basin. However, only a small part of the Radstock Formation has been examined in detail in the type area. Most of the published information (Moore and Trueman, 1937; Moore, 1938) relates to the roof measures of the one or two coal seams which were still in work in the 1930s. These were carefully searched for fossils, but most of the old workings on the Radstock seams have long been abandoned. However, it appears that *Anthraconauta* is probably rare, if not absent in the Radstock Formation south of the Farmborough Fault Belt, but becomes increasingly abundant farther north. On the other hand, *Anthraconaia prolifera* is present at Radstock, but was not found at Pensford Colliery or in the Hursley Hill Borehole, where opportunities for study were good.

Radstock Basin

Over virtually the whole of this region the strata belonging to the Radstock Formation are concealed beneath Triassic and Jurassic rocks, and knowledge of the sequence and the coals is derived from the numerous collieries, all now abandoned, sunk in the area between Kilmerdon, Clutton and Dunkerton (Figure 36a). Six main seams are more or less evenly spaced out in the lower 350 ft to 400 ft of the sequence (Figure 36d), and of these four were worked over most of the basin, although individually they rarely exceeded 27 in in thickness (Figure 37).

At the base the Nine-Inch Vein was seldom worked because of its thin section. At the Welton Colliery it reached 13 in and was worked over a small area near the shaft. About 20 to 40 ft above, the Bull Vein was the lowest widely worked seam (Figure 38). It showed considerable variation of both section and thickness, being thickest in the areas where the associated measures were most fully developed. North of a line from Welton Hill to Braysdown a dirt parting split the coal, which was largely unworkable north of Clandown, though a part of the split seam was worked as far north as Dunkerton. The seam appears to have been best developed at Kilmersdon Colliery [6876 5382], in the south of the basin, where clean coal was 28 in thick. The following sections illustrate the changes in the seam when traced northwards: Welton Hill Colliery—clift roof, coal 12 in, parting 1 in, coal 12 in; Lower Writhlington Colliery—coal 5 in, parting, coal 6 in, band 3 in, coal 13 in; Clandown Colliery [6803 5592]—clift roof, coal 10 in, stone 24 in, coal 10 in, pan floor; Dunkerton Colliery—strong blue shale roof, coal 18 in, blue shale 10 in, coal 2 in, fireclay floor.

The Bull Vein is separated from the overlying Bottom Little or Little Slyving Vein by 20 to 40 ft of shale and mudstone. A thin seam of good quality hard coal, it varied from 12 to 18 in. In some of the shaft sections of the Withy Mills–Radford area it has not been recognised. The seam is succeeded by about 25 to 40 ft of mudstone with some sandstone, above which the Slyving (Slyven) Seam, a rather soft seam of good quality gas coal, occurs. The name, according to Prestwich (1871, p.58) is derived from the Middle English 'slive'—a thin band—of cannel in the middle of the seam. The seam varies in full section from about 20 to 30 in (see Figure 36b), being thickest around the margins of the basin and thinnest along a median area running north-north-west to south-south-east. In the east of the area, the seam splits along the cannel parting to form the Two Coal Slyving of parts of the Clandown and Camerton [685 580] workings. East of a line joining Huish [6974 5405], Ludlows and Clandown collieries the dirt parting increases rapidly and the upper thin leaf of coal was not worked. Some splitting also appears to take place in a northerly direction, for at Fry's Bottom the Slyving Vein is recorded as being in three pieces: top coal 10 in, cannel coal 4 in, bottom coal 12 in; but the coal was worked as one seam. Other sections were recorded as follows: Welton Hill Colliery—clift roof, coal 9 in, cannel 2 in, coal 11 in, clod 6 in, on pan; Ludlows Colliery—clift roof, shab 2 in, good coal 12 in, inferior coal 4 in, stone 1 in, coal 15 in; Clandown West workings—top coal 9 in, dirt bine, coal 24 in; Tyning Colliery [6960 5520]—top coal (not worked) 10 in, dirt 10 ft, bottom coal 17 in; Braysdown Colliery—top coal 6 in, measures 31 ft, bottom

120　CHAPTER FIVE　COAL MEASURES (WESTPHALIAN)

Figure 35 Comparative vertical sections of the Radstock Formation between Hursley Hill and Radstock

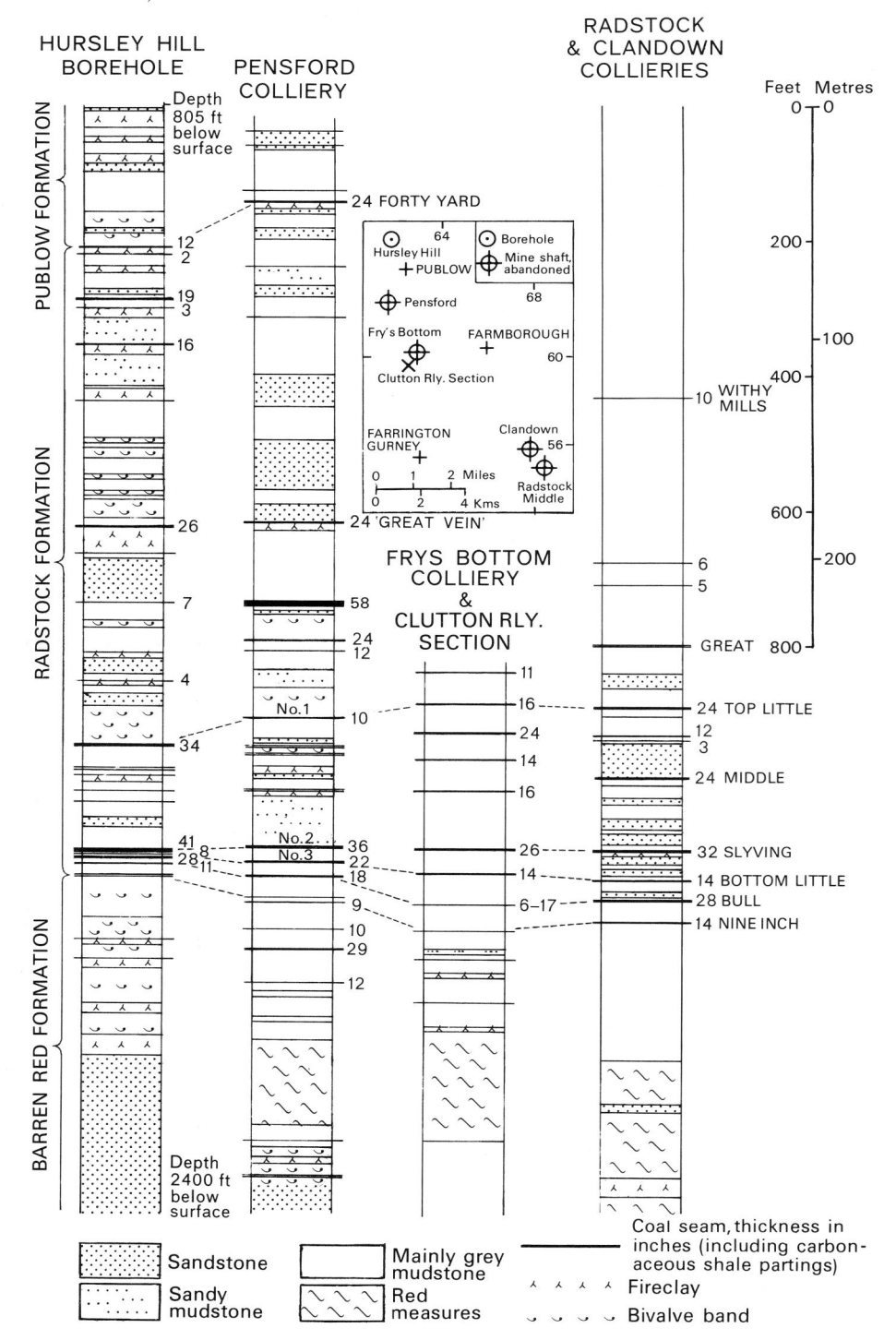

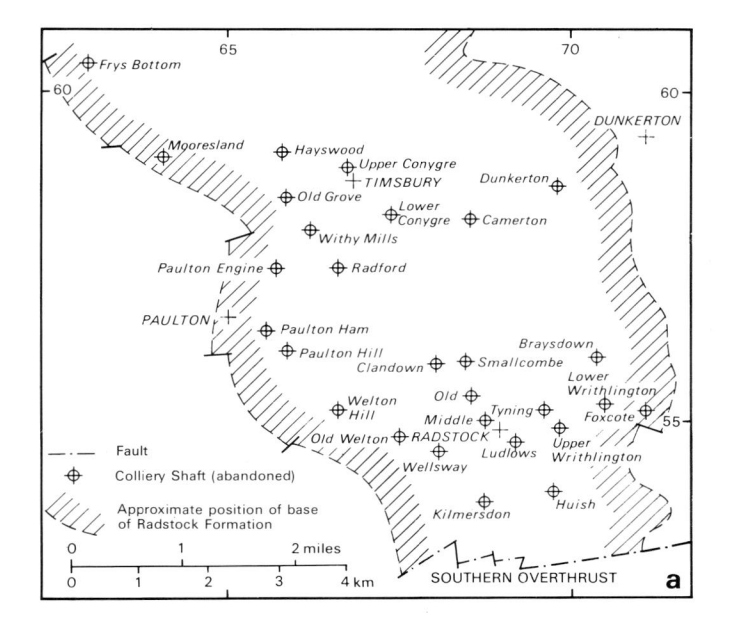

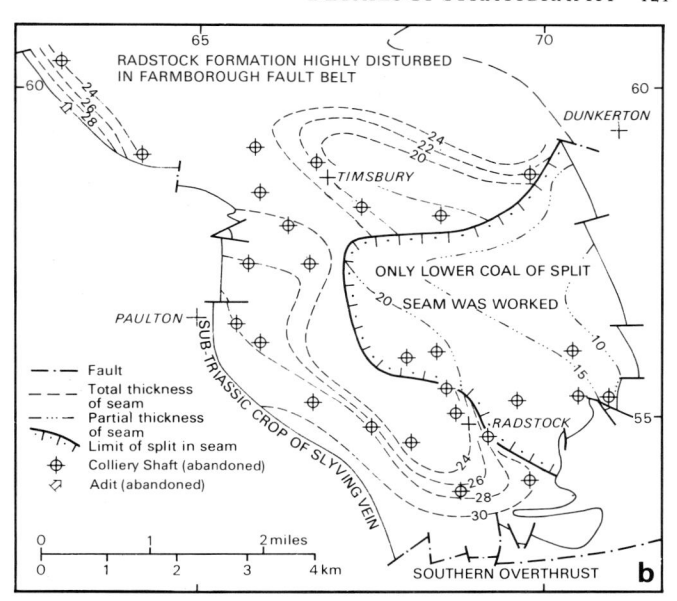

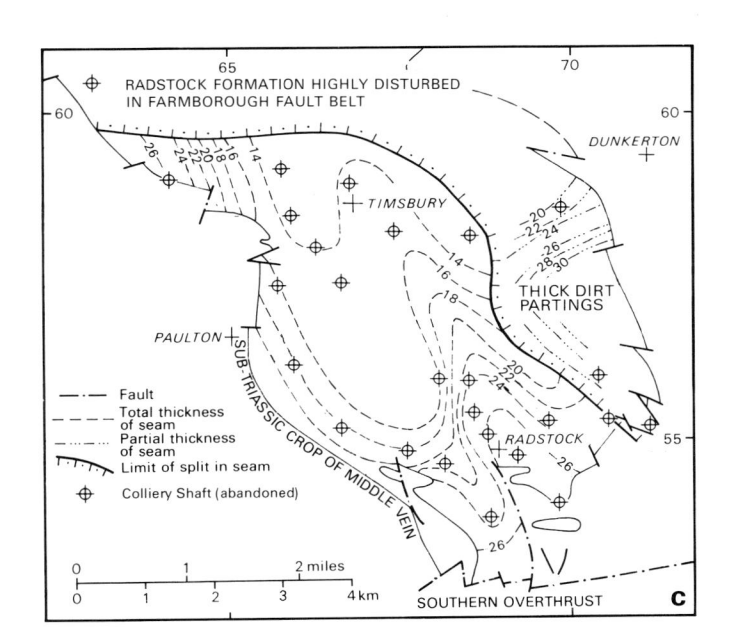

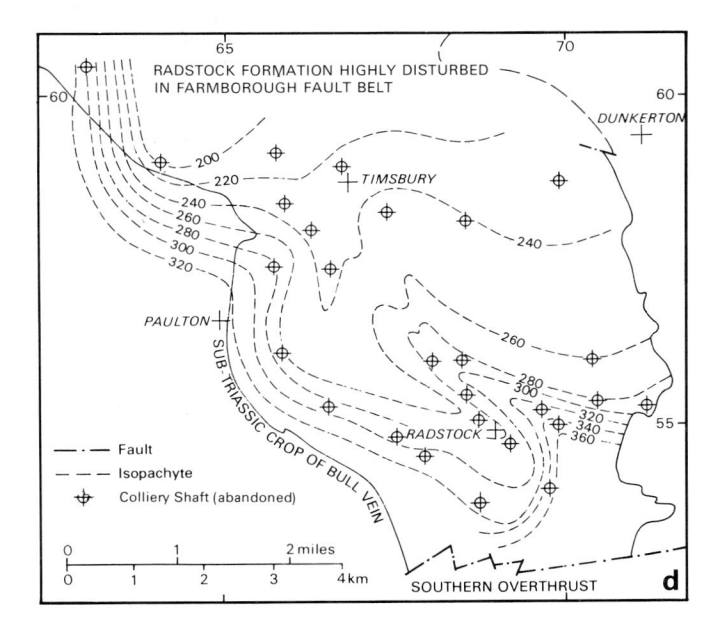

Figure 36
(a) Key map of the Radstock Basin showing approximate extent of the Radstock Formation and the names of collieries;
(b) Isopachytes of the Slyving Vein;
(c) Isopachytes of the Radstock Middle Vein;
(d) Isopachytes of the measures between the Bull Vein and the Great Vein

coal 16 in; Foxcote Colliery [7108 5518]—top coal 11 in, measures 64 ft, bottom coal 9 in; Nap Hill Adit [6275 5967]—clift roof, coal 9 in, coal 6 in, dirt 1 in, coal 15 in, black shale.

About 50 to 60 ft above the Slyving Vein, the Middle Vein is a rather 'dead' coal some 14 to 30 in thick (Figure 36c). The seam develops a north-eastern split along a line from Foxcote and Writhlington collieries through Dunkerton towards Fry's Bottom. The following sections have been recorded: Braysdown Colliery—sandstone, clift 0 to 14 in, top coal 7 in, binching ½ to 14 in, bottom coal 20 in, soft blacks, pan with a sporadic 3 in band of coal; Camerton Colliery (south side)—shale roof, coal 14 in, clod 7 in, coal 12 in; Fry's Bottom Colliery—top coal 10 in, stone 5 in, bottom coal 13 in.

The Top Little Vein, a coal of good quality, rarely exceeding 18 in, contrasts with the seams below in achieving its maximum thickness over the central part of the basin. Progressive thinning

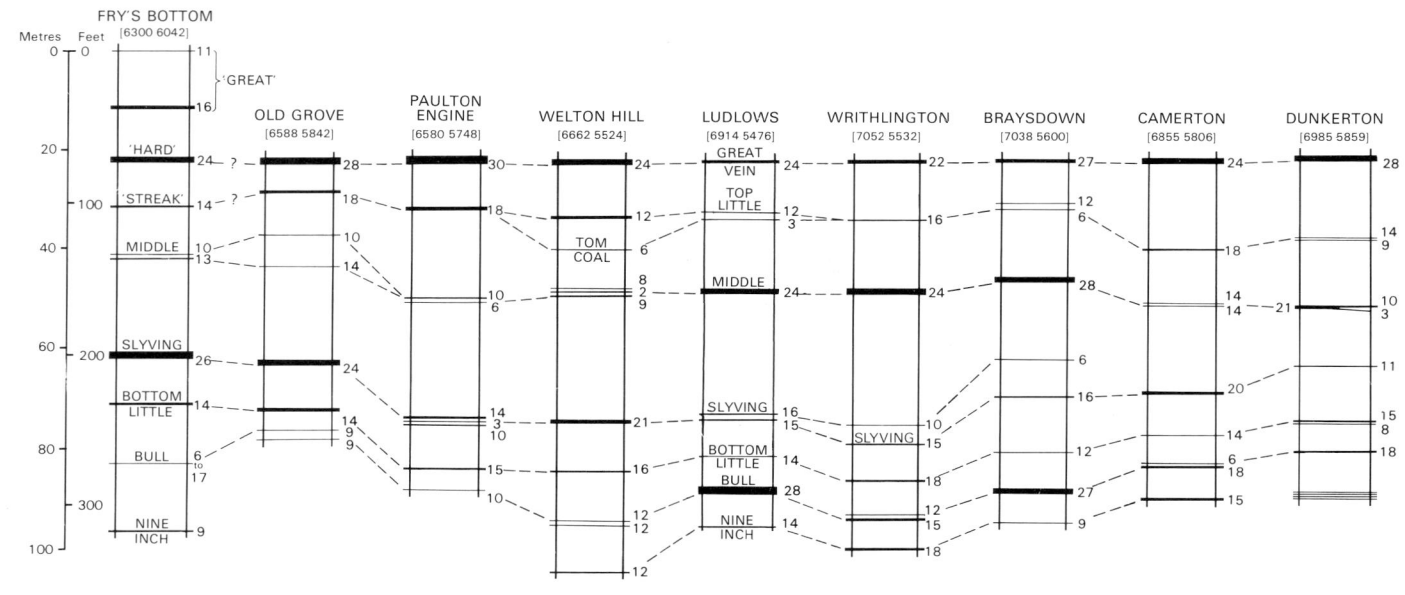

Figure 37 Diagram showing the probable correlation of the principal coals of the Radstock Formation south of the Farmborough Fault

towards the margins is combined with splitting along the eastern edge, as seen in the Camerton (Longside) workings: clift roof, coal 14 in, binching ½ to 9 in, coal 10 in. Farther east, in the Dunkerton Colliery workings, the dirt parting thickens to 10 in.

About 30 ft above the Top Little Vein, the Great Vein (Figure 37 and 38) was the best seam worked in the Radstock Formation, consisting for the most part of a single leaf of coal 24 to 26 in thick with a good roof.

The identity of the so-called 'Great Vein' of Fry's Bottom Colliery is difficult to establish. According to all published accounts the seam consisted of: top coal (good but tender) 11 in, measures 43 ft 8 in, bottom coal (inferior) 16 in. At a distance of 40 ft below this 'bottom coal' lies the 'Hard Vein', some 24 in thick. It is unusual for a persistent seam like the Great Vein, which has maintained a constant thickness throughout the Radstock Basin and is 26 to 28 in thick in the adjacent collieries south of Fry's Bottom, to split into two bands of coal separated by over 43 ft of measures. Moreover, the 'Hard Vein' of Fry's Bottom is a single coal, 24 in thick, lying the same distance (37 ft 10 in) above the 'Streak Vein' (equated with the Top Little Vein) as does the Great Vein above the Top Little Vein over the remainder of the basin (see Figure 37). Thus it is possible that the Fry's Bottom 'Hard Vein' may be the Great Vein, whilst the two coals called 'Great Vein' of Fry's Bottom are higher seams which are represented farther north by coals above the No. 2 seam of Pensford Colliery.

In general two seams, approximately 70 to 100 ft apart, were worked in the faulted ground north of Camerton Colliery (Figures 39 and 40). From their position they were termed 'Top Little' and 'Great' Veins, respectively, and it was considered that they represented the Radstock Veins of the same names with the seam sections changed by splitting and incoming of dirt partings. This view was accepted until late in the life of Camerton Colliery when an exploratory branch proved what appeared to be the true Great Vein, a one-piece coal 28 in thick, some 9 ft above the two-coal seam known as the 'Great Vein' (Figure 40). The so-called 'Top Little Vein' in the Northside Camerton laps is recorded as having the following general section: clift, top coal 12 to 18 in, blacks 9 to 18 in, bottom coal 3 to 18 in, blacks 12 to 36 in, pan; a section measured in the second lap was: clift, dirt 1½ in, top coal 20 in, dirt 3 in, bottom coal 17 in, the latter being locally replaced by slickensided and polished rashings. At Priston Colliery [6919 5942] the locally called 'Top Little Vein' which lies 65 to 70 ft below the 'Great Vein' has the following section: clift roof, shale 4 in, coal 24 to 36 in, black 6 to 24 in, pan.

At Camerton the 'Great Vein' in the Northside second lap workings (Figure 40) showed the section: clift roof, clod 2 to 10 in, coal 12 to 13 in, binching 0 to 3 in, coal 10 to 13 in, pan. In the second lap it was clift roof, rashings ½ in, coal 11 in, binching 3 in, coal 16 in, binching on pan. At Priston Colliery the section was: clift roof, shab 12 in, parting 1 in, coal 15 in, blacks on pan. These sections may be compared with those of the Middle Vein at Fry's Bottom Colliery given by Anstie (1873, p.78) as: top coal 10 in, stone 5 in, bottom coal 13 in. It would thus appear that the coals worked as 'Top Little' and 'Great' in the region of the Framborough Fault Belt at Priston may really be the correlatives of the Slyving and Middle veins of the Radstock Basin proper (Figure 38, 39 and 40).

At Priston Colliery a dip branch was driven north, intersection seven coal seams as listed below. The coals contain many shaly partings and it has not been possible to correlate them, though they are thought to lie within the main coal group of the Radstock Formation.

	Thickness ft in
No. 1 Seam (Firestone): clift roof, shab 2½ in, coal 6½ in, dirt 1½ in, coal 5 in, black shale 7 in, listy coal 10 in, shaly coal 14 in, pan	3 10½
Measures	56 0
No. 2 Seam: clift roof, coal 9 in, parting, coal 3 in, binching 3 in, coal 4 in, pan	1 7
Measures	34 0
No. 3 Seam: hard clift roof, coal 5 in, binching 3 in, shaly coal 9 in, parting, coal 11 in	2 4
Measures	44 0
No. 4 Seam: clift roof, good coal 7 in, dirt 2½ in, coal 3½ in, coaly shale 8 in, dirt 1 in, coal 3½ in, stony coal 13 in, pan	3 2½
Measures	45 0
No. 5 Seam: clift roof, shale 20 in, coal 10 in, coaly shale 7½ in, blacks 7½ in, pan	3 9
Measures	15 0
No. 6 Seam: no details	— —
	208 9

DETAILS OF STRATIGRAPHY 123

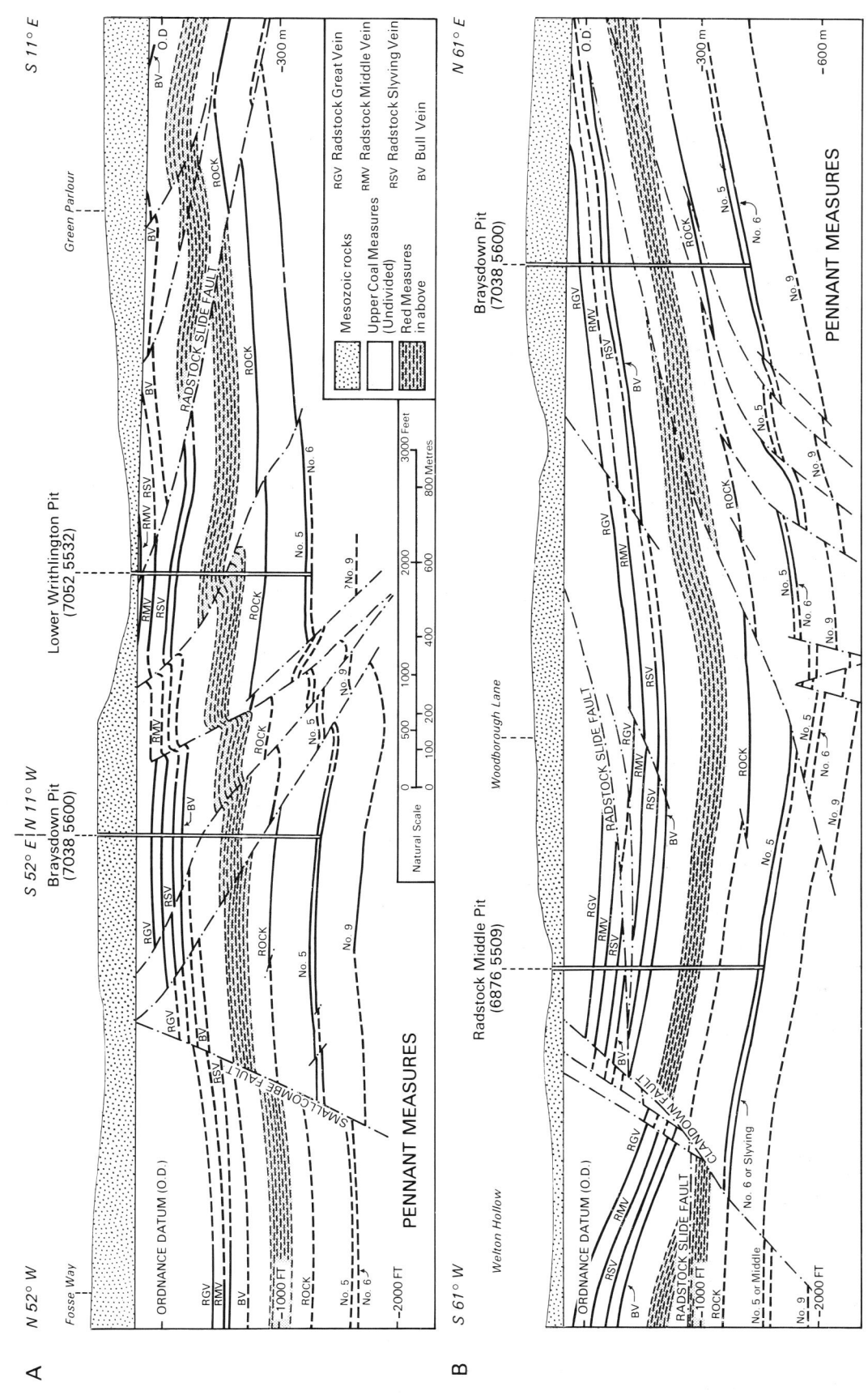

Figure 38 Sections showing the stratigraphy and structure of the productive Upper Coal Measures in the Radstock Basin

124 CHAPTER FIVE COAL MEASURES (WESTPHALIAN)

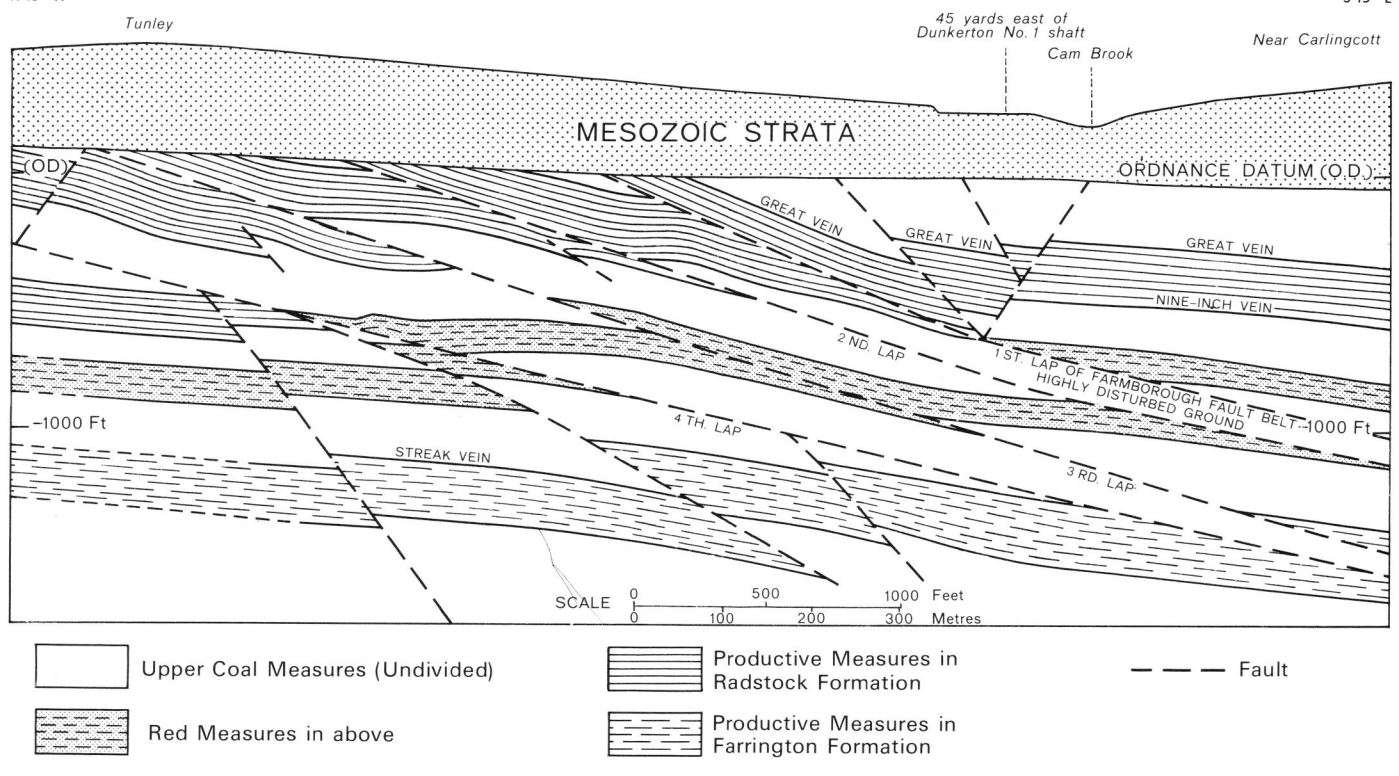

Figure 39 Section through the Farmborough Fault Belt between Dunkerton and Tenley showing faulting of the Upper Coal Measures

A seventh seam was reached, but there are no details of its position or section.

Little is known of the measures above the Great Vein in the Radstock Basin. Two thin seams were found in the Clandown and Radford [6662 5740] shafts. The Ore Seam lies 222 ft above the Great Vein of Clandown with two shaly coaly bands in the intervening strata (Buckland and Conybeare, 1824, p.278, sect.22). A 4 in coal is said to occur 90 ft above the Great Vein in the Radford Shaft. The Withy Mills Seams, taken as the top of the Radstock Formation, is 10 to 12 in thick and is separated from the Great Vein by a little over 400 ft of measures. It takes its name from Withy Mills Colliery [6620 5794] where it was intersected at a depth of 150 ft.

Pensford Basin

North of the Farmborough Fault Belt Supra-Pennant Measures crop out in the low-lying area between Chelwood, Hursley Hill and Compton Dando, drained by the River Chew and its tributaries. South of this area the intensely disturbed ground of the Farmborough Fault Belt is concealed by the Triassic and Jurassic strata in the uplands between Stowey and Marksbury Plain. To the north and east of the Coal Measures outcrops are bounded by the Mesozoic terrain extending from Winford to Hursley Hill and Burnett.

In the Bromley Horst the crop of the upper part of the Barren Red Formation is marked by red shales and fireclays about 230 to 300 yd east of Bromley Colliery shaft. The Pensford seams must crop still farther to the east, but the belt of exposed measures is narrow and the surface is covered with debris from old workings. The crops of the Pensford Nos. 2 and 3 seams shown on six-inch Geological Sheet ST 66 SW have therefore been calculated.

The most complete sections of the Radstock Formation in the Pensford Basin are those given in Pensford Colliery shaft and the Hursley Hill Borehole (Figure 35). Additional information has come from the examination of cross measures headings or branches at the colliery: the combined section of the Coronation Branch east of the shafts and the Jubilee Branch west of the shafts is given below.

Coronation Branch — measurement commenced 325 ft east of centre of North Shaft.

	Thickness		Depth	
	ft	in	ft	in
Streaky sandstone and mudstone	5	6	5	6
Streaky sandy mudstone with nonmarine bivalves	0	9	6	3
Coal, shale and fireclay	17	6	23	9
Mudstone and sandy mudstone with plants	13	10	37	7
Carbonaceous mudstone and coal	2	0	39	7
Pale grey fireclay and clunch	3	0	42	7
Measures (not seen)	10	0	52	7
Grey mudstone with siderite	1	0	53	7
Carbonaceous shale and dirty coal	1	0	54	7
Mudstone with siderite bands and plants	1	3½	65	10½
Rashings	0	1½	66	0
Sandy mudstone with *Calamites*	3	1	69	1
Hard sandstone	3	0	72	1
Measures (not seen)	10	0	82	1
Sandy and sideritic mudstone	5	8	87	9
Streaky sandstone and mudstone	9	8	97	5
Slick grey mudstone	0	2	97	7

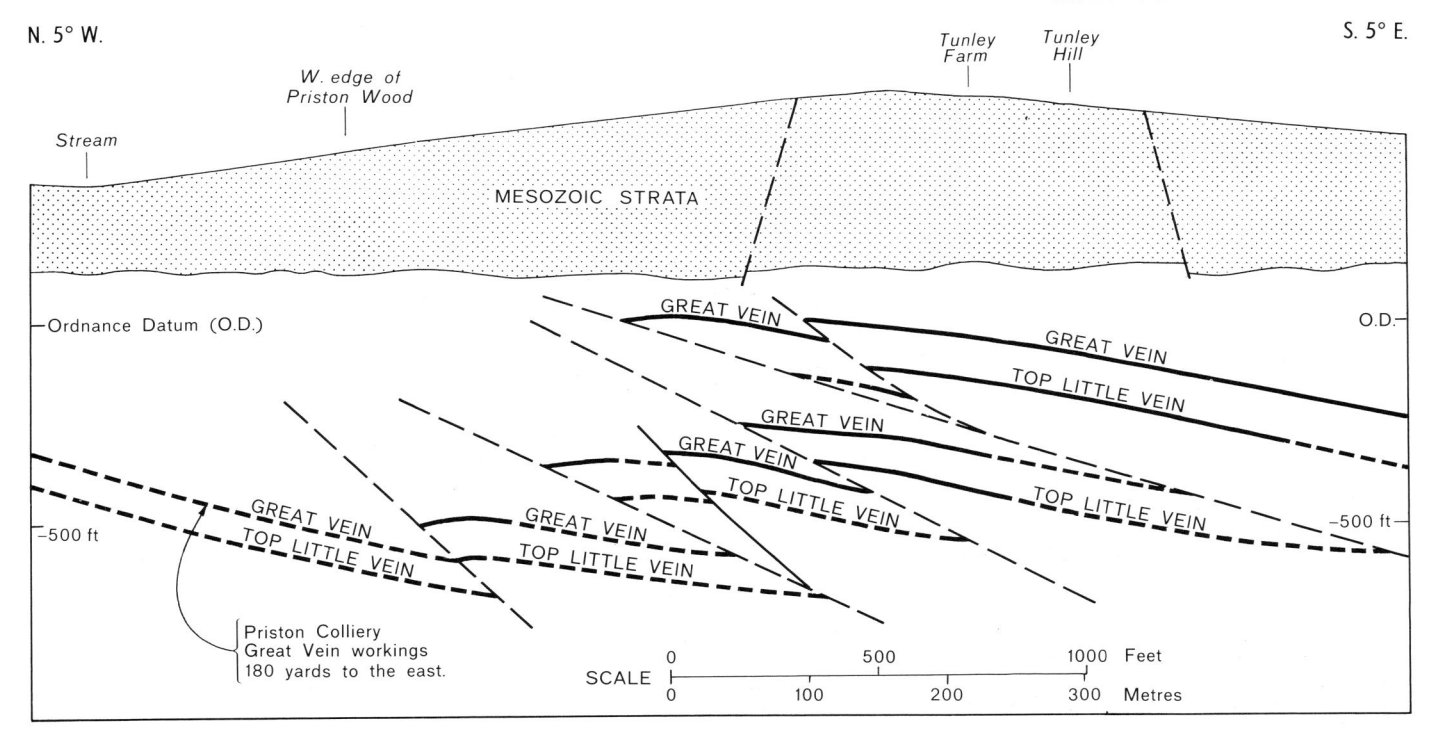

Figure 40 Section through the Farmborough Fault Belt proved in the Northside workings of Camerton Colliery showing faulting in the Radstock Formation

Hard sandstone	2	6	100	1
Interbedded mudstone and sandstone	13	11	114	0
Laminated grey mudstone with non-marine bivalves	0	6	114	6
Mudstone with sideritic layers	11	11	126	5
Laminated mudstone with nonmarine bivalves	0	4	126	9
Ironstone	0	1	126	10
Rashings and silty mudstone with ironstone bands	1	8½	128	6½
Laminated grey mudstone with non-marine bivalves	1	4	129	10½
Hard sandy ironstone-banded mudstone	9	6	139	4½
Laminated mudstone full of nonmarine bivalves; *Anthraconauta* cf. *calcifera*, *Carbonita spp.*	2	0	141	11½
Pensford No. 1 Seam (Gob)	1	1	143	0½
Medium grey fireclay	0	10	143	10½
Tom Coal: dirty coal	0	6	144	4½
Pale grey fireclay and mudstone with plants	4	0	148	4½
Grey mudstone with plants	1	9	150	1½
Carbonaceous shale, coal and rashings	3	7	153	8½
Fireclay	2	6	156	2½
Carbonaceous mudstone	4	5½	160	8
Fireclay and clunch passing down into mudstone with plant remains	3	0	163	8
Mudstone and streaky sandy mudstone with plant remains	17	6½	181	2½
Coal and rashings	3	4½	184	7
Fireclay	1	0	185	7
Sandy and sideritic mudstone	4	5½	190	0½
Laminated mudstone with bivalves	2	7	192	7½
Sulphury coal and rashings	3	3	195	10½
Fireclay	1	3	197	1½
Sandy and streaky sideritic mudstone with plant remains	7	8	204	9½
Mudstone with sideritic and coaly streaks	4	3	209	0½
Pale grey soft fireclay	5	0	214	0½
Grey mudstone with plants	3	9	217	9½
Sandy fireclay and mudstone	5	6	223	3½
Grey sandy silty mudstone with plants	19	6	242	9½
Coal and carbonaceous shale	2	11½	245	9
Fireclay, coal and shale	1	3	247	0
Sandy mudstone and clunch with rootlets	9	4	256	4
Coal and carbonaceous shale	2	3	258	7
Fireclay with ironstone nodules	2	0	260	7
Streaky sandy mudstone with plants	14	6	275	1
Dark grey laminated mudstone	0	4	275	5
Coal and rashings	0	9	276	2
Grey clunch with rootlets	1	6	277	8
Sandy mudstone with some sideritic layers and plant remains	13	0	290	8
Coal, rashings and carbonaceous shale	2	0	292	8
Sandy, streaky mudstone with some sideritic bands and *Calamites* stems and other plants	31	1	323	9
Dark grey slightly shaly mudstone with plant remains	0	4	324	1
Pensford No. 2 Seam: coal 1½ in, dirt ½ in, coal 7½ in, dirt 3 in, coal 4 in, dirt and coal 2 in, coal 1½ in, dirt ½ in, coal 3½ in, dirt and coal ¾ in, coal 8 in, dirt 4 in	3	0½	327	1½
Pale grey fireclay	2	0	328	11½
Tom Coal	0	4½	329	4
Fireclay and carbonaceous mudstone with *Neuropteris sp.* and some sideritic bands	19	6	348	10

	ft	in	ft	in
Carbonaceous shale in roof of coal				
Pensford No. 3 Seam: blacks 18 in, coal 22 in	3	2	352	0
Jubilee Main Branch				
Mudstone and fireclay	16	0	368	0
Pensford No. 4 Seam: hard clean coal 10 in, soft binching 3 in	1	1	369	1
Fireclay	3	0	372	1
Mudstone, mainly hard and sandy	23	0	395	1
Hard mudstone with ironstone bands, bivalves at base	8	0	403	1
Pensford No. 5 Seam: coal 8 in, soft coaly shale and fireclay 12 in, blacks and fireclay 30 in	4	2	407	3
Hard mudstone, not well exposed	10	0	417	3
Hard mudstone and sandstone passing down into mudstone; *Anthraconauta phillipsi*, *Carbonita spp.*, *Spirorbis sp.*	7	0	424	3
Mudstone and greys passing down into clift	20	0	444	3
Mudstone with ironstone nodules	5	0	449	3
Fracture cleaved shale	1	0	450	3
Hard mudstone with ironstone nodules	0	7	450	10
Pensford No. 6 Seam: shale and blacks 2 in, coal 3 in, binching 1 in	0	6	451	4
Fireclay	3	0	454	4
Hard sandy mudstone and sandstone passing down into mudstone with plants	23	0	477	4
Pensford No. 7 Seam: soft shaly coal with dirt bands	2	6	479	10
Ganiserised mudstone passing down into clift	35	0	514	10
Mudstone	10	0	524	10
Pensford No. 8 Seam: coal and dirt	0	9	525	7
Ganiserised mudstone	2	0	527	7
Shale with ironstone nodules and plants	2	6	530	1
Coaly streak	0	3	530	4
Barren Red Formation (see pp. 117–118)	—	—	—	—

West of Pensford Colliery the Forty Yard Seam is probably marked by the coal seen in the bed of Salter's Brook [6213 6202] 817 yd south-south-east of Pensford Colliery shaft. The crop is cut by the fault bounding the northerly margin of the Bromley Horst, and continues northwards to a position about 470 yd west of Pensford Church. To the west only the upper part of the Radstock Formation is exposed, the lower beds, together with the crops of the Pensford No. 3 and No. 2 seams, being concealed by Triassic rocks. North of Pensford the Radstock Formation continues as far north as Belluton, but beyond only the overlying barren beds of the Publow Formation continue as far as the foot of Hursley Hill.

Evidence of coal working survives at many places in the deep valley known as Fry's Bottom between East Chelwood Bridge [6300 6185] and Perry's Pit [6267 6008]. The measures exposed in the floor and sides of the valley show intense structural disturbance, reflecting the presence of the Farmborough Fault Belt which passes immediately north of Fry's Bottom Colliery [6300 6043], and it may well be that the crops shown on geological maps are over simplified. Other isolated areas where one or more thin coals have been worked are at Common Wood [655 625], east of Hunstrete, and to the south end of Compton Dando. The Common Wood seams are thought by some to belong to the Radstock Formation and to include the Slyving Vein, though there is no definite evidence to support this.

North of Fairy Hill [645 655] in the deep valley called Wooscombe Bottom, evidence of the presence of one or more seams has been found. These coals appear to lie above the two-foot seam worked at Chewton Keynsham Engine Pit [6520 6620] and may be situated in the upper part of the Radstock Formation. North of Fairy Hill the measures are faulted against Blue Lias, but in spite of the structural disturbance, the coal (or coals) was worked on both sides of the Chew Valley, especially in the area near Burnett [658 656].

A tract of Mesozoic rocks over 2 miles wide separates the exposed Supra-Pennant Measures at Hursley Hill from those at Brislington. South of Brislington there is evidence of a solitary coal above the Farrington Formation. This may be the Pensford No. 2 seam (or Nos. 2 and 3 seams combined) cropping in the low ground between Flowers Hill Farm [6620 6975] and Flowers Hill (see above, p. 118); if so, this is the only evidence for the existence of the Radstock Formation in south-east Bristol. The seam may correlate with the two-foot coal of Chewton Keynsham Engine Pit (see above).

Publow Formation

Above the Withy Mills Seam of the Radstock Basin and the Forty Yard Seam of Pensford lies the incomplete Publow Formation, which represents the highest Carboniferous rocks of Somerset, truncated upwards by Permo-Triassic erosion. First recognised and defined in the Pensford Basin (Kellaway, 1970), the formation consists for the most part of grey mudstones, silty mudstones and siltstones with some sandstones and a few sporadic thin coals of 2 ft or less and dug here and there at crop on a very limited scale. Little is known of the formation in the Radstock Basin, where it is almost wholly concealed beneath Mesozoic rocks, though it is known to reach about 425 ft in the Clandown–Camerton area.

Pensford Basin

Although the Publow Formation is exposed over much of the Pensford Basin, detailed knowledge derives mainly from the Hursley Hill Borehole [6180 6565] (Figure 35) drilled in 1951 for the National Coal Board at the northern edge of the exposed Coal Measures. Here over 1000 ft 6 in of beds were proved above the presumed position of the Forty Yard Seam, (Kellaway, 1970, p. 1047) summarised as follows:

	Thickness		*Depth*	
	ft	in	ft	in
Coarse grey-brown sandstone	39	3	39	3
Grey mudstone and fireclay; *Arthropleura* at base	19	9	59	0
Coal	0	6	59	6
Mudstone and fireclay; *Lepidoderma (Eurypteris)* cf. *wilsoni* at 68 ft	22	0	81	6
Muddy sandstone and sideritic mudstone, ironstone at base	15	7	97	1
Sandstone and mudstone	54	5	151	6
Fireclay	2	8	154	2
Mudstone with ironstone bands and nodules	38	11	193	1
Grey sandy mudstone with plants	29	11	223	0
Alternations of sandstone and mudstone	38	8	261	8
Mudstone with ironstone; plants	27	1	288	9
Fireclay and mudstone	15	9	304	6
Streaky sandstone and sandy mudstone	21	6	326	0
Dark carbonacous mudstone	1	0	327	0
Coal (dirty)	3	5	330	5
Fireclay and carbonaceous mudstone	7	1	338	4
Coal (dirty)	2	4	340	8
Fireclay with thin mudstone bands	15	2	355	10
Sandstone and micaceous mudstone; plants	17	8	373	6
Fireclay and carbonaceous mudstone with coal streak	0	10	374	4
Micaceous sandstone and sandy mudstone	15	2	389	6

Description	ft	in	ft	in
Well-bedded mudstone with ironstone bands; *Anthraconaia* aff. *pringlei* at 392 ft	11	3	400	9
Mudstone and fireclay with ironstone nodules	33	4	434	1
Alternations of sandstone and mudstone; plants	32	0	466	1
Coal	1	0	467	1
Clunch	3	3	470	4
Sandstone and mudstone	31	11	502	3
Fireclay	7	9	510	0
Micaceous sandy mudstone with ironstone bands	3	2	513	2
Fireclay passing down into mudstone with thin bands of sandstone	37	7	550	9
Mudstone with ironstone nodules; well-preserved plants	35	7	586	4
Grey sandstone	2	9	589	1
Mudstone and fireclay	28	5	617	6
Streaky sandstone and mudstone	8	3	625	9
Alternations of grey sandy mudstone, streaky sandstone and fireclay	148	9	774	6
Streaky sandstone and mudstone with slump structures	27	6	802	0
Streaky sandstone, sandy mudstone, and fireclay	31	10	833	10
Mudstone with ironstone nodules	13	10	847	8
Clunch passing into fine-grained sandstone and sandy mudstone with ironstone nodules	19	7	867	3
Mudstone and fireclay; plants	11	3	878	6
Rashings	2	6	881	0
Mudstone and sandy clunch	17	0	898	0
Sandstone and sandy mudstone with ironstone nodules	7	9	905	9
Mudstone; well-preserved plants	26	6	932	3
Fireclay passing into sandy mudstone and sandstone with ironstone nodules	16	2	948	5
Sideritic mudstone; plant debris	6	7	955	0
Smooth mudstone with sideritic bands; *Anthraconauta* aff. *phillipsii*, *A. tenuis*, *Carbonita sp.*	16	2	971	2
Carbonaceous mudstone, fireclay and dirty coal	1	8	972	10
Black mudstone with bivalves	0	8	973	6
Hard ganisteroid mudstone	5	7	979	1
Current-bedded sandstone and sandy mudstone	10	2	989	3
Smooth mudstone; *Anthraconauta* aff. *phillipsii*, *Euestheria sp.*, plants	4	7	993	10
Sideritic mudstone with thin sandy streaks; plants	12	2	1006	0
Coal (dirty)	1	0	1007	0

In an unweathered condition the rocks were mostly grey, buff or drab brown, but on exposure the non-grey colours became accentuated, with red-brown, pink and biscuit-coloured tints predominating. Nonmarine bivalves and associated fossils of the *Anthraconauta tenuis* Zone are particularly abundant in smooth mudstone bands throughout the Publow Formation, and particularly towards the base. The plants indicate a Westphalian D age.

Over much of the Pensford Basin the Publow Formation is poorly exposed and knowledge is confined to sporadic crop-workings in thin coals. The formation appears to reach a maximum thickness of nearly 2000 ft in the Compton Dando area, but apart from the 118 ft 7 in of measures proved in the top of the Pensford Shaft (Figure 35) there are few other sections of note. The main ridge on which Pensford Colliery was situated is composed of easterly dipping mudstone with substantial beds of hard current-bedded Pennant-type sandstone. These rocks are well exposed in the disused railway cutting [617 633] about 430 yd south-south-west of Pensford Church. They were formerly regarded by some authors as part of the Pennant Measures, but their position relative to Rydons Pit [6090 6300] led Anstie (1873) to regard them as being well above the Radstock seams. The floor of the Pensford Syncline roughly coincides with the deep gorge-like valley of Salter's Brook east of Pensford Colliery shafts. This fold continues northwards towards Pensford village and may correspond with a north–south syncline which has been identified on the surface near Parsonage Farm, Publow. East of Salter's Brook the rocks of The Common and the western side of Publow Leigh have a gentle westerly dip.

Part of the plateau surface of Publow Leigh is composed of thin red Triassic sandstone resting on the Coal Measures. A thin (unnamed) coal seam is partly concealed by the Triassic sandstone and there is evidence of working. From the levels taken underground in Pensford Colliery it would appear that this coal is about 1070 ft above the Pensford No. 2 Seam, that is, about 80 ft above the Forty Yard Seam and within the Publow Formation.

About 400 yd east of Leigh Farm an anticlinal fold with north–south strike separates the outlier of the Publow Leigh seam from the coal seam which has been worked near Birchwood House [6335 6347] and thence to Lords Wood [6356 6278]. This coal is almost certainly the Publow Leigh seam, noted above, exposed on the eastern flank of the anticline. It is overlain by a sandstone with an erosive base which rests in some places on the black shale roof of the coal. Elsewhere the coal and its roof shale have been removed by erosion during the deposition of the sandstone. The distribution of the old shallow pits and crop workings suggests that pits are numerous where the roof shale is intact, but are sparsely scattered or absent where the sandstone rests directly on the coal seam or its seatearth.

South of Lord's Wood and north of Chelwood village the Supra-Pennant Measures include a coal seam which has been worked on the sides of the deep valleys leading northward from Fry's Bottom and Hunstrete Plantation. This coal may also be the correlative of the Publow Leigh Seam.

South-east of Compton Dando in the Tuckingmill area [655 638] the Coal Measures appear to be almost horizontal or have very low westerly dips. At least one coal has been worked here, possibly in measures which are at the top of the Radstock Formation or the base of the Publow Formation. The seam appears to be subject to 'washouts' at the base of a sandstone and in this respect is very like the Publow Leigh Seam.

The massive quartzose sandstones of Compton Dando and Compton Common bear a strong general resemblance to those of the Publow Formation at Pensford and Hursley Hill. Among these rocks there are occasional lenses of very hard coarse quartzitic sandstone which are found at Hunstrete, at Compton Dando and near Hursley Hill. They appear to be characteristic of the Publow Formation and have not been seen below the level of the Forty Yard Seam or its correlatives.

CHAPTER 6

Triassic

In the Bristol district rocks regarded as Triassic in age comprise reddish arenaceous beds overlain by red mudstones and marls, in turn succeeded by pale and dark grey and greenish argillaceous and calcareous deposits. Younger beds overlap or overstep older ones in the proximity of contemporary topographical features formed largely of Carboniferous rocks, against which all formations display some modification, usually a coarsening, producing a marginal conglomeratic facies. Minor deposits of unusual character include the infills of karst and solution features developed on the contemporaneous surface of Carboniferous Limestone outcrops. The reddish rocks in particular show every sign of having been deposited under hot desert conditions in a number of largely interconnected, low-lying terrestrial basins which came into being following the uplift and erosion that accompanied the Hercynian–Variscan orogeny.

A revision of the lithostratigraphical nomenclature of the British Triassic sequences was published by Warrington et al. (1980). In addition, the position of the Triassic–Jurassic boundary was taken at the lowest occurrence of ammonites of the genus *Psiloceras*; this is always above the base of the Blue Lias, the lowest beds of which are thus regarded as Triassic in age (Cope et al., 1980; Warrington et al., 1980).

These revisions in nomenclature postdate the publication of the Bristol district special sheet, which employs the names used on the published 1 to 10 000 scale maps. The following account uses both sets of names; their equivalence is shown in Tables 3 and 4 and is indicated in brackets in the text.

GENERAL DESCRIPTION

South of the Mendip Hills up to 1500 ft of Triassic rocks are preserved beneath the superficial deposits of the Somerset Levels. They are poorly exposed in the Wedmore Inlier (Green and Welch, 1965) where Keuper Marl (Mercia Mudstone) and Rhaetic, together with Lower Lias strata, emerge from beneath the alluvium. Extensive low-lying tracts of peat and alluvium separate the Wedmore outcrops from the Triassic breccias and conglomerates which were deposited on the Carboniferous Limestone on the southern flanks of the Mendips.

Owing to the discontinuous nature of the outcrop of the Triassic rocks and the lack of evidence as to the precise position of their base, it is difficult to determine their exact thickness and the nature of the facies changes on the

Table 3 Stratigraphical classification of the Triassic rocks of the Bristol district

Sheets 250, 251, 264, 265, 280, 281 and Bristol District Special Sheet	Published six-inch and 1:10 000 sheets	Classification after Warrington et al., 1980		
		Group	Stage	System
Blue Lias & White Lias	Blue Lias [1]	Lias	Hettangian	JURASSIC
	White Lias [1]	Penarth Group	Rhaetian	
Rhaetic Cs	Rhaetic Cs			
Tea Green Marl	Tea Green Marl		Norian	TRIASSIC
Keuper Marl (including thin sandstones) Dolomitic Conglomerate	Keuper Marl (including thin sandstones) Dolomitic Conglomerate	Mercia Mudstone Group (including marginal conglomerates)	Carnian	
[2] sandstone	[2] sandstone			

[1] Basal Blue Lias and very thin White Lias shown as combined formation in marginal areas
[2] Redcliffe Sandstone (Kellaway, 1991 and this memoir)
Cs conglomeratic and sandy facies of Rhaetic

Table 4 Stratigraphical classification of the late Triassic and Lower and Middle Jurassic

Margin of Bristol district map (1962), based on mapping in 1938–39, 1943–52	Major subdivisions used in this text and in Donovan and Kellaway, 1984	Lithostratigraphy after Cope et al., 1980a, 1980b, and Warrington et al., 1980	Biostratigraphy Substage	Stage
		Upper Cornbrash		Callovian
Great Oolite	Cornbrash	Lower Cornbrash		
Forest Marble	Forest Marble including Hinton Sands and Upper Rags	Forest Marble		
Great Oolite Limestone	Great Oolite { Bath Oolite, Twinhoe Beds, Combe Down Oolite } Frome Clay	Great Oolite	Upper	Bathonian
Fuller's clay/limestone/clay Earth (mainly Fuller's Earth Rock clay) clay	Upper Fuller's Earth Clay (including Fuller's Earth Bed) Fuller's Earth Rock Lower Fuller's Earth Clay Fulliconus Limestone	Upper Fuller's Earth Clay (including Fuller's Earth Fuller's Earth Rock { Rugitela Beds, Ornithella Beds, Milborne Beds } Lower Fuller's Earth Clay { acuminata Beds, clay, knorri clays & fullonicus limestone }	Middle / Lower	
	Upper Inferior Oolite	Anabacia Limestone Coralline Beds Doulting Stone Dundry Freestone Coral Bed & Upper Maes Knoll Trigonia Grit Conglomerate	Upper	Bajocian
Inferior Oolite (limestone)	Middle and Lower Inferior Oolite	Elton Farm Limestone Grove Farm Beds	Lower	
		Barns Batch Beds Iron Shot Limestone		Aalenian
Upper Lias Midford Sands Clay Junction Bed	Midford Sands (0–18m), Cephalopod Bed (lsts.) & Upper Lias Clay (0–30m), Cotteswold Sands Junction Bed (0–3m)	Limestone Midford Sands Cephalopod Bed & clays Cotteswold Sands Junction Bed Junction Bed Clay & limestone		Toarcian
Middle Lias	Marlstone Rock Bed	Marlstone Rock Bed		Pliensbachian
Dyrham Silts	Dyrham Silts (to 9m)	Dyrham Silts sandstones, clays & shales (condensed Radstock sequence)		
Lower Lias Lower Lias Clays	Lower Lias Clay* (to 124m)	Lower Lias Clays (condensed Radstock sequence)		Sinemurian
Blue Lias†	Blue Lias* (including Saltford Shales) (to 40m)	Blue Lias (including Saltford Shales)		Hettangian
White Lias	White Lias (to 6m)	Penarth Group Lilstock Formation (Langport Member and Cotham Member) Westbury Formation	TRIASSIC	Rhaetic
Rhaetic	Rhaetic { Cotham Beds, Westbury Beds }			
Tea Green Marl	Tea Green Marl	Blue Anchor Formation		
Keuper Marl	Mercia Mudstone	Mercia Mudstone		

* Many local subdivisions identified on the Radstock sheet.

† There is a massive conglomeratic and shelly facies on Broadfield Down (up to 7.5m thick).

southern slopes of the Mendips. Elsewhere, as in the higher parts of the central Mendips, the local absence of the Tea Green Marl (Blue Anchor Formation) and the transgressive relationship of the black shales of the Westbury Formation (Green and Welch, 1965, fig. 2a) are additional complications. In general, however, the rocks of the Mercia Mudstone Group show progressive northerly attenuation towards the central Mendips and then thicken again towards the Vale of Wrington and the Chew Valley. The attenuation is probably attributable to a combination of depositional thinning and erosion of the upper beds prior to the deposition of the Penarth Group.

In parts of the eastern and central Mendips and the uplands of Broadfield Down, as well as locally along the margins of the Bristol coalfield, strata of upper Triassic (Rhaetic) and lower Jurassic age rest unconformably on Palaeozoic formations. In these areas the Mercia Mudstone Group is represented only by infillings of fissures and solution cavities in the Carboniferous Limestone.

North of the Mendips widespread removal of the Mesozoic cover makes it difficult to estimate the original

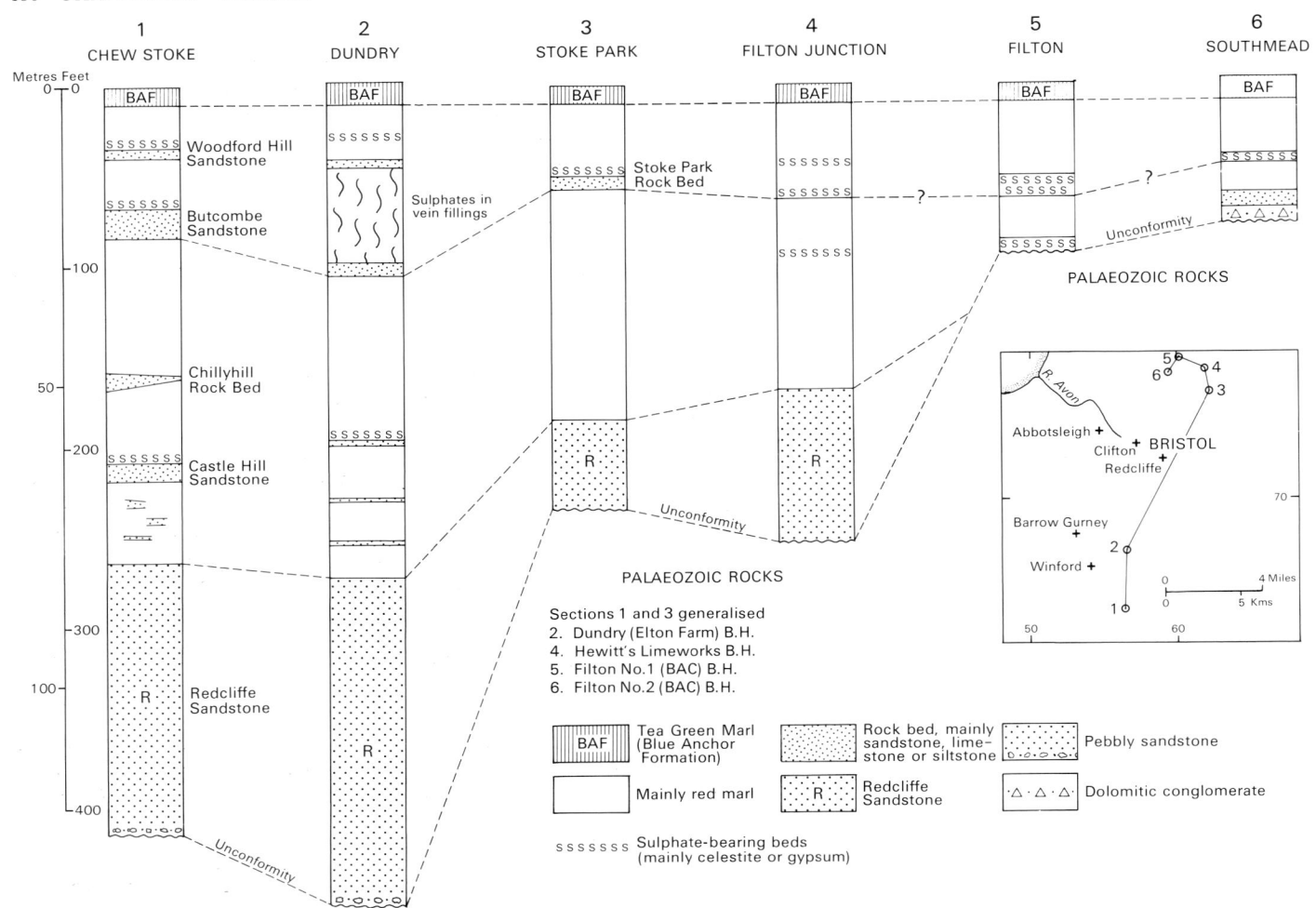

Figure 41 Comparative vertical sections in the Triassic rocks of the Bristol district

thickness of the Triassic deposits, except where the highest beds are preserved. In the Vale of Wrington, between Banwell and Butcombe, about 700 to 800 ft of rocks are present, but 2½ miles to the east in the Chew Stoke area there are only some 400 ft (Kellaway *in* Green and Welch, 1965, p.66). East of Chew Magna the lower part of the red Mercia Mudstone becomes increasingly arenaceous and the total thickness is not more than 100 ft along the western flank of the faulted ridge of Upper Coal Measures extending from Belluton and Pensford to Red Hill [623 605], Temple Cloud and Farrington Gurney. This fault-controlled hinge-line appears to have been an important influence on Triassic sedimentation, extending southwards to Stratton-on-the-Fosse and Shepton Mallet. The same zone of attenuation is traceable northwards from Belluton towards Brislington and east Bristol and thence north-north-eastwards to Frampton Cotterell, Tytherington and Purton. To the east of this line a thin red Mercia Mudstone sequence comprising red mudstone with a sandy or conglomeratic base seldom exceeds 100 ft, and is overlain by the thin green mudstones of the Tea Green Marl (Blue Anchor Formation) in the Radstock and Coalpit Heath basins.

Though widely developed in the Bristol and Somerset Coalfield the Tea Green Marl is seldom more than about 12 ft in thickness. It consists mainly of greenish mudstones with occasional harder, green, grey or yellowish, silty, calcareous sandstones. The base of the formation is seldom well defined. Green spots and blotches in the underlying red mudstones become more common and tend to merge as they pass into the overlying strata. About 30 ft below its base a hard greenish grey mudstone and siltstone corresponds to the Stoke Park Rock Bed of the Bristol area (Figure 41). The upper part of this bed and the lower beds of the overlying red and green marl contain celestite and gypsum which have been called the Celestine Bed (Kellaway and Welch, 1948), and have been described by Nickless et al. (1976) as the Severnside Evaporite Bed. This term is inappropriate (p.155) and the deposit has been re-named after Yate, the principal area of occurence.

The west-facing sub-Triassic declivity situated on the hinge-line separating the eastern and western parts of the Bristol Coalfield, though primarily an erosion feature, was probably in part controlled by faulting. The Redcliffe Sandstone Formation, consisting of highly ferruginous red sandstone, occurs to the west of this line (Figures 41 and 42) and is one of the older Triassic formations preserved in the Bristol district. It accumulated in an elongated depression with a floor sloping gently south-south-westwards from Latteridge, through Winterbourne, Stapleton, Redcliffe and Bedminster to Dundry. In Triassic times, therefore, it appears that red sand, derived largely from weathering of Coal Measures sandstones, was transported southwards towards Bristol and Dundry, being intercalated with muds and clays in the Chew Valley area (Figure 42). The boundary of the Redcliffe Sandstone with the Mercia Mudstone tends to be transitional and beds of red sandy marl are intercalated locally with the sandstones which pass into red marl in the western part of the vale of Wrington. At Long Ashton, Redland and other localities on the western margin of its depositional basin, the Redcliffe Sandstone passes laterally into the coarse clastic deposits long known as the Dolomitic Conglomerate, composed mainly of limestone debris and boulders derived from the Carboniferous Limestone. At Bristol, however, brecciconglomerates composed mainly of quartzitic sandstone fragments in a dark red, ferruginous matrix pass laterally into Redcliffe Sandstone and red, sandy mudstone, on the lower slopes of Kingsdown, St Michael's Hill, Brandon Hill and Clifton Wood.

West of Bristol Triassic rocks are widely distributed in the Severn Estuary north-east of Clevedon, and they overlie Coal Measures throughout the Avonmouth Basin. The classic Aust Cliff section (Plate 10) (Hamilton, 1977) shows about 126 ft of Keuper Marl and Tea Green Marl (Mercia Mudstone), overlain by the Rhaetic consisting of black shales of the Westbury Beds and the pale marls of the Cotham Beds with a nonsequence at the base of the Blue Lias, the White Lias being absent (Reynolds, 1947; Welch and Trotter, 1961). At Avonmouth the beds above the Keuper Marl have been removed by erosion, but at the eastern margin of the Avonmouth Basin the Westbury Beds and Cotham Beds are preserved between Hallen and Compton Greenfield. Here the total thickness of the Mercia Mudstone Group is about 200 to 225 ft. Sandstones, which may in part at least be contemporaneous with the Redcliffe Sandstone, occur in the lower part of the succession.

DETAILS OF STRATIGRAPHY

Redcliffe Sandstone Formation

Over much of the area east of the Whitefaced Fault and of a line extending southwards towards Farrington Gurney the Triassic sequence seldom exceeds 200 ft in thickness and only the basal beds are sandy or pebbly, the bulk of the succession comprising red or green calcareous mudstone and siltstone. West of the Whitefaced Fault at Hambrook, and of the Flowers Hill Fault at Brislington, the lowest part of the Trias is composed of dark red calcareous and highly ferruginous sandstones, typically developed in the eastern and southern parts of Bristol, notably at Stapleton, Easton and Redcliffe. The name Redcliffe Sandstone, which does not appear on the published one-inch geological sheets but which has been used informally for some time, is introduced here for these beds (Figures 41 and 42). At Harry Stoke, west of Hambrook, the sandstone is 90 ft thick, but to the north it thins and is last seen at Tockington as a thin calcareous sandstone interbedded with red sandy marl beneath the main mass of red Mercia Mudstone.

The brilliant red sandy soil developed on the crop of the Redcliffe Sandstone is recorded in the names of such districts as Redcliffe and Redfield now included in the Bristol urban area. A section showing the sandstone resting unconformably on Pennant Sandstone is still visible in the railway cutting west of St Anne's Park Station [620 722], but otherwise exposures are scarce in this part of Bristol. The sandstone can still be seen in the riverside cliffs at Redcliffe Parade, west of St Mary Redcliffe [5883 7230], but the best section in Bristol is on the south side of the New Cut between Bathurst Basin and Ashton Gate [581 721], where some 40 ft are exposed. South of the New Cut red sandstone underlies much of the densely populated area of Bedminster where it is known to be up to 175 ft thick, resting on Coal Measures. Many of the old colliery shafts at Bedminster and Ashton Gate were sunk through the sandstone into the underlying Coal Measures.

Until recently the Redcliffe Sandstone was an important aquifer for Bristol (Richardson, 1930, pp.226–37), but many of the old wells and boreholes are now disused. Bristol Castle and the old walled town were built on a promontory of Redcliffe Sandstone,

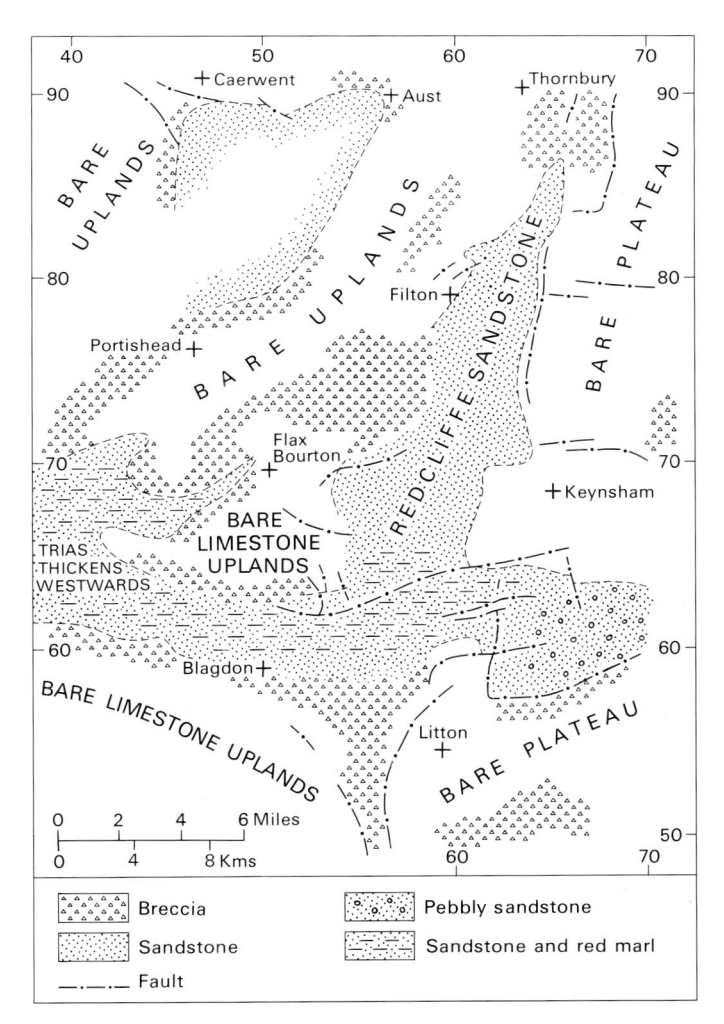

Figure 42 Sketch map showing the distribution of the Redcliffe Sandstone and homotaxial formations in areas east of the Severn

bounded on three sides by salt marshes and tidal waters, and were supplied with potable water from wells sunk in massive red sandstone. The sandstone also provided the foundations for Bristol Bridge, which was, in the Middle Ages, the lowest crossing point on the Avon.

West and south-west of Bedminster in the low ground extending from Long Ashton to Cambridge Batch and Flax Bourton the Redcliffe Sandstone passes into interbedded sandstone, sandy marl and conglomerate and loses its identity. South of Bedminster it extends beneath the red Mercia Mudstone under Dundry Hill and southwards towards the Chew Valley. The Dundry (Elton Farm) Borehole [5636 6589] proved 480 ft of Triassic rocks of which the lowest 206 ft consisted of fairly massive calcareous and ferruginous sandstone with a little interbedded red mudstone, resting unconformably on Middle Coal Measures; the basal 5½ ft contained pebbles of Coal Measures sandstone. Farther south, at Chew Magna, the Redcliffe Sandstone is represented by sandy marls and sandstones about 150 ft thick.

West of Broadfield Down, the No. 1 Well at Chelvey Pumping Station [4736 6797] proved 312 ft of Triassic rocks of which the topmost 100 ft consisted of red mudstones and the lowest 30 ft of marginal Dolomitic Conglomerate; the remainder was sandstone with some pebbly beds that may well be homotaxial with the Redcliffe Sandstone. Thick beds of Triassic marl, sandstone and conglomerate are also present in the western part of the Severn Tunnel (Figure 19). At the eastern end 50 ft of red and yellow sandstones are overlain by red gypsiferous marls similar to those seen at Aust Cliff. At Sudbrook in the Middle Shaft at the western end of the tunnel a basal conglomeratic sandstone is separated from a higher yellow sandstone by 65 to 68 ft of red marl. This is unusual, for nowhere at the surface does such a thickness of intervening red marl appear (Welch and Trotter, 1961, p.117).

In the Avonmouth, Clapton and Nailsea basins the red marls and sandstones in the lowest part of the Triassic tend to pass laterally into, or interdigitate with, Dolomitic Conglomerate. The resulting pattern of sedimentation is complex, with lenticular beds of sandstone and marl interlocked with, or intersected by, fan-shaped spreads of breccia and by channel-fillings composed of sandy and pebbly material.

In spite of its ferruginous character the Redcliffe Sandstone is strongly calcareous, though cementation is irregularly developed. Decalcified Redcliffe Sandstone is soft and friable when dry and quickly breaks down when saturated with water. Waterlogged, red, ochreous sand produced by digging or tunnelling in saturated Redcliffe Sandstone can become thixotropic as a result of vibration and is then difficult to handle. The use of the red calcareous sandstone for land restoration purposes also presents problems as it readily decalcifies. This process can cause subsidence if the sandstone is penetrated by migrating groundwater.

No fossils have been found in the Redcliffe Sandstone.

Keuper Marl

The Keuper Marl (Mercia Mudstone) sequence consists mainly of dominantly red mudstones, normally calcareous and with green mottles and bands. Thicknesses vary greatly and in areas such as west Bristol, Chipping Sodbury and the central Mendips it is missing altogether. This variability is well illustrated at Aust, where trial borings behind the cliff were described by Whittard (1948). The cliff section (Plate 10) shows some 115 ft of red Keuper Marl whereas immediately east of the cliff thicknesses of up to 184 ft were proved. In the Dundry (Elton Farm) Borehole the corresponding thickness is 260 ft and a comparable thickness is present at Chew Stoke (Kellaway *in* Green and Welch, 1965, p.66). The greatest thickness recorded north of the Mendips is about 700 ft at Banwell Moor.

Though water-laid and predominantly of silt grade (Hamilton, 1977), the mudstones are poorly laminated and well jointed with a curving or conchoidal fracture. Beds of hard, sandy mudstone or siltstone (skerries), which are seldom more than 20 ft thick, occur sporadically. These are usually grey or mottled red and greenish grey, interbedded with softer grey or green mudstone or clays and are normally quite strongly calcareous or dolomitic, passing locally into fine-grained limestone or conglomerate. Beds of impure, dolomitic, silty sandstone with small amounts of calcite and sulphates also occur. Some of the calcareous rocks contain voids or cavities due to the leaching out of carbonates and sulphates. Porous rocks of this kind are permeable to groundwater and may be structurally weakened by the production of solution cavities.

Typical sections through composite rock beds (or skerry bands) have been given by Woodward (1876, p.59) and Green and Welch (1965, pp.71–72). The Stoke Park Rock Bed of the Bristol Coalfield produces a very strong physical feature that can be recognised and mapped over wide areas (Figure 41). This and other hard rock beds intercalated with the softer red marls are of great value in determining the local stratigraphy and structure, particularly in areas such as the Chew Valley where a number of hard rock beds are developed (Kellaway *in* Green and Welch, 1965, p.66). Siliceous sandstones are comparatively rare in the red Mercia Mudstone but are known to occur locally in the Radstock Basin where they have been termed 'firestones'.

In the Radstock Basin the Keuper Marl (Mercia Mudstone) is some 200 ft thick and consists mainly of red and green mudstone with only a few persistent hard bands. The basal beds are composed of dark red sandy mudstone with a few scattered pebbles resting on a thin layer of Dolomitic Conglomerate. About six miles north of the Mendips conglomeratic sandstones with fragments of Carboniferous Limestone occur at the base of the red Keuper Marl in the Pensford–Chelwood area, resting unconformably on Coal Measures. The sandstones thicken southwards at the expense of the mudstones; at Radstock they are 12 to 24 ft thick and on the northern margins of the Mendips they pass into massive Dolomitic Conglomerate, notably around Chilcompton and Stratton-on-the-Fosse. Further to the north-west in the Yeo Valley between Bourne and the Blagdon Reservoir, 'pale conglomerates' again occur intercalated with red marls and sandstones. One of these bands has yielded fish remains, including *Saurichthys apicalis*, as well as the bivalve *Modiolus minimus* (Woodward, 1876, p.80; Green and Welch, 1965, p.71).

Similar facies changes involving the passage of red marls and sandstones into conglomeratic rocks characterise the uplands surrounding the Nailsea, Avonmouth and Coalpit Heath basins. Correlation of individual beds is difficult over any distance, but some breccia beds, for example the Westclose Conglomerate, can be traced for considerable distances (Green and Welch, 1965, p.68).

Although the red and green mudstones of the local Keuper Marl are generally similar to their counterparts in the English Midlands and central Somerset, their associated evaporite deposits show significant differences. In particular, the Bristol evaporites contain strontium sulphate (celestite) (Figure 41) associated with some barium sulphate (baryte); gypsum and secondary calcite and dolomite also occur fairly widely. Halite as such has not been detected, though the gypsiferous red marls of the Avonmouth Basin and Aust Cliff include greenish grey mudstones with clay pseudomorphs after halite crystals.

Celestite is present in exploitable quantity mainly in the marginal areas of the shallow Triassic sedimentary basin surrounding and overlying the Bristol and Somerset Coalfield. One such area, around Yate and Westerleigh, has for many years been an important world source of the mineral (Baker, 1902; Sherlock and Hollingworth, 1938; Nickless et al., 1976). Primary celestite forms

Plate 10 Triassic and Jurassic rocks in the section at Aust Cliff [5645 8920], south of the Severn Bridge.

The upper part of the red Keuper Marl is overlain by pale Tea Green Marl and then the black shales and thin limestones of the Westbury Beds. The succeeding Cotham Beds are pale grey mudstones with limestone bands; the overlying flaggy limestones lie at the base of the Blue Lias. A normal fault cutting the section throws down about 10 ft to the south.

reddish-coloured nodules up to 6 in in diameter, consisting of platy crystals, embedded in red marl and forming impersistent seams roughly parallel with the bedding. The tabular (orthorhombic) white or blue crystals, commonly exhibited in museums, are found in irregular drusy cavities and veins in red marl or Dolomitic Conglomerate, nearly always exhibiting evidence of recrystallisation and secondary emplacement. Locally such deposits may extend downwards into joints and fissures in underlying Triassic or Palaeozoic rocks (Baker, 1902, pp.163–164).

Replacement of primary gypsum (or anhydrite) by celestite has been observed at many localities (Nickless et al., 1976), some deposits (e.g. at Barrow Gurney) being an intimate mixture of gypsum and celestite. Baryte and barytocelestite are less common, though there is some evidence to suggest that the barium content of the evaporites increases at the depositional margins of the basins. Baryte is the dominant mineral in the veins filling fissures in Carboniferous Limestone uplands adjacent to the sedimentary basins in which the evaporites were laid down.

Very small amounts of galena and blende are present in the evaporites or in vughs developed in the hard greenish grey calcareous and dolomitic mudstones associated with the evaporite beds. Thus at Pill [5227 7613] hard grey mudstones intercalated with red marl contain galena and blende in association with sulphates (mainly baryte), and at Abbots Leigh [545 751] small cubic crystals of galena have been found embedded in celestite nodules.

The main evaporite horizon to the north of the River Avon is associated with the Stoke Park Rock Bed, which at Stoke Park [622 773], north of Stapleton, consists of hard greenish grey sandy and silty dolomitic mudstone. Celestite-bearing mudstones are present immediately above the Stoke Park Rock Bed at Sims Hill [6272 7776] and Stoke Gifford [6328 7962]. The sulphate-bearing mudstones and the underlying rock-bed are included in the 'Severnside Evaporite Bed' as defined by Nickless et al. (1976)[1]. In north Bristol the Stoke Park Rock Bed and its associated celestite

---

1 It is proposed to discontinue the use of the term 'Severnside' and substitute Yate (see p.155).

deposits occupy a position about 40 to 50 ft below the base of the Tea Green Marl. The general correspondence of the celestite and gypsum-bearing marls of the Coalpit Heath and Avonmouth basins and their probable extension southwards to Bitton has long been recognised (Kellaway and Welch, 1948, p.47). South of the Bristol Avon, however, correlation of the evaporite deposits is less certain as gypsum and celestite occur at more than one horizon in the Triassic rocks.

A detailed investigation of areas north-east of Bristol, where there are prospects of future development, is described by Nickless et al. (1976). Other occurrences in the Chepstow (sheet 250), Malmesbury (sheet 251) and Wells (sheet 280) districts have been described by Welch and Trotter (1961), Cave (1977) and Green and Welch (1965) respectively. The urban area of Bristol and the region extending from Bristol to the Chew Valley and the Mendips is not adequately covered by these accounts, however, and is therefore considered in greater detail below.

The record made at Compton Dundon by William Smith in 1815 and reproduced by Moore (1867, p.457) shows that celestite ('strontia') is present in the Tea Green Marl south of the Mendips, but that the principal evaporite bed lies in the red mudstones 200 ft below and consists almost entirely of gypsum (Figure 41). The Banwell Moor Borehole [3995 6087] also proved gypsiferous marls about 200 ft below the Tea Green Marl (Green and Welch, 1965, pp.198–199). It is possible that this horizon may correlate with the gypsum-bed of Compton Dundon and Aust Cliff and it may also mark the position of the celestite-bearing strata associated with the Butcombe Sandstone of the Chew Valley and Vale of Wrington and with the Westclose Hill Conglomerate of Easton (Green and Welch, 1965, fig.7).

In the Chew Valley (Figure 41) the red Keuper Marl includes several sandstone beds in its upper part. Celestite is known to occur within or just above the Woodford Hill, Butcombe and Castle Hill sandstones, the best developed being found above the Woodford Hill Sandstone some 25 ft below the Tea Green Marl. The probable relationships of the red Keuper Marl strata of Chew Stoke, the Dundry Borehole and Stoke Park (north of Bristol) are shown in Figure 41, and an extension of this correlation to the Vale of Wrington and the Mendips was illustrated by Green and Welch (1965, fig.7, p.67).

Scattered occurrences of celestite are known in the Chew Valley and at localities around the southern and eastern flanks of Broadfield Down. Of these, Regilbury [5276 6285] and Winford [5424 6460] appear to have been the richest deposits, though there is little published information. Both of the areas where workable deposits occur are situated at or near the passage of red marl into Dolomitic Conglomerate. Their situation and geological environment are comparable with that of the deposits of celestite proved in the 19th century in Clifton and with those worked at Abbots Leigh prior to 1914. The Abbots Leigh deposits were among the richest ever worked in the Bristol district, possibly less extensive but otherwise comparable with the massive nodular and coarsely crystalline celestite worked in the highly productive Stanshawes Court area, south of Yate [715 815]. Boulder-size masses of celestite up to half a ton in weight were extracted from yellow and red mudstone overlying Dolomitic Conglomerate in the ground north-west of Leigh Court [544 751].

At Abbots Leigh, as in some other areas where the celestite is located within or immediately above Dolomitic Conglomerate, the correlation of the celestite-bearing strata is doubtful. Comparison of marginal deposits such as those of Filton and Southmead (Figure 41), Cotham, Clifton and Westbury-on-Trym, suggests that these may equate with more than one individual celestite-bearing horizon in the basinal sediments. However, all the known celestite occurrences in the Bristol district are younger than the Redcliffe Sandstone and the majority are probably of an age which is intermediate between that of the Butcombe Sandstone and the Tea Green Marl (Figure 41).

At Barrow Gurney a celestite-bearing horizon occurs about 5 to 9ft below the Tea Green Marl, and Nickless et al. (1976, fig.21, p.71) give a number of localities where it has been proved, though not worked. The crop is marked by a feature formed of hard red and green silty mudstone which can be traced from Barrow Gurney by way of Redwood Farm [5265 6906] north-eastwards towards Wild Country Lane [5345 6930].

The succession established in the Barrow Gurney area is almost identical with that of Winford where a bed of nodular granular celestite up to about 2 ft thick was formerly exposed in the banks of a sunken lane leading to Dundry [5405 6512]. Here the evaporite-bed is situated above a sandstone about 20 to 30 ft below the Tea Green Marl. This sandstone makes a small but distinct feature on the valley side that can be followed eastwards as far as Court Farm [5543 6495].

Beyond Court Farm a rock-bed occupying a similar stratigraphical position gives rise to a plateau area on the spur north of Lane End Farm [555 643]. The same rock-bed can be traced along the western face of Limekiln Hill and thence by way of Hillgrove [5719 6389] into the steep-sided gully which joins the Winford branch of the River Chew at the west end of Chew Magna village. East of the gulley a sandstone feature about 40 to 45 ft below the Tea Green Marl can be traced for about a quarter of a mile before dying out. The red Keuper Marl is cut by a fault in the area north of Stanton Drew. East of this fault it has not been possible to trace the celestite-bearing rock-bed but an 'anomalous concentration of strontium' recorded by Nickless et al. (1976, p.73) is situated in a position where it might be expected to occur if the Winford evaporite bed continues in this direction.

Beneath the Winford rock-bed and its evaporite horizon a more massive and continuous rock-bed can be traced from the valley floor south-east of Winford by way of Littleton Court [5527 6455] and Lane End Farm [5558 6423] to Bitham's Wood [561 642] but appears to die out north of Chew Magna. Another bed of sandstone and hard green mudstone, some 10 to 15 ft below, first appears at Limekiln Hill, north of Port Bridge, and is traceable to Halfway House [5943 6360], Hautville's Quoit [6017 6380] and Amercombe Cottage [6163 6492] north-east of Belluton. It is possible that these two beds represent the Butcombe Sandstone. The underlying red marl is sandy and shows rapid easterly attenuation at Belluton due to overlap against the rising surface of the Coal Measures of Pensford and Publow.

East of Winford only traces of celestite have been found. The richest deposits crop out to the south of the church and these may be of the same age as those of Regilbury [5270 6290]. Occurrences described by Nickless et al. (1976) at Knowle Hill [583 613] are likely to be at the horizon of the Woodford Hill Sandstone.

Appreciable quantities of celestite have been reported within or adjacent to the urban area of Bristol, notably at Clifton and Abbots Leigh. In addition to these localities, of which only one (at Abbots Leigh) has been worked, quantities of celestite have been observed at a number of other localities in west-central Bristol including Kingsdown, Cotham and Redland. The occurrences in Clifton are mainly in red marl resting on Dolomitic Conglomerate or in vughs within the conglomerate itself, as at the entrance to Clifton Down Tunnel [5737 7417]. One of the most interesting localities is situated on Clifton Down above the well-known road-side section in Bridge Valley Road [5640 7376]. An account of this section was given by De la Beche (1846, pp.246–247, fig.26) who recognised the presence of fine-grained dolomitic limestone and breccia resting on the coarse red calcareous conglomerate seen in the road-side cutting. These upper beds have since been found to contain interstitial celestite and are overlain by sulphate-bearing red or yellow mudstones marking the position of the main evaporite horizon at

Clifton. The rich celestite deposits in Clifton were described by Collie (1879) and Tawney (1878). In the adjoining districts of Cotham and Kingsdown it would appear that the main celestite horizon is situated about 20 to 30 ft below the Tea Green Marl. Since it can be traced almost continuously from Clifton to Cotham and thence to Narroways Hill and Stoke Park it would appear that the celestite bed of Clifton is likely to equate with the Stoke Park Rock Bed.

Tea Green Marl

In the Bristol district the Tea Green Marl (Blue Anchor Formation) at the top of the Keuper Marl seldom exceeds 12 to 15 ft in thickness. Hard, greenish grey, sandy, calcareous and dolomitic mudstones with interbedded impure limestone and dolomite bands dominate the sequence, which is in general terms similar to the underlying red Keuper Marl in all but colour. The striking change in appearance is well seen in the famous Aust Cliff section near the M4 Severn Bridge. The base of the formation is poorly defined; the transition from dominantly red to dominantly pale green mudstone occurs over about 3 ft. Definition of the upper limit of the formation, though difficult south of the Mendips, presents no problems farther north, where greenish grey mudstones are succeeded abruptly by dark grey-black shales of the Westbury Beds. In the Wedmore area the Tea Green Marl also includes beds of dark grey or black shale, which are generally distinguishable from shales of the Westbury Beds by their harder nature and powdery texture.

Soils produced by the weathering of the Tea Green Marl tend to be drier and less sticky than those of the red Keuper Marl below and the Westbury Beds above. The presence of hard greenish grey sandy and calcareous rock beds gives rise to a characteristic rib-feature, already remarked upon by Welch and Trotter (1961, p.114). At some localities, such as Patchway, north of Filton, the beds form substantial plateau areas, and for this reason the outcrop is shown separately on the published six-inch and 1:10 000 maps.

South of the Mendips the Tea Green Marl is up to 130 ft thick in the Wedmore inlier (Green and Welch, 1965, fig.2, p.68). The uppermost 50 to 65 ft may be correlated with the Grey Marl of the west Somerset coastal sections (Richardson, 1911; Whittaker, 1973; Warrington and Whittaker, 1984) and of Uphill (Kellaway and Oakley, 1934; Whittaker and Green, 1983). The Grey Marl is absent or only poorly developed north of the Mendips, as for example at Chewton Keynsham (Kellaway and Welch, 1948, p.47).

Divergent views are held as to the origin of the green coloration of the Tea Green Marl. The presence of a passage zone of green-veined or mottled red mudstone at the base suggests post-depositional reduction of the iron oxides, possibly during the deposition of the black pyritic shales of the Westbury Beds which rest nonsequentially on the Tea Green Marl (Hamilton, 1977). There are also indications that erosion of consolidated green mudstone occurred locally prior to black shale deposition. In a number of widely separated areas in the central Mendips, as well as on Broadfield Down, the Westbury Beds rests unconformably on red Keuper Marl, or on Palaeozoic rocks which contain vertical fissures filled with red and green marl. Rounded and subangular lumps of hard green mudstone and sandstone in the basal Westbury Beds conglomerates were derived from the Tea Green Marl and show that the latter was consolidated to some degree before erosion. This hiatus may represent the time during which the Grey Marl was deposited farther south. The reducing conditions responsible for the green coloration of the Tea Green Marl may therefore have been operative before the deposition of the Westbury Beds. Cave (1977, p.70) has suggested that the parallelism between the base of the Tea Green Marl and the top of the Yate Evaporite Bed in the northern part of the Bristol Coalfield and the Tortworth area indicates a primary origin for the colour. The change from red to green is accompanied by a reduction in the amount of total iron present and Hamilton (1977, p.115) has suggested that the iron oxides responsible for the coloration are strongly adsorbed on the sedimentary particles. This would account for the relative stability of the colour of the green mudstones and sandstones under a wide range of conditions, including prolonged weathering.

Baryto-celestite has been recorded in the Tea Green Marl (Hamilton, 1977) and a celestite-bearing horizon has been noted within and just below the formation at Locking and Banwell.

Dolomitic Conglomerate

Although some early writers considered the possibility of a Permian or Middle Triassic age for the strongly magnesian rocks found in the Bristol area (Etheridge, 1870), it has long been recognised that both the dolomitised and undolomitised breccias and conglomerates are a strongly diachronous facies. The name 'Dolomitic Conglomerate' was first used by Buckland and Conybeare (1824) and replaced the old miners' and quarrymen's terms 'Millstone' or 'Millgrit'. It was adopted for use on the Geological Survey maps of the Mendips and Bristol district (De la Beche, 1846; Woodward, 1876) and was applied to all the massive Triassic breccias and conglomerates (Plate 11) without reference to the degree of secondary dolomitisation which they have undergone. Strongly dolomitised breccias and conglomerates can usually be recognised by their buff, yellow or orange-brown colour. Less highly magnesian varieties are reddish brown, red and green or grey-green in colour. The term 'marginal deposits' has been proposed (Warrington et al., 1980) but the breccias and brecciconglomerates have originated in several different ways and in various situations. Accordingly, the original name, Dolomitic Conglomerate, is retained here in view of its widespread use in all previous publications, including the maps of the Geological Survey.

Records claiming that there are rock fragments showing glacial striae in the Dolomitic Conglomerate (Valpy in Woodward, 1887, p.231) should be treated with reserve, no satisfactory supporting evidence for such an origin having been found. Some 'striated' fragments are tectonically slickensided; others such as those noted by Sollas (1880) at Portskewett may have originated during gravitational movements of scree-like deposits prior to final consolidation. In this sense not all the limestone debris found at the base of the Keuper Marl is of marginal origin. In some instances, as in the Radstock Basin and at Avonmouth and Clapton isolated masses of Carboniferous Limestone resting on slides or thrust planes (Figure 46) may have broken up or degraded by Triassic weathering in situ. This accounts for some of the isolated patches of coarse conglomerate which are encountered in the southern part of the Radstock Basin, and at Clapton, Avonmouth and other localities.

Much of the exposed Dolomitic Conglomerate is of late Triassic age and was formed when the post-Variscan uplifted surface was already deeply dissected (Plate 12). The effect of post-Carboniferous tectonic activity was to induce rifting and fissuring followed by in-situ break-up of the Palaeozoic rocks over wide areas; exposures at Holwell near Nunney [7275 4501] and at Small's Quarry on Hartcliffe Rocks near Winford [5305 6628] illustrate these processes. At Small's Quarry a north–south-aligned wedge-shaped fissure at least 20 ft deep was filled with red mudstone in which cobbles of hard quartzitic sandstone (?Cromhall Sandstone) were embedded. Resting on the fissure was an angular or subangular limestone breccia of Dolomitic Conglomerate type.

The formation of the massive deposits of scree and boulders which make up the Dolomitic Conglomerate was probably initiated and periodically renewed by contemporaneous tectonic activity, producing extensive fan-like deposits by wasting of elevated rock outcrops, aided by downslope movement accelerated from time to time by flash floods in a generally desert climate.

Plate 11 Dolomitic Conglomerate in Bridge Valley Road, Clifton, Bristol.

Boulders and pebbles of Carboniferous Limestone lie within a red mudstone and sandstone matrix. The deposit lies in a hollow cut into the underlying Carboniferous Limestone in Triassic times.

The present distribution of the coarse clastic deposits is controlled by three factors; the irregular topography on which they were deposited; subsequent tectonic movements; and the considerable erosion they have undergone. In order to reconstruct the pre-Triassic surface corrections have to be made to the present sub-Triassic basement (Figure 43a) by removing the effects of post-Triassic movements. The result is shown in Figure 43b which defines the areas of Triassic sediment accumulation and the Palaeozoic 'land' areas which remained upstanding through much of late Triassic time, and which provided the source of most of the sediments. The upland areas comprised, in the main, the Lower Carboniferous and older rocks of the Mendips, of Broadfield Down, of the Westbury-on-Trym–Failand–Clevedon ridge, and of the areas bordering the Coalpit Heath Basin to the north and east. Some of the coarse breccias of the Dolomitic Conglomerate were laid down as screes and torrent fans marginal to these areas and passed distally into and interdigitated with the finer mudstones and sandy mudstones deposited in extensive lakes or playas filling the low-lying intermontane areas (Figure 42). The coarse angular rocks (Plate 11) show no sign of current or wave action, and it must be presumed that the saline playas were too shallow and the marine incursions which replenished them too weak to permit any reworking or rounding of the rock debris in the marginal scree-fan material.

Deposits infilling fissures and palaeokarst features in the sub-Triassic Carboniferous Limestone surface have yielded reptilian faunas at the following localities; Batscombe [460 550], Cromhall [704 916], Durdham Down [573 747], Emborough [623 505], Gurney Slade [624 499] and Tytherington [660 890]; early mammalian remains are associated at Emborough. Riley and Stutchbury (1840) based the genera *Thecodontosaurus* and *Palaeosaurus* on

Plate 12 Sub-Triassic unconformity at Kilkenny Bay Portishead.

Dolomitic conglomerate rests unconformably upon Black Nore Sandstone from the lower Old Red Sandstone succession.

specimens found in a quarry (now infilled) near Worrall Road on Durdham Down, Clifton. Galton and Cluver (1976) discussed the status of *Thecodontosaurus* remains, including those of *T. antiquus* from Durdham Down, and assigned them to the Anchisauridae, a slender footed family of prosauropod dinosaurs. The Durdham Down deposit has been interpreted as the infilling of a depression on the Carboniferous Limestone surface. Halstead and Nicoll (1971), however, regarded it as debris from the collapse of a cave system and noted the occurrence of previously unrecorded reptilian remains, possibly including a lacertilian and the sphenodontid *Clevosaurus*[1].

The reptilian faunas from the other Mesozoic fissure deposits in the district and from contiguous areas of the Mendips, and of South Wales, have been reviewed by Robinson (1957b) and Fraser (1985).

---

1 Named 'from *Clevum*, the Roman term for Gloucestershire' (Swinton, 1939, p.591). The Roman name was, however, Glevum, but though Robinson, 1957a, p.263; 1973, p.477) drew attention to this and amended the genus to *Glevosaurus*, the original spelling has been retained by other workers (e.g. Halstead and Nicoll, 1971, Fraser and Walkden, 1983, Whiteside, 1986).

Those from Cromhall and Tytherington are dominated by sphenodontids (*Clevosaurus, Diphydontosaurus, Planocephalosaurus, Sigmala*) which are associated at the former locality with pseudosuchians, a terrestrial crocodile, *Kuehneosaurus, Thecodontosaurus, Variodens* and sauropsids, and at the latter site with a crocodilian, a coelurosaur, a pseudosuchian and archosaurs (Robinson, 1962; Fraser, 1985; Whiteside, 1986). The Emborough fauna includes sphenodontids, archosaurs, *Kuehneosaurus* and *Variodens*, and *Kuehneotherium*, an early mammal (Fraser, 1985; Fraser and others, 1985); *Kuehneosaurus* and *Kuehneosuchus* have been recovered from the Batscombe deposit (Robinson, 1957, 1962; Fraser, 1985).

Robinson (1957) considered the fissure-faunas to be divisible into one of 'pre-Rhaetic' and another of 'Rhaeto-Liassic' age, both containing reptiles, the latter being characterised by the inclusion of early mammals. This concept is subject to revision following the discovery of Rhaetian palynomorph assemblages in the Tytherington deposit (Marshall and Whiteside, 1980), which lacks early mammals, and the recovery of *Kuehneotherium* from the Emborough deposit, which is considered to predate the Westbury Beds (Fraser et al., 1985). It is therefore probable that the faunas from fissures within the Bristol district are of late Triassic (Norian to Rhaetian)

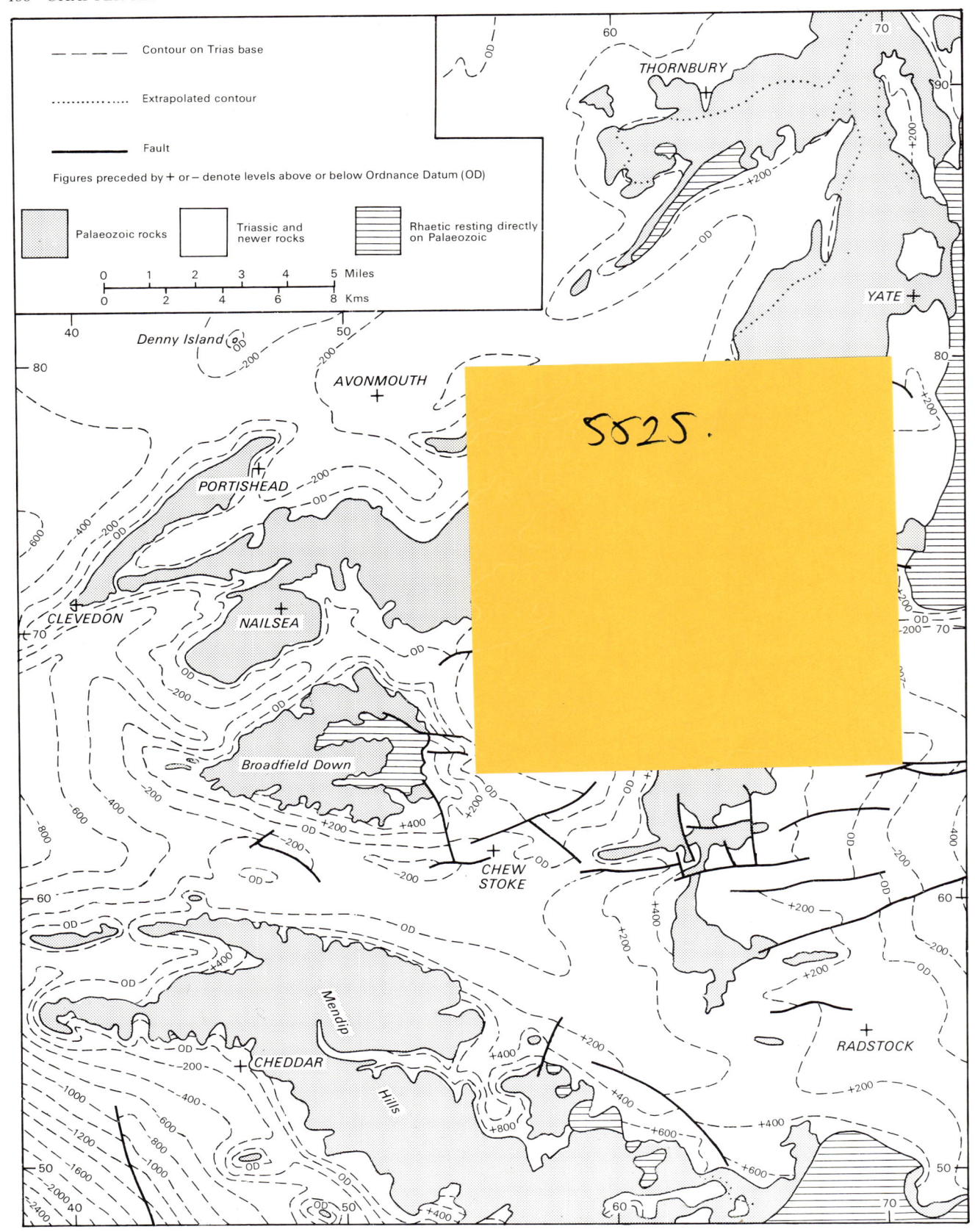

Figure 43(a) Generalised base contour map of the Triassic rocks

DETAILS OF STRATIGRAPHY 139

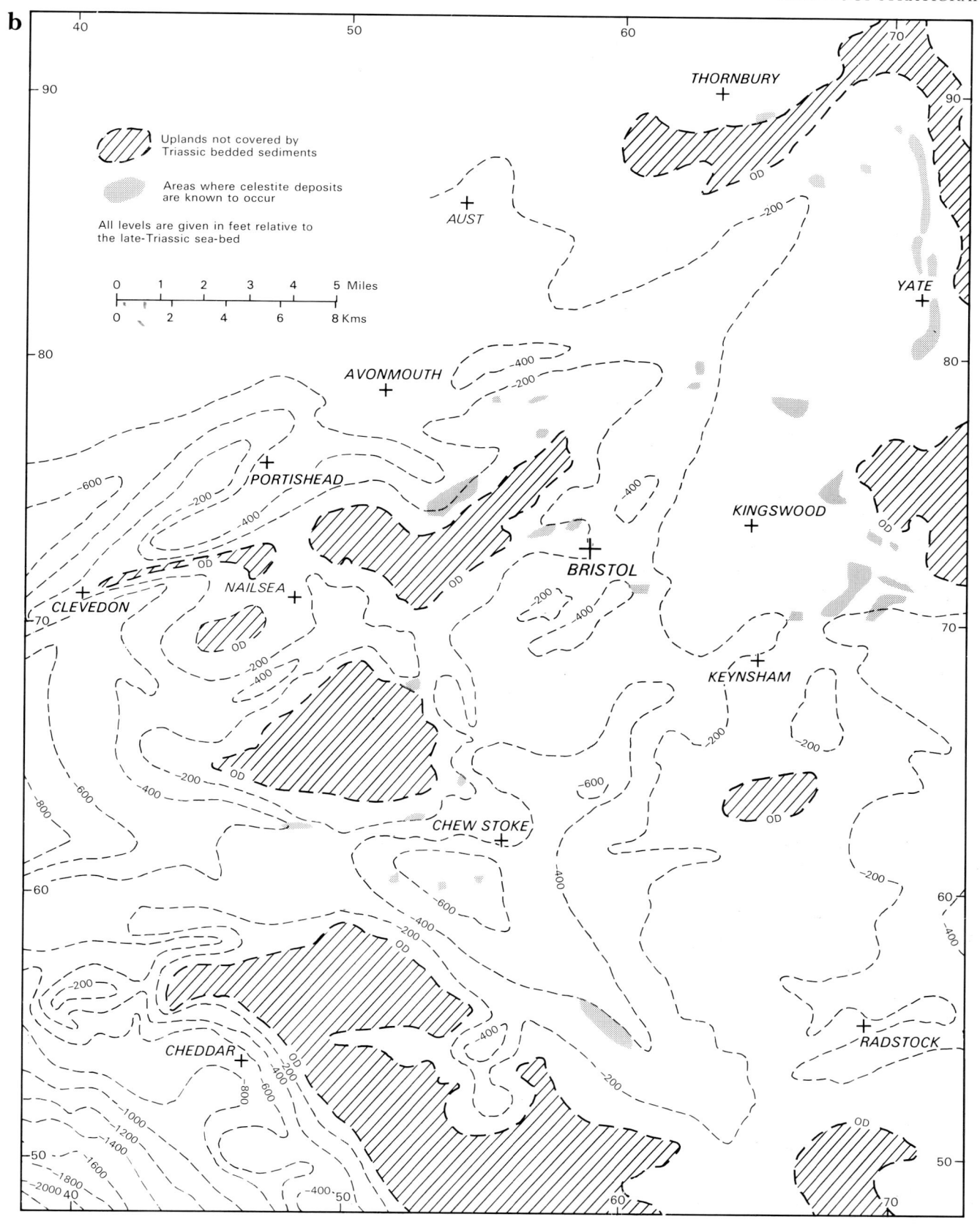

Figure 43(b) Reconstructed base-contours of the Triassic rocks after removing the effect of post-Rhaetian folding

age. Ecological and related aspects of these faunas have been considered by Marshall and Whiteside (1980), Fraser and Walkden (1983, 1984) and Fraser (1985, 1986).

Rhaetic

The closing phases of the Triassic period were marked by a sudden transition from the terrestrial desert environments to shallow-water marine conditions interspersed with brackish and freshwater episodes. The beds mapped as Rhaetic comprise the Westbury Beds overlain by the Cotham Beds (Table 4). Under the terminology of Warrington et al. (1980), these have been renamed the Westbury Formation and the Cotham Member. The last named is overlain by the White Lias (Langport Member), the base of which was taken as the base of the Jurassic system at the time of the survey.

The Westbury Beds consist mainly of dark grey or black mudstone with thin, dark grey limestone bands. The Cotham Beds include pale greenish grey or buff mudstone and clay with nodular limestone. These rocks are well seen at Aust Cliff on the east bank of the Severn, where they rest on the Tea Green Marl which in turn passes down by gradation into the Keuper Marl. At Aust Cliff, where the White Lias is absent, they are overlain by the flaggy limestones at the base of the Blue Lias. The section was first described by Buckland and Conybeare (1824). Investigations carried out in connection with the building of the Severn Bridge have provided additional information (Whittard, 1948); more recent accounts are by Welch and Trotter (1961, p.121) and Hamilton (1977). According to Welch and Trotter, the Westbury Beds are 14 ft 6 in thick and the Cotham Beds 11 ft 2 in. At Aust, as at many localities in the Bristol district, the contact between the Westbury Beds and the Cotham Beds is nonsequential. At Bristol and in the Radstock area, the White Lias is up to 21 ft in thickness (Tutcher and Trueman, 1925, fig.5) and separates the Cotham Beds from the basal Blue Lias limestones.

The thickest and most complete section of the Rhaetic so far recorded in the Bristol district was formerly exposed in the principal railway cutting [587 798] between Henbury and Filton (Tutcher, 1908). Here the total thickness of the Westbury Beds and Cotham Beds was recorded as 34 ft and the section was used as a basis for comparison with others in and around Bristol (Tutcher, 1930, p.34)[1].

Westbury Beds

The term Westbury Beds, introduced by Wright (1860, pp.377–379) for the black shales and limestones, with the basal 'Pullastra'-sandstones at Garden Cliff, Westbury-on-Severn [720 125] was adopted and amended by Richardson (1911). At Garden Cliff the black shales and associated sandstones are comparable in many respects with those at Penarth, South Wales (Etheridge, 1872, vertical sections) and those in the Sparkford Hill railway cutting in south-east Somerset (Moore, 1867; Richardson, 1911; Kellaway and Wilson, 1941, fig.11) and they mark a very sudden and widespread transgression by a shallow open sea. At Aust Cliff the sandstones are very thin or absent, as in many other sections near Bristol, but the basal thin conglomeratic bone-bed is well known for the presence of the palatal teeth of *Ceratodus* (Agassiz, 1833–1843, pp.129–138; Tutcher, 1908; Kellaway *in* Kellaway and Oakley, 1934).

Evidence of cyclic sedimentation has been recorded by Hamilton (1962). Three upwards-fining cycles were also observed in sections

---

[1] On the published six inch Geological Survey maps covering north Bristol and adjoining areas of Avon, the thin White Lias and Basal Blue Lias limestones have been mapped as a single unit (WBL), the base being the top of the Cotham Marble.

along the M5 motorway between Cribbs Causeway and Patchway Tunnel, and at Aust (Hamilton, 1977). The sandy and conglomeratic Ceratodus Bone Bed at Aust and Filton probably marks the lowest of three cycles recognised at Cribbs Causeway. In the Filton railway cutting, where *Ceratodus* was found after the publication of the original account, the Bone Bed is underlain by 11½ ft of black shales resting on the Tea Green Marl; these may represent the Grey Marl previously thought to be absent in this area. Elsewhere around Bristol the Ceratodus Bone Bed rests directly on the Tea Green Marl, as at Pylle Hill (Wilson, 1894, 1895) or is absent, as at Henleaze and Southmead (Kellaway, 1932). Tutcher (1908, 1928) considered that the local absence of the basal Westbury Beds was caused by overlap against irregularities in the underlying surface, though this and the variation in the thickness of the formation may be partly due to contemporaneous earth movements.

Lumps of greenish, calcareous sandstone and wisps of soft, green marl are commonly incorporated in the basal bone beds and black shale of the Westbury Beds, having been derived from the underlying Tea Green Marl.

The invertebrate fauna of the Westbury Beds is limited in variety though individual species (mainly bivalves) may be present in large numbers; *Eotrapezium ewaldi* is probably the commonest shell, though bands packed with *Protocardia* and *Lyriomyophoria* are also found; *Chlamys valoniensis*, *Pleurophorus elongatus* and *Rhaetavicula contorta* are other characteristic species. Gastropods include cerithiids, '*Natica*' and acteoninoids, and inarticulate brachiopods are represented by *Orbiculoidea*. The ophiuroid *Ophiolepis damesi* is present sporadically, as at Pylle Hill, Bristol (Wilson, 1891, 1894). Ostracods and the teeth and scales of elasmobranch and ganoid fishes are relatively common and echinoid fragments also occur. Insect remains, including the elytra of beetles, have also been recorded.

The palatal teeth of the dipnoan fish *Ceratodus* are well known from Aust, where bones and teeth of the marine reptiles *Ichthyosaurus* and *Plesiosaurus*, as well as bones of the terrestrial carnivorous dinosaur *Avalonia*, have also been found (Hamilton, 1977). 'Labyrinthodont remains' recorded from Aust (Reynolds, 1947) have been shown to be bones of the large fish *Birgeria* (Savage and Large, 1966).

Divergent views have been expressed as to the origin of the sandy, phosphatic and pebbly deposits described as bone beds. Phosphatised bones, mostly fragmentary and worn or abraded, together with fish teeth and scales form an appreciable part of these deposits. Invertebrate fossils are less common though thin-shelled *Mytilus* occur in the Ceratodus Bone Bed at Aust and Charlton (Tutcher, 1908). Scattered bones and phosphatic nodules ('coprolites') and thin layers with fish scales and teeth occur at several levels in the Westbury Beds, but the palatal teeth of *Ceratodus* and the largest concentration of vertebrate remains are almost invariably found in the principal bone bed at or near the base. This is overlain by dark shales of consistent thickness lying beneath a thin grey limestone—the lower Pecten Bed.

At Aust the Ceratodus Bone Bed is a pale greenish grey calcareous or slightly sandy conglomerate with thin sandstone partings. At other localities in and around Bristol it is represented by current-bedded pyritic sandstone with fish teeth and scales and abraded reptilian bones. Well-rounded smooth white, pink and grey quartz pebbles are a common constituent in many sections. As a general rule sands are confined to the lower part of the Westbury Beds. Where the strata are strongly attenuated, however, as at Henleaze [5824 7720], the black shales pass locally into lenticular masses of soft sandstone or clay with lenses of gritty limestone full of *Lyriomyophoria* (Kellaway, 1932, p.288).

Pebbles, including well-rounded quartz, worn phosphatic nodules and mudstone and limestone clasts, occur embedded in the laminated black shales. The occurrence of reptilian bones and fish remains in Mesozoic rocks has been commonly linked with the presence of isolated pebbles or scattered groups of pebbles composed of rocks of non-local origin (see e.g. Kellaway and Wilson, 1941, p.156). This has led to the belief that such pebbles may be 'gastroliths' or stomach stones similar to those found in the stomachs of some modern vertebrates (Wickes, 1904, 1910). Some of the pebbles may be drop-stones from drifting masses of weed such as may have abounded in shallow land-locked seas devoid of strong currents. This view is supported by the presence of thin-shelled *Mytilus* in the coarse conglomeratic bone beds (Tutcher, 1908). Although the pebbles in the bone beds suggest a high-energy environment of sedimentation, their association with the fragile *Mytilus* indicates a much quieter mode of deposition. Indeed, drifting weed could have carried the stones and also provided a habitat for *Mytilus* with its byssal mode of attachment.

In the Wedmore inlier a local arenaceous member, the Wedmore Stone, is present within the Westbury Beds (Green and Welch, 1965, pp.69, 82–84). Reptilian remains recovered from a bed resting on the Wedmore Stone near Wedmore (Sandford, 1894) were described by Seeley (1898) and assigned to the taxa *Avalonia sanfordi* and *Picrodon herveyi*. Galton (1985) has reassessed this material and has designated the postcranial remains as the holotype of *Camelotia borealis*, a new member of the herbivorous prosauropod dinosaur family Melanorosauridae.

Cotham Beds

The name Cotham Beds (Richardson, 1911), derived from the district of Cotham in Bristol, refers to strata called 'Upper Rhaetic' in some earlier accounts. It comprises 10 to 12 ft of pale greenish grey marl or soft calcareous shaly mudstone with some bands and nodules of impure limestone and, more rarely, of ripple-marked sandstone, resting nonsequentially on the Westbury Beds. Clasts of black shale derived by erosion from that formation occur in the basal beds. Tutcher (1908) described 1½ ft of argillaceous limestone at the base of the Cotham Beds at Charlton, west of Filton as the 'Estheria Bed'. Others (Richardson, 1903; Kellaway, 1932) have recognised similar lithologies elsewhere separated by thin beds of pale grey clay from the surface of the black shale of the Westbury Beds. In the north western suburbs of Bristol between Clifton and Filton this facies comprises greenish grey limestone with marine, brackish and freshwater faunas and floras. Well-preserved remains of the bryophytes *Naiadita lanceolata*, considered to have been a submerged freshwater species, and *Hepaticites solenotus* occur (Harris, 1938). These and other elements of the flora and fauna point to a shallow fresh or brackish-water environment, in which extensive areas of calcareous muds and silts were subject to periodic flooding by sea water. *Chlamys valoniensis* is found in the thin laminae resulting from these incursions.

Another indication of the presence of mudflats subject to periodic drying out is provided by the septarian nodules of earthy limestone found scattered through the clays of the Cotham Beds. These show vertical contraction cracks filled with crystalline calcite on which was deposited acicular radiating and felted masses of pale blue and white barytocelestite. As with the occurrences in the Keuper Marl, the localities where this mineral is most abundant, for example at Redland (Wickes, 1901) and Clifton (Collie, 1879, p.292), lie near the margin of the late Triassic depositional basin. Desiccation cracks, ripple markings and sporadic trace-fossils resembling worm trails also indicate deposition in very shallow lagoonal areas, and sabkha-type conditions, under which crystallisation of barytocelestite took place, probably prevailed for short periods.

At the top of the Cotham Beds lies the thin lenticular limestone first described by Owen (1754) as Cotham Marble. This crops out on high ground at Cotham House, Bristol [5836 7393]. It has an irregular top and an internal arborescent structure which is responsible for the alternative name of 'Landscape Marble'. It ranges in thickness from ½ to 9 in, but has nonetheless a wide distribution, being known from Shepton Mallet in the south to Tortworth in the north, and from Bath in the east to the Severn Estuary in the west. William Smith (MS) described the 'Cotham Stone' as marking the base of the White Lias at Radstock. Its origin was for a long time a matter for speculation, but it is now considered to be algal (Hamilton, 1961; Mayall and Wright, 1981; Wright and Mayall, 1981).

Although the Cotham Marble is known to have cropped out at Cotham House, Bristol, no detailed section through the Cotham Beds is known. Cotham House has recently been demolished, but blocks of Cotham Marble, showing the curious vermiform upper surface, can still be seen in ornamental pillars at the entrance of Cotham Park [5862 7403]. The only reliable record of the succession in the Cotham outlier (Tawney, 1878) was obtained during the construction of Oakfield Road Pumping Station, Clifton [5762 7372]. This is situated at the southern (faulted) extremity of the outlier, approximately 850 yd west-south-west of the site of Cotham House. A section not far away at Redland (Wickes, 1901) showed 8 in of Cotham Marble overlain by thin White Lias (Langport Member). The Cotham Marble can be identified in the area of the Tortworth Inlier at least as far north as Chase Hill [7413 8882] where it was recorded by Richardson (1904).

Evidence of penecontemporaneous chanelling of the partially lithified algal limestone is provided by the presence of the so-called 'Crazy Cotham' which is developed locally as channel-fillings on the Landscape Marble (Hamilton, 1961). Apart from fish scales, some small turbinate gastropods (Wilson, 1894) and a small form of *Modiolus* there are few marine organisms preserved in the Landscape Marble. The commonest fossil, the bivalve *Meleagrinella decussata*, occurs in appreciable numbers, both in the Landscape Marble and in the shaly clays underlying and overlying it. In areas such as Cotham where the Cotham Marble is up to 9 in thick with as many as three 'landscape' layers (Hamilton, 1961) it was used for the construction of ornamental walls and rockeries and was also cut and polished.

White Lias

The white limestones and marls which succeed the Cotham Beds are identified as White Lias (Langport Member) on all of the component six-inch geological maps of the Bristol district, though on the special sheet and the relevant one-inch sheets (264, 265, 280 and 281) they are included with the Blue Lias and classified as Jurassic. The White Lias is now included in the Triassic following Warrington et al. (1980). The White Lias consists of pale grey and cream-coloured limestones, marls and clays which assume even lighter hues when weathered, a feature which accounted for the name given to it by quarrymen, later adopted by Stachey (1719) William Smith (1799) and defined by Tutcher and Trueman (1925). It ranges up to about 20 ft in thickness and the thicker sections in the Avon valley and near Radstock show a clear division into two parts: a lower consisting of thin rubbly limestones with clay partings, and an upper with stronger, more regular limestones separated by clays not more than about 2 in thick. The limestones of the upper part are hard, fine-grained and uniform in texture, with conchoidal fracture, and are usually bounded by flat bedding planes, in contrast to those of the succeeding Blue Lias. The top limestone bed is known locally as the 'Sun Bed' (Arkell and Tomkeieff, 1953, p.116) and may reach 1½ ft in thickness. It commonly shows vertical U-shaped burrows usually referred to as *Diplocraterion*. The upper beds were formerly extensively quarried for building stone, especially in the Avon valley between Keynsham and Bath and in the area extending southwards to Radstock.

A small-scale isopachyte map of the White Lias by Kellaway and Welch (1948, fig.15a) showed that deposition took place in two basins, respectively north and south of the Mendips. The thickest sedimentation occurred in a narrow belt extending east–west through Radstock, 19 ft being recorded in the Hemington Borehole [7247 5296] and 20 ft at Buckland Colliery [7446 5079]. The thick belt continues eastwards at depth, for 27 ft 3 in were recorded in a borehole at Westbury (Wilts.), 8 miles east of Buckland (Pringle, 1922). The White Lias is reduced to between 12 and 18 in at Dundry Hill and north of Bristol, and it dies out altogether about one mile north-west of Filton.

The fine-grained, white limestones forming the upper part of the White Lias are nearly devoid of fossils though trace fossils are often conspicuous in the Sun Bed, and there is a restricted fauna of bivalves, gastropods and ostracods in the lower part. Twelve or thirteen genera of bivalves are recorded. *Modiolus*, represented by *M. hillanus* and small shells often recorded as *M. minimus*, is locally abundant. A few genera, including *Astarte*, *Plagiostoma* and *Pleuromya*, make their earliest appearance. More sporadically echinoids and simple corals occur. The bennettitalean leaf *Otozamites obtusus* represents the land flora. Most of the recorded genera and species of marine organisms occur also in the Blue Lias. The fauna indicates that the sea was very shallow with the sea-bed suffering intermittent emergence, corresponding with Hallam's interpretation (1960) of the White Lias in Devon.

Some of the higher limestones, including the Sun Bed, commonly show very fine lamination and current-bedding where weathered. That these deposits had hardened and compacted before the deposition of the Blue Lias took place is shown by a section seen in Harris's Quarry on Marksbury Plain. Here differential compaction of the Blue Lias has taken place over ridges and hollows in the dissected upper surface of the Sun Bed. At Aust Cliff, pockets of conglomerate formed of White Lias limestone are seen from time to time beneath the basal Blue Lias limestone. Desiccation or shrinkage cracks are found locally.

The foregoing account of the White Lias is based on that given in the memoir on the Lower Jurassic rocks of the Bristol district by Donovan and Kellaway (1984, pp.5–6).

Important sections of the Westbury Beds, Cotham Beds and White Lias are described in the Geological Survey's memoirs covering the Monmouth (233) and Chepstow (250) sheets (Welch and Trotter, 1961), the Malmesbury (251) Sheet (Cave, 1977) and Wells and Cheddar (280) Sheet (Green and Welch, 1965). Many of the older published accounts have inadequate or outmoded locality descriptions; the National Grid references for these and other important sections are set out below.

Sheet 264 (Bristol)

Locality	Grid Ref	Reference
Brislington (West Town Lane, now known as Hazlebury Road)	[6135 6920]	Kellaway, 1935
Charlton (Railway Cutting)	[587 798]	Tutcher, 1908
Clifton (Oakfield Rd. Pumping Station)	[5762 7372]	Tawney, 1878
Cribbs Causeway–Patchway	[573 807–584 815]	Hamilton, 1962
Elton Farm Borehole, Dundry	[5636 6689]	Ivimey-Cook, 1978
Henleaze (Northumberland Ho. Es.) (area includes Fallodon Road)	[577 762–579 762]	Kellaway, 1937
Knowle (Red Lion Hill)	[6085 7030]	Kellaway, 1937
Pylle Hill (Railway Cutting)	[597 719]	Wilson, 1894
Redland (New Clifton, at junction of Lincoln Road and Cranbrook Road)	[5805 7560]	Wickes, 1901
Winford (Grove Farm)	[5387 6613]	Trueman, 1936

Sheet 265 (Bath)

Bitton (Rye Down Lane)	[6775 7052]	Tutcher, 1923
Willsbridge (Railway Cutting)	[6695 7070]	Moore, 1867

Sheet 280 (Wells)

Emborough (Old Down)	[626 509]	Savage, 1962

MINERALISATION AND METASOMATISM AFFECTING TRIASSIC ROCKS

Both the massive Dolomitic Conglomerate beds and the skerries of the red Keuper Marl which extend into the basinal areas have been subjected to widespread diagenetic and postdiagenetic changes, including haematitisation, the development of quartz veins and geodes and secondary dolomitisation. The deposition of iron oxides, crystalline quartz, nonferrous sulphides (notably sphalerite and galena), and the emplacement of baryte, barytocelestite and other minerals in fissured Palaeozoic bedrocks are all related to Permo-Triassic metasomatism. Mass silicification, which has strongly affected the Triassic rocks of the central Mendips as well as underlying Palaeozoic and overlying Jurassic rocks, is a much younger phenomenon. Silicified late Triassic and Lower/Middle Jurassic rocks are now described collectively under the name Harptree Beds (Green and Welch, 1965, pp.94–95; Donovan and Kellaway, 1984, pp.19–20), and are referred to subsequently (p.152).

Haematitisation

The Triassic age of the haematite of the Bristol district was demonstrated by Etheridge (1870, pp.183–186); subsequent contributions were made by Grenfell (1873), Sibly and Lloyd (1927), O. T. Jones (1931) and Trotter (1942). The general distribution of the haematite ores in the Bristol district was described by Cantrill et al. (1919) and much additional material has been published by Welch and Trotter (1961), Cave (1977), Green and Welch (1965) and Kellaway (1967).

Oxidation and red staining of grey Coal Measures and other Carboniferous rocks overlain unconformably by Triassic sediments has given rise to a highly ferruginous zone at the contact, and red-stained fissures have been recorded at depths of 2243 ft below the base of the Trias at Harry Stoke, north of Bristol (Kellaway, 1967). Locally, as at Winford and Hartcliffe Rocks, this ferruginous zone is so rich that the rocks have been worked as iron ore. All stages in the conversion of both calcareous and magnesian Dolomitic Conglomerate into massive iron ore could formerly be seen in sections on the south side of Hartcliff Rocks [534 661], with veins of haematite extending downwards into the Carboniferous Limestone.

The source of the iron is considered to have been the sulphides and carbonates of iron present in the unweathered Palaeozoic formations; these appear to have been attacked by oxidising ground waters which penetrated the bedrock at depth and also permeated the basal Triassic sediments.

In late Triassic times erosion of the uplands and infilling of the basinal areas led to a slowing down of the circulation of ground water and this was accompanied by a relative rise in sea level with the creation of sabkhas and mudflats at or near sea level. This process culminated in a general submergence of much of the region beneath the transgressive waters of the Rhaetian sea and the establishment of reducing conditions beneath the sea floor. The contrast between the mineralisation processes associated with the oldest and youngest Triassic rocks in the Bristol district is most striking. The dark purplish red haematitic and siliceous breccia seen at the southern end of Brandon Hill, Bristol [5788 7266] is typical of many localities where the older Triassic sandstones and conglomerates rest on strongly pyritic mudstones and quartzitic sandstones of Namurian or Westphalian age. At the opposite extreme are the lead veins found on Clifton Down where the black shales of the Westbury Beds rest unconformably on Carboniferous Limestone. In these veins the metallic minerals are all sulphides, generally galena, marcasite or sphalerite. Crystalline quartz is conspicuous by its paucity and the commonest gangue minerals are baryte or calcite. In steeply inclined vein-fillings extending down into the underlying Carboniferous rocks the normal paragenesis is haematite and quartz deposited first, with galena, baryte and calcite forming a later infilling of the central zone of the vein or fissure.

Quartz veins and geodes

Geodes of banded agate lined with quartz crystals were formerly cut and polished at Bristol under the name of 'Bristol Diamonds'. The presence of these semiprecious 'potato-stones' was known to early travellers including John Evelyn (1654) and Celia Fiennes (1698). The largest known collection of 'Bristol Diamonds' is preserved in the walls of the 18th century Grotto at Goldney House, Clifton [5743 7278]. This construction has been described by Stembridge (1969) and is of considerable historical interest.

On the eastern side of the Boyd valley to the east of Bristol quartz geodes occur in association with celestite near the depositional margin of the Stoke Park Rock Bed and its related celestite-bearing marls. Many of the trial bores made by the Geological Survey between Wapley and Bitton show similar occurrences. Celestite nodules with quartz cores have been described by Nickless et al. (1976, p.47) from Wapley Common, and quartz-calcite nodules from Dulcote, near Wells, have been shown by Jobbins (in Green and Welch, 1965, p.168) to be pseudomorphs after baryte or celestite.

Buckland and Conybeare (1824) recorded a section at Sandford Hill in the western Mendips where contemporary excavations were made to recover nodules of agate and crystalline quartz. Similar deposits of agate and quartz geodes have been observed at several localities in the Bristol and Somerset Coalfield, notably at Flax Bourton [503 687] and in west Bristol, more particularly in the area between Henbury Church and Passage Road, Brentry [570 788]. Here the Triassic rocks consist of hard red mudstone and yellow Dolomitic Conglomerate overlain by red marls with pink or reddish coloured nodules composed of tightly packed laths or tiny plates of celestite. Partially recrystallised clasts of quartzitic sandstone were observed in a number of temporary excavations exposing the underlying yellow Dolomitic Conglomerate. However, some quartz veins and geodes showed no direct relationship with any surviving siliceous clasts and it is possible that these may have been replacements of anhydrite or celestite.

Recent work on the forms of quartz associated with evaporites (Folk and Pittman, 1971) shows that length-slow chalcedony develops preferentially when replacement of sulphates by silica takes place. Using this approach, Tucker (1976) has shown that some quartz geodes ('Bristol Diamonds') may be replacements of primary anhydrite. Thus the presence of quartz geodes within the Dolomitic Conglomerate may have resulted from the alteration of reworked anhydrite nodules or anhydrite-replaced pebbles of Carboniferous rocks. From an examination of the quartz geodes Tucker (1976) concluded that the mean annual temperature in later Triassic ('Keuper') times may have been about 22°C rising seasonally to 35°C or more. This interpretation of the climate from mineralogical evidence agrees broadly with the conclusions based on palaeontology, more especially with reference to the reptilian faunas. It also accords with the reconstruction of the conditions which gave rise to the formation of sabkhas or playa-like depositional basins in which the evaporites accumulated.

Other aspects of megaquartz formation include the deposition of haematite-quartz linings and vein fillings which are developed locally in the Carboniferous rocks. Fissures lined with euhedral red, yellow or colourless quartz extend to depths of at least 200 to 300 ft below the base of the Triassic rocks, as at St Vincent's Rocks, Clifton [5654 7313]. The only sulphates detected in these fillings are traces of baryte.

Replacement of sulphates by silica appears to have occurred intermittently between the time of deposition of the Butcombe and Woodford Hill sandstones and that of the Tea Green Marl (Blue Anchor Formation), and may well have been related to the general amelioration of the desert climate at the close of the Triassic with the onset of open-sea conditions.

Dolomitisation

The skerry beds so common in the red Keuper Marl (Mercia Mudstone) show innumerable small euhedral crystals of dolomite embedded in a calcareous, arenaceous or muddy matrix. Whittard and Smith (1944, p.66) suggested that the dolomite was an original constituent precipitated from saline solutions. The same authors described a 5½ ft bed of sandy dolomitic mudstone (?the Stoke Park Rock Bed) lying 38 ft beneath the green mudstones of the Tea Green Marl at Wickwar, and containing clasts of the underlying Silurian limestone. The magnesian rock contained 40.6 per cent of carbonate of which dolomite made up 34.2 per cent in the form of tiny idiomorphic crystals.

Most unaltered Triassic sandstones, breccias and conglomerates in the Bristol district and the Mendips are predominantly grey or grey-green with wisps of red and green marl or with a red or pinkish tinge due to the presence of ferric oxides. The changes induced by secondary dolomitisation include hydration of ferric oxides to limonite, giving tints of yellow or brown. On the whole these colours, like those of the red and green marls (Hamilton, 1977) have been remarkably stable and resistant to subsequent weathering.

When secondary dolomitisation of Triassic conglomerate has taken place both the fine-grained groundmass and calcareous clasts have been converted to dolomite and magnesian rock (Green and Welch, 1965, p.65). This process was intensified where the Triassic conglomerate is in direct contact with Carboniferous dolomite or where it contains a high proportion of Carboniferous dolomite fragments. Optimum conditions for secondary dolomitisation are found where dolomite-rich conglomerates are overlain by or flank sabkha deposits and have been penetrated by highly saline water. Thus the hillock known as the Tump at Kings Weston, Bristol [5400 7773] is a shattered mass of Black Rock Dolomite resting on Coal Measures and enveloped by dolomite-rich conglomerates which are roughly comparable in age to the Stoke Park Rock Bed. As a result of secondary dolomitisation almost the entire mass of magnesian limestone forming The Tump has been converted to structureless or cavernous purplish grey and yellow dolomite.

No fundamental changes of colour have been produced by Quaternary weathering and oxidation though the yellow coloration of the Dolomitic Conglomerate and some of the sandy and dolomitic mudstones has been heightened by post-Triassic weathering. This is due in part to Quaternary decalcification which has resulted in the removal of calcitic cement, and to a relative increase in the proportion of soft ochreous matter mixed with dolomite and insoluble sand, silt and clay.

Secondary Triassic dolomitisation of conglomerate, sandstone and sandy mudstone is essentially a very late post-diagenetic feature which can, in some instances, be seen to postdate cementation and jointing of the rock. It is probably due to subsurface movement of ground water migrating under hydrostatic pressure through permeable strata (refluxion). Primary deposition of dolomite crystals during the formation of the greenish grey rock beds in the red Keuper Marl took place in shallow marine embayments or on tidal mudflats and may have been independent of the refluxion process. There are however, many occurrences where magnesian rocks with primary dolomite have been affected by secondary dolomitisation.

PALYNOLOGY

The following account has been provided by Dr G. Warrington.

No palynomorphs have been recovered from beds older than the Mercia Mudstone Group, a sample from the Redcliffe Sandstone Formation at 992ft 5 in in the Dundry (Elton Farm) Borehole having proved barren.

Samples from the red Keuper Marl (Mercia Mudstone) were examined from twelve horizons between 663 ft 10 in and 834 ft in the Dundry (Elton Farm) Borehole; one sample, from 730 ft 9 in, was productive but yielded only *Tasmanites*, a prasinophyceaen alga, and bisaccate miospores indicative of an early Mesozoic age. Samples from around the Severnside Evaporite Bed were examined from boreholes to the east of Bristol and from an outcrop near West Leigh Court [545 751] (Warrington, 1977). Only one, from a borehole near Westerleigh, was productive, yielding *Classopollis*, a miospore indicative of a late Triassic or younger Mesozoic age.

Samples from outcrops in the upper part of the red Keuper Marl at Long Ashton [5445 6950], Bedminster Down [575 704] and Stanton Drew [604 629] and from the Butcombe Sandstone at Butcombe [514 614] and the Woodford Hill Sandstone at Chew Stoke [544 618] (Figure 41) also proved barren. Beds immediately below the Butcombe Sandstone near Butcombe Farm [507 600] and from within that unit at Blagdon reservoir [5045 6040] yielded miospore assemblages including *Camerosporites secatus*, *Ellipsovelatisporites plicatus*, and *Ovalipollis pseudoalatus* with, from the former locality, *Duplicisporites granulatus* and *Enzonalasporites vigens*; these indicate a Carnian age (Table 3).

The Tea Green Marl (Blue Anchor Formation) was sampled from boreholes east of Bristol (Warrington, 1977) and from three horizons between 634 ft 11 in and 643 ft 8 in in the Dundry (Elton Farm) Borehole (Figure 44). The sample from 634 ft 11 in yielded miospores and organic-walled microplankton, associated with test linings of foraminifera, in an assemblage similar to those recorded from the formation in its type area in west Somerset (Warrington and Whittaker, 1984) and which is also comparable with those from the overlying Westbury Formation in the same borehole (Figure 44).

Palynomorph assemblages have been recovered from the Penarth Group in the Dundry (Elton Farm) Borehole, from a borehole at Wapley (Warrington, 1977) and from outcrop at Chilcompton (Warrington, 1984). Marshall and Whiteside (1980) have recorded palynomorphs typical of the Westbury Formation–Cotham Member sequence from fissure deposits at Tytherington [660 890]. The Penarth Group is documented here from the Dundry (Elton Farm) Borehole, where it was examined in conjunction with the underlying Tea Green Marl and the overlying basal (Triassic) beds of the Lias.

Assemblages from the Westbury Formation are similar to those of the Tea Green Marl and comprise miospores with, in most instances, organic-walled microplankton and test linings of foraminifera. Miospores generally dominate these assemblages and comprise associations characterised by *Rhaetipollis germanicus*, *Ricciisporites tuberculatus*, *Ovalipollis pseudoalatus* and circumpolles, principally *Classopollis*. Organic-walled microplankton are important and in one instance are the dominant constituent; they comprise the dinoflagellate cyst *Rhaetogonyaulax rhaetica* and sporadic representatives of *Micrhystridium*, an acanthomorph acritarch. A slight increase in the diversity of the miospore component is evident in the top 2 ft of the formation and this trend is continued into the succeeding Lilstock Formation.

Assemblages from the lowest 3 ft of the Cotham Member are dominated by *Classopollis* and other miospore taxa which also characterise the Westbury Formation, but the associations are augmented by *Porcellispora*, the spore of *Naiadita*, a bryophyte (Harris, 1938), and representatives of trilete spores including *Calamospora*, *Convolutispora*, *Cornutisporites*, *Limbosporites*, *Perinosporites*, *Stereisporites*, *Triancoraesporites* and *Zebrasporites* (Figure 43). Specimens of *Porcellispora* from the Cotham Member at Pylle Hill, Bristol, were illustrated by Sollas (1901) in an account that constitutes the earliest published record of spores from British Triassic rocks. Har-

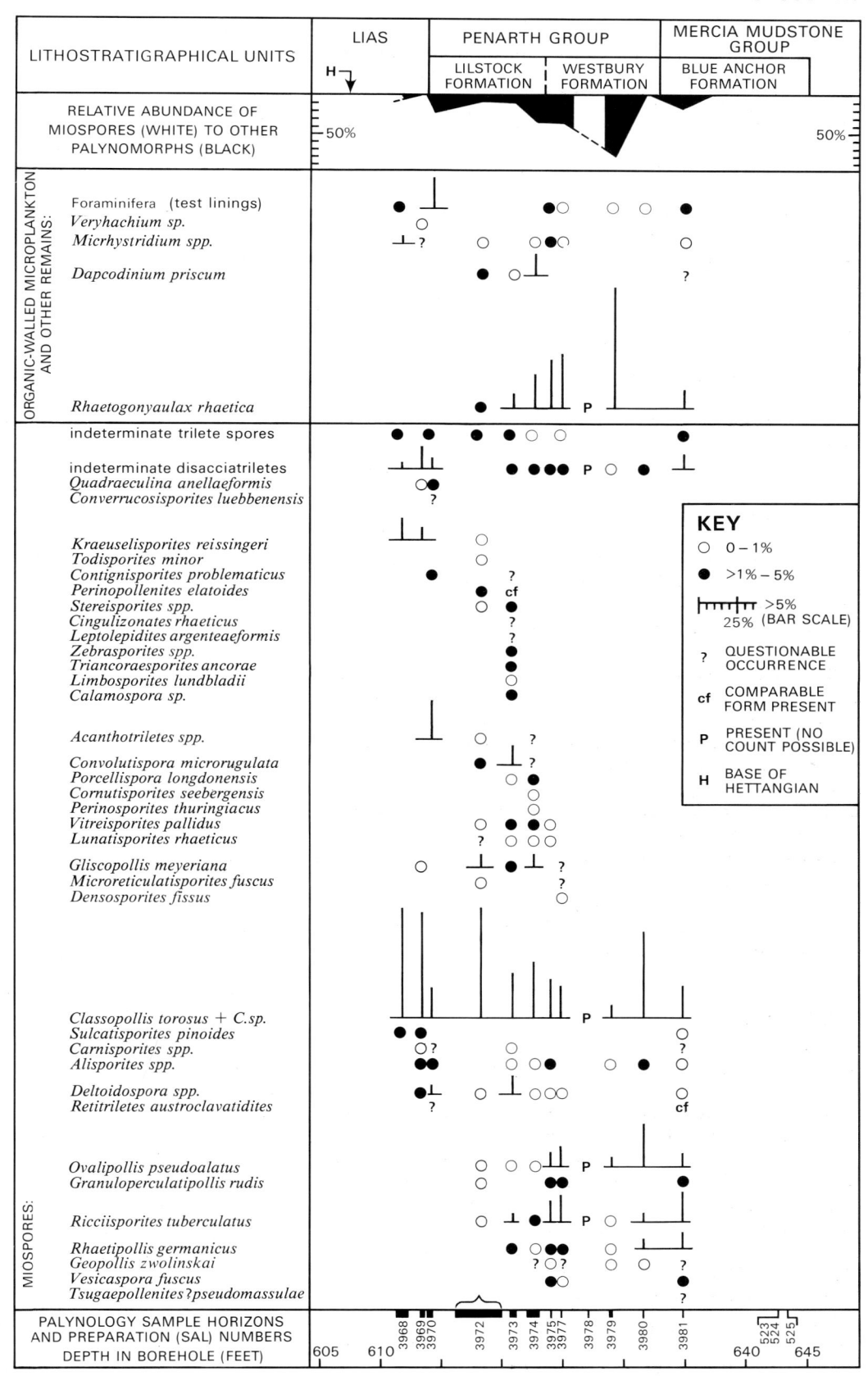

Figure 44 Palynomorphs from the Triassic rocks of the Dundry (Elton Farm) Borehole

ris (1938) described and illustrated *Naiadita* material from the Bristol district in more detail and demonstrated its bryophytic affinity.

Organic-walled microplankton associations from the lowest 3 ft of the Cotham Member are generally less abundant than those from the Westbury Formation. They are dominated by the dinoflagellate cysts *Dapcodinium priscum* and *Rhaetogonyaulax rhaetica*; acanthomorph acritarchs (*Micrhystridium*) occur sporadically.

Core recovery from the upper part of the Cotham Member was poor in the Dundry (Elton Farm) Borehole and a preparation from these beds represents a drilled interval of 3 ft 11 in. The assemblage recovered is less diverse than those from lower down and is dominated by miospores, principally *Classopollis* and *Gliscopollis meyeriana*. Specimens of *Rhaetipollis germanicus* do not persist to this level, which is marked by the highest occurrences of *Ricciisporites tuberculatus* and *Ovalipollis pseudoalatus* and the lowest occurrence of *Kraeuselisporites reissingeri*. The associated organic-walled microplankton include a small number of dinoflagellate cysts and sporadic acanthomorph acritarchs.

A preparation from beds that may represent the White Lias (Langport Member) yielded miospores of limited diversity associated with some remains of foraminifera. The formal base of the Jurassic is now taken at the lowest occurrence of the ammonite *Psiloceras*; this is at 607 ft in the Dundry (Elton Farm) Borehole and consequently a few feet of beds in the basal (Blue) Lias sequence above the White Lias are of Triassic age. Palynomorph assemblages from these beds are of limited diversity and are dominated by miospores, principally *Classopollis* and *Kraeuselisporites reissingeri*. Organic-walled microplankton, including the polygonomorph acritarch *Veryhachium* and the acanthomorph *Micrhystridium*, and the remains of foraminifera are also present.

The palynomorph succession from the topmost Tea Green Marl (Blue Anchor Formation) to the basal Lias in the Dundry (Elton Farm) Borehole is comparable with others recorded from the same sequence elsewhere in the Bristol and adjoining districts (see, for example, Orbell, 1973; Warrington, 1977, 1984), and is indicative of a Rhaetian (late Triassic) age. The change in character of the assemblages evident within the Lilstock Formation reflects environmental changes and does not serve to identify the Triassic–Jurassic boundary.

The majority of the palynomorph preparations comprise miospores and other remains of land plant origin, in association with organic-walled microplankton and remains of foraminifera that are indicative of a marine depositional environment. The changes in the relative abundances of the palynomorphs (Figure 44) reflect environmental and biogeographical changes associated with a late Triassic transgression that resulted in the establishment of marine conditions throughout much of the British Isles by Hettangian times (Warrington, 1981).

CHAPTER 7
Jurassic

The Lower Jurassic rocks of the Bristol district have been described in a separate memoir (Donovan and Kellaway, 1984). Only a few spurs and outliers of Middle Jurassic rocks, lying principally in the south-west, therefore remain to be described.

MIDDLE JURASSIC

The lowest part of the Middle Jurassic succession (Table 4) is the Inferior Oolite, which crops out between Twerton, near Bath and Mells, near Frome; it also forms a large outlier, some 4 miles by 1¾ miles, capping Dundry Hill, where the rocks are much affected by superficial cambering and landslipping (Figure 45). It comprises a sequence of very variable limestones, seldom more than 50 ft thick, and is very incomplete when compared with the 300 ft or so of limestones developed in the Cotswolds to the north-east. The comparatively complete Cotswolds succession has been divided into Lower, Middle and Upper divisions. Whilst the Lower and Middle Inferior Oolite correspond in general to the Aalenian Stage and Lower Bajocian Substage respectively, deposition of the Upper Inferior Oolite continued through the Upper Bajocian Substage into the oldest part of the Bathonian. By reference to work on the ammonites, particularly by S. S. Buckman (1893, 1895) in Dorset and the mid-Cotswolds, a sequence of 11 zones and 24 subzones has been recognised and these have been used to provide evidence of the major changes in sedimentation, associated with earth movements and intervals of erosion which have given rise to significant nonsequences and unconformities during both the Aalenian and Bajocian stages. The main lithostratigraphical and biostratigraphical units recognised within the district are shown in Table 5.

The Inferior Oolite of Dundry Hill was described by Buckman and Wilson (1896) and that of the Bath–Doulting area by L. Richardson (1907). The important structural and depositional changes which culminated in the major Upper Bajocian transgression, as seen in the Bristol–Mendip region, were summarised by Kellaway and Welch (1948, pp.73–76). Briefly, the existing sediments were folded, uplifted and eroded in Lower Bajocian times. In places these were preserved to a greater or lesser extent in gentle downwarps, but in others, especially in the west and in the Mendips area, the erosion cut down through the Middle and Lower Inferior Oolite, removing part or all of the Lias. The latter condition is well seen in the classic section at Vallis Vale and elsewhere where nearly horizontal Upper Inferior Oolite was described nearly 150 years ago by de la Beche (1846, pp.287–292, figs. 41–45) as resting unconformably upon Carboniferous Limestone.

The base of the Upper Inferior Oolite in the eastern Mendips is commonly conglomeratic and fragments of volcanic rocks indicate that the Silurian volcanics of Beacon Hill, north-east of Shepton Mallet, were probably undergoing erosion at this time. Clasts of hard grey andesite found in the Inferior Oolite Limestone in Merehead Quarry in 1973 (Alabaster and Wilson, 1975, pp.73–75) were remarkable for their freshness, being much less weathered than any of the Silurian rocks exposed in situ at the present time. Traces of similar lavas are also reported in the Upper Old Red Sandstone of the eastern Mendips, and it is feasible that the Bajocian material may have been recycled via the Upper Palaeozoic, though the freshness of the clasts makes this unlikely.

The term Great Oolite was first used by R. Warner (1811) for the massive oolitic limestones, as well as the clays below and above, in the vicinity of Bath, though it may have been derived from a manuscript of William Smith (Worssam *in* Donovan and Hemingway, 1963, p.151). The rocks crop out along the eastern margin of the district in four hilly promontories projecting westwards from the main Cotswold plateau north and south of Bath: Lansdown Hill; Duncorn Hill; Huddox Hill and White Ox Mead; and the hills east of Kilmersdon. Exposures are poor and there are few sections of any significance within the boundaries of the sheet, but in the strip of ground immediately to the east a number of boreholes drilled in recent years in Horsecombe Vale, Combe Hay, Twinhoe, Baggridge and the Frome area has given very complete information concerning the sequence and in particular of the southwards passage of the Great Oolite into the Frome Clay. These boreholes have been described in two papers by Penn et al. (1979) and Penn and Wyatt (1979), which form the basis of the following summary.

The lithostratigraphical sequence of the Great Oolite in the Lansdown Hill to Kilmersdon area of the Bristol district and its correlation with the Bath and Frome areas immediately to the east is shown in Table 6.

Details

INFERIOR OOLITE

Dundry Hill

Parsons (1979) compares the nomenclature used in Table 5 with that of Buckman and Wilson (1896) and other authors, and has given details of the location and stratigraphy of sections to be seen on Dundry Hill. Slight modifications were made by Parsons when it was used in Cope et al. (1980, p.10, fig. 3b). The Lower and Middle Inferior Oolite limestones are thin, 'condensed' and more or less 'ironshot' and they occur only on the western part of the hill. They accumulated slowly in a broad shelf of shallow sea into which very little extraneous sediment was introduced, but in which precipitation of iron was frequent, and they contain an abundant molluscan fauna. The sequence was preserved by a local downwarp from the erosion, which prior to the transgression of the Upper Bajocian sea had cut down on the west of the hill into the Lower Inferior Oolite and on the east into the Upper Lias. The oldest beds are the ironshot limestones, seen by Buckman and Wilson (1896, p.677) near Castle Farm [c.551 671] at the west end of the hill.

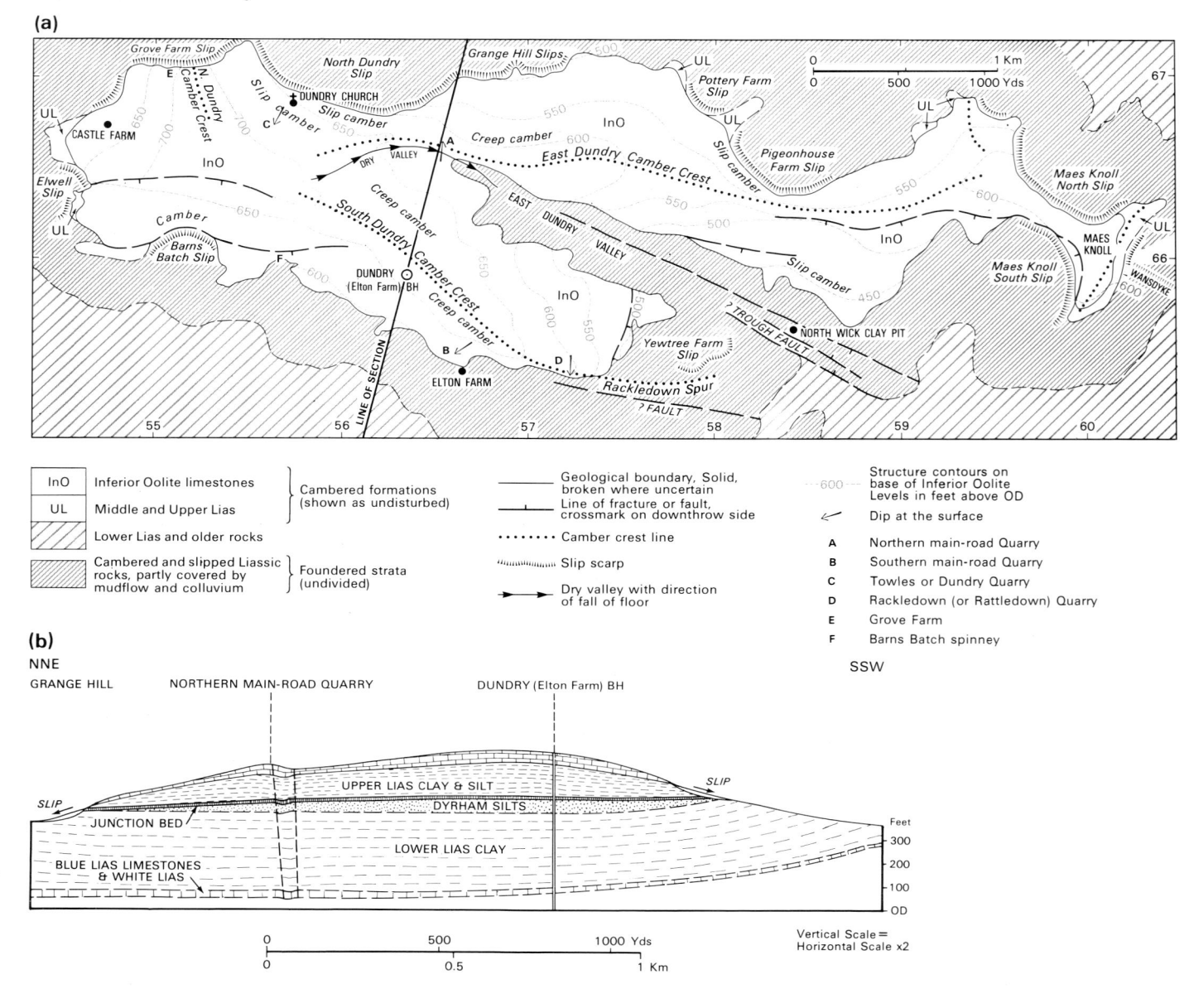

Figure 45 Dundry Hill, showing
(a) the effect of cambering, faulting and landslipping on the structure of the Inferior Oolite limestone plateau
(b) structural inversion and marginal attenuation produced by cambering

These are overlain by the Barns Batch Beds, hard sandy and oolitic limestones, exposed south of Dundry, and thickening to nearly 10 ft at [572 654] near Rackledown Farm (also known as Rattledown). The upper part of this unit contains abundant pleurotomariid gastropods. The overlying Grove Farm Beds comprise up to 6½ ft of nodular ironshot limestones and marls seen at the farm of that name [551 671] on the west of the hill and at Barns Batch spinney [557 659].

The Elton Farm Limestone, up to 6½ ft thick, includes the 'Lower White Ironshot', the 'Upper White Ironshot' and the 'Ironshot' of Buckman and Wilson (1896), terms introduced to record the location of the ammonite faunas, not as discrete layers (Parsons, 1979, p.136). The best exposure at present is in the Southern Main Road quarry [567 655] at Elton Farm. The beds of crystalline limestone contain variable amounts of limonite ooliths and are very fossiliferous; individual layers are separated by limonite-stained partings. The 'Brown Ironshot', above which occurs the main Upper Bajocian gap, is truncated by a smooth flat erosion surface with many borings and encrusted oysters.

The basal bed of the Upper Inferior Oolite—the Maes Knoll Conglomerate—is an ironshot limestone with a base containing limonite- and serpulid-coated clasts of blue-grey limestone and rolled Toarcian ammonites first described by Buckman and Wilson (1896, p.686, table 4). The bed is about 2¼ ft thick and of late *garantiana* Zone age; it can be seen on the west side of Dundry Hill at Grove Farm [551 671], at South Main-Road quarry and on Maes Knoll on the east. The succeeding Dundry Freestone has been extensively sought on Dundry Hill both from quarries and from underground galleries, but is now virtually worked out. It was used in the construction of St Mary Redcliffe in Bristol. The building stone is about 12 ft thick at Dundry but it thins rapidly eastwards. It can be seen on Dundry Down [552 666] where 15 ft of massive bioclastic limestones, including the Freestone, are overlain by 4 ft of more thinly bedded limestones.

The youngest beds of the Inferior Oolite seen on Dundry Hill were called the Coralline Beds by Buckman and Wilson (1896, p.672). The pale bioclastic limestones with numerous coral fragments are said to be over 20 ft thick (Donovan, 1958, p.132); 1 ft of disturbed limestone can be seen on Barns Batch Spinney and rather more in similar tumbled masses (up to 5 ft) at Rackledown and South Main-Road quarries.

Cambering and landslipping have affected the rocks forming the upper part of Dundry Hill (p.145). In particular, the Inferior Oolite limestone has been converted to loose rubble in many areas and any interpretation of the stratigraphy has to take these factors into account.

Eastern and southern outcrops

In the area to the east of the 'Malvern hinge-line' which extends roughly north–south through the middle of the Bristol and Somerset Coalfield, a long and complex period of earth-movements and erosion resulted in the remarkable regional planed and bored surface during late Lower and early Upper Bajocian times which was later to be transgressed by the Upper Bajocian deposits (Upper Inferior Oolite). No Lower or Middle Inferior Oolite is known anywhere within this area. At Tog Hill and Freezing Hill, southeast of Wick, Upper Inferior Oolite rests directly upon Upper Lias, and the Lower Inferior Oolite is not seen again to the south for some 25 miles across the eastern Mendips until the Cole syncline near Bruton is reached.

Table 5 Stratigraphical classification of the Inferior Oolite of the Bristol district

Stage/Substage		Zone	Subzone*	Sequence in Bristol special sheet district		
				Dundry	E and SE outcrops	
Bathonian	Lower (part)	*Zigzagiceras (Z.) zigzag*	*Oppelia (Oxycerites) yeovilensis* *Morphoceras macrescens* *Parkinsonia convergens*	Lower Fuller's Earth		
				Coralline Beds 20 ft +	Anabacia Limestones	Upper Inferior Oolite
Bajocian	Upper	*Parkinsonia parkinsoni*	*Parkinsonia bomfordi* *Strigoceras truelli*	Dundry Freestone 0–20 ft	Doulting Stone Upper Coral Bed Upper Trigonia Grit	
		Strenoceras (Garantiana) garantiana	*Parkinsonia acris* (3 older subzones)	Maes Knoll Conglomerate 0–2 ft		
		Strenoceras subfurcatum	(3 subzones)	〰〰〰〰〰〰〰〰		
		Stephanoceras humphriesianum	(3 subzones)			
	Lower	*Emileia (Otoites) sauzei*		Brown Ironshot Elton Farm Limestone 0–6 ft	Witchellia Bed Limonitic Bed ovalis Bed	Middle Inferior Oolite
		Witchellia laeviuscula	*Witchellia laeviuscula* *Sonninia (Fissilobiceras) ovalis*			
		Hyperlioceras discites		_ _ _ _ _ bivalve bed _ _ Grove Farm Limestone (Beds) 0–6 ft	Grey limestones and marls	
Aalenian		*Graphoceras concavum*	*Graphoceras formosum* hor. *Graphoceras concavum*			
		Ludwigia murchisonae	(2 younger subzones) *Ludwigia murchisonae* *Ludwigia haugi*	〰〰〰〰〰〰〰〰 Barns Batch Limestone (Beds) 0–6 ft	Sands and oolitic limestones	Lower Inferior Oolite
		Leioceras opalinum	*Tmetoceras scissum* *Leioceras opalinum*	Ironshot limestone 0–1¼ ft		

* Subzones listed only where beds of this age may occur in Bristol district   〰〰 non-sequence

At Doulting, just to the south of the Mendips and beyond the southern margin of the district, deposition of Upper Inferior Oolite began with the Doulting Conglomerate of grey crystalline limestone with pebbles, serpulids and many terebratuloids. This is of *subfurcatum* Zone age (Parsons, 1975). It is overlain by crinoidal and bioclastic limestones totalling about 45 ft in thickness. This sequence of limestone includes the Doulting Stone, a freestone about 20 ft thick which has been widely used for high-grade building, for example in Wells Cathedral and Glastonbury Abbey. Above are some 11 ft of rubbly white-brown oolitic Anabacia limestones with small solitary corals *'Anabacia' = Chomatoseris*) scattered throughout; its top surface is bored and iron-stained and the upper 1 ft contains early Bathonian ammonites (Torrens, 1969, p.B19).

In the north-eastern part of the Mendips at Vallis Vale and Whatley the Upper Inferior Oolite rests directly upon the Carboniferous Limestone. At Mells, Woodward (1894, p.93) described a railway cutting with 26 ft of Inferior Oolite resting on Lias. Whilst most of the fossils listed are characteristic of the Upper Inferior Oolite, there is a reference to *'Ammonites humphresianus'* which if correctly identified would indicate the presence of Middle Inferior Oolite. The record has never been confirmed and it is possible that the specimen was a *remanié* one. The base of the Upper Inferior Oolite at Doulting, Nunney, Limpley Stoke, Welton, Maes Knoll and many localities around Bath contains an abundance of derived fossils (Hawkins *in* Savage, 1977, p.119) including ammonites of Toarcian, Aalenian and early Bajocian age which testify to the magnitude of the mid-Inferior Oolite erosion.

Northwards from Wellow the shelly limestones of the Upper Trigonia Grit are developed as a lateral equivalent of the Maes Knoll Conglomerate of Dundry. At Midford, south of Bath, these are about 5 ft thick, overlain by the Upper Coral Bed, nearly 8 ft thick, and 12 ft of massive Doulting Stone; at the top 11 ft of Anabacia Limestones are capped by 6 in of fossiliferous rubbly limestone with terebratuloids.

Descriptions of all the principal lithological subdivisions of the Inferior Oolite in the Bristol district and the eastern and central Mendips and south Cotswolds have been given by Richardson (1907, 1910), Kellaway and Welch (1948), Green and Welch (1965) and Cave (1977). Accounts of sections which can be seen in the field have been provided by Murray and Hancock (*in* Savage, 1977, pp.140–165) and of Bath and the Mendips respectively by Hawkins (*in* Savage, 1977, pp.119–132) and Savage (1977, pp.85–100). Older descriptions include those of de la Beche (1846) and Woodward (1876, 1894).

Great Oolite

Fuller's Earth The Lower Fuller's Earth consists of about 40 ft of predominantly calcareous mudstones with sporadic thin bands of argillaceous limestone, giving nine recognisable units which can be grouped into three cycles of sedimentation. Three of the units are especially recognisable and so are useful for correlation. Each is named after a fossil species which is particularly abundant at that level. They are, in ascending order, the knorri Clays, with abundant *Catulina knorri* (Unit 2 of Penn et al., 1979); the acuminata Bed, a silty calcareous shell-fragmental mudstone with common *Rhynchonelloidella smithi* amd *Praeexogyra acuminata* (Unit 6); and the echinata Bed, a shell-fragmental mudstone with harder calcareous bands and *Meleagrinella echinata* (Unit 8). At the base of the formation is a thin pale grey fine-grained limestone — the *Fullonicus* Limestone — with a Bathonian fauna resting nonsequentially on the Anabacia Beds of the Inferior Oolite.

The Fuller's Earth Rock makes a strong feature in the vicinity of Bath. It consists of interbedded rubbly shelly and marly limestones with subsidiary shell-fragmental calcareous mudstones, in all about 11 ft in thickness at Baggridge and thickening to about 16½ ft at Horsecombe Vale. It is composed of two distinct units. The lower comprises up to 3 ft of fine-grained compact limestone with an abrasive sandy texture; it may contain dark-skinned (due to pyrite with some phosphate), reworked pebbles of limestone resting on an irregular eroded surface. The upper unit is thicker and can be further divided into two sequences of rubbly marly shell-detrital limestone, each with a distinctive fauna. The lower, the Ornithella Beds, is dominated by serpulids, *Rhynchonelloidella smithi, Ornithella bathonica*, and by attached, cemented and burrowing bivalves. The upper subunit, the Rugitela Beds, locally succeeds the *Ornithella* Beds nonsequentially; ornithellids are absent, *R. smithi* is very abundant, some *Rugitela bullata* occur and *Catinula matisconensis* is also prominent.

The area south of Bath is the type area of the Upper Fuller's Earth, comprising at most some 90 ft of mudstones with some limestones. A lower unit, dominated by mudstones with few limestones, thins southwards from about 48 ft at Bath to 11½ ft at Baggridge, while becoming increasingly silty and calcareous. The upper part is also mainly argillaceous but contains significantly more limestone bands, the proportion increasing northwards. South of Baggridge Hill the Upper Fuller's Earth as a whole becomes much thinner, and it is only about 1¾ ft at Frome.

The commercially exploited Fuller's Earth Bed has a sharp base, is about 8 ft thick, and occurs between 10 and 30 ft below the top of the Upper Fuller's Earth. It is a bluish grey homogeneous structureless clay, restricted to the area of thickest Upper Fuller's Earth and has a waxy texture and subconchoidal fracture when dry and hard. The highest grade 'earth' comes from the lowest part of the bed where it contains over 80 per cent montmorillonite and has a low calcite content. It is commonly capped by a thin limestone, passing upwards into silty calcareous mudstones. The commercial bed cannot be traced south of Baggridge No. 2 Borehole [7407 5602]. It was worked underground by the pillar-and-stall method in the area between Duncorn Hill in the west and Combe Hay and Midford to the south-east and east.

A volcanic origin for the earth was suggested by Kerr (1932) and Grim (1933), and was confirmed by Jeans et al. (1977). The latter authors concluded that the bed was formed by the deposition and rapid argillisation in shallow marine waters of volcanic ash of trachytic composition. Abundant pyroclastic minerals, such as sanidine, and rock particles occur in the sand grade fraction of the deposit, together with fresh glass shards preserved in nodules and so escaping the early alteration of the ash. Penn et al. (1979) discussed possible sources and concluded that the bed was not a direct ash fall but was produced after penecontemporaneous reworking of ash deposited elsewhere. The very restricted area in which it is known, the presence of deposit-feeding and burrowing bivalves as well as benthonic foraminifera and ostracods which survived during accumulation and also the occurrence of small-scale cross-bedding and channels with fine shale detritus, all indicate some degree of redistribution and admixture with bioclastic carbonate materials.

Great Oolite The Great Oolite or Great Oolite Limestone was so named because the limestones of the Bath area were of superior quality and thickness compared with those of the Inferior Oolite. To the north of a line drawn roughly east–west through the north end of Baggridge Hill and south of White Ox Head the Great Oolite separates the clays of the Upper Fuller's Earth from the Forest Marble; to the south of this line the limestones no longer exist, having been replaced first by detrital and argillaceous limestone and then by the Frome Clay (see below). The limestones make up the plateaux on either side of the River Avon at Bath, though the exact details of the sequence are commonly obscured by extensive landslipping on the steep hillsides.

The Great Oolite comprises three major formations in the Bath area: at the base the Combe Down Oolite is overlain, first by the Twinhoe Beds, and then by the Bath Oolite (Table 6). The locally

Table 6 Stratigraphical classification of the Great Oolite of the Bristol district

	Bristol special sheet[1] Lansdown Hill to Kilmersdon	Bath area[2]	Frome area[2]
		Cornbrash	Cornbrash
	Forest Marble	Forest Marble (including Upper Rags and Bradford Clay horizons)	Forest Marble (including Bradford Clay horizon)
Great Oolite	Great Oolite Limestone	Great Oolite ⎧ Bath Oolite ⎨ Twinhoe Beds ⎩ Combe Down Oolite	Frome Clay[3]
	Fuller's Earth ⎧ clays ⎨ limestone ⎨ Fuller's Earth Rock ⎩ clay	Upper Fuller's Earth Fuller's Earth Rock Lower Fuller's Earth	Upper Fuller's Earth Fuller's Earth Rock Lower Fuller's Earth

1 Classification used on the side margin of the one-inch Bristol special sheet; there is no Cornbrash exposed within the district
2 Classification after Penn and Wyatt (1979, p.26)
3 Beds in the Frome area which correspond to the limestones of the Great Oolite (excluding Upper Rags) were mapped as Upper Fuller's Earth. The term Upper Fuller's Earth is now confined to beds underlying the Great Oolite at Bath and to their attenuated equivalents at Frome (Penn and Wyatt, 1979, pp.27, 33)

conglomeratic base of the Combe Down Oolite contains phosphatised pebbles of Fuller's Earth limestones and rests on an extensive plane of erosion cutting down into successive horizons in the upper part of the Upper Fuller's Earth. The extent of this nonsequence suggests a regional erosive phase rather than local channelling during oolith formation. The deposits in the transition zone to the Frome Clay (Twinhoe Ridge and Baggridge Hill) show alternation of facies reflecting changes in the depositional environment. The intercalations of grit-sized shale fragments with ooliths in the Twinhoe area may represent the northerly extension of clays into which ooliths and shell-debris were washed from nearby banks. The abundance of mudstone clasts in some beds suggests contemporaneous erosion of clays, perhaps in tidal channels intersecting the marginal slopes of the shelf sea.

The Combe Down Oolite consists of fairly uniform, massive, cross-bedded oolites up to 57½ ft thick south of Bath with a well-defined top surface which was extensively bored (Penn and Wyatt, 1979). Green and Donovan (1969) described the Twinhoe Beds and the Bath Oolite of the same area. The Twinhoe Beds comprise three main facies: at the base a locally impersistent basal marly ironshot pisolitic limestone is overlain by marly pisolitic shell-fragmental limestone and then by fine-grained detrital limestones. These total about 20 ft at Combe Hay. South of the Twinhoe ridge these limestones pass into fine-grained argillaceous limestone overlying 3 ft of rubbly marly shelly limestone. The overlying Bath Oolite consists mainly of relatively pure oolites, some shell-fragmental limestones and a local coral bed at the base. The Oolite is generally poorly fossiliferous except for some of the coarser limestones which are shelly. In the area south of Bath it ranges from 16 to 30 ft thick. The joint thickness of the Twinhoe Beds and the Bath Oolite is relatively uniform at about 50 to 60 ft, though constituent beds vary widely. The Twinhoe Beds thicken southwards at the expense of the Bath Oolite, the one facies passing into the other.

The Frome Clay comprises the argillaceous strata lying between the Upper Fuller's Earth and the Forest Marble in the area south of the Twinhoe ridge. It was formerly classified as part of the Upper Fuller's Earth and as such it was mapped in the Bristol district in the area east of Kilmersdon. An overstepping relationship to the Upper Fuller's Earth has been demonstrated by Penn and Wyatt (1979, fig.14), who also traced it northwards into the Great Oolite of the Bath area. In the Frome area it rests on a reduced sequence of Upper Fuller's Earth, its base being marked by the lower smithi limestone, the lateral equivalent of the base of the Combe Down Oolite. A second and higher smithi limestone occurs higher in the formation, which has been traced into the base of the Twinhoe Beds.

Forest Marble   Within the Bristol district the Forest Marble occurs only on the high ground on either side of the main road from Radstock to Frome (A362), between Terry Hill and Mills Down. Here the sequence is predominantly argillaceous throughout. Just east of the present district, The Forest Marble, a name originally applied by William Smith to distinctive shell-detrital limestones lying between the Great Oolite and the Cornbrash, is now taken to include all strata lying between the Bath Oolite and the Cornbrash. These comprise a very variable sequence, generally about 100 ft thick, of predominantly clays and mudstones irregularly interbedded with limestones of varying type, silts, sands and sandstones. At the base in the area around Bath occurs 20 to 55 ft of shell-fragmental oolites and coralline limestones — the Upper Rags — a division which was formerly included in the Great Oolite (Green and Donovan, 1969). The Upper Rags can be recognised on the Baggridge Hill plateau, but from here southwards the division passes rapidly into blocky pale grey silty calcareous mudstones. The Forest Marble above the Upper Rags horizon comprises mainly poorly fossiliferous calcareous slightly silty mudstones, interbedded with calcareous sandstones, sandy limestones, and some crystalline limestones. Lateral variation is both rapid and considerable and correlation even of closely spaced sequences is difficult. In general terms limestone developments are more marked in the upper part of the sequence, where they are associated with lenses of sand and sandstone, particularly in the higher parts of the calcareous succession.

Harptree beds

In the Mendips, and on Broadfield Down south of Bristol, there are occurrences of strata silicified in situ and referred to as the Harptree Beds after their principal development between East Harptree and Oakhill. They include attenuated Upper Triassic to Middle Jurassic strata which have been largely converted to chert. In places, the underlying Carboniferous Limestone has also been affected. On Broadfield Down a similar process has affected the conglomeratic Blue Lias limestones of Felton Hill [520 650] (Donovan and Kellaway, 1984, pp.19–20). Locally the decalcified cherty rock occurs as doggers up to 5 ft long, in a soft ochreous matrix. From a quarry [620 493] between Binegar and Gurney Slade, a block of this rock yielded *Garantiana* sp., an Upper Bajocian ammonite (Green and Welch, 1965, pp.94–95, 108–109). This occurrence, combined with the absence of silicified pebbles in any of the local Jurassic rocks, indicates a post-Middle Jurassic age for the silicification, which may date from Albian–Cenomanian times (Kellaway, 1991, pp.246–248 and fig. 14.2).

CHAPTER 8
Economic geology

Various nonmetallic and metallic minerals have been exploited over the centuries in the Bristol district, but apart from the quarrying of limestone there is little activity at the present time. Coal has been by far the most important over the past 150 to 200 years, but mining has now ceased; the last colliery (Kilmersdon) closed in 1973. Among the metallic minerals, the ores of iron, manganese, lead and zinc have all been exploited in the past and evaporite deposits (gypsum and celestite) have also been worked. Hard rocks (sandstones and limestones) have been extensively quarried for building stone and aggregate, as well as clays for brickmaking and pottery and sand for the manufacture of glass. Although most of these have ceased to be exploited, some of the industries founded on them still persist though they are now dependant on imported raw materials.

Amongst the most important assets of the Bristol district are the major water supplies deriving from the Carboniferous Limestone of the Mendips and Broadfield Down. The limestone scenery is itself an important element in the local economy since the gorges and limestone caverns, notably those of Cheddar, are important to the tourist industry.

IRON ORE

Iron ores are widely distributed in the Bristol district and there is an extensive literature dealing with them: Etheridge (1870), Grenfell (1873a, b), Anstie (1873), Woodward (1876), Cantrill et al. (1919) and Green and Welch (1965). The haematites and goethites of the Bristol district closely resemble those of the Forest of Dean and South Wales with which they have a community of origin (Kellaway, 1967; Trotter, 1942). Associated with the ores in some areas, as for example at Wick and Winford, are earthy forms of hydrated iron oxide, produced either by alteration of Triassic deposits or emplaced in cavities in the Carboniferous Limestone. Hematite and goethite are commonly found filling fissures and joints in the limestone or in Millstone Grit and Coal Measures sandstones. Locally, as at Hartcliff Rocks [533 663] and Brandon Hill, Bristol [5790 7268] the Dolomitic Conglomerate has been converted into massive iron ore by the replacement of limestone by haematite. A similar replacement body, partly in Carboniferous Limestone and partly in Dolomitic Conglomerate was described by Grenfell (1873a) at the Iron Mine at Clifton.

One of the largest deposits of haematite was encountered in the Coalpit Heath Basin of the Bristol coalfield, where the ore occurs in Pennant Sandstone (Cantrill et al., 1919). About 111 000 tons were extracted from the Frampton Cotterell Iron Mine between 1862 and 1874, from vertical fissures in the sandstone down to a depth of 480 ft. The workings were abandoned in 1875. Three pits working iron ore in Pennant Sandstone at Downend, in north-east Bristol, are said to have been operated until 1900. Traces of the blooming slags have also been found at Brislington, Bristol and other places where haematite and goethite were easily accessible. Iron smelting was carried out in Iron Age times in a cave at Rowberrow Warren near Burrington Combe in the Mendips.

Some clay-ironstone was worked from the Coal Measures in Ashton Vale and Kingswood in former times but neither this nor the haematite deposits gave rise to a major industry in the Bristol district.

MANGANESE

Manganese ore has been worked at a number of scattered localities, mainly in the form of pyrolusite or wad. It occurs mainly in the Dolomitic Conglomerate or in fissures and cavities in the Carboniferous Limestone. It was used to produce a black or dark colour in pottery, and also at Wadbury to harden steel. Manganese workings are said to have been in operation on Leigh Down, south of the Avon Gorge in 1756. Buckland and Conybeare (1824) and Woodward (1876) list a number of localities where the ore was dug, but none of the deposits was large enough to sustain prolonged working.

EARTH PIGMENTS (OCHRES)

Winford and Wick were formerly two of the most important centres of pigment manufacture, based initially on the availability of red and yellow ochre. Bedded deposits of red ochre are generally older than those which yield brown, yellow or orange coloured pigments. All are of late Triassic origin, but some have been redeposited. Yellow ochre is associated with the strongly dolomitised facies of the Dolomitic Conglomerate and was commonly found in association with lead and iron ores, notably calamine, in fissures and cavities in the Carboniferous Limestone.

At Winford, where iron ore was worked from shallow shafts, beds of highly ferruginous purplish red mudstone and Dolomitic Conglomerate were also dug opencast and ground to make the reddle used for marking sheep. At Wick, red and yellow ochre is commonly found in fissures in Carboniferous Limestone, some containing galena, but the principal workings, all long-since abandoned, exploited roughly stratified bands and lenses of fine-grained ferruginous conglomerate and mudstone of Triassic age. As at Winford, these red, purple, brown and yellow strata are succeeded by yellow and green mudstones (Tea Green Marl). The overlying pyritic black shales of the Westbury Beds constitute an unconformable cover and are not affected by the colour changes observed in the underlying rocks.

One area worked for red ochre is at Compton Martin on the north face of the Mendips (Green and Welch, 1965, p.165); here, as at other localities to the north, the contact between Triassic and Carboniferous rocks is strongly haematitised, red and purple colours being dominant. On the uplands of Broadfield Down and the Mendips, particularly in the lead-zinc orefield, yellow ochre has been worked sporadically over wide areas (Woodward, 1876, p.106). Some of these deposits occur in swallets in silicified Lower Lias limestones (Harptree Beds). The silification postdates the formation of the ochreous rocks, which have been rendered useless as a source of pigment.

LEAD AND ZINC

Among the oldest known industries in the post-Bronze Age history of the Bristol district is the lead mining for which the Mendips were once famous. The presence of lead fishing weights in Iron Age settlements near Glastonbury and the evidence of the use of lead by the Romans are well documented (Woodward, 1876, pp.167–175; Gough, 1930; Dobson, 1931; Buchanan and Cossons, 1969). The geology of the central Mendip orefield has already been described in some detail (Green, 1958; Green and Welch, 1965) but much less material is available regarding the lead mines of Broadfield Down and Bristol. Like the central Mendips, the uplands of Broadfield Down still retain patches of black shale and greenish grey silty clay of the Westbury Formation and Cotham Member, as well as the conglomeratic Brockley Down Limestone, resting unconformably on the Carboniferous limestones in which the majority of the lead veins were emplaced. Where the Mesozoic strata still survive the old lead mines were sunk through these strata to work the mineralised veins below; in some instances lead ore was also recovered from the Mesozoic strata themselves. Elsewhere lead was worked at many localities from open diggings and relatively shallow pits along the outcrops of veins on the bare limestone surface. The output of many of the mines on Broadfield Down is unknown as is the date of their operation. Some were worked before the 18th century, a number may be much older. Unlike the Mendips, however, there are no indications of lead having been smelted on any appreciable scale on Broadfield Down.

Extensive areas of rough pitted and worked ground on Broadfield Down have been cleared for cultivation during the present century. Nevertheless the evidence of old workings is still to be seen in the Wrington Warren, at Downside, at the head of Brockley Combe, in Bourton Combe and elsewhere. Some of the workings were still open in the wooded areas in and around Bourton Combe at the time of the survey, about 1950. The mineralised zones in the Carboniferous Limestone show the same general relationship to the base levels of the Upper Triassic and Lower Jurassic as those described in the Ashton Park–Leigh Woods–Clifton areas (Kellaway, 1967). Some of the more prominent veins which have been identified on the limestone uplands adjacent to the Avon Gorge are shown on the published 1:10 560 sheet ST 57 SE and on the Bristol district special sheet.

In the Carboniferous Limestone ridge extending from Penpole Point and Kings Weston to Henbury and Brentry (Upper Knole) several old lead mines are known to have existed in former times. Some of these workings were of very limited extent, particularly at the western end, but there were more extensive workings near Westbury-on-Trym and at Brentry. These include Pen Park Hole, one of the best documented lead mines in the Bristol district (Tratman, 1963). There were several openings, the main one being located [585 793] on the Ordnance Survey 1:10 000 map ST 57 NE (Palmer *in* Tratman, 1963). Much of the evidence of this mine dates from accounts published in the 17th and 18th centuries notably by Southwell (1863) and the Rev. A. Catcott in his *Treatise on the Deluge* (1768). Later accounts by his brother, G.S. Catcott (1792) and S. Rudder (1779) are important, while that of Nicholls (1880) includes material compiled from manuscripts of A. Catcott and from other sources, some of which are preserved in Bristol Central Library. Nicholls regarded Pen Park Hole as a Roman lead mine, but it was later claimed by Kerslake (1883) as being of Saxon date. Both these assessments were later shown to be wrong and a late Tudor date of about 1590 is suggested by Palmer *in* Tratman, 1963. Supporting evidence for this conclusion came in 1935–36 when old lead workings were encountered in the nearby Brentry Quarry, from which an Elizabethan coin was recovered.

Saxon mining for lead was referred to in the Saxon charter relating to Stoke Bishop, but it would appear that this could not have been Pen Park Hole. One possible area may have been the highly mineralised belt at the northern end of Durdham Down, extending along the margin of the open space north and south of the White Tree [573 757]. There is evidence for old workings in this area and galena has been seen in excavations for roadworks.

Between 1934 and 1936 small quantities of lead ore were recovered from Brentry Quarry and some of this was sent to smelting works at Bedminster. All the veins occur in steeply dipping or vertical Carboniferous Limestone, cut by steeply inclined faults or thrusts, along which mineralisation took place. Most of the veins seen in Brentry Quarry are thin and impersistent, and occur for the main part in the Clifton Down Limestone in the southern part of the quarry. One lenticular vein consisting of galena with calcite, barytes, (or possibly barytocelestite) ankerite and marcasite with a little sphalerite achieved a width of 15 ft at a depth of about 60 ft from the ground surface. Much ochreous material was present towards the surface due to oxidation of the marcasite. In the upper part of the veins seen in the quarry galena was the dominant nonferrous sulphide; the lower parts contained less galena and more sphalerite. At floor level in the quarry some parts of the thicker veins consisted almost entirely of sphalerite. The significance of this occurrence was underlined by the presence of extensive but shallow deposits of sphalerite and 'calamine' (probably a mixture of weathered smithsonite and cerussite) associated with some galena and ochreous barytes in excavations for drains on the Henbury Estate in 1949–50, west of Passage Road. Red mudstone with small nodules of pink platy celestite occurred above the mineralised sandy calcareous mudstone which passes laterally into conglomerate. The tendency for sphalerite and calamine to be concentrated at the stratigraphical level of the Stoke Park Rock Bed is not surprising, but the downward passage of the galena veins of Brentry Quarry into sphalerite-rich deposits at the same topographical level as the Henbury deposits, and the alignment of the lead and zinc deposits of the Henbury Estate with the north-east–south-west-trending lead veins of Pen Park Hole, suggest a correlation of level as well as orientation.

Observations throughout the area west of Bristol suggest that most of the lead veins die out downwards and very few extend much beyond 100 ft below the base of the black shales of the Westbury Beds. The richest veins are found mostly at higher levels and though the depth range of the Mendip lead veins may have been somewhat greater than those of Broadfield Down and Bristol, all the workings are relatively shallow in relation to their very wide extent (Kellaway, 1967). This accounts for the relatively small annual outputs which have been recorded from the Mendips and elsewhere in the 17th and 18th centuries in contrast to the very extensive areas of old 'gruffy-ground' which formerly existed on many of the Carboniferous Limestone uplands.

North of Bristol the Carboniferous Limestone uplands show less evidence of lead mineralisation though galena is been found at Almondsbury where a lead mine was recorded by Catcott (1775). Lead mines formerly existed at Chipping Sodbury and Bury Hill, Wick. Numerous lead veins, mostly rather thin, have been exposed from time to time during the quarrying of Carboniferous Limestone in the Wick Inlier.

Working of sphalerite and calamine appears to have commenced in the 16th century, one of the most important and earliest sites of calamine mining being on Worle Hill near Weston-super-Mare. Woodward (1876) describes this and other occurrences of smithsonite in Broadfield Down and the Mendips. It is known that many of the occurrences of 'calamine' both in the Mendips and at Wrington and elsewhere are mineralogical assemblages of fairly complex composition. One such complex vein seen near Maesbury Castle was recorded by Ponsford (*in* Green and Welch, 1965, pp. 164–165).

The history of the brass industry in Bristol has been described by Buchanan and Cossons (1969, pp.115–127). Its development depended on a unique combination of mineral resources, that is, local supplies of calamine and coal, plus the availability of water power in the Avon valley and the capability of importing copper from Cornwall through the Port of Bristol. Much of the calamine was found in the Dolomitic Conglomerate in the form of flats and irregular veins, or secondary concentrations in the solution cavities. Some of the calamine in the waste tips of old lead mines in the Mendips and Bristol district may also have been recovered. Shipham and Rowberrow became two of the most important centres of calamine production, but the full extent of the old calamine workings, mainly concentrated in the Mendips and Broadfield Down, may never be known.

COPPER

Copper veins at Henbury were mentioned by Catcott (1775) and were said to have been worked about half a mile west of the village. This would appear to be on the side of the prominent hill crowned by Moorgrove Wood which is situated on an outlier of Keuper Marl, Tea Green Marl and Westbury Beds clays. In a shallow valley on the south side of the hill a white celestite-rich breccia with specks of malachite was noted during the six-inch survey. The site [804 787] is now built up, but it may well have been the site of 'Pope's Work', and the copper veins may well form part of a continuous belt of mineralised fissures extending eastwards to Pen Park Hole (see above). The celestite-malachite-bearing horizon is thought to equate with the Stoke Park Rock bed about 45ft below the top of the red Keuper Marl.

CELESTITE

Celestite ($SrSO_4$) has been produced over a wide region extending from the Mendips to the northern part of the Bristol Coalfield, and over the last century a high percentage of total world production has come from the area surrounding Bristol. Production is now confined to the area of Yate, but it was formerly worked at a number of other localities, notably at Abbots Leigh, west of Bristol.

Deposits of celestite occur at several stratigraphical levels in the upper part of the Keuper Marl but are best developed immediately above the Stoke Park Rock Bed, situated about 40 to 50 ft below the Tea Green Marl at Bristol. The mineral assumes a number of forms, the primary deposit being small reddish nodules of crystalline celestite or irregular nodular masses of fine-grained granular celestite which roughly follow the bedding of the Keuper Marl. Where secondary leaching has taken place the recrystallised mineral forms thin joint and fissure fillings in the underlying red marl. Elsewhere, recrystallisation has produced drusy cavities in red marl filled with white and pale blue tabular (orthorhombic) crystals of strontium sulphate. Crystalline celestite is also concentrated in voids in the Dolomitic Conglomerate and in the sandy rocks which are transitional to the Keuper Marl. Where the higher beds of the Keuper Marl rest unconformably on Upper Carboniferous rocks, as in the area around Yate and Stanshawe's Court, quite large quantities of celestite are emplaced along the contact with the Keuper Marl or extend downwards as veins and stringers following joints and other discontinuities in the underlying bedrock.

An assessment of celestite resources north-east of Bristol was made by the Geological Survey (Nickless et al., 1976), the principal celestite deposits being referred to as the 'Severnside' Evaporite Bed. The name Severnside is now applied to a major industrial site on the banks of the Severn north of Avonmouth. This is remote from the celestite deposits of the Coalpit Heath Basin where the main concentration of celestite in the Coalpit Heath Basin is probably related to the Stoke Park Rock Bed (p.133). It is therefore proposed to substitute Yate for 'Severnside' as applied to the principal deposits of the eastern side of the Coalpit Heath Basin. In case of doubt as to the stratigraphical position any evaporite bed can take the name of the underlying skerry band. The method of showing the presence of celestite on the published six-inch (or 1:10 000) Geological Survey maps involves no commitment to specific correlation and is therefore unaffected by the proposed change. Further details for Malmesbury (Sheet 251) and Wells and Cheddar (Sheet 280) districts have been given by Cave (1977) and Green and Welch (1965) respectively. Celestite is used mainly in pyrotechnics, as well as in the electronics, glass, ceramics and paint industries.

Figure 46 Probable distribution of Triassic celestite deposits prior to dissection

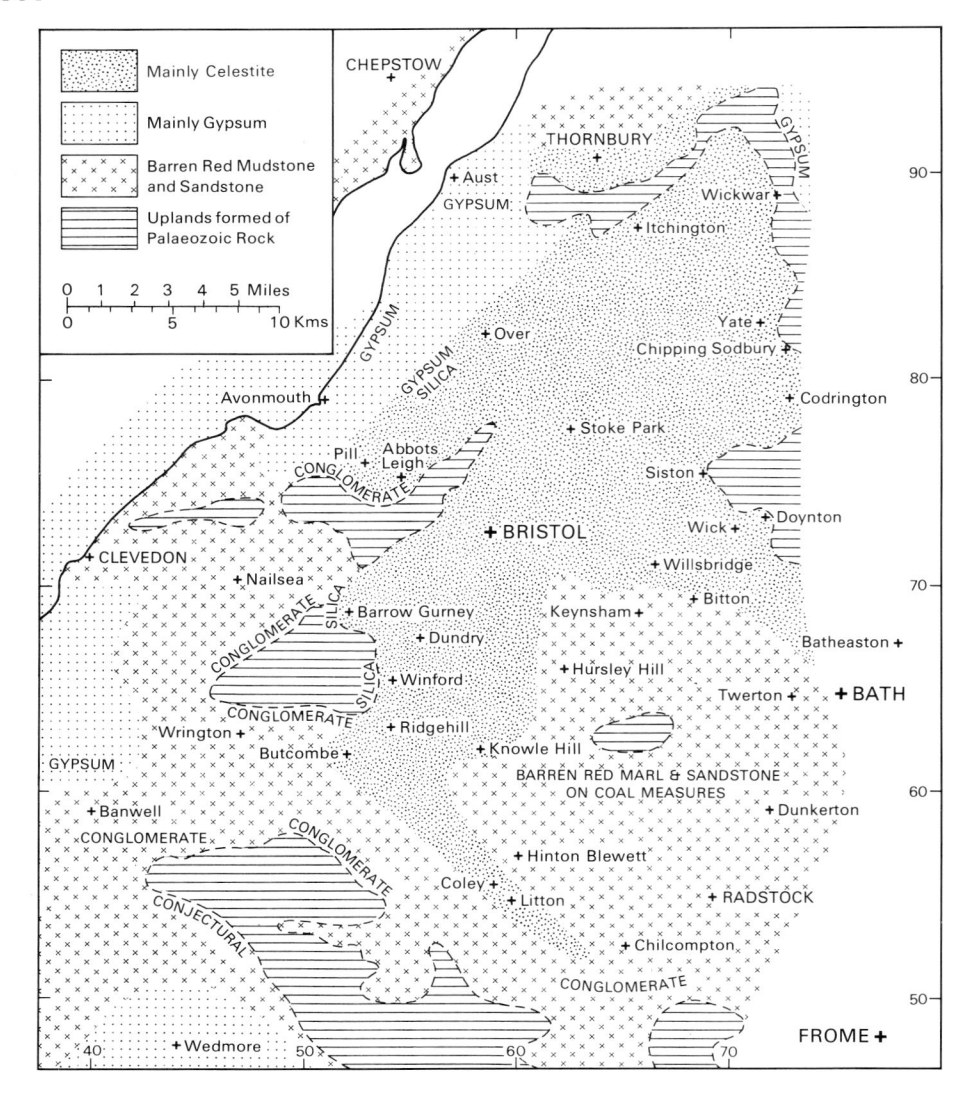

ROADSTONE AND AGGREGATE

Limestone quarrying has now become the most important extractive industry in the Bristol district. Although there is some incidental working of Liassic limestone in the overburden of the Carboniferous Limestone at Lulsgate Quarry on Broadfield Down, almost all hard rock quarrying is concentrated in the Carboniferous Limestone in the peripheral areas surrounding the Bristol and Somerset Coalfield. Geologically speaking, the resources of high grade limestone, dolomite and sandstone exceed any amount that is likely to be required in the foreseeable future. In practice their workability is dependent on environmental considerations, not least being the need to protect the underground water resources.

Although they relate mainly to Somerset and also cover areas to the south of the Mendips, the reports published by the County Planning Department of Somerset County Council provide much useful information on quarrying in the region south of Bristol.[1]

In the Mendips, production is mainly concentrated in a small number of large Carboniferous Limestone quarries, all the main limestone formations being used. On Broadfield Down active quarrying is largely concentrated on the steep north-western face at Backwell and in the north-east at Lulsgate and Hartcliff Rocks. Other areas where quarrying is significant are Failand and on the Clevedon–Portishead ridge at Weston-in-Gordano.

Limestone quarrying was carried out actively in the Avon Gorge until the outbreak of war in 1939. The extensive quarries here on both sides of the Avon provided a convenient source of stone for the inhabitants of Bristol and for those in adjoining areas in reach of water transport. Quarrying was also active until 1939 in the ridge extending from Penpole Point, Shirehampton, to Brentry. The last quarry in work was situated on Henbury Hill [566 782], excavated to a depth of a hundred feet in vertical Clifton Down Limestone with a bedding plane of the Concretionary Beds forming its north-western side (Plate 6). Much of the gorge of the River Trym at Blaise Castle was formerly quarried for limestone, but the workings have long been overgrown.

---

[1] *Quarrying in Somerset* (1971) and Supplement No. 1, *Hydrology and rock stability — Mendip Hills* (1973).

In the northern part of the Bristol Coalfield limestone has been quarried at Elberton, Grovesend near Tytherington, Cromhall and Wickwar (Welch and Trotter, 1961; Cave, 1977; Savage, 1977). In the Wick Inlier massive quarrying has taken place, the main excavation in the Clifton Down Limestone west of the Wick Fault having been taken down below the level of the River Boyd.

Namurian and Dinantian quartzitic sandstones have been worked as 'Quartzite' on a small scale at Cromhall, Wick, Winford and other places but the cost of crushing and treating this abrasive material has inhibited its use.

BUILDING STONE

A wide variety of local stone has been used for building in the Bristol district and only the more important are mentioned here.

Carboniferous Limestone has been extensively used, especially for walling and it is still obtainable from working quarries. It has not been used to any great extent for ashlar, the rock in general proving difficult to trim.

In former days Pennant sandstone was used on a very large scale for shaped building stone, kerbs and paving, but of the many large quarries that were worked all have now been abandoned. The suburbs of Bristol built during the 19th and early 20th centuries show acres of grey or reddish Pennant stone houses with lintels and sills of Bath Stone. Big sandstone quarries were concentrated in the Avon Valley between Hanham and Netham, and in the Frome Valley above Stapleton.

Dolomitic Conglomerate was worked near Draycott under the name of 'Draycott Marble' and used for building and paving stones, chimney pieces, etc. Kingsweston House [539 773], built between 1710 and 1725 by Sir John Vanbrugh for Sir Robert Southwell, used a beautiful ashlar of fine-grained yellow Dolomitic Conglomerate tinged with pink, known as 'Penpole Stone', which probably came from old quarries in woodland 300 yd south-west of the house.

The hard, white, fine-grained limestones of the White Lias were once widely employed as a building stone in the Radstock and adjacent areas. The uppermost bands provide a tough, creamy white stone almost immune to frost action. Conglomeratic and shelly limestones of the Lower Lias were extensively used locally in the Mendips. On Broadfield Down, the Brockley Down Limestone was quarried at Felton and Downside (Donovan and Kellaway, 1984, pp.19,48).

White Lias and Blue Lias limestones have been widely used for building houses and walls in the Avon valley between Bristol and Bath and in the central part of the Bristol and Somerset Coalfield between Stoke Gifford and Radstock. They vary greatly in durability according to their stratigraphical position, the flaggy limestone of the Blue Lias (widely used for paving) is hard and resistant to weathering; the overlying, nodular Blue Lias limestones are generally less satisfactory.

Dundry Stone from the Inferior Oolite of Dundry Hill was famous as a building stone but it has long since been worked out. It was partly quarried and partly mined from galleries in the area of Dundry village. The ornate tower of Dundry Church was built of Dundry Stone in 1484 by the Merchant Venturers as a beacon for sailors, and it is clearly visible from the Severn Estuary off Clevedon. The most famous building constructed of Dundry Stone is St Mary Redcliffe in Bristol, built mostly in the 13th to 15th centuries and founded on Redcliffe Sandstone.

The Great Oolite limestone has been worked in small quarries on Lansdown and the western end of the Odd Down plateau, but the main workings in Bath Stone on Combe Down are situated east of the area covered by the Bristol district special sheet.

SAND AND GRAVEL

Resources of river and gravel have always been poor in the basin of the Bristol Avon. The most important deposits of river gravel are situated in the Avon valley between Keynsham and Twerton. Most of these are now worked out. The principal source of gravel used in the Bristol district is marine gravel dredged in the Bristol Channel.

GLASS SAND

Bristol and Nailsea are two famous names in the history of glass manufacture, yet the apparent absence of any suitable local glass sand has always been presented as something of a mystery. A Venetian glass worker is said to have been operating in Bristol in 1651 but the oldest known glassworks is almost certainly that of Stanton Wick, situated midway between Sutton Court and Chelwood and dating from 1660. This was probably well supplied with fuel from numerous small coal-pits working the Bromley and Pensford seams at shallow depth. There is evidence that Triassic sandstones were worked on the north side of the Chelwood–Bishop Sutton road [6140 6170; 6160 6140] and at Chelwood. Most of this sandstone is red in colour, though some is white and pale buff. It is likely that this is the rock referred to by John Strachey of Sutton Court in his MS of about 1730, preserved in the Somerset Record office at Taunton in which he states that 'in these parts ... sand for glass is made out of the red stone ...'. In view of the prevailing red colour of the sandstone it is not surprising that much of the glass found at the site of the old works is bottle glass of a greenish colour.

Triassic sandstones, notably the massive Redcliffe Sandstone (p.132), are widely distributed in the western parts of the Bristol district, including the Severn Estuary and the south side of the Nailsea Basin and so it is not difficult to visualise these as a source of much of the bottle glass formerly made in Bristol and Nailsea. To the east of Bristol there are no Triassic sandstones of any consequence, neither is there any evidence of glass-making.

Much of the area where the Redcliffe Sandstone crops out in Bristol has long since been built over, and little evidence remains of sandstone workings. It has been suggested, however, that the so-called 'caves' or subterranean chambers excavated in the sandstone at Redcliffe originated as underground workings for sand-rock used in the early glassworks, and subsequently adapted for storage purposes. Redcliffe was one of the main centres of glass manufacture in Bristol. Of all the glass cones shown in old pictures of Bristol only the basal part of the Prewett Street cone remains

(Buchanan and Cossons, 1969, p.89), and this has now been converted into a circular restaurant. In all, about 15 glassworks were in operation in the middle of the 18th century, nearly all of them concentrated in the central area of the city adjacent to the port.

The history of the glassworks at Nailsea, founded about 1788, has been described by Sir Hugh Chance (1968). The glassworks was adjacent to a number of old coalpits which were in work during the 18th and early to mid 19th century, but by 1873 the coal reserves were exhausted, obliging the glassworks to close. For the latter part of its life the Nailsea works imported much sand from the Isle of Wight and Wareham, although the price of 'local sand' was still being recorded by the glassworks in 1836. An old, overgrown sand pit found at Flax Bourton is said by local inhabitants to have once supplied sand to the glassworks at Nailsea. Other deposits of very similar sand, possibly of glacial origin, are present on Court Hill, Tickenham and other localities between East Clevedon and Flax Bourton (Hawkins and Kellaway, 1971).

POTTERY AND BRICK CLAY

Earthenware has been made in the Bristol area since time immemorial and the use of Coal Measures clays from Kingswood Forest was mentioned in the Great Pipe Roll of Bristol Castle at the beginning of the 13th century. The earliest pottery of which there is any specific mention appears to have come into operation before 1652. Situated at Brislington, it manufactured Bristol delft and remained active until 1770. The history of the pottery industry has been summarised by Buchanan and Cossons (1969, pp.148–151). One pottery at Fishponds remains as the sole representative of this local industry.

Of greater significance in recent times is the brick and pipe-making industry. Some brickyards, notably those at Stoke Gifford [629 798] and Malago Vale [579 699], used Keuper Marl from beds below the Stoke Park Rock Bed. Others attempted, without much success, to use the red marl above the Stoke Park Rock Bed. At Welton, near Radstock, the Lower Lias Clay (*Prodactilyoceras davoei* Zone) was dug at a brickpit at Broadway Lane [667 563], but difficulty was experienced owing to the presence of small calcareous and pyritic nodules and belemnites in the clay.

At the outbreak of war in 1939 there were no less than five large clay pits using Coal Measures mudstones (either Middle Coal Measures or the lower part of the Upper Coal Measures) in the area between east Bristol (St George) and Warmley. The Warmley works manufactured salt-glazed pipes, the remaining clay pits provided material for brickmaking. All of these works have since closed. Only one major brickworks now remains open in the Bristol district, at Cattybrook, near Almondsbury, where mudstones of the Middle Coal Measures are exploited.

COAL

Coal has undoubtedly been the main mineral exploited in the Bristol district over the past two and a half centuries and it has made a significant contribution to the development of Bristol as a major centre of population, supported by its port and a complex of associated industries. The distribution of the main coals throughout the Coal Measures sequence, as well as of the principal collieries that exploited them, are described in detail in Chapter 5. The historical development of coal-working in the Bristol and Somerset Coalfield is outlined below.

There is archaeological evidence that coal was used at Bath in Roman times and documentary evidence of coal working in Bristol and Somerset in the early 13th century. Coal must have been used in ever-increasing amounts from that time forward, but it was not until the 17th century and the decline of the forests as a source of fuel and charcoal that large-scale coal mining became widespread, not only in Bristol and Somerset, but throughout Britain.

Among various important uses of the local coal from the 17th century onwards were sugar refining and the manufacture of pottery and glass. Many of the seams were used for specialised purposes locally and the seam names are indicative of this: for example, Limekiln Vein, Smith's Coal, New Smith's Coal, and others. Some of the coals which were developed in preference to others were not always the thickest or the best for domestic use. Anstie (1873) records that the Parrot Vein of the Golden Valley Pit at Bitton, south-east of Bristol, had a high reputation as a smith's coal. This coal, seldom exceeding 24 in and commonly only 18 in, had by 1873 been worked to a depth of 1020 ft below the surface. Thicknesses of 24 in or less would normally be regarded as unworkable in most coalfields, but the older coal miners of the Bristol and Somerset Coalfield were experts at thin-seam working and many preferred working in a prone position. They used short-handled picks and wore felt or leather caps to protect their heads; and, with the exception of the Nettlebridge area, firedamp was normally absent and naked lights could be used.

Methods of working seams were specialised, and varied greatly from one part of the coalfield to another, in accordance with the geological conditions. Thus in the steeply inclined or vertical coals in the southern part of the Radstock basin, methods of stoping akin to those used in the tin mines of Cornwall were employed. Workings in the Kingswood Anticline, where dips were steep and variable, used gugs or self-acting inclines. Extraction by pillar-and-stall working was restricted to a few areas where dips were low and fairly regular; elsewhere most working was done from longwall faces. Drainage in some inland areas presented serious problems and steam engines were brought into use for this purpose by 1727. Elsewhere as at Kingswood the waters were drained by tunnels leading to adits situated alongside the Avon or its tributary streams. One such tunnel draining to the Syston Brook was about 2½ miles in length.

The first workings in medieval times must have commenced with shallow diggings along the crops of the seams, especially in the area east and west of Kingswood, where the thick seams of the Lower and Middle Coal Measures come to surface along the flanks of the Kingswood Anticline. The seams would have been wrought entirely by hand and on a small scale, but as mining methods improved and demand increased, deeper pits producing larger outputs were developed throughout the 19th century. Even in the Nettlebridge valley the early miners would have been able to make use of the thick, structurally disturbed seams exposed at the southern end of the Radstock Basin, and in later years small pits were developed, even in these strongly contorted measures in which the torn, crumpled and folded seams were worked at Vobster, Luckington and Edford (Anstie, 1873, fig.13).

North of the Nettlebridge valley the greater part of the Radstock basin is concealed by Mesozoic rocks but another ancient centre of coalmining probably developed on the exposed measures at High Littleton, Clutton and Pensford. Evidence of very old crop workings in thin seams in the Upper Coal Measures occurs in many areas, both in the Bromley horst and in the much larger tract of exposed Coal Measures between Pensford and Compton Dando. Some of the seams exposed in this area are stratigraphically higher than those worked in Pensford Colliery. Judged by modern standards many of these are worthless, but the thoroughness with which they have been dug or tested indicates that they have been exploited wherever it was worth doing so. From the evidence provided by Strachey's accounts of 1719, 1725 and 1727, it will be observed that in the Bishop Sutton–Stanton Drew area workings already extended beneath an unconformable cover of Triassic and Jurassic rocks in the early part of the 18th century (p.7). Crop workings of the kind seen at Bromley, Stanton Wick (see below, p.110) and Upper Stanton Drew are likely to be much older, as are some of the very ancient workings on exposed coal seams in the Upper Coal Measures at Newton St Loe, Corston, Queen Charlton, Burnett, Compton Dando and Brislington, in the valleys of the Chew and Avon.

On the western side of the coalfield from Harptree to Barrow Gurney the Lower and Middle Coal Measures are also concealed and there is still considerable doubt as to the quality and structure of the coal seams. Between Barrow Gurney and Bristol, however, the Lower and Middle Coal Measures contained much workable coal. Mining was prosecuted in this area with great vigour throughout the 19th and into the first decade of the 20th century.

On the Old Series Geological Survey map (sheet 35) the area of Bedminster and Ashton was shown as being completely concealed by Triassic rocks and this led to speculation as to how the miners located this productive area (Anstie, 1873, p.9). The primary six-inch geological survey showed, however, that a considerable area of exposed (but red-stained) Coal Measures exists in the low-lying ground at Ashton Vale. This indicates that, as in other areas, coal working commenced on the exposed seams and then extended laterally beneath the Triassic cover as the workings became deeper and more extensive. The Ashton seams which underlie the Bedminster coals are nowhere exposed at the surface and there are indications that these may have been detected long after the Bedminster coals were first worked.

Other isolated areas of exposed Coal Measures in the western part of the coalfield include the Clapton-in-Gordano inlier and the Nailsea basin. Both of these have been worked, the former only on a limited scale, the latter more extensively. One of William Smith's reports, prepared in 1811, dealt with the prospects of the Nailsea coal basin where John Robert Lucas had established his crownglass factory in 1788. This was supplied with coal from an adjacent pit, and one of the factors leading to the closure of the Nailsea glassworks in 1874 was exhaustion of the supplies of suitable coal.

In general there are very few areas of exposed Coal Measures in the coalfield which do not show some evidence of having been tested as a source of fuel. Even such small areas as the exposed Coal Measures at Cattybrook (Prestwich, 1871), and at Emborough and Ebbor Rocks (Green and Welch, 1965, p.57) provide evidence of ancient coal mining or exploration.

The Bristol Channel and the Severn with its tributaries played a major role in the development of the coalfield prior to the coming of the railways in the 19th century. About 1682 Celia Fiennes saw the pack horses laden with coal at Kingswood, the 'shipps carrying coales' in the harbour at Bristol and the coal barges or trows being unloaded at Creech St Michael whence it was carried on horses to Taunton. The low cost of water transport combined with easy access to the Bristol Channel and the industrial Midlands were important assets (Hudson, 1965).

Above Bristol the Avon is tidal to Hanham Mills but beyond this point it was not navigable to Bath which suffered in consequence. In 1727 therefore, a series of locks was constructed creating the Avon Navigation linking Bath to the Port of Bristol. This gave improved access to some of the coal mines at Hanham, Bitton, Newton St Loe and other places, but it also demonstrated the vulnerability of the local coal mines to outside competition and some pits lost their captive markets. Construction of the Kennet and Avon Canal and the Somerset Coal Canal assisted some of the mines, notably in the landlocked Radstock Basin, but in other areas it tilted the balance against the use of local coal and the coming of the railways did not reverse this general trend.

Annual production for 1781–90 has been estimated as about 140 000 tons. By 1870 it was 1 057 000 tons, a figure which was improved upon in 1920 when 1 505 360 tons was produced. This was never exceeded in the years which followed as failing reserves led to loss of markets and closure of the pits.

Working continued throughout all the main areas of the coalfields during the 1914–18 war on a very active scale, but the progressive closure of the Ashton, Bedminster and Kingswood pits through exhaustion greatly reduced output from the northern part of the coalfield. The need for coal during and immediately after the war of 1939–45 ensured continuance of the working pits, but thereafter the increasing use of fuel oil, further exhaustion of known reserves and the need to adopt modern working methods in areas of great structural complexity imposed an ever-increasing strain.

The condition of the coalfield in 1946 was described in detail in the Regional Report produced by the Ministry of Fuel and Power. Although some progress had been made by the Geological Survey from 1942 onwards, many important stratigraphical and structural problems were still unsolved. Following the establishment of the National Coal Board, co-operation between the Board and the Geological Survey led to more rapid progress.

It was hoped that it would be possible to develop the thick coals known to exist in the Lower and Middle Coal Measures at the southern margin of the Radstock basin. The redevelopment of Strap Pit (later adapted for use by New

Rock Colliery) was designed to do this, but it was found that the measures beneath the Pennant Sandstone were so distorted that mechanised working was impossible.

Two other areas investigated at this time were the area north of Avonmouth and the Harry Stoke–Downend area north of the Kingswood Anticline. The seams in the Avonmouth basin correlate with those of Coalpit Heath and are of similar quality. The structure of the coal basin is relatively simple and is now known from boreholes in sufficient detail for making a preliminary assessment. Unfortunately the alluvial flats, beneath which the coalfield is situated, are at a very low level (c.21 ft above OD) i.e. about high water mark. It was almost certain therefore that mining subsidence would give rise to severe flooding or incur heavy pumping costs, and this fact led to the abandonment of plans for new pits.

The Harry Stoke area north of the Kingswood Anticline has a relatively simple geological structure. All the principal coal seams of the Middle Coal Measures are present throughout the area and some of the Lower Coal Measures seams may be developed at the southern margin. Unfortunately all the coal crops are concealed beneath Triassic strata and are not easily accessible from the surface. Calculations of thickness and composition were based on the borehole results, but the workability of coals by machine-mining methods is not determinable on such evidence.

With the object of proving the condition of the principal coals a Drift Mine was opened at Harry Stoke in 1955. The seams were found to have a very regular dip to the east-south-east of about 20° and were expected to improve in thickness and quality as the workings approached the Kingswood Anticline. It was found however that the variable thickness of the Lower Five Coals and the weakness of the roof shale of the Kingswood Great Vein presented serious problems. In view of the difficulty of adapting machine-mining methods to these conditions, and the very low cost of fuel oil at that time, the Drift Mine was closed in 1963.

Working of the coals of the Upper Coal Measures of the Radstock Basin continued at Writhlington and Kilmersdon collieries until 1973. This marked the demise of the Bristol and Somerset Coalfield.

The quality of many of the seams worked in the Bristol and Somerset Coalfield has been regarded by some as very unsatisfactory, but it must be remembered that the chemical composition of many of the coals, including those most extensively worked in former times, is unknown. The Red Ash Vein of Hanham Colliery was supplied to Bristol Gas Works for many years and other pits were largely dependent on sales to town gas as far distant as Reading. The impact of North Sea gas would have caused the demise of the coalfield even if competition with imported oil had not already done so. Analytical data are available for seams sampled in later years (*NCB Reg. Surv. Rep.*, 1946, p.17; Adams *in* Kellaway, 1967, p.153) and some additional information prepared by Mr. H.F. Adams is also given.

In general the volatile content of dry ash-free coal is high, being in the range of 23–38 per cent the lowest values being found at Pensford, the highest in the southern part of the Radstock Basin. Many of the coals are strongly caking with a moderate to low ash fusion temperature. The calorific value of the coals is generally good even when the ash content is high. This combination of properties may account for the excessively strong clinkering which takes place when some of the coals are burnt.

The following information, provided by Mr H. F. Adams of the Scientific Department (Coal Survey), National Coal Board, in 1959, relates to the seams in the Lower and Middle Coal Measures of the Harry Stoke and Kingswood areas. At this time working had ceased in the Kingswood Anticline. One of the last coals in the Kingswood group of seams to be worked was the Two Feet at Speedwell Colliery: this contained about 28 per cent of volatile matter[1], 7 to 9 per cent of ash and 1 to 1.2 per cent of sulphur. Farther north the Five Coals, Great Vein and Gillers Inn seams were proved at depths of 1200 ft and over in Harry Stoke B Borehole; they were proved at shallower depths in D, E and F boreholes and subsequently in the Harry Stoke Drift Mine. Ash content ranged from 4 to 9 per cent in the Five Coals and Gillers Inn and a little higher in the Great Vein; Sulphur was about 1 per cent in the Great Vein and up to 1.5 in the other two seams; and volatile content ranged from 31 to 34 per cent, and all seams had strongly caking properties at depth in B Borehole, though the Great Vein was only weakly caking at shallower depths. In the Downend Borehole the Gillers inn was strongly caking, with under 25 per cent of volatile matter and low ash and sulphur.

Several of the seams between the Kingswood group and the base of the Coal Measures had thin or dirty sections in Harry Stoke B Borehole. Two of the cleaner seams had ash contents of 9.4 and 6.8 per cent with volatile matter about 33 per cent. Sulphur contents were 9.4 and 6.8 per cent. To the south-west of the Kingswood Anticline the Ashton Great Vein of the Ashton Park Borehole was a medium caking coal with 25 per cent volatile matter, 6 per cent ash and under 1 per cent sulphur. The Ashton Little was an inferior seam of very high sulphur content.

FULLER'S EARTH

Fuller's earth (montmorillonite) is worked at Combe Hay, south of Bath from a bed 6 to 10 ft thick lying 13 to 26 ft below the base of the Great Oolite (p.129). The workings are from adits [727 611; 728 617] driven beneath the hillside and are by pillar-and-stall method. The lower part of the bed is of higher quality and is mined separately for the foundry and pharmaceutical industries. The upper part of the bed was used for agricultural purposes. Further details are given by Highley (1972), Hawkins (*in* Savage, 1977, pp.121,124) and Penn and Wyatt (1979).

---

1 Volatiles have been calculated on a dry ash-free basis.

CHAPTER 9

Pleistocene and Recent

Unlike the Midlands and South Wales, the Bristol district is characterised by only scattered patches of Drift deposits, confined very largely to the lower-lying areas of the valleys and the coastal margins. It is rarely possible to correlate these fragmentary sediments in terms of the nomenclature now commonly used in correlating the Quaternary deposits of the British Isles (Mitchell et al., 1973); indeed there are doubts as to the applicability of this classification to the nonglacial deposits of southern England. For this reason the following account describes the deposits under broad chronological headings, giving only approximate correlations with the stage nomenclature.

EARLY TO MID-PLEISTOCENE (Before about 600 000 yr BP)

Of the Pleistocene deposits in the Bristol district those of Westbury sub-Mendip (Bishop, 1974) are almost certainly the oldest. They consist of siliceous sands and gravels, overlain by limestone conglomerates and breccias and a deposit of dark red-brown earth, rich in small mammal remains (Rodent Earth), filling an elongated trough-like fissure in the Carboniferous Limestone [506 504], the surface of which lies at about 800 ft above OD. The trough has clearly been widened by solution, but because its long axis trends parallel to the slope of the ground it may well be of neotectonic origin. Dating of the deposits by their vertebrate remains, which include *Hyaena brevirostris* in the siliceous sands and *Pitymys gregaloides* and *Arvicola cantiana* in the (younger) Rodent Earth (Bishop, 1974), indicates that both predate the Elster Glaciation of north-west Europe, thought by some to equate with the Anglian Stage of Britain. Petrological evidence supports the faunal, suggesting that the deposits were laid down under cold conditions which later ameliorated.

Sands and gravels preserved in a depression or channel in the Clevedon–Failand limestone ridge at Court Hill, between Clevedon and Tickenham, are thought to be next in order of age. The sections were originally described by Trimmer (1853, p.284), and similar ones were exposed during the construction of the M5 motorway in that general area (Hawkins and Kellaway, 1971, p.285). All these deposits are now considered to be of glacial origin. Most of the channel-fill at Court Hill [4365 7225] consists of sand and boulder-gravel. The clastic material consists primarily of Black Rock Limestone, and Black Rock Dolomite, and Triassic conglomerate and sandstone; Cretaceous flint and Carboniferous chert make up only a minor element. A red stony clay (possibly till) lines the channel on its eastern side and the bouldery sands are banked against the western side. Roughly graded foreset beds suggest periodic flows of water capable of carrying coarse gravel and small boulders. The presence of so much material of local origin, nearly all subangular or rounded, with only a small fraction of nonlocal detritus, is hard to explain in view of the size of the channel cut in the limestone ridge. The crest of the Carboniferous Limestone ridge stands at about 300 ft above OD on Tickenham Hill, east of the motorway, and is only slightly below this level on Court Hill to the west; the deepest part of the floor of the channel lies at about 163 ft above OD. It is possible that the Court Hill gully originated as a tunnel through the limestone ridge or as a narrow glacial channel. Collapse of the roof and sides of such a tunnel or channel would produce much coarse local debris. Sand, filling fissure- and joint-systems in the limestone beneath the motorway between Court Hill and the Tickenham channels (Hawkins and Kellaway, 1971, fig.5), appears to have been introduced by water under pressure.

Underlying the extensive area of estuarine alluvium south of Clevedon there are further deposits that are now regarded as being of glacial origin. They were exposed during the construction of the M5 motorway and excavations associated with sewerage and drainage improvement schemes in and around the village of Kenn. The deposits were shown (incorrectly) on the Bristol district special sheet as 'Burtle Beds (Sand and Gravel)', rising as a low mound through the alluvium around the village; they were then regarded as marine because of the local abundance of shells of *Macoma baltica* (Welch, 1956). The full succession, now interpreted as comprising glacial till and gravels overlain by pre-Ipswichian fluvial and Ipswichian marine interglacial deposits, succeeded by Flandrian postglacial sediments (Andrews et al., 1984), has been described in detail by Gilbertson and Hawkins (1978). The bulk of the glacial deposits consist of coarse outwash sands and gravels, well or poorly sorted, and containing large cobbles and boulders; the term Kenn Gravels has been given to these. Resting in hollows in the surface of the gravels, but in one locality interdigitating with them, is a reddish purple boulder clay of no great thickness, containing many cobbles and boulders of erratic rocks, some over a ton in weight. Its best exposure was at Kennpier [428 698], and for this reason it has been called Kennpier Till. The stone content of the gravels and the till is similar: Carboniferous Limestone, Upper Greensand cherts, Jurassic limestones and Cretaceous flints are very common; Devonian red quartzites, sandstones and conglomerates, Pennant sandstones and yellow Triassic sandrock with barytes less so.

The Kennpier Till is generally thought to be contemporaneous with the glacial deposits at Court Hill; both it and the Kenn Gravels carry a much more varied suite of erratic rocks than the Court Hill gravels, suggesting that they may be younger or that the provenance of the two may be different. The presence in the Kenn deposits of yellow Triassic sandstone with baryte, typical of the Clevedon area, and hard grey Pennant sandstone and coal, possibly from South Wales, suggests a westerly or north-westerly derivation; this is supported by the absence of Inferior Oolite limestone, which forms the top of Dundry Hill to the east, and of quartzitic sandstone and Hotwells Limestone from the outcrops between the Avon Gorge and Cambridge Batch. Much work has recently been carried out on the Kenn deposits, and the dates derived from amino-acid racemisation (Andrews et al., 1984) confirm previous conclusions that the glaciations involved are early, possibly contemporaneous with the Anglian glaciation of East Anglia (Hawkins and Kellaway, 1971), or even with an earlier Beestonian glaciation.

Another occurrence which may relate to a possible early glaciation of the Nailsea Basin and Broadfield is the fissure and cavity fillings at Backwell [503 683] of finely laminated (?varved) clay, contorted as a result of subsidence (Hawkins in Savage, 1977). These deposits were seen at an altitude of about 460 ft above OD during preparatory stripping of the ground for quarrying between Farleigh Combe and Bourton Combe. There is, however, no evidence as to their exact origin or date.

MID TO LATE PLEISTOCENE (About 600 000 to 10 000 yr BP)

Another deposit interpreted as being of glacial origin is the sand and cobble gravel of Portishead Down [450 751]. These gravels form a knoll at about 300ft above OD at the head of a narrow valley leading from Portishead Down eastwards into the Gordano valley. The deposit contains a variety of rocks including cobbles of hard sandstone (possibly Old Red Sandstone) and fragments of flint, some quite fresh. It is seldom exposed and the form of its base is not known. The erratic content differs considerably from that of both the Court Hill gravels and Kenn deposits though the rounded sandstone cobbles are similar to those in the 100-Foot Shirehampton Terrace gravels (p.162). The surprising feature, however, of this deposit is its height, nearly 100 ft above the level of the Fifth Terrace at Alvington west of the Severn (Welch and Trotter, 1961, p.132), thought to be late-Wolstonian in age, and 120 to 140 ft above the Fourth Terrace of the Wye in the meander of an abandoned course of the Wye at Bigsweir (Welch and Trotter, 1961, p.134). Possibly the sandstone cobbles were brought in by outwash from a glaciation later than that at Court Hill and Kenn, presumably the Wolstonian, and then recycled. If the gravels are morainic, then ice banked up against the western side of Portishead Hill may have produced a glacially dammed lake in the Gordano valley. Seen from the sea the Portishead ridge has the appearance of glaciated terrain, the valley at Walton-in-Gordano having all the attributes of an overflow channel. Drainage from a Gordano lake would have been via the East Clevedon gap, and the meltwater from the Wolstonian ice might have washed away much of the material deposited during any earlier glaciations. Further Wolstonian deposits are represented by terrace gravels, and are described separately below.

Fluvial deposits overlying the till sequence at Kennpier, at 10 to 15 ft above OD, have been dated on the basis of amino-acid studies (Andrews et al., 1984) as belonging to amino-acid Group IV (c.400 000–600 000 BP).

At Swallow Cliff [325 661], on the Woodspring Promontary between Clevedon and Weston-super-Mare just west of the Bristol district special sheet margin, a '10-Foot' raised beach and its associated deposits have been described by Gilbertson and Hawkins (1977). The beach platform, at c.41 ft above OD, is usually correlated with others at similar heights around the coasts of southern and western Britain and has been regarded as Ipswichian in the general sense. The shelly raised beach deposits have been dated (Andrews et al., 1984) as belonging to amino-acid Group II, considered to be 120 000 to 130 000 years in age (Ipswichian). However, silts and gravels underlying the beach deposits show evidence of both cold and warm periods, and the trimming of the elevated platform may mark a much earlier period of raised sea level. Other possible Ipswichian deposits appear to be restricted to river terraces.

In the colder Devensian the local deposits are very varied but are patchy in their distribution. They include river gravels, mudflows and poorly sorted head or colluvium deposits formed largely by the degradation of hard and soft rocks: coversands of aeolian origin are also present in some areas. There is no evidence for the late-Devensian ice-sheet that affected parts of South Wales and the Welsh Borderlands.

Lower Severn terraces

There are no recognisable river terraces on the east bank of the Severn Estuary within the confines of the Bristol district. The elaborate series of terraces on the west bank between Lydney and Rogiet have been described by Welch and Trotter (1961, pp.132–134), who grouped them into five major units, broadly correlated with those of the Middle Severn and Warwickshire Avon. The highest (Fifth) terrace at Alvington is thought to represent the Woolridge Terrace (Wills, 1938, p.171), now considered to be late-Wolstonian in age (Shotton in Mitchell et al., 1973, p.19, table 3). The successively lower terraces would appear to range from Ipswichian to Devensian. There are, however, anomalies in the correlation of the Lower Severn terraces with those of the tributary rivers.

Bristol Avon terraces

The terraces of the Avon east of the Hanham gorge can be classified into three main groups (Chandler et al., 1976) — a Twerton (No. 3) Terrace, a middle or Saltford (No. 2) Terrace and a Flood Plain (No. 1) Terrace. Below Hanham no terraces can be recognised in the Pennant sandstone gorge that extends to Netham, and only the Flood Plain Terrace has been identified on the low ground extending through the centre of Bristol. Some gravelly deposits yielding Palaeolithic implements have been described (Fry, 1956) but do not appear to form part of any regular terrace formation.

The highest recognisable terrace below the Avon Gorge occurs on the left bank near Leigh Court at about 100 ft above OD, roughly opposite the mouth of River Trym at Sea Mills. From the exit of the Gorge the terrace flat extends to Ham Green and Pill, being deeply dissected by narrow valleys joining the Avon at Chapel Pill and Pill. On the opposite bank of the Avon, at Shirehampton, a well-marked flat marks the only surviving area of the 100-Foot Terrace north of the river. Several authors (Fry, 1956; Brown, 1957;

Apsimon and Boon, 1960) have recorded Palaeolithic implements in the associated deposit.

Situated at about 50 ft above OD is a lower terrace known as the 50-Foot Terrace at Shirehampton. This is not known with certainty above the Horseshoe Bend, but it is well developed at Sheephouse Farm on the south bank of the river opposite Avonmouth Dock, where it is incised in Mercia Mudstone. A broad gravel deposit at Sheepway, a mile or so south-west, rises at its northern end to 47 ft above OD and is possibly a degraded remnant of the same terrace. Degradation by solifluction has also affected the 100-Foot Terrace at Lodway and Pill so that there is some difficulty in identifying individual terraces.

The presence of two major gorges on the Bristol Avon (at Clifton and Hanham) creates difficulty in the correlation of the Avon terraces with those of the Severn. One method of solving this problem is to compare the terraces of the Upper Avon between Hanham and Twerton with those of the Stroud Frome. The correlation has been set out by Chandler et al. (1976) and links the three terraces of the Avon above Bristol with those of the Middle Severn as follows: the No.1 (Flood Plain) Terrace of the Avon at 10 ft above the fluviatile flood plain is equivalent to the Worcester Terrace of the Severn (late-Devensian); the No. 2 (Saltford) Terrace at 30 to 50 ft represents the Severn Main Terrace (mid-Devensian); and the No. 3 (Twerton) Terrace at 90 ft equates with the Kidderminster Terrace (Ipswichian or very early Devensian). Welch and Trotter (1961, pp.132–133) use the following correlation for certain of the terraces of the Lower Severn: the Second (50-Foot) Terrace is equivalent to the Kidderminster Terrace of the Middle Severn; the Fourth Terrace (at 160–170 ft above OD) to the Bushley Green Terrace, and the Fifth Terrace (at 200 ft above OD) to the Woolridge Terrace (no older than Wolstonian); the Third (100-Foot Terrace) appears to have no correlative in the Middle Severn.

Recent work on the Bristol Avon terraces and the high level Drift deposits near Bath in relation to Quaternary tectonism have placed in doubt many of these altimetric correlations (Kellaway, 1991). The palaeolith-bearing terraced gravels of Shirehampton may be related to the former course of the Wye prior to the main phase of downcutting the Avon Gorge. The terraces of the Bristol Avon above Hanham are much younger, postdating the downcutting of the Avon Gorge at Clifton. Allowance has also to be made for the possible effect of ice-damming of the lower Wye (or Wye-Severn system) at the time of the glaciation of Clevedon, Kenn and Flax Bourton.

Welch and Trotter (1961) pointed out the difficulties associated with altimetric correlation in areas where fluviatile flood plains merge with estuarine ones, and the conclusions set out above are clearly fraught with doubt. Despite these and other difficulties, still largely insoluble in the absence of palaeontological or other forms of dating, Chandler et al. (1976) described the No. 1, No. 2 and No. 3 terraces of the upper Avon as Late Devensian, Middle Devensian and Ipswichian respectively. This leaves the question of the age of the 100-Foot Terrace at Shirehampton with its contained Palaeolithic implements unanswered.

In addition to the river terraces, some of which are known to pass beneath the alluvial deposits of the River Severn, there are several areas where Pleistocene deposits lie beneath an alluvial cover of variable thickness. Among these are the glacial deposits of Kenn Moor (p.161) and extensive deposits of sand and gravel of various ages, most having been proved by boreholes, tunnels or deep excavation. Those deserving mention include coarse gravels north of Avonmouth which were proved in the approach cutting on the English side of the Severn tunnel. The position of these gravels is shown in Richardson's (1887) sections of the tunnel, and Lucy (in Morgan, 1887) described the large range of clasts which were found at the time the tunnel was driven. A summary of this and other information has been given by Welch and Trotter (1961, p.136). Leese and Vernon (1961) gave an account of the boring programme carried out by the Imperial Chemical Industries Limited at the Severnside site north of Avonmouth [about 544 850], an area which includes the approach cutting and portal of the tunnel. In the tunnel cuttings and in the borings the base of the gravel shows considerable variation, ranging from about 20 to 42 ft below OD. Wills (1938, pp.228–229) suggested that these gravels may be a continuation of the Worcester Terrace of the Middle Severn, which downstream from Tewkesbury and Gloucester lies beneath the modern alluvium. Leese and Vernon (1961), however, drew attention to the fact that the Severnside deposits lie on a highly irregular surface with deep holes and rises, rather than on a cut bench. Several observers have recorded ice-scratched stones from the gravels, both from the Severn tunnel cutting and from the gravel banks south of the English Stones reef where Severnside gravel is being reworked by the tidal river. Unlike the drift deposits of Gwent north of the estuary, which are dominated by relatively local Welsh Old Red Sandstone and Carboniferous rocks, the Severnside gravels contain erratics presumed to have come from north-west England and the Welsh border country; if this provenance is correct then some at least of the Severnside deposits must be late-Devensian, when north-western erratics first entered the lower Severn valley in the Severn Main Terrace gravels.

Sand and gravel has also been found beneath alluvial deposits in the Portbury Dock area roughly midway between Avonmouth and Portishead.

Head or Colluvium

The terms Head and Colluvium are applied to coarse granular, unsorted or poorly sorted, detrital matter that has moved under the influence of gravitational forces assisted in some instances by freeze-thaw processes. Scree is essentially a Head formed by rock fragments falling from a cliff or rolling down slope in a dry condition. At the other end of the scale Head includes mudflows and other fine-grained material which has been mobilised under saturated or supersaturated conditions, with or without the assistance of freeze-thaw processes.

Some of the largest deposits of Head in the Bristol district occur on low ground adjacent to the Mendips and other limestone uplands. Many Head deposits are too thin and patchy to be mappable on any but the largest-scale maps, and some are little more than a veneer of weathered material. Others, however, are substantial deposits up to 20 ft in thickness and form an important element in the local geology. To this class belong the large exposures of loam with limestone rubble which mask the Triassic rocks between Cheddar and Axbridge, at Draycott, Rodney Stoke and Westbury, south of the Mendips. Other large areas occur at Woodborough between Shipham and Winscombe, and at Churchill and Longford in the Vale of Wrington (Green and Welch, 1965, pp.112–116). Similar tracts of stony loam overlie the Triassic rocks at the south-western margin of the Nailsea Basin at Yatton, Brockley and Flax Bourton.

In Ashton Park, Bristol (Kellaway, 1967, p.53), and along the slopes bordering the limestone uplands between Long Ashton and Wraxall, stony loam, generally of a reddish brown colour, covers much of the ground. Where the hard reddish quartzitic sandstones of Namurian or Dinantian age crop out on the slope, stony and sandy ferruginous soils are produced. The less stony varieties resemble soils formed in situ on red sandy mudstone or sandstone of Triassic age. Considerable expanses of red or reddish brown sandy loam of this kind are found at Brislington, Hambrook and other parts of the coalfield north of the Avon. Their resemblance to weathered red sandy Triassic mudstone is so close that it is likely that some areas of Head may have escaped detection. In cases of uncertainty, where neither jointing nor bedding can be seen, it is necessary to search any section carefully for included pebbles of post-Triassic rocks, before concluding that undisturbed Triassic strata are exposed.

All the larger Head deposits, including those which flank the estuarine deposits flats at Easton-in-Gordano and Portbury, predate the alluvial mudstone of the main river valleys and the Severn Estuary. Some, such as those which mantle the exposed Triassic marl supporting the patch of Terrace Gravel at Sheephouse Farm, Easton-in-Gordano [808 774], postdate the formation of the nearby terrace (? = Kidderminster Terrace) and predate deposition of the Estuarine Alluvium. Since the extensive belt of head at Easton-in-Gordano was formed by the degradation of all the younger Terrace gravels as well as the Triassic bedrock, it must also be Devensian in age.

Archaeological evidence has provided much information about conditions in late Pleistocene and early Flandrian times. A great deal of this has come from the investigation of caves and fissures discovered by lead miners and ochre-diggers in the 18th century, their discoveries being later followed up by skilful investigators like Buckland and Boyd Dawkins. Much of the descriptive work relating to the Upper Palaeolithic period as recorded in the limestone caves of the Mendips and the Bristol district generally can be found in the Proceedings of the University of Bristol Spelaeological Society. A summary of this work is given by Dobson (1931) and Grinsell (1958), and detailed bibliographies have been compiled by Donovan (1954, 1964). Conditions at the close of the late-Devensian glaciation were favourable to the use of the Mendip caves by late Palaeolithic man (Green *in* Green and Welch, 1965, pp.116–119). For this reason many of the local Upper Palaeolithic deposits which provide evidence for dating purposes are cave deposits. There are, however, some locations where remains of Upper Palaeolithic man, or of the animals which lived contemporaneously with him, are found on open sites. In most instances, as at Bleadon (Palmer, 1934), Brean Down near Weston-super-Mare (Apsimon et al., 1961), and Clevedon (Palmer, 1934), these deposits consist of Head or scree banked against cliffs and covered by more recent debris or blown sand. Such sections are particularly useful as they are likely to range from Pleistocene to Recent. Although scree formation continues to the present day it is noticeable that nearly all the major deposits of Head or colluvium are of Pleistocene age.

RECENT (FLANDRIAN): 10 000 BP to present

By far the most important of the Flandrian deposits are the very extensive spreads of estuarine mud, peat, silt and sand which form the moors or levels bordering the Bristol Channel and Severn Estuary. Only those areas adjoining the Severn Estuary above Newport and Weston-super-Mare are relevant to present consideration. The northern part of this district is covered by the Chepstow and Monmouth Memoir (Sheets 233 and 250) and has been described by Welch and Trotter (1961). South of the Mendips appreciable parts of the Somerset levels fall within the Wells (280) sheet described by Green and Welch (1965). The main emphasis of the present account is therefore concentrated on the alluvial flats around the mouth of the Bristol Avon and the areas to the south, at Nailsea and Kenn.

Deposition of the muds, silts, sands and peats is thought to have commenced at a time of low sea-level about 9000 yr BP. The origin of the deposits has received considerable attention and, though there is disagreement as to the precise degree of sediment transportation from landward sources, there is little doubt that the Severn Estuary acts as a sediment trap (Collins, 1987). Some of the fine-grained sediments may have been introduced by rivers, but other material, including the shells of foraminifera, was transported from the open sea to the west (Murray and Hawkins, 1976).

Perhaps the best introduction to a study of the conditions which prevail under the levels and marshes is the excellent section by Charles Richardson (1887) showing the structure of the alluvial deposits in the approaches to the eastern entrance of the Severn Tunnel. Taken with the information regarding this area as proved in the Severnside borings by Imperial Chemical Industries (Leese and Vernon, 1961), the disposition of the older (Pleistocene) gravels and the Flandrian sands, muds and peats at the margin of the English Stones can be seen in detail.

South of the Bristol Avon, borings and excavations at Portbury Dock (Hawkins, 1968) have enabled the base of the Flandrian deposits to be contoured. Some of the many boreholes made in the Avonmouth Dock area are shown, together with sub-drift levels, on the published six-inch map ST 57 NW of the Geological Survey. Alluvial deposits form a substantial proportion of the material filling the buried channel of the Bristol Avon and other rivers.

The transition from estuarine to fluviatile alluvium is always difficult to define, and is seldom if ever accompanied by any significant change in the physical characters of the deposits. In most rivers it is usually taken at the upper limit to which ordinary tides flow; in the case of the Avon this is taken near the entrance to the Hanham gorge.

Among other aspects of the study of Flandrian deposits are

palaeobotanical studies of the peats, notably that of Godwin (1943), though most of these studies are concerned with areas outside the present district. Radiocarbon dating is an important element in correlation of peat beds and the determination of the rate of postglacial changes of sea level (Hawkins and Kellaway, 1971). Though now in need of updating, the NERC account *The Severn Estuary and the Bristol Channel. An assessment of present knowledge* (1972) is of value.

APPENDIX 1

Abstracts of selected shafts, wells and boreholes

Full logs are held in the National Geosciences Data Centre at BGS, Keyworth; those published in full, elsewhere, are indicated.
 * Indicates shafts and boreholes logged by officers of the British Geological Survey.

1:25 000 Sheet ST 58

1 **Avonmouth No. 1 (Chittening Farm) Borehole*** Geological one-inch sheet 250. NGR 5355 8185. BGS Ref. No. ST 58 SW/10. Height above OD 20.5 ft. Bored for National Coal Board by John Thom, Ltd in 1953. Published, Welch and Trotter, 1961, pp.143–144.
 Estuarine alluvium to 20 ft; Quaternary deposits to 50 ft; Triassic (red marl) to 166 ft 3 in; Supra-Pennant Measures to 465 ft 8 in.

2 **Avonmouth No. 2 (Severn Farm) Borehole*** Geological one-inch sheet 250. NGR 5444 8305. BGS Ref. No. ST 58 SW/11. Height above OD 22–24 ft. Bored for National Coal Board by John Thom, Ltd in 1953. Published, Welch and Trotter, 1961, pp.144–146.
 Estuarine alluvium to 30 ft; Quaternary deposits to 57 ft; Triassic (red marl) to 159 ft 6 in; Supra-Pennant Measures to 367 ft 6 in.

3 **Avonmouth No. 3 (Willow Farm) Borehole*** Geological one-inch sheet 264 NGR 5436 8115. BGS Ref. No. ST 58 SW/12. Height above OD 21.5 ft. Bored for National Coal Board by John Thom, Ltd in 1953–54.
 Estuarine alluvium to 68 ft 6 in; Quaternary deposits to 72 ft 6 in; Triassic (red marl) to 170 ft 1 in; Supra-Pennant Measures to 405 ft 3 in.

4 **Avonmouth No. 4 (Brook Farm) Borehole*** Geological one-inch sheet 250. NGR 5554 8258. BGS Ref. No. ST 58 SE/1. Height above OD 20.5 ft. Bored for National Coal Board by John Thom, Ltd in 1954. Published, Welch and Trotter, 1961, pp.146–148.
 Estuarine alluvium to 30 ft 6 in; Triassic (red marl) to 182 ft 10 in; Supra-Pennant Measures to 729 ft.

5 **Avonmouth No. 5 (Stowick Farm) Borehole*** Geological one-inch sheet 250. NGR 5378 8259. BGS Ref. No. ST 58 SW/13. Height above OD 21.5 ft. Bored for National Coal Board by John Thom, Ltd in 1954. Published, Welch and Trotter, 1961, pp.148–149.
 Estuarine alluvium to 33 ft; Triassic (red marl) to 173 ft 9 in; Supra-Pennant Measures to 370 ft.

6 **Avonmouth (Chittening No. 2, Washingpool Farm) Borehole*** Geological one-inch sheet 264. NGR 5347 8152. BGS Ref. No. ST 58 SW/1. Height above OD 25 ft. Bored for Bristol Dock Board by H. Brown & Co. in 1928.
 Estuarine alluvium to 52 ft 3 in; Triassic (red marl) to 158 ft; Supra-Pennant Measures to 793 ft 4 in.

7 **Portskewett Borehole No. 103*** Geological one-inch sheet 250. NGR 5127 8848. BGS Ref. No. ST 58 NW/15. Height above OD 23 ft. Bored for Central Electricity Generating Board (Midlands) by Soil Mechanics Ltd in 1970.
 Quaternary deposits to 29 ft 6 in; Triassic (red marl and Dolomitic Conglomerate) to 129 ft 11 in; (? Middle) Coal Measures to 196 ft 2 in; Carboniferous Limestone Series (Cromhall Sandstone) to 298 ft 1 in.

8 **Portskewett Borehole No. 105*** Geological one-inch sheet 250. NGR 5128 8838. BGS Ref. No. ST 58 NW/16. Height above OD 23 ft. Bored for Central Electricity Generating Board (Midlands) by Soil Mechanics Ltd in 1970.
 Quaternary deposits to 26 ft 6 in; Triassic to 105 ft 8 in; (?Middle) Coal Measures to 187 ft; Carboniferous Limestone to 295 ft 3 in.

9 **Portskewett Borehole No. 109*** Geological one-inch sheet 250. NGR 5139 8842. BGS Ref. No. ST 58 NW/17. Height above OD 23 ft 6 in. Bored for Central Electricity Generating Board (Midlands) by Soil Mechanics Ltd in 1970.
 Quaternary deposits to 25 ft 7 in; Triassic (red marl and Dolomitic Conglomerate) to 119 ft 11 in; (?Middle) Coal Measures to 282 ft 10 in; Carboniferous Limestone to 296 ft 5 in.

10 **Severnside Turbo-drill Borehole** Geological one-inch sheet 250. NGR 5487 8400. BGS Ref. No. ST 58 SW/41. Height above OD about 22 ft. Bored by Bristol Siddeley, Whittall Tools Ltd in 1966. Published, Fry, 1971, p.69.
 Quaternary deposits (mainly Estuarine alluvium) to 80 ft; Triassic (red marl and sandstone) to 170 ft; Pennant Measures to 760 ft; indeterminate strata to 790 ft; Carboniferous Limestone Series to 1510 ft; Upper Old Red Sandstone to 1860 ft; Lower Old Red Sandstone to 3340 ft.

1:25 000 sheet 68

11 **Frog Lane Pit, Coalpit Heath** Geological one-inch sheet 265. NGR 6870 8155. BGS Ref. No. ST 68 SE/4. Height above OD 208 ft. Published, Prestwich, 1871, p.53.
 Supra-Pennant Measures to 625 ft (Hard Vein at 480 ft, High Vein at 625 ft); Pennant Measures to 655 ft.

12 **Patchway Turbo-drill Borehole** Geological one-inch sheet 264. NGR 6073 8092. BGS Ref. No. ST 68 SW/54. Height above OD about 190 ft. Bored by Bristol Siddeley, Whittle Tools Ltd in 1965.
 Rhaetic to 20 ft; Tea Green Marl to 35 ft; Keuper Marl to 265 ft; Dolomitic Conglomerate to between 285 and 300 ft; Upper Cromhall Sandstone to 325 ft.

13 **Rangeworthy (Old Wood) Colliery Shaft** Geological one-inch sheet 251. NGR 6997 8515. BGS Ref. No. ST 68 NE/1. Height above OD about 220 ft. Sunk about 1870. Abandoned 1888. Published, Anstie, 1873, pp.27–28.
 Middle Coal Measures (to ?Smith's Coal) to 310 ft.

14 **Stoke Gifford No. 1 (Hambrook Brickyard) Borehole*** Geological one-inch sheet 264. NGR 6346 8053. BGS Ref. No. ST 68 SW/10. Height above OD 158.2 ft. Bored for National Coal Board by John Thom, Ltd in 1953–54.
 Mercia Mudstone to 83 ft; Dolomitic Conglomerate to 95 ft; Pennant Measures (Downend Formation) and Middle Coal Measures to 620 ft 6 in.

15 **Stoke Gifford No. 2 Borehole*** Geological one-inch sheet 250. NGR 6235 8066. BGS Ref. No. ST 68 SW/12. Height above OD 163.5 ft. Bored for National Coal Board by John Thom, Ltd in 1954.
 Lower Lias and Rhaetic to 25 ft; Tea Green Marl and Keuper Marl to 216 ft; Redcliffe Sandstone to 329 ft 6 in; Middle Coal Measures to 1057 ft.

16 **Stoke Gifford No. 3 (Leyland Court) Borehole*** Geological one-inch sheet 250. NGR 6345 8219. BGS Ref. No. ST 68 SW/11. Height above OD 186.2 ft. Bored for National Coal Board by John Thom, Ltd in 1954. Published, Welch and Trotter, 1961, pp.149–153.
 White Lias and Rhaetic to 16 ft; Tea Green Marl to 28 ft; red Mercia Mudstone to 310 ft; Pennant Measures

(Downend Formation) to 493 ft 6 in; Middle Coal Measures to 1438 ft.

17 **Tytherington No. 1 Borehole*** Geological one-inch sheet 250. NGR 6643 8823. BGS Ref. No. ST 68 NE/28. Height above OD about 275 ft. Bored by Bristol Avon River Authority in 1972.
    Carboniferous Limestone Series to 484 ft.

18 **Winterbourne Borehole*** Geological one-inch sheet 264. NGR 6461 8010. BGS Ref. No. ST 68 SW/1. Height above OD 140 ft. Bored by Vivian's Boring and Exploration Co. Ltd in 1915–17. Published, Cantrill and Smith, 1919, pp.53–57.
    Triassic (marl) to 11 ft; Pennant Measures to 1485 ft; Middle Coal Measures to 2338 ft 6 in.

19 **Yate Deep Borehole*** Geological one-inch sheet 251. NGR 6975 8253. BGS Ref. No. ST 68 SE/11. Height above OD 225 ft. Bored by Vivian's Boring and Exploration Co. Ltd in 1918–20. Published, Pringle, 1921, pp.92–95.
    Pennant Measures (Mangotsfield Formation to 695 ft 9 in, Downend Formation to 1211 ft) Middle Coal Measures to 2203 ft 4 in. Average dip c.30°.

1:25 000 sheet 78 (Western part)

20 **Chipping Sodbury R.D.C. No. 1 (Broad Lane) Borehole*** Geological one-inch sheet 251. NGR 7024 8382. BGS Ref. No. ST 78 SW/10. Height above OD about 220 ft. Bored by Geotechnical Engineering Ltd.
    Middle Coal Measures (Crofts End Marine Band near top) to 99 ft 8 in.

21 **Westerleigh No. 1 Borehole** Geological one-inch sheet 265. NGR 7061 8127. BGS Ref. No. ST 78 SW/1. Height above OD about 245 ft. Bored for Coalpit Heath Co. by Vivian's Boring and Exploration Co. Ltd in 1912.
    Mercia Mudstone to 11 ft. Pennant Measures to ?905 ft 5in; Middle Coal Measures to 1441 ft 6 in.

22 **Westerleigh No. 2 Borehole** Geological one-inch sheet 251. NGR 7077 8185. BGS Ref. No. ST 78 SW/2. Height above OD about 250 ft. Bored 1913.
    Mercia Mudstone to 9 ft; Pennant Measures to ?201 ft 6 in; Middle Coal Measures to 713 ft.

23 **Yate (Limekilns Lane) Borehole*** Geological one-inch sheet 251. NGR 7066 8589. BGS Ref. No. ST 78 NW/5. Height above OD about 225 ft. Bored for Geological Survey in 1968. Published, Cave 1977, pp.307–314.
    Dolomitic Conglomerate to about 15 ft; Lower Coal Measures to 202 ft 9 in; Millstone Grit Series to 703 ft.

1:25 000 sheet 47

24 **Clapton-in-Gordano Borehole.** Geological one-inch sheet 264. NGR 4692 7408. BGS Ref. No. ST 47 SE/58. Height above OD about 35 ft. Published, Anstie, 1873, p.55.
    Dolomitic Conglomerate to 21 ft; Pennant Measures to 510 ft.

25 **Clapton-in-Gordano (Portishead Waterworks Co.) Borehole** Geological one-inch sheet 264. NGR 4748 7435. BGS Ref. No. ST 47 SE/60. Height above OD about 30 ft. Bored by John Thom, Ltd in 1948.
    Pennant Measures to 602 ft.

26 **Clevedon (Clevedon and Portishead Laundry Co. Ltd) Borehole** Geological one-inch sheet 264. NGR 4022 7056. BGS Ref. No. ST 47 SW/96. Height above OD about 20 ft. Published, Richardson, 1928, pp.166–167.
    Estuarine alluvium to 40 ft; Triassic to 216 ft.

27 **Clevedon (Clevedon Water Co. Pumping Station) Borehole** Geological one-inch sheet 264. NGR 4296 7167. BGS Ref. No. ST 47 SW/97. Height above OD about 50 ft. Published, Richardson, 1928, p.167.
    Head deposits to 10 ft; Mercia Mudstone to 267 ft 5 in.

28 **Portishead Generating Station Borehole** Geological one-inch sheet 264. NGR 4705 7721. BGS Ref. No. ST 47 NE/19. Height above OD 113 ft. Bored about 1935. Published, Smith and Reynolds, 1935, pp.521–527.
    Lower Limestone Shale to 250 ft; Portishead Beds (Upper Old Red Sandstone) to 375 ft; Thrust; Lower Limestone Shale to 570 ft; Portishead Beds to 674 ft.

29 **Portishead Middle Bridge No. 1 Borehole** Geological one-inch sheet 264. NGR 4715 7550. BGS Ref. No. ST 47 NE/20. Height above OD 23 ft. Bored for Portishead Water Co. Published, Richardson, 1928, p.172.
    Estuarine alluvium to 44 ft; Triassic (red marl and sandstone) to 250 ft.

30 **Portishead Railway Station Borehole** Geological one-inch sheet 264. NGR 4716 7700. BGS Ref. No. ST 47 NE/22. Height above OD 25 ft. Bored for ?Great Western Railway Co. by Le Grand, Sutcliff and Gell in 1930.
    Probably Portishead Beds (Upper Old Red Sandstone) to 550 ft.

31 **Portishead Wireless Station Borehole** Geological one-inch sheet 264. NGR 4500 7578. BGS Ref. No. 47 NE/18. Height above OD 372 ft. Bored in 1927. Published, Richardson, 1928, p.272–273.
    Portishead Beds (Upper Old Red Sandstone) to 361 ft.

32 **Tickenham Mill Borehole** Geological one-inch sheet 264. NGR 4545 7173. BGS Ref. No. ST 47 SE/61. Height above OD about 40 ft. Bored for Clevedon Water Co. in 1937.
    Triassic (red marl and sandstone) to 70 ft; Pennant Measures to 141 ft.

1:25 000 sheet 57

33 **Avonmouth Docks, Cooperative Wholesale Society's Flour Mills Borehole** Geological one-inch sheet 264. NGR 5137 7858. BGS Ref. No. ST 57 NW/2. Height above OD 25 FT. Published, Richardson 1930, pp.236–237.
    Quaternary deposits to 60 ft; Triassic (red marl and sandstone) to 133 ft; Pennant Measures to 478 ft.

34 **Avonmouth Docks, Hosegood Industries, Borehole A** Geological one-inch sheet 264. NGR 5130 7778. BGS Ref. No. ST 57 NW/5. Height above OD 30 ft. Bored by C. Isler and Co. Ltd in 1938.
    Estuarine alluvium to 85 ft, Triassic (red marl and sandstone) to 95 ft; Dolomitic Conglomerate to 159 ft; ?Pennant Measures and Upper Old Red Sandstone to 255 ft.

35 **Avonmouth Docks, Spillers Ltd Borehole** Geological one-inch sheet 264. NGR 5122 7812. BGS Ref. No. ST 57 NW/4. Height above OD 30 ft. Bored by C. Isler and Co. Ltd in 1933–34.
    Estuarine alluvium to 57 ft; Quaternary gravel to 70 ft; Triassic (red marl and sandstone) to 115 ft; Dolomitic Conglomerate to 125 ft; ?Pennant Measures to 181 ft 9 in.

36 **Avonmouth, Imperial Smelting No. 6 Borehole*** Geological one-inch sheet 264. NGR 5330 7926. BGS Ref. No. ST 57 NW/28. Height above OD 22 ft. Bored by C. Isler and Co. Ltd, and Timmins and Co. in 1951.
    Estuarine alluvium to 54 ft; Triassic (red marl and sandstone) to 171 ft; Supra-Pennant Measures to 191 ft 6 in.

37 **Avonmouth, Imperial Smelting No. 7 Borehole*** Geological one-inch sheet 264. NGR 5294 7859. BGS Ref. No. ST 57 NW/29. Height above OD 22 ft. Bored by C. Isler and Co. Ltd in 1951.
    Estuarine alluvium to 51 ft; Triassic (red marl and sandstone) to 177 ft; Supra-Pennant Measures to 186 ft.

38 **Avonmouth, Penpole No. 3 Borehole*** Geological one-inch sheet 264. NGR 5295 7760. BGS Ref. No. ST 57

NW/23. Height above OD 26 ft. Bored National Smelting Co. Ltd by C. Isler and Co. Ltd in 1950.

Drift to 5 ft 6 in; Triassic (red marl and sandstone) to 127 ft 4 in; Pennant Measures to 183 ft.

39 **Bristol, Ashley Down, Ashley Hill No. 33 Borehole*** Geological one-inch sheet 264. NGR 5968 7553. BGS Ref. No. ST 57 NE/11. Bored by Bristol City Engineer's Department in 1956.

Blue Lias to 101 ft; White Lias to 103 ft 6 in; Rhaetic to 118 ft; Tea Green Marl to 127 ft 6 in; Keuper Marl to 245 ft.

40 **Bristol, Ashton Gate, Duckmore Road No. 1 Borehole** Geological one-inch sheet 264. NGR 5712 7149. BGS. Ref. No. ST 57 SE/86. Height above OD about 25 ft. Bored by Ground Engineering Ltd in 1961.

Quaternary deposits to 34 ft; Triassic (red marl and sandstone) to 59 ft; Middle Coal Measures to 216 ft.

41 **Bristol, Ashton Gate, Raleigh Road (W. D. and H. O. Wills No. 4 Factory) Borehole** Geological one-inch sheet 264. NGR 5753 7188. BGS Ref. No. ST 57 SE/56. Height above OD about 30 ft. Bored by C. Isler and Co. Ltd in 1939.

Quaternary deposits to 10 ft; Triassic (red marl and sandstone) to 118 ft; Middle Coal Measures to 147 ft 6 in.

42 **Bristol, Ashton Park (Geological Survey) Borehole*** Geological one-inch sheet 264. NGR 5633 7146. BGS Ref. No. ST 57 SE/73. Height above OD 60 ft. Bored by John Thom, Ltd in 1952–53. Published, Kellaway, 1967, pp.49–153.

Head to 5 ft 6 in; Triassic (red marl and sandstone) to 130 ft; Lower Coal Measures to 708 ft 6 in; Millstone Grit Series to 1315 ft 2 in; Carboniferous Limestone Series to 2195 ft, ending in Clifton Down Limestone.

43 **Bristol, Ashton Vale, Frayne's Pit** Geological one-inch sheet 264. NGR 5673 7121. BGS Ref. No. ST 57 SE/71. Height above OD 30 ft. Section on abandoned mine plan.

Quaternary deposits to 35 ft; Mercia Mudstone to 85 ft; Dolomitic Conglomerate to 100 ft; Middle and Lower Coal Measures to 1060 ft.

44 **Bristol, Ashton Vale New Pit** Geological one-inch sheet 264. NGR 5656 7137. BGS Ref. No. ST 57 SE/72. Height above OD 40 ft. Section from Abandoned Mine Plan No. 5033.

Walling to 12 ft; Quaternary deposits to 34 ft; Triassic (red marl and sandstone) to 98 ft; Middle and Lower Coal Measures to 783 ft.

45 **Bristol, Ashton Vale, South Liberty Downcast (Great Engine) Shaft** Geological one-inch sheet 264. NGR 5648 7013. BGS Ref. No. ST 57 SE/78. Height above OD 55 ft. Section on Abandoned Mine Plan 8837.

Triassic (red marl and sandstone) to 100 ft; Middle and Lower Coal Measures to 1316 ft.

46 **Bristol, Ashton Vale, Starveall Pit** Geological one-inch sheet 264. NGR 5652 7083. BGS Ref. No. ST 57 SE/75. Height above OD about 40 ft. Published Prestwich, 1871, p.56.

Middle and Lower Coal Measures to 1166 ft 4 in (Bedminster Great Seam at 297 ft 8 in; Ashton Top Seam at 932 ft 6 in; Ashton Great Seam at 1007 ft 8in; Ashton Little Seam at bottom).

47 **Bristol, Bath Street Brewery Borehole** Geological one-inch sheet 264. NGR (approx.) 5911 7295. BGS Ref. No. ST 57 SE/41. Height above OD about 30 ft. Published, Richardson, 1930, pp.229–231.

Made ground to 27 ft; Gravel to 34 ft; Triassic (marl and sandstone) to 91½ ft; Lower Coal Measures to 576 ft.

48 **Bristol, Bathurst Wharf Borehole** Geological one-inch sheet 264. NGR (approx.) 5875 7210. BGS Ref. No. ST 57 SE/54. Height above OD about 36 ft. Published, Richardson, 1930, pp.231–232.

Made ground to 7 ft; Quaternary deposits to 49 ft; Triassic (red marl and Redcliffe Sandstone) to 137 ft; Coal Measures to 164 ft.

49 **Bristol, Cheese Lane Borehole*** Geological one-inch sheet 264. NGR 5973 7270. BGS Ref. No. ST 57 SE/43. Height above OD 35 ft. Bored by C. Isler and Co. Ltd in 1947.

Quaternary deposits to 18 ft 6 in; Triassic (red marl and sandstone) to 102 ft; Coal Measures to 118 ft.

50 **Bristol, Lewins Mead (Bristol United Breweries) Borehole** Geological one-inch sheet 264. NGR 5880 7337. BGS Ref. No. ST 57 SE/37. Height above OD 30 ft. Published, Moore, 1941, pp.279–292; Kellaway, 1967, pp.62–63.

Triassic (Redcliffe Sandstone) to 30 ft; Lower Coal Measures and Millstone Grit Series to 301 ft.

51 **Bristol, Old Market Street (Roger's Brewery) Borehole** Geological one-inch sheet 264. NGR 5953 7304. BGS Ref. No. ST 57 SE/83. Height above OD about 40 ft. Published Richardson, 1930, p.235.

Made ground to 37 ft; Triassic (Redcliffe Sandstone) to 152 ft. Middle Coal Measures to 300 ft.

52 **Bristol, Temple Gate Borehole** Geological one-inch sheet 264. NGR 5940 7227. BGS Ref. No. ST 57 SE/47. Height above OD about 30 ft. Published Richardson, 1930, p.232.

Triassic (Redcliffe Sandstone) to 138 ft.

53 **Bristol, Wilder Street (Messrs Carter) Borehole** Geological one-inch sheet 264. NGR 5928 7381. BGS Ref. No. ST 57 SE/30. Height above OD about 48 ft. Bored by C. Isler and Co. Ltd in 1887.

Dug well to 30 ft; Triassic (Redcliffe Sandstone) to 87 ft; Lower Coal Measures to 100 ft; ?Millstone Grit Series to 121 ft 8 in.

54 **Bristol, Pithay (J. S. Fry and Son Ltd) Borehole** Geological one-inch sheet 264. NGR (approx.) 5896 7324. BGS Ref. No. ST 57 SE/39. Height above OD about 35 ft. Bored by Thomas Matthews, Ltd in 1897–99.

Quaternary deposits to 38 ft 6 in; Triassic (Redcliffe Sandstone) to 92 ft; Lower Coal Measures to 271 ft 6 in.

55 **Bristol, River Street, St Jude's Borehole** Geological one-inch sheet 264. NGR 5972 7350. BGS Ref. No. ST 57 SE/85. Height above OD about 36 ft. Published, Richardson, 1930, pp.233–234.

Triassic (mainly Redcliffe Sandstone) to 129½ ft; Lower or Middle Coal Measures to 136½ ft.

56 **Bristol, St. Phillips (Avonside Paper Mills) Borehole** Geological one-inch sheet 264. NGR (approx.) 5984 7275. BGS Ref. No. ST 57 SE/44. Height above OD about 33 ft. Published, Richardson, 1930, pp.227–228.

Made ground to 16 ft; Triassic (Redcliffe Sandstone) to 125 ft.

57 **Bristol, Bedminster, Argus Pit** Geological one-inch sheet 264. NGR 5818 7101. BGS Ref. No. ST 57 SE/66. Height above OD about 80 ft. Section from Abandoned Mine Plan No. 3738 and Hippisley MS.

Quaternary deposits to 18 ft 6 in; Triassic (Redcliffe Sandstone) to 152 ft; Middle and Lower Coal Measures to 1733 ft; (Bedminster Great Vein at 1493 ft; Bedminster Little Vein at 1546 ft).

58 **Bristol, Bedminster, Dean Lane Pit** Geological one-inch sheet 264. NGR 5835 7168. BGS Ref. No. ST 57 SE/62. Height above OD 50 ft. Section from Abandoned Mine Plan No. 5178.

Quaternary deposits and Triassic (Redcliffe Sandstone) to 76 ft 6 in; Middle and Lower Coal Measures to 601 ft 5 in (Bedminster Great Vein at bottom).

59 **Bristol, Bedminster, Malago Pit** Geological one-inch sheet 264. NGR 5816 7107. BGS Ref. No. ST 57 SE/65. Height above OD about 80 ft. Section from Abandoned Mine Plan 3738.

Quaternary deposits to 18 ft; Triassic to 134 ft; Lower and Middle Coal Measures to 1500 ft (Bedminster Top Vein at 1270 ft 8 in; Bedminster Great Vein at 1407 ft 10in, Bedminster Little Vein at 1478 ft 1 in).

60 **Bristol, Bedminster Smelting Works (Mill Lane) Borehole** Geological one-inch sheet 264. NGR (approx. 5865 7155). BGS Ref. No. ST 57 SE/149. Height above OD about 30 ft. Published, Richardson, 1930, pp.228–229.

Triassic (marl and sandstone) to 52 ft; Middle Coal Measures to 315 ft.

61 **Bristol, Filton No. 2 (Sports Ground) Borehole** Geological one-inch sheet 264. NGR 5932 7848. BGS Ref. No. ST 57 NE/10. Height above OD 209 ft. Bored for Bristol Aeroplane Co. in 1944b. Published, Whittard and Smith, 1944b, pp.521–529.

Blue and White Lias to 45 ft; Rhaetic to 63 ft; Tea Green Marl to 75 ft; Mercia Mudstone to 136 ft 3 in; Dolomitic Conglomerate to 140 ft 6 in; Carboniferous Limestone Series to 420 ft (Hotwells Limestone to 253 ft; Middle Cromhall Sandstone to 267 ft 4 in; Clifton Down Limestone to 420 ft).

62 **Bristol, Hotwells, Cumberland Basin No. 1 Borehole*** Geological one-inch sheet 264. NGR 5713 7228. BGS Ref. No. ST 57 SE/53. Height above OD 28 ft. Bored for South Western Gas Board by F. G. Clements, Ltd.

Estuarine alluvium to 49 ft; Triassic (?Dolomitic Conglomerate) to 63 ft.

63 **Bristol, Kingsdown, Hospital Precinct No. 1 Borehole** Geological one-inch sheet 264. NGR 5893 7390. BGS Ref. No. ST 57 SE/110. Height above OD 122.5 ft.

Made ground to 10 ft; Triassic (red marl) to 30 ft; Triassic (sandstone and Dolomitic Conglomerate) to 110 ft; Millstone Grit Series to 153 ft.

64 **Bristol, Kingsdown, Hospital Precinct No. 6 Borehole** Geological one-inch sheet 264. NGR 5865 7350. BGS Ref. No. ST 57 SE/109. Height above OD 108 ft. Bored by Elmat, Ltd in 1963.

Made ground to 7 ft; Dolomitic Conglomerate to 45 ft; Millstone Grit Series to 75 ft.

65 **Bristol, Kingsdown, Hospital Precinct No. H1 Borehole** Geological one-inch sheet 264. NGR 5845 7352. BGS Ref. No. ST 57 SE/108. Height above OD 220 ft. Bored by Geotechnical Engineering, Ltd in 1965.

Made ground and Quaternary deposits to 8 ft; Triassic (red marl and sandstone) to 21 ft 9 in; Dolomitic Conglomerate to 73 ft 3 in: Millstone Grit Series to 85 ft.

66 **Henbury, Blaise Castle No. 1 Borehole*** Geological one-inch sheet 264. NGR 5553 7853. BGS Ref. No. 57 NE/1. Height above OD 185 ft. Bored for Port of Bristol Authority by E. Timmins and Sons, Ltd in 1931–34.

Soil to 1 ft; Carboniferous Limestone Series (Clifton Down Group) to 1363 ft 6 in. Folded with steep dips.

67 **Henbury, Blaise Castle No. 2 Borehole*** Geological one-inch sheet 264. NGR 5522 7829. BGS Ref. No. ST 57 NE/2. Height above OD about 180 ft. Shaft 38 ft 9 in, rest bored for Bristol Port Authority by E. Timmins and Sons, Ltd in 1945–48.

Head and Triassic to 6 ft; Carboniferous Limestone Series (Clifton Down Limestone) to 726 ft.

68 **Kings Weston, B Borehole*** Geological one-inch sheet 264. NGR 5442 7762. BGS Ref. No. ST 57 NW/34. Height above OD about 210 ft. Bored by Bristol City Engineer's Department in 1959.

Head and ?Triassic to 10 ft; Pennant Measures to 210 ft.

69 **Kings Weston, D Borehole*** Geological one-inch sheet 264. NGR 5463 7783. BGS Ref. No. ST 57 NW/35. Height above OD 209 ft. Bored by Bristol City Engineer's Department.

Head and ?Triassic to 9 ft; Carboniferous Limestone Series (Clifton Down Group) to 239 ft; Kingsweston Conglomerate (Upper Carboniferous) to 273 ft; Pennant Measures to 316 ft. Strata inverted.

70 **Kings Weston, J Borehole*** Geological one-inch sheet 264. NGR 5478 7712. BGS Ref. No. ST 57 NW/37. Height above OD 162 ft 6 in. Bored by Bristol City Engineer's Department.

Triassic (red marl) to 30 ft; Dolomitic Conglomerate to 73 ft 6 in; Upper Old Red Sandstone to 140 ft.

71 **Kings Weston, K Borehole*** Geological one-inch sheet 264. NGR 5495 7648. BGS Ref. No. ST 57 NW/38. Height above OD 16 ft 6 in. Bored by Bristol City Engineer's Department.

Triassic (red marl) to 62 ft; Dolomitic Conglomerate to 148 ft; Upper Old Red Sandstone to 155 ft.

72 **Kings Weston, L Borehole*** Geological one-inch sheet 264. NGR 5504 7636. BGS Ref. No. ST 57N. Height above OD 73 ft. Bored by Bristol City Engineer's Department.

Triassic (red marl and Dolomitic Conglomerate) to 100 ft.

73 **Kings Weston, Home Farm Borehole** Geological sheet 264. NGR 5420 7784. BGS Ref. No. ST 57 NW/26. Height above OD 100 ft. Bored by C. Isler and Co. in 1935.

?Head to 13 ft; Pennant Measures to 291 ft.

74 **Lawrence Weston, Aust Farm Borehole** Geological one-inch sheet 264. NGR 5451 7809. BGS Ref. No. ST 57 NW/27. Height above OD about 50 ft. Bored about 1934.

Dolomitic Conglomerate to 25 ft; Pennant Measures to 191 ft.

75 **Long Ashton, By-pass 'A' Borehole** Geological one-inch sheet 264. NGR 5500 7014. BGS Ref. No. ST 57 SE/97. Height above OD 59 ft. Bored for Somerset Country Council by G. K. N. Foundations, Ltd in 1965.

Triassic (red marl) to 81 ft; Lower Coal Measures to 120 ft.

76 **Long Ashton, Gore's Pit** Geological one-inch sheet 264. NGR 5589 7015. BGS Ref. No. ST 57 SE/80. Height above OD 45 ft. Sunk before 1840.

Triassic (red marl) to 28 ft 6 in; Middle Coal Measures to 64 ft 6 in.

77 **Wraxall, Watercress Farm Borehole** Geological one-inch sheet 264. NGR 5001 7057. BGS Ref. No. ST 57 SW/9. Height above OD 65 ft. bored by C. Isler and Co. in 1903. Published, Richardson, 1928, pp.99–100.

Quaternary deposits to 10 ft; Triassic (red marl) to 30 ft; Dolomitic Conglomerate to 45 ft; Lower Coal Measures to 237 ft; ?Millstone Grit Series to 490 ft.

1:25 000 sheet 67

78 **Bitton, Golden Valley New Pit** Geological one-inch sheet 265. NGR 6860 7103. BGS Ref. No. ST 67 SE/16. Height above OD about 130 ft. Sunk before 1873. Abandoned 1898. Published, Anstie, 1873, pp.36–37, fig.8.

Through Triassic into Pennant Measures (Millgrit Vein at 507 ft; Rag Vein at 557 ft; Parrot Vein at 912 ft).

79 **Bitton, Golden Valley Old Pit** Geological one-inch sheet 265. NGR 9037 7080. BGS Ref. No. ST 67 SE/21. Height above OD about 115 ft. Sunk before 1871. Abandoned 1898. Section from Abandoned Mine Plan No. 3878.

Through Triassic into Pennant Measures (Millgrit Vein at 75 ft; Rag Vein at 150 ft; Parrot Vein at 450 ft).

80 **Bitton, Ryedown Lane Borehole** Geological one-inch sheet 265. NGR 6775 7052. BGS. Ref. No. ST 67 SE/19.

Height above OD 265 ft. Bored about 1922. Published, Tutcher, 1923, p.278.
    White Lias to 7 ft 10 in; Rhaetic to 24 ft.

81 **Bristol, Baptist Mills, Pennywell Road Pit**   Geological one-inch sheet 264. NGR 6021 7398. BGS Ref. No. ST 67 SW/4. Height above OD about 40 ft. Section from Abandoned Mine Plan. Published, Prestwich, 1871, p.55.
    Redcliffe Sandstone to 90 ft; Middle and Lower Coal Measures (Kingswood Great Vein at 280 ft; Kingswood Little Vein at 390 ft; Easton Red Ash at 945 ft).

82 **Bristol, Barton Hill No.1 Borehole***   Geological one-inch sheet 264. NGR 6112 7291. BGS Ref. No. St 67 SW/14. Height above OD 48 ft.
    Made ground and Quaternary deposits to 5 ft; Triassic (red marl) to 15 ft; Triassic (Redcliffe Sandstone) to 56 ft; Pennant Measures to 145 ft.

83 **Bristol, Barton Hill No.3 Borehole***   Geological one-inch sheet 264. NGR 6107 7278. BGS Ref. No. ST 67 SW/16. Height above OD 40 ft.
    Made ground and Quaternary deposits to 5 ft; Redcliffe Sandstone to 57 ft; Pennant Measures to 90 ft.

84 **Bristol, Barton Hill No.5 Borehole***   Geological one-inch sheet 264. NGR 6097 7291. BGS Ref. No. ST 67 SW/18. Height above OD 57 ft.
    Made ground to 2 ft; Redcliffe Sandstone to 63 ft; Pennant Measures to 80 ft.

85 **Bristol, Burchells Green, Belgium Pit**   Geological one-inch sheet 264. NGR 6352 7433. BGS Ref. No. ST 67 SW/12. Height above OD about 280 ft. Section from Abandoned Mine Plan No. 11831.
    Head to 4 ft 6in; Middle Coal Measures to 408 ft.

86 **Bristol, Chester Park, Lodge Engine Pit**   Geological one-inch sheet 264. NGR 6391 7452. BGS Ref. No. ST 67 SW/13. Height above OD about 310 ft. Published, Prestwich, 1871, p.55. Middle Coal Measures.

87 **Bristol, Clay Hill, Deep Pit**   Geological one-inch sheet 264. NGR 6256 7458. BGS Ref. No. ST 67 SW/9. Height above OD 245 ft. Sections unreliable.
    Coal Measures highly disturbed.

88 **Bristol, Crew's Hole Pit**   Geological one-inch sheet 264. NGR 6254 7313. BGS Ref. No. ST 67 SW/54. Height above OD about 50 ft. Published Prestwich, 1871, p.55. Coal Measures.

89 **Bristol, Easton Area Redevelopment No.1 Borehole**
Geological one-inch sheet 264. NGR 6029 7435. BGS Ref. No. ST 67 SW/69. Height above OD 44.5 ft. Bored by G. Wimpey and Co. Ltd in 1963.
    Made ground to 10 ft; Redcliffe Sandstone to 120 ft 3 in; Middle Coal Measures to 200 ft 5 in.

90 **Bristol, Easton Area Redevelopment No.2 Borehole**
Geological one-inch sheet 264. NGR 6035 7405. BGS Ref. No. ST 67 SW/70. Height above OD 42 ft. Bored by G. Wimpey and Co. Ltd in 1962.
    Made ground to 3 ft 6 in; Redcliffe Sandstone to 111 ft; Middle Coal Measures to 200 ft.

91 **Bristol, Easton Area Redevelopment No.3 Borehole**
Geological one-inch sheet 264. NGR 6010 7390. BGS Ref. No. ST 67 SW/71. Height above OD 39.5 ft. Bored by G. Wimpey and Co. Ltd in 1962-63.
    Made ground to 8 ft; Redcliffe Sandstone to 118 ft; Middle Coal Measures to 267 ft.

92 **Bristol, Easton Area Redevelopment No.4 Borehole**
Geological one-inch sheet 264. NGR 6010 7375. BGS Ref. No. ST 67 SW/72. Height above OD 38.7 ft. Bored by G. Wimpey and Co. Ltd in 1962-63.
    Made ground to 7 ft; Redcliffe Sandstone to 109 ft 6 in; Middle Coal Measures to 201 ft 7 in.

93 **Bristol, Easton Area Redevelopment No.5 Borehole**
Geological one-inch sheet 264. NGR 6013 7322. BGS Ref. No. ST 67 SW/73. Height above OD 54.3 ft. Bored by G. Wimpey and Co. Ltd in 1963.
    Made ground to 11 ft; Redcliffe Sandstone to 121 ft 3 in; Middle Coal Measures to 200 ft 3 in.

94 **Bristol, Easton Area Redevelopment No.6 Borehole**
Geological one-inch sheet 264. NGR 6041 7378. BGS Ref. No. ST 67 SW/74. Height above OD 53.7 ft. Bored by G. Wimpey and Co. Ltd in 1962-63.
    Made ground to 5 ft 6 in; Redcliffe Sandstone to 102 ft; Middle Coal Measures to 212 ft 6 in.

95 **Bristol, Easton Area Redevelopment No. 7 Borehole**
Geological one-inch sheet 264. B.G.R. 6056 7370. BGS Ref. No. ST 67 SW/75. Height above OD 58.6 ft. Bored by G. Wimpey and Co. Ltd in 1963.
    Made ground to 3 ft; Redcliffe Sandstone to 104 ft 4 in; Middle Coal Measures to 200 ft.

96 **Bristol, Easton Area Redevelopment No. 8 Borehole**
Geological one-inch sheet 264. NGR 6059 7346. BGS Ref. No. ST 67 SW/76. Height above OD 72.2 ft. Bored by G. Wimpey and Co. Ltd in 1963.
    Made ground to 22 ft 6 in; Redcliffe Sandstone to 102 ft 6 in; Middle Coal Measures to 200 ft.

97 **Bristol, Easton Colliery**   Geological one-inch sheet 264. NGR 6065 7389. BGS Ref. No. ST 67 SW/7. Height above OD 50 ft. Published, Anstie, 1873, p.31.
    Redcliffe Sandstone to 75 ft; Middle and Lower Coal Measures to 1920 ft. Structurally distributed.

98 **Bristol, Easton, St Mark's Road No. 15 Borehole**
Geological one-inch sheet 264. NGR 6087 7450. BGS Ref. No. ST 67 SW/36. Height above OD 41.7 ft.
    Made ground to 1 ft; Redcliffe Sandstone to 55 ft 4 in; Pennant Measures (near base) to 85 ft.

99 **Bristol, Filton No. 1 (Fairlawn House) Borehole**
Geological one-inch sheet 264. NGR 6007 7946. BGS Ref. No. ST 67 NW/12. Height above OD 249 ft. Bore for Bristol Aeroplane Co. about 1941. Published, Whittard and Smith, 1943, pp.434-50.
    Lower Lias to 55 ft; Rhaetic to 68 ft; Tea Green Marl to 86 ft; Mercia Mudstone to 169 ft. Carboniferous Limestone Series to 694 ft. Fault at 211 to 222 ft.

100 **Bristol, Filton Junction, Hewitt's Lime Works Borehole**
Geological one-inch sheet 264. NGR 6145 7900. BGS Ref. No. ST 67 NW/2. Height above OD 178 ft. Bored for Bedminster Coal Co. Ltd in 1901.
    Blue and White Lias to 12 ft 1 in; Rhaetic to 31 ft 9 in; Tea Green Marl to 43 ft 3 in; Mercia Mudstone to 200 ft 4 in; Redcliffe Sandstone to 284 ft 7 in; Lower Coal Measures and Millstone Grit Series to 503 ft 5 in.

101 **Bristol, Filton Laundry Borehole**   Geological one-inch sheet 264. NGR 6025 7909. BGS Ref. No. ST 67 NW/1. Height above OD 270 ft. Bored in 1923-24. Published, Richardson, 1930, p.234; Whittard and Smith, 1943, p.447.
    Lower Jurassic and Triassic rocks to 199 ft; Carboniferous Limestone Series to 397 ft.

102 **Bristol, Hanham Colliery Downcast Shaft**   Geological one-inch sheet 264. NGR 6373 7204. BGS Ref. No. ST 67 SW/2. Height above OD about 180 ft. Section from Abandoned Mine Plan No. 8953.
    Walling above rock-head 12 ft; Pennant Measures to 900 ft.

103 **Bristol, Harry Stoke 'A' Borehole***   Geological one-inch sheet 264. NGR 6226 7905. BGS Ref. No. ST 67 NW/3. Height above OD 197 ft. Bored for National Coal Board by Craelius Co. in 1949.
    Triassic (mostly red marl) to 195 ft 10 in; Redcliffe

Sandstone to 304 ft. Middle Coal Measures to 761 ft 1 in; Lower Coal Measures 1291 ft 9 in; Millstone Grit Series to 1338 ft 11 in.

104 Bristol, Harry Stoke 'B' (Hambrook) Borehole*
Geological one-inch sheet 264. NGR 6321 7816. BGS Ref. No. ST 67 NW/7. Height above OD 151 ft. Bored for National Coal Board by Craelius Co. in 1950.
Triassic (red marl) to 10 ft; Pennant Measures (Downend Formation) to 475 ft 6 in; Middle and Lower Coal Measures to 2246 ft 9 in; Millstone Grit Series to 2260 ft 5 in.

105 Bristol, Harry Stoke 'C' (Downend) Borehole*
Geological one-inch sheet 264. NGR 6504 7677. BGS Ref. No. ST 67 NE/1. Height above OD 229.5 ft. Bored for National Coal Board by John Thom, Ltd in 1955.
Quarry fill to 34 ft; Pennant Measures (Downend Formation) to 1591 ft 2 in; Middle Coal Measures to 3244 ft 9 in; Lower Coal Measures to 3464 ft 3 in.

106 Bristol, Harry Stoke 'D' Borehole* Geological one-inch sheet 264. NGR 6253 7849. BGS Ref. No. ST 67 NW/6. Height above OD 210 ft. Bored for National Coal Board by Craelius Co. in 1950.
Triassic (red marl) to 148 ft; Redcliffe Sandstone to 235 ft 6 in; Middle Coal Measures to 651 ft 9 in.

107 Bristol, Harry Stoke 'E' Borehole* Geological one-inch sheet 264. NGR 6234 7857. BGS Ref. No. ST 67 NW/4. Height above OD 193 ft. Bored for National Coal Board by Craelius Co. in 1951.
Triassic (red marl) to 151 ft 5 in; Redcliffe Sandstone to 237 ft 3 in; Middle Coal Measures to 405 ft 10 in.

108 Bristol, Harry Stoke 'F' Borehole* Geological one-inch sheet 264. NGR 6220 7837. BGS Ref. No. ST 67 NW/5. Height above OD 209 ft. Bored for National Coal Board by Craelius Co. in 1951.
Triassic (red marl) to 165 ft 1 in; Redcliffe Sandstone to 267 ft; Middle Coal Measures to 875 ft; Lower Coal Measures to 910 ft.

109 Bristol, Kingswood, Coronation Pit, Cadbury Heath
Geological one-inch sheet 264. NGR 6656 7243. BGS Ref. No. 67 SE/7. Height above OD about 150 ft. Sunk 1901–02.
Pennant Measures to about 140 ft.

110 Bristol, Moorfields, Russell Town Avenue No.3 Borehole
Geological one-inch sheet 264. NGR 6098 7347. BGS Ref. No. ST 67 SW/65. Height above OD 63 ft. Bored 1962.
Redcliffe Sandstone to 34 ft; ?Middle Coal Measures to 120 ft.

111 Bristol, Netham, Old Pylemarsh Pit Geological one-inch sheet 264. NGR 6174 7298. BGS Ref. No. ST 67 SW/51. Height above OD 70 ft. Published, Prestwich, 1871, p.55; Anstie, 1873, p.40.
Pennant Measures to 480 ft.

112 Bristol, Potterswood, Douglas Road Borehole
Geological one-inch sheet 264. NGR 6456 7340. BGS Ref. No. ST 67 SW/110. Height above OD about 325 ft. Published, Richardson, 1930, pp.192–193.
Middle Coal Measures to 433 ft.

113 Bristol, St. George, Air Balloon Pit Geological one-inch sheet 264. NGR 6309 7337. BGS Ref. No. ST 67 SW/55. Height above OD 230 ft. Published, Prestwich, 1871, p.55.
Pennant Measures and Middle Coal Measures 606 ft.

114 Bristol, St. George, Whitehall Pit Geological one-inch sheet 264. NGR 6181 7380. BGS Ref. No. ST 67 SW/8. Height above OD 115 ft.
?Pennant Measures and Middle Coal Measures to 1104 ft, ground heavily disturbed.

115 Bristol, Redfield, Victoria Avenue Borehole Geological one-inch sheet 264. NGR 6149 7332. BGS Ref. No. ST 67 SW/50. Height above OD 81 ft. Published, Richardson, 1930, pp.232–233.
Redcliffe Sandstone to 8 ft; ?Middle Coal Measures to 206 ft 6 in.

116 Bristol, Ridgeway, Castles Pit Geological one-inch sheet 264. NGR 6263 7563. BGS Ref. No. ST 67 NW/16. Height above OD 175 ft. Published Prestwich, 1871, p.55.
Pennant Measures (to Hen Vein) 600 ft.

117 Bristol, St. Phillip's Marsh, Great Western Pit
Geological one-inch sheet 264. NGR 6091 7249. BGS Ref. No. ST 67 SW/5. Height above OD 42 ft.
Made ground to 15 ft; Quaternary deposits to 32 ft; Walled shaft (probably mainly Triassic) to 244 ft 8 in, Pennant Measures and Middle Coal Measures to 951 ft 1 in.

118 Bristol, St. Phillip's Marsh, J. Lysaghts Borehole
Geological one-inch sheet 264. NGR 6159 7262. BGS Ref. No. ST 67 SW/3. Height above OD about 30 ft. Bored by J. W. Titt and Co. Ltd in 1952.
Quaternary deposits to 39 ft; Redcliffe Sandstone to 88 ft; Pennant Measures to 200 ft.

119 Bristol, Soundwell High Pit Geological one-inch sheet 264. NGR 6495 7519. BGS Ref. No. ST 67 NW/9. Height above OD about 310 ft. Published, Prestwich, 1871, p.54.
Middle and Lower Coal Measures to 1296 ft.

120 Bristol, Speedwell Pit Geological one-inch sheet 264. NGR 6323 7442. BGS Ref. No. ST 67 SW/10. Height above OD 277 ft. Published, Anstie, 1873, p.32.
Middle and Lower Coal Measures (disturbed) to 1452 ft.

121 Mangotsfield, Brandybottom Pit Geological one-inch sheet 264. NGR 6823 7715. BGS Ref. No. ST 67 NE/7. Height above OD 229 ft. Published, Prestwich, 1871, p.54.
Supra-Pennant Measures and Pennant Measures to 674 ft.

122 Oldland, California Colliery No.1 Shaft (Downcast), Oldland Common Geological one-inch sheet 265. NGR 6654 7140. BGS Ref. No. ST 67 SE/11. Height above OD about 140 ft. Sunk, after 1873. Data from Abandoned Mine Plan No. 4617.
Pennant Measures to about 522 ft.

123 Oldland, California Colliery No.2 Shaft (Upcast), Parkwall Geological one-inch sheet 265. NGR 6648 7175. BGS Ref. No. ST 67 SE/10. Height above OD about 130 ft. Sunk, before 1873.
Pennant Measures (Rag Vein at 160 ft; Buff Vein at 365 ft; Parrot Vein at 480 ft).

124 Oldland Common, Cowhorn Pit Geological one-inch sheet 265. NGR 6699 7195. BGS Ref. No. ST 67 SE/17. Height above OD about 115 ft. Published, Prestwich, 1871, p.56.
Pennant Measures (Millgrit Vein at 21 ft; Rag Vein at 35 ft; Buff Vein at 249 ft; Parrot Vein at 370 ft 6 in).

125. Oldland, Hole Lane Colliery Geological one-inch sheet 265. NGR 6779 7175. BGS Ref. No. ST 67 SE/15. Height above OD 160 ft. Published, Prestwich, 1871, p.56.
Pennant Measures (Millgrit Vein at 300 ft; Rag Vein at 348 ft; Buff Vein at 576 ft; Parrot Vein at 660 ft.

126 Pucklechurch, Parkgate Pit Geological one-inch sheet 265. NGR 6835 7876. BGS Ref. No. ST 67 NE/6. Height above OD about 180 ft. Sunk 255 ft 8 in; bored 325 ft.
Triassic (red marl) to 28 ft; Supra-Pennant Measures to 580 ft 8 in.

127 Pucklechurch, Parkfield Colliery Geological one-inch sheet 265. NGR 6885 7777. BGS Ref. No. ST 67 NE/20. Height above OD 231 ft.
Triassic (red marl) to 65 ft; Supra-Pennant Measures to 596 ft; Pennant Measures to 782 ft.

128 **Warmley, Crown Colliery** Geological one-inch sheet 265. West Pit, NGR 6716 7354; South Pit, NGR 6724 7343; Engine Pit, NGR 6725 7350. BGS Ref. No. ST67SE/2. Height above OD about 150 ft. Depths about 500 ft.
   Middle Coal Measures.

129 **Warmley Tower, Goldney Pit** Geological one-inch sheet 265. NGR 6700 7258 BGS Ref. No. ST67SE/8. Height above OD about 140 ft. Information from F. C. Sadler's 1 chain to 1 inch plan.
   Middle Coal Measures to 567 ft.

130 **Westerleigh, New Engine Pit** Geological one-inch sheet 265. NGR 6780 7939. BGS Ref. No. ST 67 NE/8. Height above OD 179 ft. Published, Buckland and Conybeare, 1824, p.244; Woodward, 1876, p.55.
   Head to 6 ft; Supra-Pennant Measures and Pennant Measures (Rag Vein at 317 ft 4 in; Hollybush and Great Vein at 502 ft 10 in) to 508 ft 10 in.

1:25 000 sheet 36 (Eastern part)

131 **Banwell Moor Borehole** Geological one-inch sheet 280. NGR 3995 6087. BGS Ref. No. ST 36 SE/1. Height above OD 19 ft. Bored for Weston-super-Mare U.D.C. in 1929. Published, Green and Welch, 1965, pp.198–199.
   Quaternary deposits to 56 ft 6 in; Triassic (red marl) to 570 ft 6 in; Dolomitic Conglomerate to 603 ft; ?Pennant Measures to 1018 ft.

1:25 000 sheet 46

132 **Claverham, Court de Wyck Tannery Borehole** Geological one-inch sheet 264. NGR 4495 6623. BGS Ref. No. ST 46 NW/73. Height above OD 20 ft. Bored by F. G. Clements, Ltd in 1933.
   Quaternary deposits to 15 ft; Triassic (red marl) to 300 ft.

133 **Flax Bourton, Chelvey Pumping Station Borehole** Geological one-inch sheet 264. NGR 4736 6797. BGS Ref. No. 46 NE/5. Height above OD 50 ft. Bored 1871–72. Published, Richardson, 1928, pp.212–214.
   Triassic (red marl) to 307 ft; Dolomitic Conglomerate on Coal Measures to 327 ft.
   A second well proved Triassic rocks to 312 ft and Coal Measures to 400 ft.

134 **Wrington, West Hay Borehole** Geological one-inch sheet 264. NGR 4601 6341. BGS Ref. No. ST 46 SE/1. Height above OD about 105 ft. Published, Whitaker and Woodward, 1895, p.345; Richardson, 1928, p.48.
   Quaternary deposits to 9 ft; Triassic (red marl) to 93 ft; ?Carboniferous Limestone to 99 ft.

135 **Yatton, Bishop's Farm No.2 Borehole** Geological one-inch sheet 264. NGR 4487 6559. BGS Ref. No. ST 46 NW/71b. Height above OD about 67 ft.
   Triassic (red marl and sandstone) to 265 ft; Dolomitic Conglomerate to 270 ft 9 in; Carboniferous Limestone Series to 280 ft.

1:25 000 sheet 56

136 **Barrow Gurney, Barrow Court Borehole** Geological one-inch sheet 264. NGR 5142 6841. BGS Ref. No. ST 56 NW/27. Height above OD about 360 ft. Bored by F. G. Clements, Ltd in 1941.
   Lower Lias and Triassic to 290 ft; Carboniferous Limestone Series to 302 ft.

137 **Barrow Gurney, Gratwicke Hall (Home Farm) Borehole** Geological one-inch sheet 264. NGR 5181 6820. BGS Ref. No. ST 56 NW/31. Height above OD 390 ft. Well sunk about 1914, boring by F. G. Clements, Ltd in 1943.
   Dug well to 190 ft; Triassic (red marl and Dolomitic Conglomerate) to 272 ft; Lower Coal Measures (with Ashton Vale Marine Band) to 291 ft.

138 **Bedminster, Old Engine Pit** Geological one-inch sheet 264. NGR 5610 6999. BGS Ref. No. ST 56 NE/1. Height above OD 55 ft. Published, G. S. Hor. Sect. No. 106 (1875).
   Triassic to 100 ft; Lower Coal Measures to 450 ft.

139 **Chew Stoke Borehole 'A'** Geological one-inch sheet 280. NGR 5713 6156. BGS Ref. No. ST 56 SE/1. Height above OD about 148 ft. Bored for Bristol Waterworks Co. in 1938. Published, Moore and Trueman, 1939, pp.66–68.
   Triassic (red marl and sandstone) to 45 ft; Middle Coal Measures to 106 ft.

140 **Dundry, (Elton Farm) Borehole*** Geological one-inch sheet 264. NGR 5636 6689. BGS Ref. No. ST 56 NE/3. Height above OD 703 ft. Bored for Geological Survey by John Thom, Ltd Published, Ivimey-Cook, 1978.
   Inferior Oolite to 36 ft 4 in; Upper Middle and Lower Lias to 615 ft 3 in; White Lias and Rhaetic to 632 ft 11 in, Mercia Mudstone to 887 ft 2 in; Redcliffe Sandstone to 1093 ft 2 in, Middle Coal Measures to 1500 ft 10 in.

141 **Winford No. 1 Borehole*** Geological one-inch sheet 264. NGR 5573 6375. BGS Ref. No. ST 56 SE/5. Height above OD 198 ft. Bored for National Coal Board by John Thom, Ltd in 1955.
   Triassic (red marl and sandstone) to 229 ft; Lower Coal Measures to 787 ft.

142 **Winford No. 2 Borehole*** Geological one-inch sheet 264. NGR 5636 6343. BGS Ref. No. ST 56 SE/6. Height above OD 200 ft. Bored for National Coal Board by John Thom, Ltd in 1955.
   Triassic (red marl and sandstone) to 200 ft; Middle and Lower Coal Measures to 1538 ft 2 in.

143 **Winford, Crown Inn Borehole*** Geological one-inch sheet 264. NGR 5401 6374. BGS Ref. No. ST 56 SW/1. Height above OD 400 ft. Bored by F. G. Clements, Ltd in 1950.
   Triassic (red marl and Dolomitic Conglomerate) to 81 ft 6 in; Carboniferous Limestone Series (Upper Cromhall Sandstone) to 100 ft.

144 **Winford, Orthopaedic Hospital Borehole** Geological one-inch sheet 264. NGR 5341 6565. BGS Ref. No. ST 56 NW/29. Height above OD 480 ft. Well dug to 160 ft 9 in by W. Bryant, bored by F. G. Clements, Ltd in 1927–28.
   Triassic (sandstone and Dolomitic Conglomerate) to 11 ft 6 in; Carboniferous Limestone Series to 326 ft.

1:25 000 sheet 66

145 **Bromley Colliery** Geological one-inch sheet 280. NGR 6061 6173. BGS Ref. No. ST 66 SW/3. Height above OD 305 ft.
   Supra-Pennant Measures to 503 ft 2 in.

146 **Farmborough Pumping Pit** Geological one-inch sheet 281. NGR 6576 6011. BGS Ref. No. 66 SE/1. Height above OD 430 ft.
   Blue Lias, White Lias, Rhaetic and red marl in faulted ground to 86 ft; Mercia Mudstone to 199 ft; Dolomitic Conglomerate to 256 ft; Supra-Pennant Measures to 1440 ft.

147 **Fry's Bottom Colliery** Geological one-inch sheet 280. NGR 6300 6042. BGS Ref. No. ST 66 SW/1. Height above OD about 450 ft. Section from Abandoned Mine Plan No. 3033.
   Triassic (red marl) thin and Supra-Pennant Measures (Radstock Formation) to below 562 ft.

148 **Hursley Hill Borehole*** Geological one-inch sheet 264. NGR 6180 6565. BGS Ref. No. ST 66 NW/2. Height above OD 220.6 ft. Bored for National Coal Board by Craelius Co. in 1951.

Supra-Pennant Measures to 2400 ft (Pensford No. 1 Seam at 1724 ft; Pensford No. 2 Seam at 1883 ft 7 in; Pensford No. 3 Seam at 1889 ft 5 in.

149 Pensford Colliery, North Shaft Geological one-inch sheet 264. NGR 6178 6269. BGS Ref. No. ST 66 SW/2. Height above OD 340 ft.

Supra-Pennant Measures to 1497 ft (Forty Yards Seam at 120 ft 7 in; Pensford No. 1 Seam at 814 ft 6 in; Pensford No. 2 Seam at 1001 ft 1 in; Pensford No. 3 Seam at 1034 ft 8 in).

150 Priston Borehole Geological one-inch sheet 281. NGR 6981 6063. BGS Ref. No. ST 66 SE/2. Height above OD 251.9 ft. Bored 1909.

Quaternary deposits to 13 ft 9 in; Triassic (Tea Green Marl and red marl) to 159 ft 8 in; conglomeratic sandstone to 210 ft 10 in; Supra-Pennant Measures to 912 ft 3 in.

151 Somerdale, Keynsham J.S. Fry and Sons, Ltd Borehole Geological one-inch sheet 264. NGR 6545 6941. BGS Ref. No. ST 66 NE/23. Height above OD about 40 ft. Published, Tutcher, 1928, pp.56–57.

Rhaetic to 12 ft; Tea Green Marl to 24 ft; Mercia Mudstone to 208 ft; Pennant Measures to 277 ft.

152 Stockwood, Grove Farm Borehole* Geological one-inch sheet 264. NGR 6267 6870. BGS Ref. No. ST 66 NW/1. Height above OD 280 ft. Bored by Bristol City Engineer's Department in 1951.

Lower Lias to 18 ft; White Lias to 25 ft; Rhaetic to 44 ft 10 in; Tea Green Marl to 54 ft 6 in; Mercia Mudstone to 85 ft.

1:25 000 sheet 76 (Western part)

153 Twerton Colliery No. 1 shaft Geological one-inch sheet 265. NGR 7149 6456. BGS Ref. No. ST 76 SW/27. Height above OD about 145 ft. Sunk before 1848, Abandoned about 1875. Section from Abandoned Mine Plan No. 802.

Lower Jurassic, Triassic and ?Middle Coal Measures to 642 ft.

154 Twerton Colliery No. 2 shaft Geological one-inch sheet 265. NGR 7128 6420. BGS Ref. No. 76 SW/28. Height above OD 90 ft. Sunk before 1875. Published, Richardson, 1928, p.207.

Quaternary deposits to 8 ft 6 in; Lower Lias (including White Lias) to 150 ft 6 in; Rhaetic to 173 ft 6 in, Triassic (Tea Green Marl and red marl) to 359 ft 6 in; ?Middle Coal Measures touched.

1:25 000 sheet 35 (Eastern part)

155 Banwell Spring Borehole Geological one-inch sheet 280. NGR 3987 5921. BGS Ref. No. ST 35 NE/18. Height above OD 19 ft. Bored for Weston-super-Mare U.D.C. in 1931.

Mercia Mudstone to 61 ft; Dolomitic Conglomerate to 66 ft 3 in; Carboniferous Limestone Series (Clifton Down Limestone) to 369 ft 1 in.

1:25 000 sheet 55

156 Bishop Sutton, New Pit Geological one-inch sheet 280. NGR 5828 5924. BGS Ref. No. ST 55 NE/1. Height above OD 227.6 ft.

Triassic (red marl) to 120 ft; Dolomitic Conglomerate to 132 ft. Supra-Pennant Measures to 739 ft 10 in (Peacock Vein at 643 ft 7 in).

157 West Harptree, Gurney Farm Borehole Geological one-inch sheet 280. NGR 5630 5810. BGS Ref. No. ST 55 NE/2. Height above OD about 200 ft. Bored for W. Pritchard Morgan and Co. by John Thom, Ltd. in 1911-12.

Quaternary deposits to 8 ft; Triassic (red marl) to 179 ft 9 in; Triassic (Dolomitic Conglomerate) to 209 ft 6 in; Supra-Pennant Measures to 622 ft 4 in.

1:25 000 sheet 65

158 Camerton Colliery New Pit Geological one-inch sheet 281. NGR 6855 5806. BGS Ref. No. ST 65 NE/9. Height above OD 256.65 ft.

Triassic (marl and sandstone) to 84 ft 7 in; Supra-Pennant Measures (Radstock Formation to 788 ft 9 in; Barren Red Formation to 1420 ft 11 in; Farrington Formation to 1697 ft 8 in).

159 Camerton Colliery Old Pit Geological one-inch sheet 281. NGR 6813 5793. BGS Ref. No. ST 65 NE/8. Height above OD about 257 ft. Published, Prestwich, 1871, p.57.

Triassic to 144 ft; Supra-Pennant Measures (Radstock Formation) to 968 ft 4 in.

160 Clandown Colliery Geological one-inch sheet 281. NGR 6803 5592. BGS Ref. No. ST 65 NE/20. Height above OD 453 ft. Published, Buckland and Conybeare, 1824, p.278.

Lower Lias to 16 ft; White Lias to 28 ft; Rhaetic to 52 ft; Tea Green Marl to 64 ft; Triassic (red Mercia Mudstone) to 262 ft 4 in; Dolomitic Conglomerate to 274 ft 4 in; Supra-Pennant Measures (Publow Formation to 683 ft 4 in; Radstock Formation to 1380 ft).

161 Clandown, Smallcombe Colliery Geological one-inch sheet 281. NGR 6832 5590. BGS Ref. No. ST 65 NE/21. Height above OD 402 ft.

Lower Lias and Rhaetic to 54 ft; Triassic (red and green marl) to 195 ft; Triassic (Dolomitic Conglomerate) to 228 ft; Supra-Pennant Measures (Publow Formation, Radstock Formation into top of Barren Red Formation) 1074 ft.

162 Clutton, Rudge Pit Geological one-inch sheet 280. NGR 6286 5858. BGS Ref. No. ST 65 NW/5. Sunk ?1865.

Supra-Pennant Measures (Mangotsfield Formation) to 834 ft.

163 Dunkerton Colliery No. 1 Shaft Geological one-inch sheet 281. NGR 6985 5859. BGS Ref. No. ST 65 NE/10. Height above OD 244.83 ft. Published, Richardson, 1928, p.51.

Blue Lias and White Lias to 32 ft 3 in; Rhaetic to 61 ft 3 in; Tea Green Marl to 75 ft 3 in; Mercia Mudstone to 222 ft 6 in; Dolomitic Conglomerate to 257 ft; Supra-Pennant Measures to 681 ft 10 in.

164 Farrington Colliery Geological one-inch sheet 280. NGR 6415 5554. BGS Ref. No. ST 65 NW/10. Height above OD 418.7 ft. Published, Greenwell and McMurtrie, 1864, p.27; G.S. Vert.Section, sheet 49, No. 10 (1873).

Redcliffe Sandstone to 19 ft 6 in; Dolomitic Conglomerate to 31 ft 6 in; Supra-Pennant Measures (Barren Red and Farrington formations to 857 ft 6 in).

165 Farrington Gurney Borehole Geological one-inch sheet 280. NGR 6280 5526. BGS Ref. No. ST 65 NW/8. Height above OD 428 ft. Published, Moore, 1936, pp.220–222.

Triassic (red marl) to 60 ft; Dolomitic Conglomerate to 150 ft; Pennant Measures (Mangotsfield Formation) to 291 ft.

166 Greyfield Colliery New Pit, High Littleton Geological one-inch sheet 280. NGR 6400 5868. BGS Ref. No. ST 65 NW/4. Height above OD 438 ft. Published, Greenwell and McMurtrie, 1864, p.27.

Triassic to 37 ft; Supra-Pennant (Barren Red and Farrington formations) to 901 ft 5 in.

167 Kilmersdon Colliery Geological one-inch sheet 281. NGR 6876 5382. BGS Ref. No. ST 65 SE/12. Height above OD 384 ft. Pumping shaft to Bottom Little Vein at 834 ft 7 in, section continued in Winding shaft to 1454 ft 5 in.

Raised above surface 15 ft; Quaternary deposits to 19 ft 3 in; Blue Lias to 39 ft 6 in; White Lias to 60 ft; Rhaetic to 72 ft 6 in; Tea Green Marl to 82 ft 5 in; Triassic (red marl and sandstone) to 270 ft 7 in; Dolomitic Conglomerate to 318 ft 1 in; Supra-Pennant Measures (Radstock Formation to 921 ft 8 in; Barren Red Formation into Farrington Formation, disturbed at base, to 1608 ft 4 in.

168 Midsomer Norton, Norton Hill Colliery New Pit (Downcast) Geological one-inch sheet 281. NGR 6690 5409. BGS Ref. No. ST 65 SE/10. Height above OD 400 ft. Published, Greenwell and McMurtrie, 1864, p.27.

Made ground to 21 ft; Tea Green Marl to 24 ft 7 in; Triassic (red marl) to 103 ft 1 in; Dolomitic Conglomerate to 198 ft 10 in; Supra-Pennant Measures (Radstock, Barren Red and Farrington formations) to 1503 ft 7 in.

169 Midsomer Norton, Norton Hill Colliery Old Pit (Upcast) Geological one-inch sheet 281. NGR 6705 5370. BGS Ref. No. ST 65 SE/11. Height above OD 400 ft. Sunk 1843.

Rise above surface 10 ft 6 in; White Lias to 24 ft; Rhaetic to 53 ft; Tea Green Marl to 58 ft; Triassic (sandstone) to 158 ft 6 in; Dolomitic Conglomerate to 214 ft 6 in; Supra Pennant Measures (Radstock, Barren Red and Farrington formations) to 1128 ft 7 in.

170 Midsomer Norton, Old Mills Colliery Geological one-inch sheet 280. NGR 6528 5516. BGS Ref. No. ST 65 NE/16. Height above OD 380 ft.

Triassic (sandstone) to 39 ft; Dolomitic Conglomerate to 42 ft; Supra-Pennant Measures (Barren Red and Farrington formations) to 1081 ft.

171 Midsomer Norton, Old Mills Colliery, Underground Borehole Geological one-inch sheet 280. NGR 6516 5516. BGS Ref. No. ST 65 NE/15. Depth below OD 822 ft. Bored by Craelius Co. in 1953.

Upper Coal Measures (Farrington and Mangotsfield formations) to 499 ft 5 in.

172 Midsomer Norton, Old Welton Colliery, North Shaft Geological one-inch sheet 281. NGR 6754 5486. BGS Ref. No. ST 65 SE/5. Height above OD 264.63 ft. Published, Greenwell and McMurtrie, 1864, p.26. Abandoned Mine Plan No. 3617.

Triassic (sandstone) to 62 ft; Dolomitic Conglomerate to 79 ft 8 in; Supra-Pennant Measures (Radstock Formation to 488 ft 9 in; Barren Red and Farrington formations to 1650 ft).

173 Midsomer Norton, Welton Hill Colliery Geological one-inch sheet 281. NGR 6662 5524. BGS Ref. No. ST 65 NE/18. Height above OD 368.58 ft. Published, Prestwich, 1871, p.57.

Triassic (red marl) to 106 ft; Dolomitic Conglomerate to 115 ft; Supra-Pennant Measures (Radstock Formation) to 608 ft.

174 Paulton Engine Pit Geological one-inch sheet 281. NGR 6580 5748. BGS Ref. No. ST 65 NE/6. Height above OD about 260 ft. Published, Greenwell and McMurtrie, 1864, p.26.

Triassic to 90 ft; Supra-Pennant Measures (Radstock Formation) to 609 ft 5 in.

175 Paulton Hill Pit Geological one-inch sheet 281. NGR 6588 5614. BGS Ref. No. ST 65 NE/14. Published, Conybeare and Phillips, 1822, p.429.

Inferior Oolite to 18 ft; Lower Lias Clay to 138 ft; Blue Lias to 144 ft; White Lias to 157 ft 6 in; Rhaetic to 172 ft 6 in; Triassic (red marl) to 304 ft 6 in; Dolomitic Conglomerate to 310 ft 6 in; Supra-Pennant Measures (Radstock Formation) to 672 ft 8 in.

176 Priston Colliery No. 1 Shaft Geological one-inch sheet 281. NGR 6920 5942. BGS Ref. No. ST 65 NE/12. Sunk 1914.

Made ground to 12 ft; Inferior Oolite to 59 ft; Lower Lias Clay to 210 ft; Blue Lias to 221 ft 4 in; White Lias to 234 ft; Rhaetic to 249 ft 5 in; Tea Green Marl to 273 ft; Triassic (red marl and sandstone) to 399 ft; **fault**; Supra-Pennant Measures (Radstock Formation) to 702 ft.

177 Radford Colliery Geological one-inch sheet 281. NGR 6662 5770. BGS Ref. No. ST 65 NE/7. Height above OD about 310 ft. Published, Prestwich, 1871, p.57. Abandoned Mine Plan No. 329.

Triassic to 80 ft 3 in; Supra-Pennant Measures (Publow and Radstock formations) to 1065 ft.

178 Radstock, Huish Colliery Geological one-inch sheet 281. NGR 6974 5405. BGS Ref. No. ST 65 SE/13. Height above OD 400 Ft. Published, Greenwell and Mcmurtrie, 1864, p.25.

Inferior Oolite to 6 ft 6 in; Lower Lias, Rhaetic and Tea Green Marl to 90 ft 6 in; Triassic (red marl) to 264 ft 6 in; Triassic (Dolomite Conglomerate) to 296 ft 6 in; Supra-Pennant Measures (Radstock Formation) to 505 ft 6 in.

179 Radstock, Ludlows Pit Geological one-inch sheet 281. NGR 6914 5476. BGS Ref. No. ST 65 SE/7. Height above OD 247.72 ft. Published, Greenwell and McMurtrie, 1864, p.24.

Walling to 12 ft 6 in; Triassic (mainly red marl) to 91 ft 6 in; Dolomitic Conglomerate to 118 ft; Supra-Pennant Measures (Radstock Formation to 740 ft 8 in; Barren Red and Farrington formations to 1723 ft).

180 Radstock, Middle Pit Geological one-inch sheet 281. NGR 6874 5511. BGS Ref. No. ST 65 NE/23. Height above OD 246.64 ft. Published, Greenwell and Mcmurtrie, 1864, pp.24–25.

Triassic (red marl) to 90 ft; Dolomitic Conglomerate to 117 ft; Supra-Pennant Measures (Radstock Formation, including repetition by 'Radstock Slide', to 875 ft 9 in; Barren Red Formation to 1373 ft 9 in; Farrington Formation, in shaft to 1782 ft 4 in; in Borehole to 1850 ft 10 in).

181 Radstock, Old Pit Geological one-inch sheet 281. NGR 6852 5546. BGS Ref. No. ST 65 NE/22. Height above OD 276.2 ft.

Triassic (red marl) to 102 ft; Dolomitic Conglomerate to 129 ft; Supra-Pennant Measures (Radstock Formation) to 942 ft.

182 Radstock, Wellsway Colliery Geological one-inch sheet 281. NGR 6808 5406. BGS Ref. No. ST 65 SE/6. Height above OD 349.86 ft. Published, Greenwell and McMurtrie, 1864, p.25

Lower Lias, Rhaetic and Tea Green Marl to 36 ft; Triassic (mainly red marl) to 189 ft; Dolomitic Conglomerate to 222 ft; Supra-Pennant Measures (Radstock Formation) to 754 ft 6 in.

183 Ston Easton No. 1 Borehole* Geological one-inch sheet 280. NGR 6225 5174. BGS Ref. No. ST 65 SW/4. Height above OD 671.5 ft. Bored for National Coal Board by Craelius Co. in 1952–53.

Lower Lias and Rhaetic to 18 ft 10 in; Tea Green Marl to 33 ft; Triassic (red and green marl) to 113 ft 5 in; Dolomitic Conglomerate to 259 ft 2 in; ?Middle and Lower Coal Measures (highly disturbed) to 1502 ft 5 in.

184 Ston Easton No. 2 Borehole* Geological one-inch sheet 280. NGR 6211 5158. BGS Ref. No. ST 65 SW/5. Height above OD 671.8 ft. Bored for National Coal Board by Craelius Co. in 1953.

Lower Lias and Rhaetic to 21 ft 7 in; Tea Green Marl to 25 ft 3 in; Triassic (red marl) to 111 ft 6 in; Dolomitic Conglomerate to 237 ft 10 in; Lower Coal Measures (disturbed)to 937 ft.

185 **Stratton on the Fosse, New Rock Colliery** Geological one-inch sheet 280. NGR 6479 5057. BGS Ref. No. ST 65 SW/6. Height above OD 672 ft. Published, Greenwell and Mcmurtrie, 1864, p.28.

Made Ground to 12 ft; Dolomitic Conglomerate to 152 ft; Pennant Measures (Downend Formation) to 1151 ft 9 in.

186 **Temple Cloud Pit** Geological one-inch sheet 280. NGR 6211 5816. BGS Ref. No. ST 65 NW/6. Height above OD 428 ft.

Pennant Measures to 558 ft 10 in.

187 **Timsbury, Conygre Colliery Lower Pit** Geological one-inch sheet 281. NGR 6738 5825. BGS Ref. No. ST 65 NE/4. Height above OD about 400 ft. Published, Greenwell and McMurtrie, 1864, p.27.

Triassic to 168 ft; Supra-Pennant Measures (Publow and Radstock formations to 1182 ft.)

188 **Timsbury, Conygre Colliery Upper Pit** Geological one-inch sheet 281. NGR 6671 5890. BGS Ref. No. ST 65 NE/3. Height above OD about 490 ft. Published, Greenwell and McMurtrie, 1864, p.27.

Triassic (red marl) to 180 ft; Dolomitic Conglomerate to 204 ft; Supra-Pennant Measures (Publow Formation to 384 ft; Radstock Formation to 1174 ft).

189 **Timsbury, Hayswood Colliery** Geological one-inch sheet 281. NGR 6576 5909. BGS Ref. No. ST 65 NE/1. Height above OD 525 ft.

Lower Lias and Triassic to 270 ft; Supra-Pennant Measures (Publow, Radstock Barren Red and Farrington formations) to 1143 ft.

190 **Timsbury, Old Grove Colliery** Geological one-inch sheet 281. NGR 6587 5842. BGS Ref. No. ST 65 NE/2. Height above OD about 368 ft. Published, Greenwell and McMurtrie, 1864, pp.21–24.

Triassic (red marl) to 12 ft; Supra-Pennant Measures (Radstock Formation to 534 ft; Barren Red Formation to 1051 ft; Farrington Formation to 1363 ft.)

191 **Tyning Colliery** Geological one-inch sheet 281. NGR 6960 5520. BGS Ref. No. ST 65 NE/24. Height above OD 320 ft. Published, Greenwell and McMurtrie, 1864, p.24.

Made Ground to 17 ft; Rhaetic to 24 ft 10 in; Tea Green Marl to 35 ft; Triassic (red marl and sandstone) to 202 ft; Dolomitic Conglomerate to 221 ft; Supra-Pennant Measures (Radstock Formation) to 995 ft 9 in.

192 **Withy Mills Colliery** Geological one-inch sheet 281. NGR 6620 5794. BGS Ref. No. ST 65 NE/5. Height above OD about 344 ft. Published, Greenwell and McMurtrie, 1864, pp.26–27.

Triassic to 96 ft; Supra-Pennant Measures (Publow Formation to 180 ft; Radstock Formation to 816 ft).

193 **Writhlington Upper Colliery** Geological one-inch sheet 281. NGR 6989 5500. BGS Ref. No. ST 65 NE/25. Height above OD about 260 ft. Published, Greenwell and McMurtrie, 1864, p.25.

Made ground to 15 ft; Triassic (red marl and Dolomitic Conglomerate) to 148 ft 4 in; Supra-Pennant Measures (Radstock Formation to 694 ft 5 in; Barren Red Formation to 784 ft 8 in).

1:25 000 sheet 75 (Western Part)

194 **Braysdown Colliery** Geological one-inch sheet 281. NGR 7038 5600. BGS Ref. No. ST 75 NW/3. Height above OD 472.75 ft. Published, Greenwell and McMurtrie, 1864, p.26.

Inferior Oolite to 32 ft; Lower Lias to 119 ft; Rhaetic to 148 ft; Triassic (red marl and sandstone) to 306 ft; Dolomitic Conglomerate to 324 ft; Supra-Pennant Measures (Radstock Formation to 717 ft 9 in; Barren Red Formation to 1479 ft 3 in; Farrington Formation to 1863 ft 6 in).

195 **Dunkerton Colliery No. 1 Borehole** Geological one-inch sheet 281. NGR 7001 5819. BGS Ref. No. ST 75 NW/2. Height above OD 400 ft.

Supra-Pennant Measures (Radstock Formation to 255 ft 9 in; Barren Red Formation and ?Farrington Formation to 1156 ft 10 in). Cut by five thrusts causing 216 ft repetition.

196 **Dunkerton Colliery No. 2 Borehole** Geological one-inch sheet 281. NGR 7024 5964. BGS Ref. No. ST 75 NW/1. Height above OD 405 ft.

Fullers Earth to 7 ft; Inferior Oolite to 56 ft; Lower Lias to 302 ft; Triassic to 533 ft 3 in; Supra-Pennant Measures (Radstock Formation to 624 ft; Barren Red Formation to 801 ft 5 in).

197 **Foxcote Colliery** Geological one-inch sheet 281. NGR 7108 5518. BGS Ref. No. ST 75 NW/6. Height above OD 365.75 ft. Published, Greenwell and McMurtrie, 1864, p.25.

Quaternary deposits and Inferior Oolite to 21 ft; Lower Lias and Rhaetic to 93 ft; Tea Green Marl and Triassic (red marl and sandstone) to 277 ft; Dolomitic Conglomerate to 284 ft 9 in; Supra-Pennant Measures (Radstock Formation to 378 ft 5 in; Barren Red Formation to 1057 ft 6 in; Farrington Formation to 1418 ft 9 in.

198 **Hemington Borehole*** Geological one-inch sheet 281. NGR 7247 5296. BGS Ref. No. ST 75 SW/1. Height above OD about 450 ft. Bored 1909. Published, Cantrill and Pringle, 1914, pp.98–101; Richardson,1928, pp.80–81.

Quaternary deposits and Fuller's Earth to 87 ft; Inferior Oolite to 126 ft; Lower Lias Clay to 151 ft; Blue Lias to 155 ft; White Lias to 174 ft; Rhaetic and Tea Marl Green to 197 ft; Triassic (red marl and sandstone) to 336 ft; Supra-Pennant Measures (Radstock, Barren Red and Farrington formations) to 1200 ft. Coal Measures strongly faulted.

199 **Mells Colliery** Geological one-inch sheet 281. NGR 7114 5008. BGS Ref. No. ST 75 SW/6. Height above OD 437.34 ft; Published, Moore, 1867,p.481; Geological Survey Vertical Sections, sheet 50, No. 24 (1873).

Inferior Oolite to 6 ft; Lower Lias to 27 ft 8 in; Middle Coal Measures to 341 ft 2 in.

200 **Woodborough Colliery** Geological one-inch sheet 281. NGR 7025 5556. BGS Ref. No. ST 75 NW/26. Height above OD about 365 ft.

Rhaetic to 10 ft; Triassic (Tea Marl Green and red marl) to 232 ft; Dolomitic Conglomerate to 243 ft; Supra-Pennant Measures (Radstock Formation) to 495 ft.

201 **Writhlington Lower Colliery** Geological one-inch sheet 281. NGR 7052 5532. BGS Ref. No. ST 75 NW/4. Height above OD 229.43 ft. Published Greenwell and McMurtrie, 1864, p.25.

Triassic (red marl and Dolomitic Conglomerate) to 78 ft 2 in; Supra-Pennant Measures (Radstock Formation to 390 ft; Barren Red Formation to 1129 ft; Farrington Formation to 1473 ft 6 in.)

1:25 000 sheet 64 (Northern Part)

202 **Coleford, Mackintosh Colliery** Geological one-inch sheet 281. NGR 6914 4972. BGS Ref. No. ST 64 NE/1. Height above OD 490.43 ft.

Middle and ?Lower Coal Measures to 1615 ft.

203 Coleford, Newbury Colliery Geological one-inch sheet 281. NGR 6957 4977. BGS Ref. No. ST 64 NE/2. Height above OD 457 ft. Sunk by W.Coulson, Ltd
    Middle Coal Measures to 737 ft.

204 Stratton on the Fosse, Mendip (Strap) Pit Geological one-inch sheet 280. NGR 6482 4957. BGS Ref. No. ST 64 NW/2. Height above OD 725 ft. Sunk 1862–76. Published, Prestwich, 1871, p.58.
    Dolomitic Conglomerate to 72 ft; Pennant Measures, Middle and Lower Coal Measures to 1834 ft.

205 Stratton on the Fosse, Old Rock Pit Geological one-inch sheet 280. NGR 6492 4971 BGS Ref. No. ST 64 NW/1. Height above OD about 710 ft. Published, Geological Survey Vertical Sections, sheet 50, No. 23 (1873).
    Dolomitic Conglomerate to 94 ft; Pennant Measures and Middle Coal Measures to 718 ft.

APPENDIX 2

Glossary of local mining terms used in the Bristol and Somerset Coalfield

Bastard coal   Layers of detrital coal (often with coal pebbles), commonly present in sandy shale and sandstone. Such seams are lenticular and of limited extent.
Batch   A mound or heap of spoil. A large mine tip is a 'Wark batch'.
Batt (or Bass)   A layer of dark carbonaceous shale in coal.
Bell-mould   Stem of a fossil tree standing upright in the measures above a seam of coal. When the underlying coal is extracted the unsupported sandstone cast is liable to fall suddenly into the space below.
Binching (benching)   The band of dirt below or in the seam which is taken out first in cutting the coal, or in which shot holes are drilled.
Bind   Mudstone or shale.
Blacks   Dark grey carbonaceous mudstone which often occurs between the coal seam and the fireclay.
Black Chalk   Soft black carbonaceous mudstone producing a black streak when rubbed on brick or stone.
Branch   A road cut across the measures: where horizontal it is termed a 'level branch'; where inclined to the deep it is known as a 'dip branch' or 'branch dipple'; and where inclined upwards is called a 'rise branch'.
Brasses   Iron pyrites.
Cathead   A nodule of ironstone or small spherical mass of sandstone.
Cleaves   The soft, oxidised reddened and weathered Coal Measures beneath the Red Ground (Trias). Cleaves usually extend to depths of 15 to 30 ft.
Clift (Somerset)   Hard mudstone.
Clod   A band of soft shale or mudstone above the seam, which falls after the coal is removed.
Clunch   Almost structureless hard mudstone.
Coal smut   Soft sooty black clay marking the crop of a coal seam.
Cockle   Ironstone nodule, e.g. 'cockly clift' = mudstone with ironstone nodules.
Cockroach holing   Dark grey-black holing dirt; cf. binching.
Corngrit   Three thin bands of ironshot limestone (calcarenite) which form the base of the Blue Lias in the Radstock district.
Course   A seam of coal. The word 'course' is also used in 'half-course incline', implying that the road is directed at an angle of about 45° to the strike of the seams.
Crumples   See Rumples.
Dipple   A down dip road in coal.
Drift (drift-road)   A main road driven level-course in coal.
Dung   Loose stone or mine rubbish. The Dungy Drift seam has intercalated shale or dirt layers which were encountered in the drifts.
Duns   Mudstone. (The Bristol equivalent to 'clift'.)
Firestone   A hard sandstone, generally quartzose or sub-quartzitic.
Glange   Streaky coal; often redeposited carbonaceous matter.
Greys   Sandstone. 'Glangey greys' is sandstone with coal streaks.
Gug (or Gugg)   An incline. A balance gug is (usually) a double-tracked incline or parallel incline in which the loaded tubs on one side are balanced by returning empties on the other.
Jingles (or Jingleboys)   Hard brittle shale which makes a clinking or jingling sound on falling.
Lap   A low-angle reversed fault (see Overlap).
Lism   A thin band of very dark mudstone.
Listy   Containing thin bands of fusain.
Mamstone   Red-grey sandstone in the Trias. Probably a corruption of Malmstone.
Millstone (or Millgrit)   Dolomitic Conglomerate, the basal Triassic conglomerate.
Overlap (or Overlap Fault)   A thrust fault.
Pan   Fireclay, seatearth.
Peau (or Peacock)   Iridescent colouring on pyrite (hence Peau Vein).
Pennant   Hard, blue-grey sandstone or subgreywacke.
Rashings   Carbonaceous shale generally associated with coal. Usually very soft and friable, occasionally with polished slickensided or mirror-like surfaces.
Red ground   The Triassic rocks resting unconformably on the Coal Measures.
Ricewood   Brushwood packed above the roof timbers where the roof is soft or crumbling shale.
Rumples   A bed of soft limestone in the Cotham Beds (Rhaetic), to which wells in the Radstock district were sunk.
Shab   See Clod.
Shell of coal   A thin band of coal.
Slide   A flat or low-angle fault, causing omission (by lag faulting) or repetition by overlap faulting of the coal seam.
Swamp   A syncline or downfolded area in which coal is worked. The term makes no specific reference to the presence or absence of water.
Tanger   A small post pushed over the crowntree and projecting beyond into the coal face.
Topple   A rise road in coal.
Vein   Coal seam. Many local coal seams are worked in vertical or highly folded measures and do not have the flat well-bedded appearance of coal seams in less disturbed Coal Measures.
Wark   Mine rubbish or dirt. Also, as an adjective, warky or warkey, e.g. wark—batch or heap of mine waste, and Warkey Course, the name of a coal seam with numerous thick dirt partings.
Working stone   See Clod.
Worm band   Ironstone or sideritic mudstone with worm- or grub-like structures. Characteristically (but not uniquely) developed above the Gillers Inn Vein of Kingswood.

APPENDIX 3

Biostratigraphy of the Black Rock Limestone in the Portway Tunnel, Bristol
M. Mitchell

HISTORY OF RESEARCH

The Black Rock Limestone faunas of the Avon Gorge at Bristol were divided by Vaughan (1905, pp.190–193, pls. 27,28) into the *Zaphrentis* (Z) Zone (with Horizon β at its base), Horizon γ, and the lower part of the *Syringothyris* (C) Zone in upward sequence, with Horizon γ marked by 'the co-occurrence of *Zaphrentis* and *Caninia* ('*cylindrica*') in remarkable abundance.' The Burrington Combe sequence in the Mendips was similarly divided by Reynolds and Vaughan (1911, pp.348–349, 363–368) into the *Zaphrentis* Zone, again with Horizon β at its base, and the Lower *Syringothyris* or *Caninia* (C) Zone with Horizon γ at its base. The important difference from Bristol was that at Burrington, *Caninia patula* (now referred to *Caninophyllum*) was noted (p.376) as 'diagnostic of γ' and *C. cylindrica* (now referred to *Siphonophyllia*) was recorded only from the C beds above Horizon γ. *C. patulum* was not recognised in the original Bristol paper, making correlation between the two zonal schemes difficult, but a footnote in Dixon and Vaughan (1912, p.545) recorded that '*Caninia cylindrica*, as employed in the 'Bristol Paper,' covered all the *Caninias*... these would now be differentiated into *C. cornucopiae*, *C. patula* and *C. cylindrica*'. This footnote explains the important differences between the coral terminology of Vaughan (1905) and of Reynolds and Vaughan (1911). During this short period considerable progress with the revision of coral taxonomy was made by Carruthers (1908) and Salée (1910).

Siphonophyllia cylindrica was not recorded from the Portway Tunnel section, and a check of museum material failed to trace any specimens of this species from the immediate area of Bristol. However, specimens from Burrington are preserved, and it seems likely that the 1905 Bristol records of *C. cylindrica* refer to *Caninophyllum patulum*.

The Burrington Combe section was studied in detail during the resurvey of the Wells (280) Sheet (Mitchell and Green, pp.177–197, in Green and Welch, 1965), and the range of the corals and brachiopods tabulated and discussed. The Black Rock Limestone assemblages were divided (p.182, table 1) into Lower, Middle and Upper faunas, units for which the Assemblage Biozone names *Zaphrentites delanouei*, *Caninophyllum patulum* and *Siphonophyllia cylindrica* respectively have been proposed by Ramsbottom and Mitchell (1980, p.62). Horizon β was not recognised by Mitchell and Green.

The construction in 1962–1963 of the Northern Foul Water or Portway Tunnel between Sea Mills and Gully Quarry in the northern end of the Avon Gorge (Figure 11) provided an unrivalled opportunity to restudy the lower part of the classic Avon section. The faunas collected from the Black Rock Limestone Group are listed in Figure 47, together with those from the Lower Limestone Shale Group for completeness. The nomenclature of the corals has been brought up to date to allow comparison with recent publications, but the names of the brachiopods, which are less stratigraphically useful, remain as originally identified so that they can be compared with the Burrington unrevised brachiopod faunas.

Comparison of the sequence of faunas in the Black Rock Limestone at Burrington with that of the Portway Tunnel was outlined by Mitchell (1971, p.97; 1972, p.159, fig.1) and showed that the upper part of the more complete Burrington sequence (from a more southerly, down shelf position on the southern British shelf; Lees, 1982) was not present in the Avon Gorge. This provided the first macrofossil evidence for a major faunal nonsequence in the Avon section.

Independent conodont and foraminiferal research (Rhodes and Austin, 1971, p.338; Austin, Conil and Rhodes, 1973, p.175) also suggested a substantial nonsequence at this horizon, below the Gully (=Caninia) Oolite. Three additional non-sequences were recognised in the Avon Gorge by Ramsbottom (1973, p.595, fig.8), who was able to show that the Avon was an unsatisfactory type section, paving the way for the proposal by George et al. (1976, p.6) of six regional stratotyped stages for the correlation of the British Dinantian.

Mitchell (1981, p.582, fig.1, column E) included the Bristol Portway Tunnel section in his review of the Black Rock Limestone faunas from the Bristol and South Wales area, and the present account gives the detailed fossil ranges for this Bristol section (Figure 47).

PORTWAY TUNNEL CORAL FAUNAS

The *Zaphrentites delanouei* Biozone (approximately 5350 to 6090 ft from Sea Mills Shaft) contains the typical zaphrentoid corals *Fasciculophyllum omaliusi* and *Zaphrentites delanouei*, together with *Sychnoelasma clevedonensis* (=*Zaphrentis konincki* forma α of Carruthers, 1908, p.69, pl.5, figs.1,1a) which is restricted to the upper beds of the biozone, as it is in all sections that have been studied (Mitchell, 1981, p.580).

The base of the *Caninophyllum patulum* Biozone (approximately 6090 to 6310 ft from Sea Mills Shaft) is marked by the first appearance of *Caninia cornucopiae*, *Fasciculophyllum densum* and *Sychnoelasma konincki* and of *Caninophyllum patulum greeni* (=closely septate form of Reynolds and Vaughan, 1911, pl.30, figs.6a,6b), a subspecies that is diagnostic of the lower part of this biozone. The highest record of a significant coral is that of *C. patulum greeni* (6306) and no fossils were recovered from the top beds of the Black Rock Limestone which are dolomitised.

The youngest Black Rock assemblages present in the Portway Tunnel therefore indicate the lower part of the *C. patulum* Biozone. The upper part of this biozone and the whole of the succeeding *Siphonophyllia cylindrica* Biozone are cut out by a nonsequence at the base of the Gully Oolite, although some of this missing sequence may be represented by the unfossiliferous dolomites.

The Tournaisian-Viséan boundary is currently correlated approximately with the base of the *S. cylindrica* Biozone (Mitchell and others, 1986), so the sub-Gully Oolite nonsequence at Bristol cuts out the top beds of the Tournaisian and the whole of the Viséan parts of the Black Rock Limestone sequence at Burrington, a total of about 400 ft.

Figure 47
Range diagram showing the vertical distribution of the fossils of the Lower Limestone Shale and Black Rock Groups in the Foul Water Tunnel of the Avon Gorge between Sneyd Park and the Gully

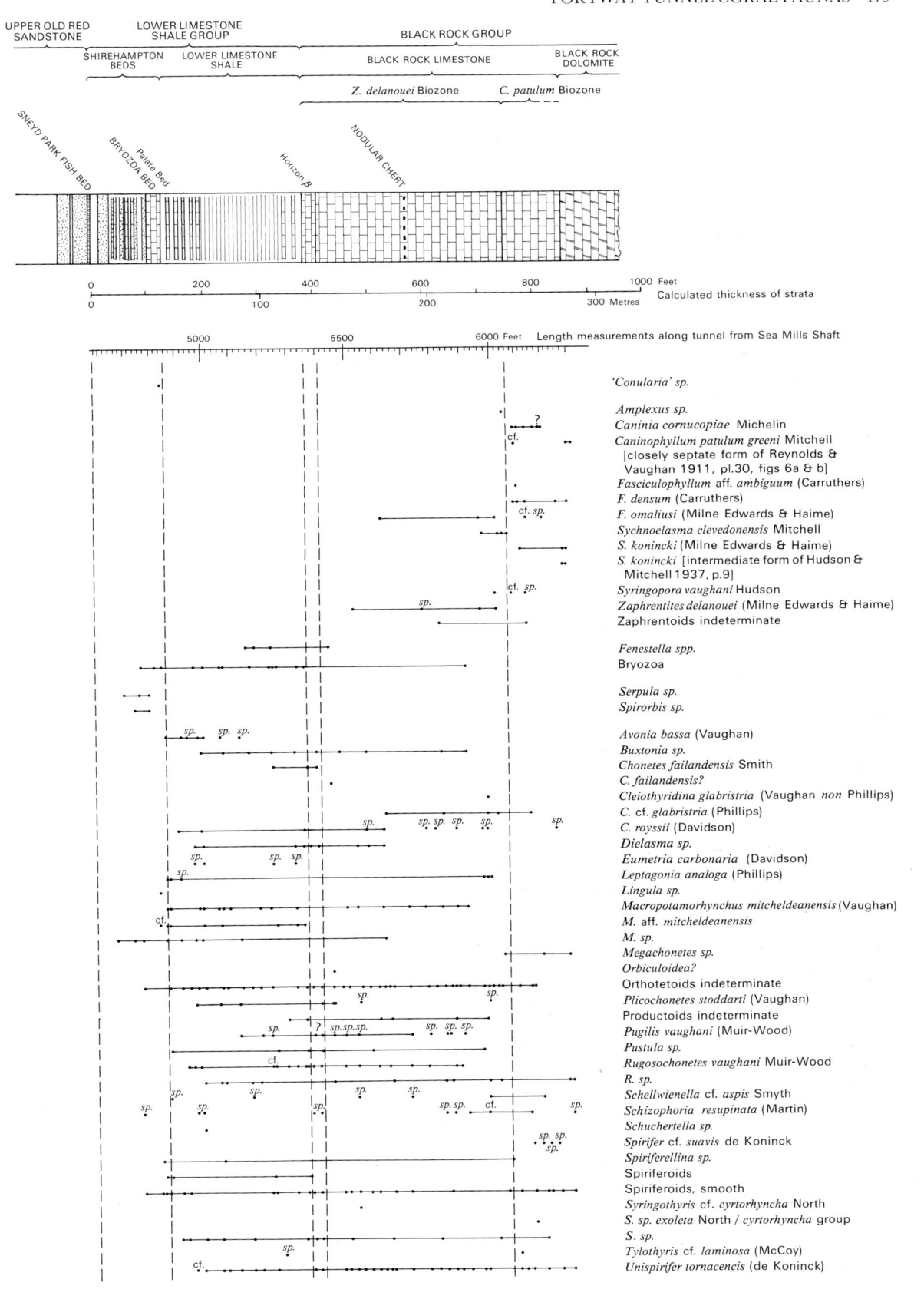

'Conularia' sp.

Amplexus sp.
Caninia cornucopiae Michelin
Caninophyllum patulum greeni Mitchell [closely septate form of Reynolds & Vaughan 1911, pl.30, figs 6a & b]
Fasciculophyllum aff. ambiguum (Carruthers)
F. densum (Carruthers)
F. omaliusi (Milne Edwards & Haime)
Sychnoelasma clevedonensis Mitchell
S. konincki (Milne Edwards & Haime)
S. konincki [intermediate form of Hudson & Mitchell 1937, p.9]
Syringopora vaughani Hudson
Zaphrentites delanouei (Milne Edwards & Haime)
Zaphrentoids indeterminate

Fenestella spp.
Bryozoa

Serpula sp.
Spirorbis sp.

Avonia bassa (Vaughan)
Buxtonia sp.
Chonetes failandensis Smith
C. failandensis?
Cleiothyridina glabristria (Vaughan non Phillips)
C. cf. glabristria (Phillips)
C. royssii (Davidson)
Dielasma sp.
Eumetria carbonaria (Davidson)
Leptagonia analoga (Phillips)
Lingula sp.
Macropotamorhynchus mitcheldeanensis (Vaughan)
M. aff. mitcheldeanensis
M. sp.
Megachonetes sp.
Orbiculoidea?
Orthotetoids indeterminate
Plicochonetes stoddarti (Vaughan)
Productoids indeterminate
Pugilis vaughani (Muir-Wood)
Pustula sp.
Rugosochonetes vaughani Muir-Wood
R. sp.
Schellwienella cf. aspis Smyth
Schizophoria resupinata (Martin)
Schuchertella sp.
Spirifer cf. suavis de Koninck
Spiriferellina sp.
Spiriferoids
Spiriferoids, smooth
Syringothyris cf. cyrtorhyncha North
S. sp. exoleta North / cyrtorhyncha group
S. sp.
Tylothyris cf. laminosa (McCoy)
Unispirifer tornacencis (de Koninck)

REFERENCES

Most of the references listed below are held in the Library of the British Geological Survey at Keyworth, Nottingham. Copies of the references can be purchased subject to the current copyright legislation.

ALABASTER, C. and WILSON, D. 1975. Volcanic clasts in the basal Inferior Oolite of East Mendip. *Proceedings of the Bristol Naturalists' Society*, Vol. 35, 73–75.

APSIMON, A. M. and BOON, G. C. 1960. An exposure of the Bristol Avon gravels at Shirehampton, near Bristol. *Proceedings of the University of Bristol Spelaeological Society*, Vol. 9, 22–29.

— DONOVAN, D. T. and TAYLOR, H. 1961. The stratigraphy and archaeology of the late-glacial and post-glacial deposits at Brean Down, Somerset. *Proceedings of the University of Bristol Spelaeological Society*, Vol. 9, 69–136.

AGASSIZ, L. 1833–1843. *Recherches sur les poissons fossiles*. Tome III, 2me Partie. Des dents de Placoides, 73–359.

ANDREWS, J. T., GILBERTSON, D. and HAWKINS, A. B. 1984. The Pleistocene succession of the Severn Estuary: a revised model based upon amino acid racemization studies. *Journal of the Geological Society of London*, Vol. 141, 967–974.

ANSTIE, J. 1873. *The coal fields of Gloucestershire and Somersetshire, and their resources.* (London.)

ARKELL, W. J. and TOMKEIEFF, S. I. 1953. *English rock terms chiefly as used by miners and quarrymen.* 139 pp. (Oxford: University Press.)

AUSTIN, R. L., CONIL, R. and RHODES, F. H. T. 1973. Recognition of the Tournaisian–Viséan boundary in North America and Britain. *Extrait des Annales de la Société Géologique de Belgique*, Tome 96, 165–188.

BAKER, B. A. 1902. Celestine deposits of the Bristol district. *Proceedings of the Bristol Naturalists' Society*, New Series, Vol. 9, 161–165.

BATHER, F. A. 1926. *William Smith. The father of English geology.* 12 pp. (Bath: Royal Literary and Scientific Institution.)

BATTEN, R. L. 1966. The Lower Carboniferous Gastropod Fauna from the Hotwells Limestone of Compton Martin, Somerset. *Monograph Palaeontolographical Society.* 109pp.

BISHOP, M. J. 1974. A preliminary report on the Middle Pleistocene mammal bearing deposits of Westbury-Sub-Mendip, Somerset. *Proceedings of the University of Bristol Spelaeological Society*, Vol. 13, 301–318.

BOLTON, H. 1907. On a marine fauna in the Basement-Beds of the Bristol Coalfield. *Quarterly Journal of the Geological Society of London*, Vol. 63, 445–469.

— 1911. Faunal horizons in the Bristol Coalfield. *Quarterly Journal of the Geological Society of London*, Vol. 67, 316–341.

BRAINE, A. 1891. *The history of Kingswood Forest including all the ancient manors and villages in the neighbourhood.* (London and Bristol.)

BROWN, J. C. 1957. Palaeolithic and other implements from the Shirehampton district. *Proceedings of the University of Bristol Spelaeological Society*, Vol. 8, 43–44.

BUCHANAN, R. A. and COSSONS, N. 1969. *The industrial archaeology of the Bristol region.* (Newton Abbot: David and Charles.)

BUCKLAND, W. 1821. Description of the Quartz Rock of the Lickey Hill in Worcestershire, and of the strata immediately surrounding it. *Transactions of the Geological Society of London*, Ser. 1, Vol. 5, 506–544.

— and CONYBEARE, W. D. 1824. Observations on the south-western coal district of England. *Transactions of the Geological Society of London*, Ser. 2, Vol. 1, 210–316.

BUCKMAN, S. S. 1895. The Bajocian of the Mid-Cotteswolds. *Quarterly Journal of the Geological Society of London*, Vol. 51, 388–462.

— 1893. The Bajocian of the Sherborne district. *Quarterly Journal of the Geological Society of London*, Vol. 49, 479–522.

— and WILSON, E. 1896. Dundry Hill: its upper portion, or the beds marked as Inferior Oolite (g^5) in the maps of the Geological Survey. *Quarterly Journal of the Geological Society of London*, Vol. 52, 669–720.

BUSH, G. E. 1928. The Avonian succession at Clevedon—a description of the coast section. *Proceedings of the Bristol Naturalists' Society*, 4th Series, Vol. 6, 392–399.

BUTLER, M. 1973. Lower Carboniferous conodont faunas from the eastern Mendips, England. *Palaeontology*, Vol. 16, 477–517.

CALVER, M. A. 1969. Westphalian of Britain. *Compte Rendu du 6me Congrès International de Stratigraphie et Géologie du Carbonifère, Sheffield, 1967*, Vol. 1, 223–254.

CANTRILL, T. C. and PRINGLE, J. 1914. On a boring for coal at Hemington, Somerset. *Summary of Progress of the Geological Survey of Great Britain for 1913*, 98–101.

CANTRILL, T. C. and SMITH, B. 1919. On a boring for coal at Winterbourne, Gloucestershire. *Summary of Progress of the Geological Survey of Great Britain for 1918*, App. 4, 53–57.

— SHERLOCK, R. L. and DEWEY, H. 1919. Iron ores: sundry unbedded ores of Durham, east Cumberland, North Wales, Derbyshire, the Isle of Man, Bristol district and Somerset, Devon and Cornwall. *Special Report on the Mineral Resources of Great Britain, Memoir of the Geological Survey of Great Britain*, Vol. 9, 87 pp. (London: HMSO.)

CARRUTHERS, R. G. 1908. A revision of some Carboniferous corals. *Geological Magazine*, Vol. 45, 20–31, 63–74, 158–171.

CATCOTT, G. S. 1792. *A descriptive account of a descent made into Penpark-Hole, Westbury-upon-Trim, Gloucester, in 1775.* (Bristol.)

CATCOTT, REV. A. 1768. *A treatise on the Deluge* (2nd edition). Part III, Section II, Proofs of the Universality of the Flood. 3, from caves, natural grottoes, swallet holes, etc., 335–356.

CAVE, R. 1977. Geology of the Malmesbury district. *Memoir of the Geological Survey of Great Britain*, Sheet 251 (England and Wales). 343 pp.

— and WHITE, D. E. 1971. The exposures of Ludlow rocks and associated beds at Tites Point near Newham, Gloucestershire. *Geological Journal*, Vol. 7, 239–254.

CHANCE, SIR HUGH. 1968. The Nailsea glassworks. Studies in glass history and design: papers read to Committee B Sessions of the VIIIth International Congress on Glass, London July, 1968. 33–39.

CHANDLER, R. J., KELLAWAY, G. A., SKEMPTON, A. W. and WYATT, R. J. 1976. Valley slope sections in Jurassic strata near Bath, Somerset. *Philosophical Transactions of the Royal Society of London*, A, Vol. 283, 527–556.

COLLIE, N. 1879. On the celestine and baryto-celestine of Clifton. *Proceedings of the Bristol Naturalists' Society*, New Series, Vol. 2, 292–300.

COLLINS, M. 1987. Sediment transport in the Bristol Channel: a review. *Proceedings Geologists' Association*, Vol. 98, Pt. 4, 367–383.

CONYBEARE, W. D. and PHILLIPS, W. 1822. *Outlines of the geology of England and Wales; and comparative views of the structure of foreign countries.* 470 pp., map, 2 plates. (London: W. Phillips.)

COPE, J. C. W., DUFF, K. L., PARSONS, C. F., TORRENS, H. S., WIMBLEDON, W. A. and WRIGHT, J. K. 1980. A correlation of Jurassic rocks in the British Isles. Part 2. Middle and Upper Jurassic. *Special Report of the Geological Society of London*, No. 15. 109 pp.

— GETTY, T. A., HOWARTH, M. K., MORTON, N. and TORRENS, H. S. 1980. A correlation of the Jurassic rocks in the British Isles. Part 1. Introduction, Lower Jurassic. *Special Report of the Geological Society of London*, No. 14. 73 pp.

COSSHAM, H. 1862. On the northern end of the Bristol coalfield. *Transactions of the Northern England Mining Institute*, Vol. 10, 97–104.

— 1865. On the geological structure of the district around Kingswood Hill, near Bristol. *Geological Magazine*, Vol. 2, 110–113.

— 1879. A lecture to the Bristol Mining School on 'Faults in the Bristol Coalfield and how caused, with special reference to lateral pressures. *Colliery Guardian*, Vol. 37, 335–373, 413–414, 452–453.

— 1885. On a discovery in the Kingswood Coalfield. *Proceedings of the Cotteswold Naturalists' Field Club*, Vol. 8, 247–253.

— WETHERED, E. and SAISE, W. 1875. The northern end of the Bristol Coalfield. *Colliery Guardian*, Vol. 30, 412–420.

COX, L. R. 1942. New light on William Smith and his work. *Proceedings of the Yorkshire Geological Society*, Vol. 25, 1–99.

— 1948. *William Smith and the birth of stratigraphy.* 8 pp., portrait. (London: International Geological Congress.)

COCKS, L. R. M., WOODCOCK, N. H., RICHARDS, R. B., TEMPLE, J. T. and LANE, P. D. 1984. The Llandovery Series of the type area. *Bulletin of the British Museum (Natural History): Geology*, Vol. 38, No. 3, 131–182.

CONIL, R., GROESSENS, E. and PIRLET, M. 1977. Nouvelle charte stratigraphique du Dinantien type de la Belgique. *Annales de la Societé Geologique du Nord*, Tome 96, 363–371

COYSH, A. W. 1926. New sections of Avonian rocks in the neighbourhood of Bristol. *Proceedings of the Bristol Naturalists' Society*, 4th Series, Vol. 6, 324–327.

CROOKALL, R. 1955. Fossil plants of the Carboniferous rocks of Great Britain. (Second section). Vol. 4, Pt. 1. *Memoir of the Geological Survey of Great Britain.*

CRUTTWELL, A. C. 1881. *On the geology of Frome and its neighbourhood.* (Frome.)

CURTIS, M. L. K. 1955. A review of past research on the Lower Palaeozoic rocks of the Tortworth and Eastern Mendip inliers. *Proceedings of the Bristol Naturalists' Society*, Vol. 29, 71–78.

— 1968. The Tremadoc rocks of the Tortworth Inlier, Gloucestershire. *Proceedings of the Geologists' Association*, Vol. 79, 349–362.

— 1972. The Silurian rocks of the Tortworth Inlier, Gloucestershire. *Proceedings of the Geologists' Association*, Vol. 83, 1–35.

— 1972. A preliminary study of the occurrence of iron ore in the new trading estate, Yate. *Proceedings of the Bristol Naturalists' Society*, Vol. 32, 163–166.

— and CAVE, R. 1964. The Silurian–Old Red Sandstone unconformity at Buckover, near Tortworth, Gloucestershire. *Proceedings of the Bristol Naturalists' Society*, Vol. 30, 427–442.

DAVIES, J. H. and TRUEMAN, A. E. 1927. A revision of the non-marine lamellibranchs of the Coal Measures and a discussion of their zonal sequence. *Quarterly Journal of the Geological Society of London*, Vol. 83, 210–259.

DE LA BECHE, H. T. 1846. On the formation of the rocks of South Wales and south western England. *Memoir of the Geological Survey of Great Britain*, Vol. 1. 531 pp.

DIX, E. 1933. The succession of fossil plants in the Millstone Grit and the lower portions of the Coal Measures of the South Wales Coalfield (near Swansea) and a comparison with that of other areas. *Palaeontographica*, Bd. 78, Abt. B, 158–202.

DIXON, E. E. L. 1907. Geology of the South Wales Coalfield, Part VIII, the country around Swansea. *Memoir of the Geological Survey of Great Britain*, Sheet 247 (England and Wales).

— 1921. Geology of the South Wales Coalfield, Part XIII, the country around Pembroke and Tenby. *Memoir of the Geological Survey of Great Britain*, sheets 244 and 245 (England and Wales).

— and VAUGHAN, A. 1912. The Carboniferous succession in Gower (Glamorganshire), with notes on its fauna and conditions of deposition. *Quarterly Journal of the Geological Society of London*, Vol. 67, 477–571.

DOBSON, D. P. 1931. The Bristol district in the Prehistoric Period. *Report of the Meeting of the British Association for the Advancement of Science, 1930, Bristol*, 344–345.

DOLBY, G. and NEVES, R. 1970. Palynological evidence concerning the Devonian–Carboniferous boundary in the Mendips, England. C. R. 6. *Compe rendu du 6me Congrès Internationale de Stratigraphie et de Géologie du Carbonifère, Sheffield 1967*, Vol. 2, 631–642.

DONOVAN, D. T. 1954. A bibliography of the Palaeolithic and Pleistocene sites of the Mendip, Bath and Bristol area. *Proceedings of the University of Bristol Spelaeological Society*, Vol. 7, 23–24.

— 1978. Lower Lias ammonites of the Elton Farm Borehole and the Dundry area, Avon, and a new species of *Aegoceras*. *Bulletin of the Geological Survey of Great Britain*, No. 69, 11–18.

— 1964. A bibliography of the Palaeolithic and Pleistocene sites of the Mendip, Bath and Bristol area. First supplement. *Proceedings of the University of Bristol Spelaeological Society*, Vol. 10, 89–97.

— and HEMINGWAY, J. E. (editors). 1963. *Lexique stratigraphique international.* No. 1 Europe. Fasc. 3aX. Jurassique. 394 pp. (Paris: Centre national de la rechèrce scientifique.)

— and KELLAWAY, G. A. 1984. Geology of the Bristol district: the Lower Jurassic rocks. *Memoir of the British Geological Survey.* 69 pp.

DUMONT, A. 1832. Mémoire sur la constitution géologique de la provence de Liège. 374 pp. (Bruxelles: Academie Royale.)

ETHERIDGE, R. 1985. On the physical structure of the northern part of the Bristol Coal Basin. *Proceedings of the Cotteswold Naturalists' Field Club*, Vol. 4, 28–49.

— 1870. On the geological position and geographical distribution of the Reptilian or Dolomitic Conglomerate of the Bristol area. *Quarterly Journal of the Geological Society of London*, Vol. 26, 174–192.

— 1872. On the physical structure and organic remains of the Penarth (Rhaetic) Beds of Penarth and Lavernock; also with description of the Westbury-on-Severn section. *Transactions of the Cardiff Naturalists' Society*, Vol. 3, 39–64.

EYLES, V. A. 1955. Scientific activity in the Bristol region in the past. 123–143 in *Bristol and its adjoining counties*. MACCINNES, C. M. and WHITTARD, W. F.(editors). 335 pp. (Bristol: British Association for the Advancement of Science.)

FOLK, R. L. and PITMAN, L. S. 1971. Length slow chalcedony. A new testament of vanished evaporites. *Journal of Sedimentary Petrology*, Vol. 41, 1045–1058.

FRASER, N. C. 1985. Vertebrate faunas from Mesozoic fissure deposits of south-west Britain. *Modern Geology*, Vol. 9, 173–300.

— 1986. New Triassic sphenodontids from south-west England and a review of their classification. *Palaeontology*, Vol. 29, 165–186.

— and WALKDEN, G. M. 1983. The ecology of a late Triassic reptile assemblage from Gloucestershire, England. *Palaeogeography, Palaeoclimatology, Palaeoecology*, Vol. 42, 341–365.

— — 1984. The postcranial skeleton of the Upper Triassic Sphenodontid *Planocephalosaurus robinsonae*. *Palaeontology*, Vol. 27, 575–595.

— — and STEWART, V. 1985. The first pre-Rhaetic therian mammal. *Nature, London*, Vol. 314, 161–163.

FRY, T. R. 1956. Further notes on the gravel terraces of the Bristol Avon, and their palaeoliths. *Proceedings of the University of Bristol Spelaeological Society*, Vol. 7, 121–129.

— 1970. Section of Lias below the Midford Sand at Bitton Hill, Bitton, Gloucestershire. *Proceedings of the Bristol Naturalists' Society*, Vol. 31, 631–634.

— 1971. Report on a turbo-drill borehole at Severnside. *Proceedings of the Bristol Naturalists' Society*, Vol. 32, p.69.

FULLER, J. G. C. M. 1969. The industrial basis of stratigraphy: John Strachey, 1671–1743, and William Smith, 1769–1839. *Bulletin American Association of Petroleum Geologists*, Vol. 53, 2256–2273.

GALTON, P. M. 1985. Notes on the Melanorosauridae, a family of large Prosauropod dinosaurs (Saurischia: Sauropodomorpha). *Géobios*, No. 18, 671–676.

— and CLUVER, M. A. 1976. *Anchisaurus capensis* (Broom) and a revision of the Anchisauridae (Reptilia: Saurischia). *Annals of the South African Museum*, Vol. 69, 121–159.

GEORGE, T. N., JOHNSON, G. A. L., MITCHELL, M., PRENTICE, J. E., RAMSBOTTOM, W. H. C., SEVASTOPULO, G. D. and WILSON, R. B. 1976. A correlation of Dinantian rocks in the British Isles. *Special Report of the Geological Society of London*, No. 7. 87 pp.

GILBERTSON, D. D. and HAWKINS, A. B. 1978. The Pleistocene succession at Kenn, Somerset. *Bulletin of the Geological Survey of Great Britain*, No. 66, 41 pp.

— — 1977. The Quaternary deposits at Swallow Cliff, Middlehope, County of Avon. *Proceedings of the Geologists' Association*, Vol. 88, 255–266.

GLANVIL, J. 1669. Observations concerning the Bath Springs. *Philosophical Transactions of the Royal Society of London*, Vol. IV, No. 49, 977–982.

GODWIN, H. 1943. Coastal peat beds of the British Isles and North Sea. *Journal of Ecology*, Vol. 31, 199–247.

GOUGH, J. 1930. The mines of Mendip. 269 pp. (Oxford: Clarendon Press.)

GREEN, G. W. 1958. The Central Mendip Lead-Zinc Orefield. *Bulletin of the Geological Survey of Great Britain*, No. 14, 70–90.

— and DONOVAN, D. T. 1969. The Great Oolite of the Bath area. *Bulletin of the Geological Survey of Great Britain*, No. 30, 1–63.

— and WELCH, F. B. A. 1965. Geology of the country around Wells and Cheddar. *Memoir of the Geological Survey of Great Britain*, Sheet 280 (England and Wales). 225 pp.

GREENWELL, G. C. 1854. Notes on the coal field of east Somerset. *Transactions of the Northern England Institute of Mining Engineers*, Vol. 2, 258–266.

— 1892. On the probability of coal being found south of the Mendips, in Somersetshire. *Transactions of the Manchester Geological Society*, Vol. 21, 596–604.

— and MCMURTRIE, J. 1864. *The Radstock portion of the Somersetshire coalfield*. 28 pp., map and 2 horizontal sections. (Newcastle-upon-Tyne: M. & M. W. Lambert.)

GRENFELL, J. G. 1873. Iron mine recently opened in the Royal York Crescent, Clifton. *Transactions of the Clifton College Scientific Society*, Part 4, 46–53.

— 1873. Clifton minerals. *Transactions of the Clifton College Scientific Society*, Part 4, p.63.

GRIM, R. E. 1933. Petrography of the Fuller's earth deposits of Olmstead, Illinois. *Economic Geology*, Vol. 28, 344–363.

HALLAM, A. 1960. The White Lias of the Devon coast. *Proceedings of the Geologists' Association*, Vol. 71, 47–60.

HALSTEAD, L. B. and NICOLL, P. G. 1971. Fossilised caves of Mendip. *Studies in Speleology*, Vol. 2, 93–102.

HAMILTON, D. 1961. Algal growth in the Rhaetic Cotham Marble of southern England. *Palaeontology*, Vol. 4, 324–333.

— 1962. Some notes on the Rhaetic sediments of the Filton By-pass Substitute near Bristol. *Proceedings of the Bristol Naturalists' Society*, Vol. 30, 279–285.

— 1977. Aust Cliff. 110–118 in *Geological excursions in the Bristol district*. SAVAGE, R. J. G. (editor). (University of Bristol.)

— and WHITTAKER, A. 1977. Coastal exposures near Blue Anchor, Watchet and St Audries Bay, north Somerset. 101–109 in *Geological excursions in the Bristol district*. SAVAGE, R. J. G. (editor). (University of Bristol.)

HANCOCK, P. L. and WILLIAMS, B. P. J. 1977. The geology of Cattybrook brick pit, Almondsbury. 79–84 in *Geological excursions in the Bristol district*. SAVAGE, R. J. G. (editor). (University of Bristol.)

HARMER, F. W. 1907. On the origin of certain cañon-like valleys associated with lake-like areas of depression. *Quarterly Journal of the Geological Society of London*, Vol. 63, 470–514.

HARRIS, T. W. 1938. *The British Rhaetic flora*. 84 pp. (London: British Museum of Natural History.)

HAWKINS, A. B. 1968. The geology of the Portbury area. *Proceedings of the Bristol Naturalists' Society*, Vol. 31, 421–428.

— and KELLAWAY, G. A. 1971. Field meeting at Bristol and Bath with special reference to new evidence for glaciation. *Proceedings of the Geologists' Association*, Vol. 82, 267–292.

HEPWORTH, J. V. and STRIDE, A. H. 1950. A sequence from the Old Red Sandstone to Lower Carboniferous, near Burrington, Somerset. *Proceedings of the Bristol Naturalists' Society*, Vol. 28, 135–138.

HIGHLEY, D. 1972. Fullers earth. *Mineral Dossier Mineral Resources Consultative Committee*, No. 3. 26 pp.

HILL, J. S. and FAIRLEY, W. 1874. The coal deposits of Great Britain and the Oolitic coal of England, Scotland, Sweden and Denmark. *Colliery Guardian*, Vol. 28, 670.

HUDSON, K. 1968. *The industrial archaeology of southern England* (2nd edition). 286 pp. (Newton Abbot: David and Charles.)

IVIMEY-COOK, H. C. 1974. The Permian and Triassic deposits of Wales. 295–321 in *The Upper Palaeozoic and Post Palaeozoic rocks of Wales*. OWEN, T. R. (editor). 426 pp. (Cardiff: University of Wales Press.)

— 1978. Stratigraphy of the Jurassic of the Elton Farm Borehole near Dundry, Avon. *Bulletin of the Geological Survey of Great Britain*, No. 69, 1–9.

JEANS, C. V., MERRIMAN, R. J. and MITCHELL, J. G. 1977. The origin of Middle Jurassic and Lower Cretaceous Fuller's earths in England. *Clay Mineralogy*, Vol. 12, 11–44.

JONES, O. T. 1925. The geology of the Llandovery district. Part I. The southern area. *Quarterly Journal of the Geological Society of London*, Vol. 81, 344–388.

— 1931. Some episodes in the geological history of the Bristol Channel region. *Report of the meeting of the British Association for the Advancement of Science*, for 1930, 57–82.

JONES, J. and LUCY, W. C. 1868. Section of the transitional beds of the Old Red Sandstone and Carboniferous Limestone at Drybrook, in the Forest of Dean. *Proceedings of the Cotteswold Naturalists' Field Club*, Vol. 4, 175–193.

KELLAWAY, G. A. 1935. Notes on a section near West Town Lane, Brislington, Bristol. *Proceedings of the Bristol Naturalists' Society*, 4th Series, Vol. 7, 565–567.

— 1932. The Rhaetic and Liassic rocks of Henleaze and Southmead. *Proceedings of the Bristol Naturalists' Society*, 4th Series, Vol. 7, 285–302.

— 1937. Further recent exposures in the Rhaetic and Jurassic rocks of the Bristol area. *Proceedings of the Bristol Naturalists' Society*, 4th Series, Vol. 8, 223–227.

— 1967. The Geological Survey Ashton Park borehole and its bearing on the geology of the Bristol district. *Bulletin of the Geological Survey of Great Britain*, No. 27, 49–153.

— 1970. The Upper Coal Measures of South West England compared with those of South Wales and the southern Midlands. *Compte Rendu du 6me Congrès International de Stratigraphie et de Géologie du Carbonifère, Sheffield, 1967*, Vol. 3, 1039–1055.

— 1971. The Bristol district. 1645–1651 in Excursion 1: Bristol-Mendip area and south-west England. MATTHEWS, S. C. *Compte Rendu du 6me Congrès International de Stratigraphie et de Géologie du Carbonifère, Sheffield 1967*, Vol. 4, 1641–1667.

— (editor). 1991. *Hot springs of Bath*. (Bath: Bath City Council.)

— and HANCOCK, P. L. 1983. Structure of the Bristol district, the Forest of Dean and the Malvern fault zone. 88–107 in *The Variscan Fold Belt in the British Isles*. HANCOCK, P. L. (editor). (Bristol: Adam Hilger Ltd.)

— and OAKLEY, K. P. 1934. Notes on the Keuper and Rhaetic exposed in a road cutting at Uphill, Somerset. *Proceedings of the Bristol Naturalists' Society*, 4th Series, Vol. 7, 470–488.

— WELCH, F. B. A. 1948. *British regional geology: Bristol and Gloucester district* (2nd edition). (London: HMSO for Institute of Geological Sciences.)

— — 1955a. The Upper Old Red Sandstone and Lower Carboniferous Rocks of Bristol and the Mendips compared with those of Chepstow and the Forest of Dean. *Bulletin of the Geological Survey of Great Britain*.

— and WILSON, V. 1941. An outline of the geology of Yeovil, Sherborne and Sparkford Vale. *Proceedings of the Geologists' Association*, Vol. 52, 131–174.

KELLING, G. 1974. Upper Carboniferous sedimentation in South Wales. 185–224 in *The Upper Palaeozoic and post-Palaeozoic rocks of Wales*. (Cardiff: University of Wales Press.)

KERR, P. F. 1932. Montmorillonite or smectite as constituents of Fuller's earth or bentonite. *American Mineralogist*, Vol. 17, 192.

KERSLAKE, T. 1883. Henbury Parish a thousand years ago. (abstract). *Gloucester Notes and Queries*, Vol. 6, 101 and 126, from *Antiquarian Magazine and Bibliographer*.

KIDSTON, R. 1894. On the various divisions of the Carboniferous rocks as determined by their fossil flora. *Proceedings of the Royal Physical Society of Edinburgh*, Vol. 12, 183–257.

KING, W. W. 1934. The Downtonian and Dittonian strata of Great Britain and north-western Europe. *Quarterly Journal of the Geological Society of London*, Vol. 90, 526–570.

LAWSON, J. D. 1955. The geology of the May Hill Inlier. *Quarterly Journal of the Geological Society of London*, Vol. 111, 85–116.

LEESE, C. E. and VERNON, W. F. 1961. Basal gravel in the alluvium near Severn Beach. *Proceedings of the Bristol Naturalists' Society*, Vol. 30, 139–143.

LEES, A. 1982. The palaeoenvironmental setting and distribution of the Waulsortian facies of Belgium and southern Britain. 1–16 in *Symposium on the palaeoenvironmental setting and distribution of US Waulsortian facies*. BOLTON, K., LANE, H. R. (editors). (El Paso: El Paso Geological Society and University of Texas.)

LOUPEKINE, I. S. 1951. Fluorite from the Carboniferous Limestone of the Avon Gorge, Bristol. *Proceedings of the Bristol Naturalists' Society*, Vol. 28, 203–220.

— 1953. Reversed faulting in the Great Quarry, Avon Gorge. *Proceedings of the Bristol Naturalists' Society*, Vol. 28, 335–342.

MARSHALL, J. E. A. and WHITESIDE, D. I. 1980. Marine influences in the Triassic 'Uplands'. *Nature, London*, Vol. 287, 627–628.

MARTYN, S. 1875. On fish remains in the Bristol Old Red Sandstone. *Proceedings of the Bristol Naturalists' Society*, New Series, Vol. 1, 141–144.

MAYALL, M. J. and WRIGHT, V. P. 1981. Algal tuft structures in stromatolites from the Upper Triassic of south-west England. *Palaeontology*, Vol. 24, 655–660.

McMURTRIE, J. 1869. A lecture on the Carboniferous strata of Somersetshire. *Proceedings of the Bath Natural History and Antiquarian Field Club*, Vol. 1, 45–60.

— 1869. Paper on the faults and contortions of the Somersetshire Coal Field. *Proceedings of the Bath Natural History and Antiquarian Field Club*, Vol. 1, 127–147.

— 1901. The geological features of the Somerset and Bristol coalfield, with special reference to the physical geology of the Somerset Basin. *Transactions of the Institute of Mining Engineers*, Vol. 20, 306–335.

MITCHELL, G. F., PENNY, L. F., SHOTTON, F. and WEST, R. G. 1973. A correlation of Quaternary deposits in the British Isles. *Special report of the Geological Society of London*, No. 4. 99 pp.

MITCHELL, M. 1971. 97–98 in *Annual Report for 1970*. INSTITUTE OF GEOLOGICAL SCIENCES. (London: Institute of Geological Sciences.)

— 1972. The base of the Viséan in south-west and north-west England. *Proceedings of the Yorkshire Geological Society*, Vol. 39, 151–160.

— 1981. The distribution of Tournaisian and early Viséan (Carboniferous) coral faunas from the Bristol and South Wales areas of Britain. *Acta Palaeontologica Polonica*, Vol. 25, 577–585.

— STRANK, A. R. E., THORNBURY, B. M. and SEVASTOPULO, G. D. 1986. The distribution of platform conodonts, corals and foraminifera from the Black Rock Limestone (late Tournaisian and early Viséan) of Tears Point, Gower, South Wales. *Proceedings of the Yorkshire Geological Society*, Vol. 46, 11–14.

MOORE, C. 1867. On abnormal conditions of secondary deposits when connected with the Somersetshire and South Wales Coal-Basin; and on the age of the Sutton and Southerndown Series. *Quarterly Journal of the Geological Society of London*, Vol. 23, 449–568.

MOORE, L. R. 1938. The sequence and structure of the Radstock Basin. *Proceedings of the Bristol Naturalists' Society*, 4th Series, Vol. 8, 267–305.

— 1941. The presence of the Namurian in the Bristol district. *Geological Magazine*, Vol. 78, 279–292.

— 1936. On a boring for coal at Farrington Gurney, near Bristol. *Proceedings of the Bristol Naturalists' Society*, 4th Series, Vol. 8, 220–222.

— and TRUEMAN, A. E. 1937. The Coal Measures of Bristol and Somerset. *Quarterly Journal of the Geological Society of London*, Vol. 93, 195–240.

— — 1939. The structure of the Bristol and Somerset coalfields. *Proceedings of the Geologists' Association*, Vol. 50, 46–67.

— — 1942. The Bristol and Somerset coalfields with particular reference to the prospects of future development. *Proceedings of the South Wales Institute of Engineers*, Vol. 57, 180–222.

MORGAN, C. LLOYD. 1885. Sub-aerial denudation and the Avon Gorge. *Proceedings of the Bristol Naturalists' Society*, New Series, Vol. 4, 171–197.

— 1885. On the mapping of the Millstone Grit, at Long Ashton, near Bristol. *Proceedings of the Bristol Naturalists' Society*. New Series, Vol. 4, 163–165.

— 1886. Contributions to the geology of the Avon Basin. III. The Portbury and Clapton District. *Proceedings of the Bristol Naturalists' Society*, New Series, Vol. 5, 1–16.

— and LUCY, W. G. 1887. The Severn Tunnel section. *Proceedings of the Bristol Naturalists' Society*, New Series, Vol. 5, 82–94.

— and REYNOLDS, S. M. 1901. Igneous rocks and association sedimentaries of the Tortworth Inlier. *Quarterly Journal of the Geological Society of London*, Vol. 52, 267–284.

MURCHISON, R. I. 1839. *The Silurian System founded on geological researches in the counties of Salop, Worcester and Stafford; with descriptions of the coalfields and overlying formations.* Vols. 1, 2. 768 pp. (London: Murray.)

MURRAY, J. W. and HAWKINS, A. B. 1976. Sediment transport in the Severn Estuary during the past 8000 to 9000 years. *Journal of the Geological Society of London*, Vol. 132, 385–398.

— and WRIGHT, C. A. 1971. The Carboniferous Limestone of Chipping Sodbury and Wick, Gloucestershire. *Geological Journal*, Vol. 7, 255–270.

NICHOLLS, F. J. 1880. Pen Park Hole. A Roman lead mine. *Transactions of the Bristol and Gloucestershire Archaeological Society*, Vol. 4, 320–328.

NICKLESS, E. F. P., BOOTH, S. J. and MOSLEY, P. N. 1976. The celestite resources of the area north-east of Bristol with notes on occurrences north and south of the Mendip Hills and in the Vale of Glamorgan: description of 1:25 000 resources sheet ST 68, and parts of ST 59, 69, 79, 58, 78, 67 and 77. *Mineral Assessment Report of the Institute of Geological Sciences*, No. 25. 83 pp.

ORBELL, G. 1973. Palynology of the British Rhaetic-Liassic. *Bulletin of the Geological Survey of Great Britain*, No. 44, 1–44.

OWEN, E. 1754. *Observations on the earths, rocks, stones and minerals, for some miles about Bristol, and on the nature of the Hot-well, and the virtues of its water.* (London.)

OWEN, G. (of HENLLYS). 1796. *History of Pembrokeshire.*

PALMER, L. S. 1934. Some Pleistocene breccias in the Severn Estuary. *Proceedings of the Geologists' Association*, Vol. 45, 145–161.

PARSONS, C. F. 1975. Ammonites from the Doulting Conglomerate Bed (Upper Bajocian, Jurassic) of Somerset. *Palaeontology*, Vol. 18, 191–205.

— 1979. A stratigraphic revision of the Inferior Oolite of Dundry Hill, Bristol. *Proceedings of the Geologists' Association*, Vol. 90, 133–151.

PENN, I. E., MERRIMAN, R. J. and WYATT, R. J. 1979. The Bathonian strata of the Bath-Frome area. 1. A proposed type-section for the Fuller's Earth (Bathonian), based on Horsecombe Vale No. 15 Borehole, near Bath, with details of contiguous strata. *Report of the Institute of Geological Sciences*, No. 78/22, 1–22.

— and WYATT, R. J. 1979. The Bathonian strata of the Bath-Frome area. 2 The stratigraphy and correlation of the Bathonian strata in the Bath-Frome area. *Report of the Institute of Geological Sciences*, No. 78/22, 23–88.

PHILLIPS, J. 1829. Illustrations of the geology of Yorkshire. 192pp. (York: printed for the author.)

— 1844. *Memoirs of William Smith, LLD.* 150 pp. (London: John Murray.)

— 1848. The Malvern Hills compared with the Palaeozoic districts of Abberley, Woolhope, May Hill, Tortworth and Usk. *Memoir of the Geological Survey.*

— 1871. *Geology of Oxford and the valley of the Thames.* 523 pp. (Oxford: Clarendon Press.)

PRESTWICH, J. 1871. Report on the quantities of coal, wrought and unwrought, in the coalfields of Somersetshire and part of Gloucestershire. 33–70 in *Report of the Royal Commission on Coal*, Vol. 1. (London: HMSO.)

PRINGLE, J. 1921. On a boring for coal at Yate, Gloucestershire. *Summary of Progress of the Geological Survey of Great Britain for 1920*, 92–95.

— 1922. On a boring for coal at Westbury, Wiltshire. *Summary of Progress of the Geological Survey of Great Britain for 1921*, 146–155.

RAMSBOTTOM, W. H. C. 1973. Transgressions and regressions in the Dinantian: a new synthesis of British Dinantian stratigraphy. *Proceedings of the Yorkshire Geological Society*, Vol. 39, 567–607.

— and MITCHELL, M. 1980. The recognition and division of the Tournaisian Series in Britain. *Journal of the Geological Society of London*, Vol. 137, 61–63.

— CALVER, M. A., EAGAR, R. M. C., HODSON, F., HOLLIDAY, D. W., STUBBLEFIELD, C. J. and WILSON, R. B. 1978. Silesian (Upper Carboniferous). *Special Report of the Geological Society of London*, No. 10. 82 pp.

REED, F. R. C. and REYNOLDS, S. H. 1908a. Silurian fossils from certain localities in the Tortworth Inlier. *Proceedings of the Bristol Naturalists' Society*, 4th Series, Vol. 2, 32–40.

— — 1908b. On the fossiliferous Silurian rocks of the southern half of the Tortworth Inlier. *Quarterly Journal of the Geological Society of London*, Vol. 64, 512–545.

REYNOLDS, S. H. 1901. Excursion to Tortworth. *Proceedings of the Geologists' Association*, Vol. 17, 150–152.

— 1905. Fish teeth and spines from the Carboniferous Limestone of the Bristol district. *Proceedings of the Bristol Naturalists' Society*, 4th Series, Vol. 2, 41–43.

— 1916. Further work on the igneous rocks associated with the Carboniferous Limestone of the Bristol district. *Quarterly Journal of the Geological Society of London*, Vol. 72, 23–42.

— 1918. The Carboniferous Limestone series of the area between Clifton and Clevedon. *Proceedings of the Bristol Naturalists' Society*, 4th Series, Vol. 4, 186–197.

— 1920. The Carboniferous Limestone of the Clifton-Westbury-King's Weston Ridge. *Proceedings of the Bristol Naturalists' Society*, 4th Series, Vol. 5, 92–100.

— 1921. The lithological succession of the Avonian at Clifton. *Quarterly Journal of the Geological Society of London*, Vol. 77, 213–243.

— 1921. A geological excursion handbook for the Bristol district (2nd edition). 45 pp. (Bristol.)

— 1924. The igneous rocks of the Tortworth Inlier. *Quarterly Journal of the Geological Society of London*, Vol. 80, 106–112.

— 1926. The effect on the Avon Section of the construction of Portway. *Proceedings of the Bristol Naturalists' Society*, 4th Series, Vol. 6, 318–323.

— 1936. The Carboniferous Limestone series (Avonian) of the Avon Gorge by A. Vaughan. *Proceedings of the Bristol Naturalists' Society*, 4th Series, Vol. 8, 29–90.

— 1947. The Aust Section. *Proceedings of the Cotteswold Naturalists' Field Club*, Vol. 29, 29–39.

— and GREENLY, E. 1924. The geological structure of the Clevedon-Portishead area, Somerset. *Quarterly Journal of the Geological Society of London*, Vol. 80, 447–466.

— and SMITH, S. 1925. The Old Red Sandstone and Avonian succession at Westbury on Trym. *Geological Magazine*, Vol. 62, 464–473.

— and VAUGHAN, A. 1911. Faunal and lithological sequence in the Carboniferous Limestone Series (Avonian) of Burrington Combe (Somerset). *Quarterly Journal of the Geological Society of London*, Vol. 67, 342–392.

RHODES, F. H. T. and AUSTIN, R. L. 1971. Carboniferous conodont faunas of Europe. *Memoir of the Geological Society of America*, No. 127, 317–352.

RICHARDSON, L. 1903. The Rhaetic and Lower Lias of Sedbury Cliff, near Chepstow, Monmouthshire. *Quarterly Journal of the Geological Society of London*, Vol. 59, 390–395.

— 1904. Note on the Rhaetic rocks around Charfield, Gloucestershire. *Geological Magazine*, Vol. 41, 532–535.

— 1907. The Inferior Oolite and contiguous deposits of the Bath-Doulting district. *Quarterly Journal of the Geological Society of London*, Vol. 63, 383–423.

— 1910. Excursion to the Frome district, Somerset. *Proceedings of the Geologists' Association*, Vol. 21, 209–228.

— 1911. The Rhaetic and contiguous deposits of west, mid and part of east Somerset. *Quarterly Journal of the Geological Society of London*, Vol. 87, 1–74.

— 1928. Wells and springs of Somerset. *Memoir of the Geological Survey of England and Wales*. 279 pp.

— 1930. Wells and springs of Gloucestershire. *Memoir of the Geological Survey of England and Wales*. 292 pp.

RILEY, H. and STUTCHBURY, S. 1840. A description of various fossil remains of three distinct Saurian animals, recently discovered in the Magnesian conglomerate near Bristol. *Transactions of the Geological Society of London*, Series 2, Vol. 5, 349–357.

ROBINSON, P. L. 1957a. The Mesozoic fissures of the Bristol Channel area and their vertebrate faunas. *Journal of the Linnean Society: Zoology*, Vol. 43, 260–282.

— 1957b. An unusual Sauropsid dentition. *Journal of the Linnean Society: Zoology*, Vol. 43, 283–292.

— 1962. Gliding lizards from the Upper Keuper of Great Britain. *Proceedings of the Geological Society of London*, No. 1601, 137–146.

ROBINSON, P. L. 1973. A problematical reptile from the British Upper Trias. *Journal of the Geological Society of London*, Vol. 129, 457–479.

RUDDER, S. 1779. A new history of Gloucestershire.

SALEE, A. 1910. Contribution à l'étude des polypiers du Calcaire Carbonifère de la Belgique. Le genre *Caninia*. *Nouvelles Mémoires de la Société Belge Géologique*, Fasc. 3. 62 pp.

SANDERS, W. 1865. Map of the Bristol Coal Fields, sheet No. 14: Bristol.

SANDFORD, W. A. 1894. On bones of an animal resembling the megalosaur found in the Rhaetic formation at Wedmore. *Proceedings of the Somersetshire Archaeological and Natural History Society*, Vol. 40, 227–235.

SAVAGE, R. J. G. 1962. Rhaetic exposures at Emborough. *Proceedings of the Bristol Naturalists' Society*, Vol. 30, 275–278.

— (editor). 1977. *Geological excursions in the Bristol district*. 196 pp. (University of Bristol.)

— and LARGE, N. F. 1966. On *Birgeria acuminata* and the absence of labyrinthodonts from the Rhaetic. *Palaeontology*, Vol. 9, 135–141.

SEAVILL, E. W. 1936. Fossil shells from the Nailsea coalfield. *Proceedings of the Bristol Naturalists' Society*, 4th Series, Vol. 8, 228.

SEELEY, H. G. 1898. On large terrestrial saurians from the Rhaetic Beds of Wedmore Hill, described as *Avalonia sanfordi* and *Picrodon herveyi*. *Geological Magazine*, Vol. 35, 1–6.

SHEPHARD, T. 1917. William Smith: his maps and memoirs. *Proceedings of the Yorkshire Geological Society*, Vol. 19, 75–253.

SHERLOCK, R. L. and HOLLINGWORTH, S. E. 1938. *Gypsum and anhydrite* (3rd edition). *Special Report on the Mineral Resources of Great Britain, Memoir of the Geological Survey of Great Britain*, Vol. 3. 98 pp.

SIBLY, T. F. 1912. The faulted inlier of Carboniferous Limestone at Upper Vobster. *Quarterly Journal of the Geological Society of London*, Vol. 68, pp.58–74.

— and LLOYD, W. 1927. Iron ores: The haematites of the Forest of Dean and South Wales (2nd edition, revised by W. Lloyd). *Special Reports on the Mineral Resources of Great Britain, Memoir of the Geological Survey of Great Britain*, Vol. 10. 101 pp.

SIMPSON, SCOTT. 1951. A new Eurypterid from the Upper Old Red Sandstone of Portishead. *Annals and Magazine of Natural History*, Series 12, Vol. 4, 849–861.

SMITH, S. 1930. The Carboniferous inliers at Codrington and Wick (Gloucestershire). *Quarterly Journal of the Geological Society of London*, Vol. 86, 331–354.

— 1942. A high Viséan fauna from the vicinity of Yate, Gloucestershire; with a special reference to the corals and to a goniatite. *Proceedings of the Bristol Naturalists' Society*, 4th Series, Vol. 9, 335–348.

— and REYNOLDS, S. H. 1929. The Carboniferous section at Cattybrook, near Bristol. *Quarterly Journal of the Geological Society of London*, Vol. 85, 1–8.

— — 1935. On a boring for water in the Lower Avonian and Old Red Sandstone at Portishead. *Proceedings of the Bristol Naturalists' Society*, 4th Series, Vol. 7, 521–527.

— and STUBBLEFIELD, C. J. 1933. On the occurrence of Tremadoc shales in the Tortworth Inlier (Gloucestershire). *Quarterly Journal of the Geological Society of London*, Vol. 89, 357–378.

— and WILLAN, G. R. 1937. A preliminary note on the geology of the Bristol Channel Islands, Steep Holme, Flat Holm and Denny Island. *Geological Magazine*, Vol. 74, 91–92.

SMITH, W. 1815. *A memoir to the map and delineation of the strata of England and Wales with part of Scotland.* (London: John Cary.)

SOLLAS, W. J. 1880. On the geology of the Bristol district. *Proceedings of the Geologists' Association*, Vol. 6, 375–391.

SOLLAS, I. B. J. 1901. Fossils in the Oxford University Museum, V: On the structure and affinities of the Rhaetic plant *Naiadita*. *Quarterly Journal of the Geological Society of London*, Vol. 57, 307–312.

STAPLES, E. H. 1917. Some effects of the master folds upon the structure of the Bristol and Somerset coalfields. *Transactions of the Institution of Mining Engineers*, Vol. 52, 187–198.

STEART, F. A. 1911. The north-western portion of the Somersetshire coalfield and the Farmborough Fault. *The Colliery Guardian*, Vol. 102, 916–918.

STEMBRIDGE, P. K. 1969. *Goldney, a house or a family.* 28 pp. (Bristol: Burleigh Press.)

STODDART, W. W. 1876. Geology of the Bristol Coalfield. *Proceedings of the Bristol Naturalists' Society*, New Series, Vol. 1, 313–334.

STRACHEY, J. 1719. A curious description of the strata observed in the coal-mines of Mendip in Somersetshire. *Philosophical Transactions*, (No. 360), Vol. 30, 968–973, Table II.

— 1725. An account of the strata in coalmines. *Philosophical Transactions*, (No. 391), Vol. 33, 395–398, plates I, II.

— 1727. *Observations on the different strata of earths and minerals (more particularly of such as are found in the coalmines of Great Britain).* (London: J. Walthoe.)

STUBBLEFIELD, C. J. 1937. Palaeontological Department Sectional Report. 80 in *Summary of Progress of the Geological Survey of Great Britain for 1936.* (London: HMSO.)

— and TROTTER, F. M. T. 1957. Divisions of the coal measures on Geological Survey maps of England and Wales. *Bulletin of the Geological Survey of Great Britain*, No. 13, 1–5.

SWINTON, W. E. 1939. A new Triassic Rhyncocephalian from Gloucestershire. *Annals and Magazine of Natural History*, Vol. 4, 591–594.

TAWNEY, E. 1878. On an excavation at the Bristol Water Works pumping station, Clifton. *Proceedings of the Bristol Naturalists' Society*, New Series, Vol. 2, 179–182.

TORRENS, H. S. (editor). 1969. International Field Symposium on the British Jurassic. Excursion No. 2. Guide for north Somerset and Gloucestershire. A. North Somerset by D. T. Donovan with contributions by J. C. W. Cope, T. A. Getty, H. C. Ivimey-Cook and H. S. Torrens. 46 pp. (Keele: Keele University Press.)

TRATMAN, E. K. 1963. Report on the investigations of Pen Park Hole, Bristol. *Publication of the Cave Research Group of Great Britain*, No. 12. 54 pp.

TRIMMER, J. 1853. On the southern termination of the Erratic Tertiaries and on the remains of a bed of gravel on the summit of Clevedon Down, Somersetshire. *Quarterly Journal of the Geological Society of London*, Vol. 9, 282–296.

TROTTER, F. M. 1942. Geology of the Forest of Dean coal and iron-ore field. *Memoir of the Geological Survey of Great Britain*. 95 pp.

TRUEMAN, A. E. 1936. Note on a boring near Winford, Somerset. *Proceedings of the Bristol Naturalists' Society*, Vol. 8, 121–123.

— 1947. Stratigraphical problems in the coalfields of Great Britain. *Proceedings of the Geological Society of London*, Vol. 103, lxv–civ.

TUCKER, M. E. 1976. Quartz replaced anhydrite nodules ('Bristol Diamonds') from the Triassic of the Bristol district. *Geological Magazine*, Vol. 113, 569–574.

TUTCHER, J. W. 1908. The strata exposed in constructing the Filton to Avonmouth railway. *Proceedings of the Bristol Naturalists' Society*, 4th Series, Vol. 2, 5–21.

— 1923. Some recent exposures of the Lias (Sinemurian and Hettangian) and Rhaetic about Keynsham. *Proceedings of the Bristol Naturalists' Society*, 4th Series, Vol. 5, 268–278.

— 1928. Note on a deep boring at Somerdale, Keynsham, Somerset. *Proceedings of the Bristol Naturalists' Society*, 4th Series, Vol. 7, 56–57.

— 1930. In *The geology of the Bristol district with some account of the physiography*. CROOKALL, R., PALMER, L. S., REYNOLDS, S. H., TRUEMAN, E., TUTCHER, J. W. and WALLIS, F. J. (editors). 59 pp. (British Association for the Advancement of Science.)

TUTCHER, J. W. and TRUEMAN, A. E. 1925. The Liassic rocks of the Radstock district. *Quarterly Journal of the Geological Society of London*, Vol. 81, 595–666.

UTTING, J. and NEVES, R. 1970. Palynology of the Lower Limestone Shale Group (Basal Carboniferous Limestone series) and Portishead Beds (Upper Old Red Sandstone) of the Avon Gorge, Bristol, England. 411–422 in Colloque sur la stratigraphie du Carbonifère. STREEL, M. and WAGNER, R. H. (editors). *Les Congrès et Colloques de l'Université de Liège*, Vol. 55. (University of Liège.)

VARKER, W. J. and SEVASTOPULO, G. D. 1985. The Carboniferous System: Part 1—Conodonts of the Dinantian Subsystem from Great Britain and Ireland. 167–189 in *A stratigraphical index of British conodonts*. HIGGINS, A. C. and AUSTIN, R. L. (editors). (Chichester: Ellis Horwood.)

VAUGHAN, A. 1905. The palaeontological sequence in the Carboniferous Limestone of the Bristol area. *Quarterly Journal of the Geological Society of London*, Vol. 61, 181–307.

— 1906. The Carboniferous Limestone series (Avonian) of the Avon Gorge. *Proceedings of the Bristol Naturalists' Society*, 4th Series, Vol. 1, 74–168.

— and TUTCHER, J. W. 1903. The Lower Lias of Keynsham. *Proceedings of the Bristol Naturalists' Society*, New Series, Vol. 10, 3–55.

VON DER BORCH, C. C. 1976. Stratigraphy and formation of Holocene dolomitic carbonate deposits of the Coorong area, South Australia. *Journal of Sedimentary Petrology*, Vol. 46, No. 4.

— and LOCK, D. 1979. Geological significance of Coorong Dolomites. *Sedimentology*, Vol. 26, No. 6, 813–824.

— — and SCHWEBEL, D. 1975. Ground-water formation of dolomite in the Coorong region of South Australia. *Geology (Boulder)*, Vol. 3, No. 5, 283–285.

WALLIS, F. S. 1922. The Carboniferous limestone (Avonian) of Broadfield Down (Somerset). *Proceedings of the Bristol Naturalists' Society*, 4th Series, Vol. 5, 205–221.

— 1928. The Old Red Sandstone of the Bristol district. *Quarterly Journal of the Geological Society of London*, Vol. 83, 760–789.

WARNER, R. 1811. *A new guide through Bath and its environs.* 174 pp. (Bath.)

WARRINGTON, G. 1977. Palynological examination of Triassic (Keuper Marl and Rhaetic) deposits north-east and east of Bristol. *Proceedings of the Ussher Society*, Vol. 4, 76–81.

— 1981. The indigenous micropalaeontology of British Triassic shelf sea deposits. 61–70 in *Microfossils from recent and fossil shelf seas*. NEALE, J. W. and BRASIER, M. D. (editors). 380 pp. (Chichester: Ellis Horwood.)

— 1984. Late Triassic palynomorph records from Somerset. *Proceedings of the Ussher Society*, Vol. 6, 29–34.

— and WHITTAKER, A. 1984. The Blue Anchor Formation (late Triassic) in Somerset. *Proceedings of the Ussher Society*, Vol. 6, 100–107.

— AUDLEY-CHARLES, M. G., ELLIOTT, R. E., EVANS, W. B., IVIMEY-COOK, H. C., KENT, P. E., ROBINSON, P. L., SHOTTON, F. W. and TAYLOR, F. M. 1980. A correlation of Triassic rocks in the British Isles. *Special Report of the Geological Society of London*, No. 13. 78 pp.

WEAVER, T. 1824. Geological observations in part of Gloucestershire and Somersetshire. *Transactions of the Geological Society of London*, Series 2, Vol. 1, 317–368.

WELCH, F. B. A. 1931(a). The Avonian Inlier at Upper Vobster (Somerset). *Geological Magazine*, Vol. 68, 421–430.

— 1931(b). On the occurrence of the D_3-Subzone in the Mendip Area. *Proceedings of the Bristol Naturalists Society*, 4th Series, Vol. 7, Pt. 4, 303–307.

— 1933. The geological structure of the Eastern Mendips. *Quarterly Journal of the Geological Society of London*, Vol. 89, 14–52.

— 1956. Note on the gravels at Kenn, Somerset. *Proceedings of the University of Bristol Spelaeological Society*, Vol. 7, p.137.

— and TROTTER, F. M. 1961. Geology of the country around Monmouth and Chepstow. *Memoir of the Geological Survey of Great Britain*, sheets 233 and 258 (England and Wales). 164 pp.

WHITESIDE, D. I. 1986. The head skeleton of the Rhaetian sphenodontid *Diphydontosaurus avonis* gen. et sp. nov. and the modernising of a living fossil. *Philosophical Transactions of the Royal Society of London*, Ser. B, Vol. 312, 379–430.

WHITTAKER, A. 1973. The central Somerset basin. *Proceedings of the Ussher Society*, Vol. 2, 585–592.

— and GREEN, G. W. 1983. Geology of the country around Weston-super-Mare. *Memoir of the Geological Survey of Great Britain*, Sheet 279 with parts of 263 and 295. 147 pp.

WHITTARD, W. F. 1931. The geology of the Ordovician and Valentian rocks of the Shelve country, Shropshire. *Proceedings of the Geologists' Association*, Vol. 62, 322–339.

— 1948. Temporary exposures and borehole records in the Bristol area. I Records of boreholes sunk for the new Severn and Wye Bridges. *Proceedings of the Bristol Naturalists' Society*, Vol. 27, 311–328.

— and SMITH, S. 1943. Geology of a recent borehole at Filton, Gloucestershire. *Proceedings of the Bristol Naturalists' Society*, 4th Series, Vol. 9, 434–450.

— — 1944. Unrecorded inliers of Silurian rocks, near Wickwar, Gloucestershire, with notes on the occurrence of a stromatolite. *Geological Magazine*, Vol. 81, 65–76.

— — 1944b. Geology of a further borehole at Filton, Gloucestershire. *Proceedings of the Bristol Naturalists' Society*, 4th Series, Vol. 9, 521–529.

WICKES, W. H. 1901. A Rhaetic section at Redland. *Proceedings of the Bristol Naturalists' Society*, New Series, Vol. 9, 99–103.

— 1904. The Rhaetic Bone Bed. *Proceedings of the Bristol Naturalists' Society*, New Series, Vol. 10, 213–227.

— 1910. Beekite. *Proceedings of the Bristol Naturalists' Society*, 4th Series, Vol. 2, 9–21.

WILLS, L. J. 1938. The Pleistocene development of the Severn from Bridgnorth to the sea. *Quarterly Journal of the Geological Society of London*, Vol. 94, 161–242.

WILSON, E. 1891. On a section of the Rhaetic rocks at Pylle Hill (Totterdown) Bristol. *Quarterly Journal of the Geological Society of London*, Vol. 47, 545–549.

— 1894. The Rhaetic rocks at Pylle Hill, Bristol, with some general considerations as to the (upper and lower) limits of the Rhaetic formation in England. *Proceedings of the Bristol Naturalists' Society*, New Series, Vol. 7, 213–231.

WINWOOD, H. H. 1892. Charles Moore, FGS, and his work; with a list of the fossil types and described specimens in the Bath Museum. *Proceedings of the Bath Natural History and Antiquarian Field Club*, Vol. 7, 232–292.

WOODLAND, A. W., ARCHER, A. A. and EVANS, W. B. 1957. Recent boreholes into the Lower Coal Measures below the Gellideg–Lower Pumpquart Coal horizon in South Wales. *Bulletin of the Geological Survey of Great Britain*, No. 13, 39–60.

— and EVANS, W. B. 1964. The geology of the South Wales coalfield, Part IV, the country around Pontypridd and Maesteg (3rd edition). *Memoir of the Geological Survey of Great Britain*, Sheet 248 (England and Wales). 391 pp.

WOODWARD, H. B. 1876. Geology of the East Somerset and the Bristol Coal-Fields. *Memoir of the Geological Survey*. 271 pp.

— 1887. *The geology of England and Wales with notes on the physical features of the country* (2nd edition). 670 pp. (London: George Phillip and Son.)

— 1893. The Jurassic rocks of Britain. Vol. III. Lias of England and Wales (Yorkshire excepted). *Memoir of the Geological Survey*. 399 pp.

— 1894. The Jurassic rocks of Britain. Vol. IV. The Lower Oolitic rocks of England (Yorkshire excepted). *Memoir of the Geological Survey*. 628 pp.

WRIGHT, T. 1860. On the zone of *Avicula contorta*, and the Lower Lias of the South of England. *Quarterly Journal of the Geological Society of London*, Vol. 16, 374–411.

WRIGHT, V. P. and MAYALL, M. 1981. Organism—sediment interactions in Stromatolites: an example from the upper Triassic of South West Britain. 74–84 in *Phanerozoic Stromatolites: case histories*. MONTY, C. (editor). 249 pp. (Berlin: Springer-Verlag.)

FOSSIL INDEX

Acaste downingiae (Murchison, 1839) 12
Acitheca polymorpha (Brongniart, 1828) 113
Actinoceras nummularium (J de C Sowerby, 1839) 11
Alethopteris
 A. lonchitica (Schlotheim, 1820) 63
Alloiopteris radstockensis Kidston, 1923 113
Amphistrophia funiculata (J de C Sowerby, 1839) 12
Anabacia 150
Angelina sedgwickii? Salter, 1859 10
Anthracoceras sp. 77, 95
 A. cf. *hindi* Bisat, 1930 90
Anthracoceratites
 A. arcuatilobis (Ludwig, 1863) 76
 A. vanderbeckei (Ludwig, 1863) 67
Anthracomya sp. 119
 A. pruvosti Tchernyshev, 1931 119
Anthraconaia
 A. lenisulcata (Trueman, 1929) 77
 A. modiolaris (J de C Sowerby, 1840) 67, 76, 77
 A. pringlei (Dix & Trueman, 1929) 107
 A. prolifera (Waterlot, 1934) 110, 119
 A. aff. *pruvosti* (Tchernyshev, 1931) 105, 112
 A. pulchella Broadhurst, 1959 77, 86
 A. pulchra (Hind, 1895) 67, 77
 A. cf. *williamsoni* (Brown, 1849) 77, 86
 A. sp. nov. cf. *williamsoni* (Brown, 1849) 86
Anthraconauta
 A. minima (Hind, 1893) 94
 A phillipsii (Williamson, 1836) 67, 84, 95, 96, 97, 101, 104, 105, 107, 112, 116, 126
 A. aff. *phillipsii* (Williamson, 1836) 110, 127
 A. tenuis (Williamson, 1927) 67, 104, 105, 107, 110, 112, 116, 127
 A. cf. *tenuis* (Williamson, 1927) 110, 119
Anthracosia
 A. similis (Brown, 1843) 67, 77
 A. regularis (Trueman, 1929) 76
Apiculatisporis variocorneus Sullivan, 1964 65
arthropods 41
Arvicola cantiana (Hinton) 161
Astarte 142
Asterolepis 17
Athyris sp. 59
Atrypa? 12
 A. reticularis (Linnaeus, 1758) 11, 12
Avalonia 140
 A. sanfordi Seeley, 1898 141
Aviculopecten 41

Avonia bassa (Vaughan, 1905) 28, 49
Axophyllum vaughani (Salée, 1913) 45

Bellerophon 33, 40, 41, 48, 49, 55, 57, 58, 60
Beyrichia sp. 13
Birgeria 140
Bothriolepis sp. 17
brachiopods 19, 27, 29–30, 32, 41–42, 45, 55
Brachyprion
 B. arenaceus (Davidson, 1871) 11
 B. waltonii (Davidson, 1848) 12
Buxtonia 45
 B. scabricula (Sowerby, 1814) 55, 64

Calamnospora 144
Camarotoechia sp. 12
Camelotia borealis Galton, 1985 141
Camerosporites secatus Leschik emend. Scheuring, 1978 144
Caneyella sp. nov. aff. *multirugata* (Jackson, 1927) 76
Caninia
 C. cornucopiae Michelin *in* Gervais, 1840 30, 178
 C. (*'cylindrica'*) Scouler *in* Griffith, 1842 178
Caninophyllum
 C. archiaci (Milne-Edwards & Haime, 1852) var. *bristolensis* (Vaughan, 1903) 30
 C. patulum (Michelin, 1846) 29, 30
 C. patulum (Michelin, 1846) *greeni* Mitchell, 1980 30
Carbonicola
 C. communis Davies & Trueman, 1927 67, 76, 77, 91
 C. pseudorobusta Trueman, 1929 76
 C. cf. *pseudorobusta* Trueman, 1929 76
 C. venusta Davies & Trueman, 1927 76
Carbonita spp. 126, 127
 C. humilis (Jones & Kirkby, 1879) 86
 C. pungens (Jones & Kirkby, 1879) 104
Catinula matisconensis Lissajous 150
Catulina knorri (Voltz) 150
Ceratodus 140
Chaetetes septosus (Fleming, 1828) 51
Chlamys valoniensis (Defrance) 140, 141
Chomatoseris 150
Chonetes failandensis S Smith, 1925 28
Classopollis 144, 145, 146
Cleiothyridina
 C. glabristria (Phillips, 1836) 29
 C. cf. *glabristria* (Phillips, 1836) 29
 C. roysii (L'Éveille, 1835) 49
Cleistopora geometrica Milne-Edwards & Haime, 1852 28, 56
Clevosaurus 137
Clisiophyllum aff. *delicatum* Smyth, 1925 45
Clonograptus sp. 10
 C. tenellus (Linnarsson, 1871) 10
Clorinda? 12
Coccosteus 17
Composita 34, 49, 50, 55, 56, 60

C. ficoidea (Vaughan, 1903) 34, 42
conodonts 19, 42
Conularia quadrisulcata Sowerby, 1821 49
Convolutispora 144
corals 19, 30, 41, 42, 43, 178, 179
Cornutisporites 144
Costistricklandia
 C. lirata (J de C Sowerby, 1839) 11, 12
Craniops implicatus (J de C Sowerby, 1839) 12
Craspedobolbina (*Mitrobeyrichia*) *clavata* (Kolmodin, 1869) 11
Crassispora kosankei (Potonié & Kremp) Bharadwaj, 1955 65
crinoids 20, 27, 31, 33, 37, 41
Ctenacanthus sp. 27
Curvirimula
 C. minima 94
 C. subovata (Dewar, 1939) 76
Cyathaxonia cornu Michelin, 1847 30
Cyathoclisia tabernaculum Dingwall, 1926 30
Cypricardinia subplanulata Reed, 1927 12
Cyrtia exporrecta (Wahlenberg, 1818) 12

Dalmanites
 D. cf. *aculeatus* (Salter, 1864) 13
 D. caudatus (Brünnich, 1781) 12
 D. weaveri (Salter, 1849) 11
Dapcodinium priscum Evitt, 1961 146
Davidsonina
 D. carbonaria (McCoy, 1855) 42, 60
 D. septosa (Phillips, 1836) 45
Daviesiella llangollensis (Davidson, 1857) 45
Delepinea
 D. carinata (Garwood, 1916) 31, 42
 D. cf. *carinata* (Garwood, 1916) 42
 D. notata (Cope, 1943) 42
Delthyris elevata (Dalman, 1828) 12
Dibunophyllum 45, 55, 58, 59
 D. bipartitum (McCoy, 1849) 45
 D. bipartitum bipartitum (McCoy, 1849) 45
 D. bourtonense Garwood & Goodyear, 1924 45, 48
Dictyoclostus multispiniferus (Muir-Wood, 1928) 29
Dictyonema flabelliforme (Eichwald, 1840) 10
Diphydontosaurus 137
Diphyphyllum lateseptatum McCoy, 1849 45
Diplocraterion 141
Diplotmema adiantoides (Schlotheim, 1820) 63
Donaldina ashtonensis (Bolton, 1907) 63
Donetzoceras (*'Anthracoceras'*)
 D. aegiranum (Schmidt, 1925) 67
 D. cambriense (Bisat, 1930) 67
Dunbarella 77, 85
 D. cf. *papyracea* (J Sowerby, 1822) 76
Duplicisporites granulatus Leschik emend. Scheuring, 1970 144

Edmondia 41
Ellipsovelatisporites plicatus Klaus, 1960 144
Emileia (Otoites) sauzei (d'Orbigny) 149
Encrinurus onniensis Whittard, 1938 11
Enzonalasporites vigens Leschik, 1955 144
Eocoelia
  E. curtisi Ziegler, 1966 11
  E. sulcata (Prouty, 1923) 12
Eomarginifera aff. *derbiensis* (Muir-Wood, 1928) 30
Eospirifer plicatellus (Linnaeus, 1758) 12
Eotrapezium ewaldi (Bornemann) 140
Euestheria sp. 127
  E. simoni (Pruvost, 1911) 96, 112
Eumetria sp. 49
Eumorphoceras 60
Euomphalus 55, 59
Eupecopteris fletti Kidston, 1925 113

Fardenia? 12
Fasciculophyllum
  F. carruthersi Hill, 1940 45
  F. densum (Carruthers, 1908) 30
  F. omaliusi (Milne-Edwards & Haime, 1851) 29, 178
Favosites spp. 11, 12
Fenestella plebeia McCoy, 1844 49
foraminifera 6, 19, 32, 146

Garantiana sp. 152
Garwoodia (Mitcheldeania) gregaria (Nicholson, 1888) Wood, 1941 37
Gastrioceras
  G. cf. *coronatum* Foord & Crick, 1897 95
  G. subcrenatum C Schmidt, 1924 62, 67, 72, 85, 91, 95
Gattendorfia subinvoluta (Münster, 1843) 19
Gigantoproductus 42
  G. giganteus (J Sowerby, 1822) 59
  G. maximus (McCoy, 1844) 45
Girvanella
  G. ducii Wethered, 1890 37
  G. nicholsoni (Wethered, 1866) Wood, 1941 37
Glassia sp. 12
  G. obovata (J de C Sowerby) 12
Gliscopollis meyeriana (Klaus) Venkatachala, 1966 146
Glyptopomus 17
goniatites 19, 41, 67, 76
Graphoceras
  G. concavum (J Sowerby) 149
  G. formosum (S S Buckman) 149
Gyronema octavium multicarinatum (Lindström, 1884) 11

Haplolasma aff. *subibicina* (McCoy) 42
Haramiya 9
Helodus sp. 49
Hemicycloleaia boltoni Raymond, 1946 104
Hepaticites solenotus Harris, 1938 141
Holoptychius 17
Howellella sp. 12

Hyaena brevirostris Gervais, 1850 161
Hyperlioceras discites (Waagen) 149

Ibrahimispores brevispinosus Neves, 1961 65
Ichthyosaurus 140
Imitoceras sp. 27

Knoxisporites dissidis Neves, 1961 65
Koninckophyllum vaughani Fedorowski, 1970 45
Kraeuselisporites reissingeri (Harris) Morbey, 1975 146
Kuehneosaurus 137
Kuehneosuchus 137
Kuehneotherium 137

Leaia 73, 112, 116, 119
  L. bristolensis Raymond, 1946 104
  L. parallela Raymond, 1946 104
  cf. *Leangella segmentum* (Lindström, 1861) 12
Leioceras opalinum (Reinecke) 149
Leptaena depressa (J de C Sowerby, 1824) 12
Leptagonia
  L. analoga (Phillips, 1836) 30, 49
  L. cf. *analoga* (Phillips, 1936) 28
Leptostrophia compressa (J de C Sowerby, 1839) 11
Levitusia humerosa (J Sowerby, 1822) 42
Limbosporites 144
Lingula 27, 45, 63, 73, 79, 81, 90
  L. mytilloides J Sowerby, 1813 63, 77, 85, 95
  sp. (*squamiformis*) Phillips, 1836 76
Lingulella sp. 10
Linoprotonia
  L. corrugatohemispherica (Vaughan *in* Dixon & Vaughan, 1911) 42
  L. hemisphaerica (J Sowerby, 1822) 45
Lioestheria 86
Lithostrotion 34, 39, 41, 42, 50, 51, 55, 56, 58
  L. araneum (McCoy, 1844) 60
  L. junceum (Fleming, 1828) 45
  L. martini (Milne-Edwards & Haime, 1851) 34, 42, 45, 48, 56, 60
  L. pauciradiale (McCoy, 1844) 45, 55, 56
  L. cf. *sociale* (Phillips, 1836) 45
Lonsdaleia 48
  L. floriformis (Martin, 1809) 43, 45, 51, 56
Loxonema sp. 12
Lugwigia
  L. haugi Douvillé 149
  L. murchisonae (J de C Sowerby) 149
Lycospora pusilla (Ibrahim) Somers, 1976 65
Lyriomyophoria 140

Macoma balthica (Linnaeus) 161
Macropotamorhynchus mitcheldeanensis (Vaughan, 1905) 49
Mariopteras acuta (Brongniart, 1831) 63
Megachonetes

M. magna (Rotai, 1931) 30, 42
M. papilionaceus (Phillips, 1836) 42
Meleagrinella
  M. decussata (Münster) 141
  M. echinata (W Smith) 150
Mendacella cf. *phiala* Whittard & Barker, 1950 11
Meristina obtusa (J Sowerby, 1818) 12
Mestognathus beckmanni Bischoff, 1957 42
Micrhystridium 144, 146
Microlestes 9
Microsphaeridiorhynchus nucula (J de C Sowerby, 1839) 12
miospores 146
Modiola sp. 49.
Modiolaris 95
Modiolus 27, 141
  M. minimus J Sowerby 132, 142
  M. hillanus (J Sowerby) 142
Morphoceras macrescens (S S Buckman) 149
Mytilus 140, 141

Naiadita 144
  N. lanceolata Buckman emend. Harris, 1938 141
Naiadites sp. 76, 85, 86
  N. flexuosus (Dix & Trueman, 1932) 91
  N. cf. *obliquus* (Dix & Trueman, 1932 86
  N. productus (Brown, 1849) 77, 85
'*Natica*' 140
nautiloids 41
Nemistium edmondsi Smith, 1928 45
Neoglyphioceras 45
Neuropteris 107
  cf. *N. antecedens* Stur, 1875 63
  N. gigantea (Sternberg, 1821) 63
  N. heterophylla (Brongniart, 1822) 63
  N. scheuchzeri Hoffmann, 1826 63

Ophiolepis damesi Wright 140
Oppelia (Oxycerites) yeovilensis Rollier 149
Orbiculoidea 90, 140
Orbiculoides sp. 63
Orionastraea 43
Ornithella bathonica (Rollier) 150
Orthoceras sp. 49
Ortonella kershopensis Garwood, 1931 37
Otozamites obtusus (Lindley & Hutton) 142
Ovalipollis pseudoalatus (Thiergart) Schuurman, 1976 144, 146

Paladin 45
Palaeosaurus 136
Palaeosmilia
  P. murchisoni Milne-Edwards & Haime, 1848 41, 42, 45, 48, 55, 56, 58, 59, 60
  P. regia (Phillips, 1836) 48
Paleocyclus 11
  P. porpita (Linnaeus, 1758) 11, 12
Parallelodon 41
Parkinsonia
  P. acris Wetzel 149
  P. bomfordi Arkell 149

P. convergens (S S Buckman) 149
P. parkinsoni (J de C Sowerby) 149
Peltocare olenoides? (Salter, 1866) 10
Pentamerus? 12
Perinosporites 144
Phaulactis cf. *angusta* (Lonsdale, 1839) 12
Phialaspis 16
Picrodon herveyi Seeley, 1898 141
Pitymys gregaloides Hinton 161
Plagiostoma 142
Planocephalosaurus 137
Planolites montanus R Richter, 1937 81
Plectodonta sp. 12
Plesiosaurus 140
Pleuromya 142
Pleurophorus elongatus Moore, 1861 140
Polygnathus bischoffi Rhodes, Austin & Druce, 1969 42
Porcellispora 144
Praeexogyra acuminata (J Sowerby) 150
Prodactylioceras davoei (J Sowerby) 158
productoids 33, 43
Proetus sp. 12
prolifera 68
Promytilus 45
Protocardia 140
Psiloceras 128, 146
Pterinea sp. 12
Pterinopecten sp. 63
Pteroretis sp. 65
Pugilis vaughani (Muir-Wood, 1928) 29
Pustula
 P. cf. *pustuliformis* Rotai, 1931 30
 P. pyxidiformis (de Koninck, 1847) 30
Pycnactis mitrata (Schlotheim, 1820) 12

Raistrickia fulva Artuz, 1957 65
Resserella elegantula (Dalman, 1828) 12
Rhacophyton 27
Rhaetavicula contorta (Portlock) 140
Rhaetipollis germanicus Schulz, 1967 144, 146
Rhaetogonyaulax rhaetica (Sarjeant) Loeblich & Loeblich emend. Harland, Morbey & Sarjeant, 1975 144, 146
Rhipidomella michelini (L'Eveillé, 1835) 29
Rhodea feistmanteli Kidston, 1923 63

Rhynchonelloidella smithi (Davidson) 150
rhynchonelloids 33
Rhynchotreta cuneata (Dalman, 1828) 12
Ricciisporites tuberculatus Lundblad, 1954 144, 146
Rugitela bullata (J de C Sowerby) 150
Rugosochonetes vaughani Muir-Wood, 1962 29
Salopina conservatrix (McLearn, 1924) 12
Sanguinolites 27, 45
Saurichthys apicalis Agassiz 132
Schartymites cornubiensis (Ramsbottom, 1970) 76
Schellwienella
 S aspis Smyth, 1930 30
 S. crenistra (Phillips, 1836) 49
Schmidites sp. 10
Schuchertella cf. *wexfordensis* Smyth, 1930 30
Secarisporites remotus Neves, 1961 65
Seminula 50, 56
Serpulites (Campylites?) perversus (McCoy, 1853) 12
Sigmala 137
Siphonophyllia 178
 S. cf. *caninoides* (Sibly, 1906) 42
 S. cylindrica Scouler *in* McCoy, 1844 29, 30, 178
 S. garwoodi Ramsbottom & Mitchell, 1980 30
Skenidioides lewisii (Davidson, 1848) 12
Sonninia (Fissilobiceras) ovalis (S S Buckman ex. Quenstedt) 149
Spelaeotriletes arenaceus Neves & Owens, 1966 65
Sphaerirhynchia davidsoni (McCoy, 1851) 12
Sphenopteris 107
spiriferoids 34, 45
Spirorbis 41, 51
Spirorbis sp. 126
Stegerhynchus
 S. borealis (von Buch, 1834) 12
 S. diodonta (Dalman, 1828) 12
 S. ?weaveri (Davidson, 1869) 11
Stephanoceras humphriesianum (J de C Sowerby) 149

Stereisporites 144
Strenoceras
 S. (Garantiana) garantiana (d'Orbigny) 149
 S. subfurcatum (Zieten) 149, 150
?Strepsodus 52
Strigoceras truellei (d'Orbigny) 149
Striispirifer plicatellus (Linnaeus) 12
Strophochonetes? 12
Sychnoelasma
 S. clevedonensis Mitchell, 1980 29, 178
 S. aff. *kentensis* (Garwood, 1912) 42
 S. konincki (Milne-Edwards & Haime, 1852) 30
Syringopora spp. 45
Syringothyris
 S. cuspidata (Martin, 1796) *cyrtorhyncha* North, 1920 29
 S. aff. *elongata* North, 1920 30

Tasmanites 144
Tentaculites
 T. anglicus Salter, 1859 11
 T. ornatus J de C Sowerby, 1839 13
Thecodontosaurus 136, 137
 T. antiquus Morris, 1843 137
Tmetoceras scissum (Benecke) 149
Tolmaia? 13
Triancoraesporites 144
Tumulites (Eumorphoceras) sp. 60, 63

Variodens 137
Vaughania vetus Smyth, 1930 28, 52, 56
Veryhachium 146

Whitfieldella sp. 12
Witchellia
 W. laeviuscula (J de C Sowerby) 149

Zaphrentis
 Z. delanouei Milne-Edwards & Haime, 1852 29, 178
 Z. konincki (Milne-Edwards & Haime, 1852) 178
Zebrasporites 144
Zigzagiceras zigzag (d'Orbigny) 149

GENERAL INDEX

Aalenian era 129, 147, 149
Abbots Leigh 5, 9, 27, 52–56, 133, 134
  celestite at 155
acuminata beds 150
Adam's Seam (Downend Formation) 99
agate 143
aggregate: economic geology of 155–157
Air Balloon Pit 99
alluvium 164
Almondsbury 43, 65, 155
Alveston 19, 70
Alvington 162
ammonites 6
Anabacia Limestone 149, 150
Anglian Glaciation 161, 162
Argus Colliery 75, 76
Argyle Drift (Nettlebridge Valley) 92
Arnold's Quarry 37, 39
Arundian 27, 31, 42
Asbian 27, 45
Ashton 73–76, 159
Ashton Gate 131
Ashton Gays Vein 73
Ashton Great Vein 73–74, 75, 78, 85, 95, 160
Ashton Little Vein 73, 95, 160
Ashton Park 52–56, 164
Ashton Park Borehole 37, 42, 43, 62, 63, 64, 73–74, 160
Ashton Top Vein 74, 75–76, 78, 79, 85
Ashton Vale 159
Ashton Vale Borehole 168
Ashton Vale Colliery 62, 73–75, 78, 79, 80
Ashton Vale Marine Band 62–64, 67, 71–72, 73, 75, 76, 78, 79–80, 85, 88, 90–91, 94, 95
Ashton Watering 56
Ashwick 41, 64, 91
Aubrey 7
Aust 5, 6
Aust cliff 5, 6, 19, 132, 133, 134, 135, 140, 142
Avening Green 12
Avon, River 163
Avon Gorge 4, 5, 9, 17, 19, 20, 23, 25, 46–48, 52
  Black Rock Limestones in 28–30
  Clifton Down Group in 31, 32, 34, 37
  Hotwells Limestones in 43
  limestones quarries in 156
  Lower Limestone Shale in 27, 28
  Pleistocene deposits in 162
  river terraces 163
  vertical sections in 21, 35
Avon Navigation 159
Avon thrust fault 4, 32, 48, 51, 52, 54–55

Avon Valley 8, 70, 141
  coal mining at 159
  surface relief 3, 4
Avonmouth 4, 104, 164
Avonmouth Basin 19, 70, 131, 132, 135
Avonmouth Boreholes 166–167
Avonmouth seams 71, 73, 116
Avonmouth Veins 104
Axbridge 94

Backwell 156
Backwell Common Pits 94
Backwell Litle Coal 95
Backwell Quarry 60
Badger Gug 107
Badger Vein 107
Baggridge 147, 150
Baggridge Borehole 150
Baggridge Hill 151
Bajocian 129, 147, 148, 149, 152
Bantam Vein 106
Banwell 3, 135
Banwell Borehole 68, 173
Banwell Moor 132
Banwell Moor Borehole 134, 172
Barlake Colliery 92–93
Barns Batch Beds 129, 148, 149
Barren Red Formation 67, 68, 72, 93, 116–118
  sections in 117, 118, 122, 123, 124, 125, 126, 127
Barrow Gurney 3, 40, 65, 91, 95, 133, 134
  coal mining at 159
Barrow Gurney Boreholes 95, 172
Barrow Hill 60
baryte 133, 142, 143, 154
barytocelestite 133, 135, 142, 154
basalts
  in Carboniferous Limestone 58, 59
  in Llandovery beds 11
  at Goblin Combe 46
Batches Wood 60
Bath 1, 3, 4, 5, 8, 141, 151
  Coal Measures at 70, 73, 90–91
  Jurassic rocks at 147, 150
Bath Oolite 150, 151
Bathonian 129, 147, 149
Bathurst Basin 131
Batscombe 136
Beachley 5, 6, 19
Beacon Hill 2, 19, 147
  Pericline 16, 17, 19, 29, 64
Bedminster 73–76, 131, 132
  coal mining at 159
Bedminster Boreholes 168–169, 171
Bedminster Deep Pit 76
Bedminster Down 144
Bedminster Great Vein 75, 76, 79, 95
Bedminster Little Vein 75, 79
Bedminster Smith's Coal 73, 75
Bedminster Toad Vein 74–75, 77, 80, 95
Bedminster Top Vein 75–76, 79
Beggar's Bush 55
Belgium Pit 84

Bellerephon Beds 33, 37, 41, 49, 52, 55, 58
Belluton 3, 126, 130, 134
Belmont Hill 37
Bendle Combe 56
Berkeley Fault 10, 13
Biddle Fault 93, 111
Big Fiery Vein 79, 80
Binching Coal 82
Binegar 41, 152
Birdcombe 58
Birnbeck Limestone 41
Bishop Sutton 95, 110, 111
  coal mining at 159
Bishop Sutton Colliery 93, 95, 109
Bishop Sutton Colliery Borehole 173
Bitham's Wood 134
Bitton 5, 73, 84, 99, 134, 142
  coal mining at 158, 159
Bitton Boreholes 169
Bitton Fault 100
Black Down 2, 3, 14, 17, 19
Black Down Pericline 2, 3, 17, 19, 28–29, 41, 43, 95
Black Nore Sandstone 14, 16
Black Rock Dolomite 27, 28, 29, 49, 51, 52, 55, 56–57, 58, 59, 60, 103, 104
  in Pleistocene deposits 161
Black Rock Limestone 56–57
  biostratigraphy 178
  in Pleistocene deposits 161
Black Rock Limestone Group 20, 24–25, 27, 28–31, 46, 48, 49, 51, 52, 54, 55, 57, 58, 59, 103
  dolomitisation of 21, 23
  isopachytes in 36
Black Rock Quarry 28, 48, 57
Black Seam 100
Black vein 76
Blackstone 91
Blagdon Reservoir 132, 144
Blaise Castle 156
Blaise Castle Boreholes 169
Bleadon 164
blende 133
Blue Anchor Formation 128–129, 130
Blue Lias 3, 8, 128–129, 131, 141, 142
  building stone from 157
Blue Pot seam 91
boreholes
  abstracts of 166–176
  map of 69
  see also under individual names
Bottom Little Vein 119
Bottom Veins 107
Bourne 132
Bourton Combe 60, 162
Boyd, River 5, 90
Boyd Valley 143
Bradford Clay 151
Bradford-on-Avon 4, 5
Bradley Brook 97
Branch Coal 92
Brandon Hill 64, 131, 143
Brandybottom Pit 112, 114
Brandybottom Pit Borehole 171

Brandybottom Pit Borehole 171
brass industry in Bristol 155
Braysdown Colliery 106–107, 119, 121
Braysdown Colliery Borehole 175
Breadstone House 10
Breadstone Shales 10
Brean Down 2, 164
Brentry 51–52, 143
 lead ore at 154
 limestone quarries at 156
Brentry Hill 51
Brentry Quarry 154
brick clay 158
Bridge Gate 90–91
Bridge Yate 4
Brigantian 27, 45
Bright's Vein 105, 106
Brinkmarsh Beds 10, 11, 12
Brislington 5, 70, 126, 130, 131, 142
 clay from 158
 coal mining at 159
 head deposits at 164
 iron ore at 153
 workings, in Farrington Formation 111, 118
Bristol
 glass manufacture at 157–158
 lead ore at 154
Bristol and Somerset Coalfield 3, 8, 28, 29, 37, 42, 60, 63, 66–68, 73, 103, 105, 135
 vertical sections in 44
Bristol Avon 3, 4, 5, 6, 90
 river terraces 162–163
Bristol Boreholes 168–169, 170, 171
Brittain's Seam 84
Broad Lane Borehole 88
Broadfield 162
Broadfield Down 7, 8, 19, 20, 23, 31, 37, 40–41, 91, 132, 135, 152
 earth pigments at 154
 Hotwells Group at 42, 43
 lead and zinc at 154, 155
 roadstone and aggregate in 156
 surface relief 3
Broadfield Down Inlier 41
Broadfield Down Volcanics 46
Broadwell Down 3
Brockley 164
Brockley Combe 3, 30, 41, 59
 lead ore at 154
Brockley Down Limestone 157
Brockley Fault 59–60
Brockley Oolite 41, 58, 59
Bromley Coals 73, 93, 118
Bromley Colliery 109, 112, 117, 118, 124
Bromley Colliery Borehole 172
Bromley Horst 110, 111, 117–118, 124, 126
 coal mining at 159
Bromley Sandstone 110, 111, 117
Bromley Seams 110, 117
Bromley Veins 110
Broomhill 97
Brown Ironshot 148, 149
Brownstones 14, 16

Bruton 149
Bryant's Pit 114
Bryozoa Beds 19, 27, 30, 46, 49, 52, 56, 57
Buckover 12
Buff Coal 73
Buff Vein 99
building stone 157
Bull Vein 119, 121
Burgh Walls 52
Burley Grove 97
Burnett 124, 126, 159
Burrington 19
Burrington Combe 178
Burrington Combe gorge 2, 15, 19, 20, 29–30, 41, 43
 vertical sections in 21
Burrington Oolite 27, 41, 42
Bursall Bridge 10
Burtle Beds 161
Bury Hill (Wick) 102, 155
Buryhill Fault 102
Butcombe 3, 60
Butcombe Court 58, 59, 61
Butcombe Sandstone 134, 143

Cadbury Camp 41, 58
Cadbury Hill 58
calamine 155
Caldecote 65
California Colliery 99
Callovian 129
Cam Brook 5
Cambrian 10
Cambridge Batch 37, 43, 56, 132
 Pleistocene deposits at 162
Camerton 107, 111, 126
Camerton Colliery 119, 122, 125
Camerton Colliery Boreholes 173
Canina bristolensis Bed 34, 48
Caninia-Oolite 31, 32
Carboniferous Limestone 6–9, 17, 19–61
 building stone from 157
 conglomerates 45–46
 drainage 4–5
 earth pigments in 153
 iron ore in 153
 manganese in 153
 section through 50
 structure 53
 surface relief 2, 3
Carnian 128, 146
Castle Hill 57
Castle Hill Sandstone 134
Castle Wood Limestone 43
Castle's Ridgeway Pit 96, 97
Cathead Vein 72, 107, 111, 116
Cattybrook 65, 66, 70, 95, 104
 brickyards at 158
 coal mining at 159
Cattybrook Clay pit 95
Celestine Bed 130
celestite 132–135, 143, 155
Ceratodus Bone Bed 140
Chadian 27, 30, 42
Chaffhouse Pit 114

Charfield 13, 14, 19
Charfield Green 10, 11, 12
Charfield Hill 12
Charlton 140, 142
Charlton Field 3
Charmborough 101
Charn Hill 97
Cheddar 94
Cheddar Gorge 2, 19, 28
Cheddar Limestone 27, 41
Cheddar Oolite 27, 41
Chelvey Pumping Station 132
Chelwood 111, 124, 127, 132
 glass sand from 157
Chepstow 20, 24, 26, 29, 37
Chepstow Shelf 23
Chesley Hill 90
Chester Park 84
Chew, River 3, 5, 124
Chew Magna 95, 130, 132, 134
Chew Stoke 132
Chew Stoke Borehole 172
Chew Valley 63, 93, 95, 132, 134
 coal mining in 159
Chewton Keynsham 135
Chewton Keynsham Engine Pit 126
Chewton Mendip 91, 93
Chick Seam (Downend Formation) 96, 97
Chief Shell Bed (Ashton Vale Marine Band) 62, 73
Chilcompton 144
Chilcompton New Pit 101
Chipping Sodbury 19, 37, 39, 132–135
 lead ore at 155
Chipping Sodbury Borehole 88, 90, 167
Church Close Vein 107
Church Farm Colliery 102
Churchill 19, 95, 164
Clandown Colliery 119, 124, 126
Clandown Colliery Boreholes 173
Clandown Fault 105, 108
Clandown West Colliery 119
Clapton Basin 103–104, 132, 135
Clapton-in-Gordano Basin 1, 14, 19, 42, 65, 66, 70, 103, 159
 surface features 3
Clapton-in-Gordano Boreholes 167
Claverham Borehole 172
Cleeve Bridge 63
Cleeve Court 59
Cleeve Fault 58, 59
Cleeve Toot 59
Clevedon 3, 5, 6, 14, 17, 19, 20, 41, 56–57, 70, 131, 135, 161, 164
Clevedon Boreholes 167
Clevedon Fault 57
Clevedon–Portishead ridge 14, 19
Clift Woods Thrust 94
Clifton 3, 4, 5, 28, 48–51, 134, 135, 136, 137, 141, 142, 163
 haematitisation at 143
Clifton Down Group 20, 23–24, 27, 31–42, 51
 isopachytes in 36

GENERAL INDEX 193

Clifton Down Limestones 20, 27, 33, 39–42, 43, 48, 50–52, 54, 55–61, 103
  conglomerates in 45
  lead ore in 154
  limestone quarries in 156, 157
Clifton Down Mudstones 20, 23, 24, 27, 28, 32–33, 37, 40, 42, 46, 48, 49, 50, 51, 52, 55, 56, 57, 58, 59–60, 61
Clifton Down Oolite 37, 41
Clifton Down Railway Tunnel 51, 52, 134
Clifton Suspension Bridge 52
Clifton Wood 131
Clutton 66, 103, 111, 117, 119
  coal mining at 159
Clutton Colliery 107
Clutton Fault 103
Clutton Ham Pit 105
coal 158–160
  principal seams 73
  in Quartzitic Sandstone 64–65
Coal Barton Pits 91
Coal Measures 1, 5–9, 66–127
  classification of 71–72
  distribution and thickness 68–71
  principal seams 73
  surface relief 3, 4
Coalpit Heath Basin 1, 5, 10, 14, 19, 20, 29, 37, 42, 66, 68, 70, 71–73, 87, 96–98, 101–102, 112–116, 117–118, 132
  celestite in 155
  surface features 3
Coalpit Heath Basin Boreholes 87
Coalpit Heath Colliery 113, 114
Coalpit Heath Fault 113, 114
Cock Seam (Downend Formation) 96, 97
Cockheap Wood 103
Codrington 19, 70
Coke Seam 100
Coking Coal 92
Coking Coal Seam 72, 99
Cole Syncline 149
Coleford 70
Coleford Boreholes 175
Coleford High Delf 112–116
Coleford High Delf seam 71, 73
colluvium 163–164
Combe Down Oolite 150, 151
Combe Hay 147, 150, 160
Compton Common 127
Compton Dando 70, 124, 127
  coal mining at 159
Compton Dundon 134
Compton Greenfield 131
Compton Martin 45
Concretionary Beds (Clifton Down Oolite) 37, 41, 48, 50, 51, 55, 56
  limestone quarries in 156
cone-in-cone structures 10
Congresbury 40, 58
Conygar Quarry 104
Conygre Colliery 111
Conygre Colliery Boreholes 175
Cook's Pit 114
Cookswood 64

Coombe Dingle 5, 52
Coombe Hill 51
copper 154–155
Coralline Beds 148, 149
Cornbrash 129, 151
Cornstone 17, 18
Cornwell Farm 61
Corston 8, 66, 90
  coal mining at 159
Cosham Colliery 86
Cotham 134, 135
Cotham beds 128–129, 131, 140, 141–142
Cotham Marble 7, 141
Cotham Member 129, 144, 146
Cotswolds 1, 3, 4, 8, 147, 150
Courceyan 27
Court Hill 94, 103, 161, 162
Crease Limestone 37
Crews Hole 99
Cribbs Causeway 140, 142
  in Clifton Down Group 31, 33, 37, 41
Crock's Bottom 91
Croft's End 82
Croft's End Brickworks 85
Croft's End Clay 78
Croft's End Marine Band 67, 71, 72, 76, 78, 83–84, 86–87, 88, 90, 93, 95
Cromhall 1, 16, 19, 20, 21, 29, 45, 63, 64, 68, 70, 73, 95, 136, 137
  quarries at 156, 157
Cromhall Sandstones 20, 23, 31, 32
Cromhall Vein 65, 88
Crow Vein 94, 95
Crown Colliery 82
Crown Hill 3
Crown Inn Borehole 60
Cuckoo Seam 99

Dabchick coal 107
Damery 10, 11
Damery Beds 11
Daniel's Wood 12
Dean Lane Colliery 74, 75, 76, 79
Deep Pit, Kingswood 72, 78, 80, 81, 82
deltaic sediments 19
Denny Island 5, 6, 19
Denny Island Fault 104
Derbyshire Limestone 8
Devensian 162, 163, 164
Devil's vein 76
Devonian 10, 14–18
Dial Hill 56
Dial Quarry 60
Dibble Vein 99
Dinantian 9, 19–61
Dirty Duck Vein 106
Dog Vein 94, 95
Dogtrap Pit 88
dolomite
  in Black Rock Limestones 20, 28–29
  in Clifton Down Group 41
dolomitic concretions 16
Dolomitic Conglomerate 5, 16, 32, 50, 51, 52, 56, 57, 58, 60, 93, 103–104, 128, 131, 132, 134, 135–140
  building stone from 157
  earth pigments in 153

  haematitisation of 142–143
  iron ore in 153
  manganese in 153
dolomitisation
  Black Rock Limestone Group 21, 30
  in Dinantian rocks 23–24
  in Triassic rocks 143–144
Doublescreen Pit 95
Doulting 150
Doulting Conglomerate 150
Doulting Stone 149, 150
Downend 85–87
  coal mining at 160
  iron ore at 153
Downend Borehole 160
Downend Formation 67, 68, 70–71, 84, 95–101
Downside 91, 154, 157
Downside Abbey 93
Downtonian 12, 16
Doxall Vein 82, 84, 85, 87
drag-marks 12
drainage 4–5
Draycott 157, 164
Draycott Marble 157
Dromley Heath 97
Drybrook Limestone 37, 45–46
Dudley Pit 114
Dulcote 143
Duncorn Hill 147, 150
Dundry 76–77, 131, 134
Dundry (Elton Farm) Borehole 76, 132, 134, 142, 144–146, 172
Dundry Freestone 149
Dundry Hill 76, 132, 142, 147–149
  building stone from 157
Dundry Outlier 3
Dundry Stone 157
Dung Ball Island 6
Dungy Drift 72, 73, 93
Dungy Seam 95
Dunkerton 124
Dunkerton branch of Somerset Coal Canal 8
Dunkerton Colliery 107, 111, 119, 121, 122
Dunkerton Colliery Boreholes 173, 175
Durdham Down 37, 48, 49, 50, 136, 137
  lead ore at 154
Dyrham Silt 129

earth pigments 153–154
East Clevedon 56–57, 57–58, 103
  glass sand from 158
East Harptree 3, 152
East Mendip Basin 23
Easton 131, 134
Easton Colliery 73, 79, 80, 81, 82
Easton Four Feet Seam 79, 80
Easton Gay's Vein 80
Easton Great Vein 81–82
Easton-in-Gordano 16, 164
Easton Red Ash Vein 79, 80, 85
Easton Seven Feet Seam 79, 80
Easton Three Feet Seam 80
Easton Two Feet Seam 79, 80

194 GENERAL INDEX

Eastville 4, 96, 97
Ebbor 43, 68, 91, 94
Ebbor Outlier 1
Ebbor Rocks 159
economic geology 153–160
Edford 159
Edford Colliery 91, 92
Eggshill Colliery 88
Elberton 37, 156
Elster Glaciation 161
Elton Farm 76
Elton Farm Limestone 129, 148, 149
Emborough 91, 93, 136, 137, 142
  coal mining at 159
Emmerson's Green 102
Engine Bottom Pit 114
Engine Common 98, 102
English Stones 5
Estheria Bed 141
evaporites 132–134

Failand 3, 14, 16, 19, 41, 49, 52–56, 136
  roadstone and aggregate at 156
Fairy Hill 126
Falfield 11
Farleigh Combe 162
Farler's Pit 95
Farmborough Borehole 172
Farmborough Fault 1, 68, 73, 93, 110
  investigation of 111–112
  in Radstock formation 119
  section through 124, 125
Farrington Colliery 107
Farrington Colliery Borehole 173
Farrington Formation 67, 68, 71–73, 93, 104, 104–116
  sections in 106, 109
Farrington Gurney 8, 105, 111, 130, 131
Farrington Gurney Borehole 173
Felton 59, 60, 157
Felton Borehole 60
Fenswood 56
Fern Rag 72, 91
Fifty Acre Plantation 54
Filton 43, 64, 134, 135, 140, 142
Firestone Coal Seam 73, 93, 95
Fishponds 73, 80, 82, 97
fissure deposits (Triassic) 135–140
Five Coals Vein 82, 160
Flandrian 161, 164–165
Flax Bourton 3, 43, 132, 164
  glass sand from 158
Flax Bourton Borehole 172
Flowers Hill 118, 126
Flowers Hill Fault 131
Fluorite Bed 34, 48
Folly Bridge 102, 113
Foot Coal 92
Forest Marble 129, 151
Forest of Dean 15, 16, 19, 20, 23–24, 26, 37, 112
  syncline 14
Forty Yard Seam 119, 126, 127
Four Foot Vein 73
Foxcote Colliery 121

Foxcote Colliery Borehole 175
Frampton Cotterell 101–102, 130
Frampton Cotterell Sandstone 102
Fraynes Colliery 74–75
Freezing Hill 149
Frenchay 97
Frog Lane Borehole 166
Frog Lane Pit 113, 114
Froglane Pit 112
Frome 1, 2, 20, 31, 41, 97, 147, 151
Frome Clay 147, 150, 151
Frome River 5, 97, 102
Fryar's Pit 114
Fry's Bottom 111
Fry's Bottom Colliery 119, 121, 122
Fry's Bottom Colliery Borehole 172
Fry's Bottom Pit 93
fuller's earth 160
Fuller's Earth 129, 149, 150, 151
Fullonicus Limestone 150

galena 133, 142, 154
Garden Cliff 140
Garden Course 100–101
Garden Course Vein 71–73, 84
Gas Coal Seam 74
Gatcombe 56
geodes 143
Gillers Inn Vein 78, 79, 80, 81, 86, 87, 160
glass sand 157–158
Glashouse Pit 95
The Glen (quarry) 50
Globe Colliery 100
Globe Fourth seam 72, 101
Gloucester 6
Goblin Combe 3, 20, 23–24, 46, 58–59
Goblin Combe Oolite 23, 27, 40–42, 48, 49, 52, 55, 56, 57–58, 59, 60, 61
goethite 153
Golden Candlestick Coal 92
Golden Valley 76, 84, 90
Golden Valley Pit 94–95
Golden Valley Pits 90, 99, 158
Golden Valley Seam 94, 95
Golden Valley Top Seam 94, 95
Goodeaves Pit 91
Gordano basin 68, 70
Gordano Valley 162
Grace's Seam 94, 103
Grandmother's Rock 19, 70, 90
Gratwicke Hall 91
Gravesend 156
Great Cart Pit 114
Great Course Coal 73
Great Course Seam 93, 95
Great Fiery Vein 74
Great Oolite 3, 129, 150–151
Great Oolite Limestone 129, 150, 151
  building stone from 157
Great Quarry (Avon gorge) 33, 34, 50, 55
Great Vein 107, 111, 112, 114, 119, 122, 124, 160
  isopachytes map 121
Great Western Colliery 98
Great Western Shaft 79

Green, The 90
Grey Marls 135
Greyfield Colliery Borehole 173
Greyfield (Grayfield ?) Colliery 105, 106, 107, 110
Grove Farm Beds 129, 148, 149
Grovesend 20
Gully Oolite 20, 23, 27, 28, 30, 31, 37, 40–42, 48, 49, 50, 51, 52, 55, 56–57, 58, 59, 60
Gully quarry 28, 30, 35, 48, 178
Gurney Slade 64, 136, 152
gypsum 132–133, 134

haematitisation of Triassic rocks 142–143
Hale Combe quarry 31
Half Moon Pit 113
halite 132
Hallen 116, 131
Ham 91
Ham Green 162
Hambrook 5, 85, 97, 102, 131, 164
Hangbeggar Pit 114
Hanham 4, 5, 70, 162, 163
  coal mining at 159
Hanham Colliery 74, 79, 80, 81, 82, 84, 99, 100, 160
Hanham Red Ash Vein 79, 80, 85
Hanham Smith's Coal 80
Hanham Two Foot Vein 81
Hanham White Ash Vein 79, 80
Hard Coal 92–93
Hard Vein 112, 113, 114, 116, 122
  structure contour map 115
Hard Vein Pit 114
Hard Venture Stone 71, 74
Hard Venture Vein 78, 79, 80
Harptree 93–94
Harptree Beds 59, 152
  earth pigments in 154
Harptree Kleppe 111
Harridge Wood 91
Harrington's Coal Works 8
Harris's Quarry 142
Harry Stoke 64, 81, 84, 85–87, 90, 131, 142
  coal mining at 160
  haematitisation in 142
Harry Stoke Boreholes 71, 77, 78, 79, 83, 84, 85, 86, 87, 88, 96, 160, 170–171
Harry Stoke Drift Mine 82, 85, 86, 160
Harry Stoke Marine Band 72, 74, 76–77, 79, 80, 85–86, 87, 88, 92
Hartcliff Rocks 60, 135, 142–143
  roadstone and aggregate at 156
Haskin's Claypit 99
Haskin's Pit 85
Hayswood Colliery 111
Hayswood Colliery Borehole 175
head 163–164
Healls Scars 59, 60
hematite 153
Hemington Borehole 175
Hen Seam 96
Hen Vein 73, 79, 84, 96–97

GENERAL INDEX 195

Henbury 3, 5, 7, 37, 43, 51–52, 65, 104, 140
  copper at 155
  lead and zinc ore at 154–155
Henbury Boreholes 169
Henbury Hill 51, 156
Henfield 112, 113
Henlease 140, 142
Henlease Bathing Lake 49, 50
Hettangian 129, 146
High Coal Seam 68
High Littleton 7
High Littleton 117, 159
High Vein 71, 73, 103, 112, 114, 116
High Vein Pit 114
Highbridge Common 75
Highbury Pit 91
Highfield Fault 90
Highridge 76
Holbrook Common 90
Holcombe 91, 92, 93, 101
Hole Vein 82
Holkerian 27
Hollybrook Brick Works 85
Hollybush Vein 112, 113, 114, 116
Holmes Rock 71
Holwell 135
Hooke 7
Hopewell Hill 81
Horsecombe Vale 147, 150
Horseshoe Farm 12
Hotwells 2, 4, 5
Hotwells Group 20, 29, 42–45, 48, 50, 51, 55, 56
  isopachytes in 36
Hotwells Limestone 27, 37, 42, 51, 52, 56, 58, 59, 60, 61, 95
  in Pleistocene deposits 162
Huddox Hill 147
Hunstrete 112, 127
Hursley Hill 3, 124, 126, 127
  vertical sections at 120
Hursley Hill Borehole 110, 118, 119, 126, 172
Hussley Hill Borehole 111

Ifton 65
Inferior Oolite 3, 5, 129, 147–150
Ipswichian deposits 161, 162
iron
  economic geology of 153
  oxides in Triassic rocks 142, 143
Iron Acton 71, 102, 112, 114
Iron Acton Fault 96
Iron Shot Limestone 129, 148, 149
Itchington 19, 37

Jingleboys 79, 81, 86
Jone's Seam 99
Jubilee Main (Pensford Colliery) 118
Jubilee Return (Pensford Colliery) 118
Jurassic 9, 60, 93, 147–152
  succession in 129

Kendelshire 102
Kenn 1, 5, 70, 161, 162, 163, 164
Kenn Gravels 161

Kenn Moor Seam 90, 99
Kennet and Avon Canal 8, 159
Kennpier Till 161, 162
Keuper Marl 95, 128–129, 132–135, 140, 141
  dolomitisation of 143–144
  haematitisation of 142–143
Keynsham 3, 5, 141
  sand and gravel from 157
Kidney Hill Fault 88, 112, 113, 114
Kilkenny Bay 16, 18
Kilmersdon 147, 151
Kilmersdon Colliery 106, 107, 119, 160
Kilmersdon Colliery Borehole 173
Kilmersdon New Great Vein 106
Kings Weston 67, 70, 95, 104, 144, 154
Kings Weston Boreholes 169
Kings Weston Conglomerate 45
Kings Weston Hill 52, 54
Kings Weston Quarries 52
Kingsdown 131, 134, 135
Kingston Seymour 6, 68
Kingswood 63, 70, 77–85
  clay from 158
Kingswood Anticline 1, 4, 64, 66, 68, 70, 71, 73, 76, 77–85, 86, 88, 90, 96–100
  coal mining in 158, 160
  and Farmborough Fault 112
Kingswood Borehole 171
Kingswood Colliery 78
Kingswood Great Vein 72, 73, 75, 79, 81–82, 83, 84, 85, 86, 87, 90, 160
  structural problems 78
Kingswood Hard Vein 82–83
Kingswood Little Vein 73, 75, 79, 80–81, 86
  structural problems 78–79
Kingswood Toad Vein 72, 82, 83, 84, 85, 87
  structural problems 78
Kingswood Two Foot Vein 80–81
knorri Clays 150
Knowle Hill 134, 142

Ladden Brook 5
Lady Bench 5, 70
Ladye Bay Fault 56, 57
Lamb Leer Fault 111
laminosa dolomite 20, 23, 48, 49, 52
Landscape Marble 7, 141
Landsdown Hill 147, 151
Langport Beds 7, 128–129, 140, 141
Lansdown 3
Lapwater Pit 114
Latterbridge 131
Lawrence Weston 37
lead 7, 154–155
Lechmore Water 91
Leigh Court 134, 162
Leigh Down 153
Leigh Farm 127
Leigh-on-Mendip 29
Leigh Woods 55
Lewins Mead Borehole 62, 168
Leyland Court 96
Lias 6, 90

Liealong Vein 84
Lilstock Formation 128–129, 144
Limekiln Hill 134
Limekiln Lane Borehole 64, 88
Limekiln Plantation 56
Limekiln Vein 158
Limonitic Bed 149
Limpley Stoke 150
Lithostrotion basaltiforme Bed 34, 48, 50
Lithostrotion Limestone 41–42
Little Avon, River 10, 12
Little Course Seam 72, 93
Little Daniel's Wood 12
Little Fiery Vein 74, 77, 79, 80, 81
Little Seam 100
Little Slyving Vein 119
Little Vein 95, 110
Little Whitfield Farm 12
Littleharp Bay 56
Littleton Court 134
Littleton-upon-Severn 19
Llandovery 11–12
Locking 135
Lodge Hill 81, 85
Long Ashton 37, 43, 52–56, 131, 132
  head deposits at 164
  palynomorphs at 144
Long Ashton Boreholes 169
Longford 164
Longwood Quarry 55
Lords Wood 127
Lower (Black Rock) Dolomite 5, 19
Lower Carboniferous rocks 9
Lower Coal Measures 67–68
  classification 71–73
  distribution and thickness 68–71
  sections in 78, 81, 82, 90, 92
Lower Conygre Colliery 111
Lower Cromhall Sandstone 23, 27, 37
Lower Drybrook Sandstone 37
Lower Five Coals 73, 78, 79, 81, 82, 84, 85, 86, 90, 160
  structural problems 78
Lower Gurney 93
Lower Inferior Oolite 147, 149
Lower Jurassic 3, 4, 5
Lower Lias 6
Lower Lias Clay 129, 158
Lower Limestone Shales 9, 15, 17, 26–28, 46, 48, 49, 52, 54, 56, 57, 58, 103
  isopachytes in 36
Lower Old Red Sandstone 10, 13, 14, 16
Lower Palaeozoic rocks 10–13
Lower Trap 11
Lower Writhlington Colliery 105, 107, 119
Luckington 9, 159
Ludlow rocks 10, 11, 13
Ludlows Colliery 105, 106, 107, 119
Luggers Hill 99
Lulsgate 60–61
Lulsgate quarry 155–156
Lyde Green 114
Lydney 5

GENERAL INDEX

Mackintosh Pit 92
Made-for-Ever Seam 101, 102
Maes Knoll 3, 150
Maes Knoll Conglomerate 148, 149, 150
Maesbury Castle 27, 155
Main Coal 91
Main Coal Seam 73
Malago Vale Colliery 75, 76
Malvern Fault Zone 2, 16, 20, 23, 41, 45, 60
  Coal Measures in 70
Malvern Hills 11, 14
manganese 153
Mangotsfield Formation 67, 68, 71, 73, 90, 97, 101–103
Mangotsfield Seams 101–102
marginal deposits in the Triassic 19–20, 135–140
marine transgressions
  in Coal Measures 67
  in Dinantian times 20
Marksbury Plain 42, 124
Marlstone Rock Bed 129
Marsh Lane Colliery 105
May Hill 13
Mayshill 113
Mayshill Pit 114
Melcombe Wood 91
Mells 3, 20, 42, 64, 66, 68, 70, 73, 91, 147
Mells Colliery 93, 101
Mells Colliery Borehole 175
Mendip Basin 23
Mendip Hills 1, 8, 13, 14, 17, 19, 20, 24–25, 26, 27, 28–29, 94, 134, 147, 149, 163
  building stone from 157
  caves in 164
  celestite in 155
  surface features 2, 3
Mendip (Strap) Pit 91
Mercia Mudstone 128, 129, 130, 131, 132
  brick clay from 158
  celestite in 155
  copper in 155
Merehead Quarry 147
Mesozoic strata 3, 5
metasomatism of Triassic rocks 142–144
Micklewood Beds 10
Middle Coal Measures 4, 67, 73–95
  brick clay from 158
  classification 71–73
  distribution and thickness 68–71
  sections in 78, 81, 82, 89, 90, 92
  Middle Cromhall Sandstone 27, 37, 42, 43, 45, 51
  section in 39–41
Middle Devonian 14
Middle Hope Peninsula 42
Middle Inferior Oolite 147, 149
Middle Jurassic 147–151
Middle Jurassic Limestone 3, 4
Middle Lias 129
Middle Pit 106

Middle Vein 107, 119, 122
Middle Vein Greys 107
Midford 150
Midford Sands 129
Midsomer Norton 70
Midsomer Norton Boreholes 174
Milbury Heath 13, 14, 16
Millgrit Vein 73, 76, 98–100
Millstone Grit 8, 60, 62–65, 95
  see also Namurian
mine shafts, map of 69
mineralisation of Triassic rocks 142–144
Mollusca Bed 43
Mona Complex (Anglesey) 14
Moons Pit 91
Mooreledge 95
Moorewood Colliery 91, 92, 93
Mount Hill 85
Mount Hill Brickworks 85
Mountain Limestone 9, 25, 28
Muxen Seam 99

Nailsea 164
  glass manufacture at 157–158
Nailsea Basin 1, 3, 19, 65, 66, 68, 70, 71, 73, 76, 103, 132, 162, 164
  coal mining in 159
Nailsea Coalfield 94–95
Nailsea Heath Colliery 94
Naish House Fault 58
Naishcombe Hill Fault 90
Namurian 62–65
  vertical sections in 44
Nap Hill Adit 121
Narroways Hill 135
Nempnett 3
Netham 99, 162
Netham Vein 84
Nettlebridge 66, 68, 70, 73
Nettlebridge Valley 91–93, 100–101
  coal mining in 158–159
New Cheltenham Pit 79
New Engine Pit 112, 113
New Golden Valley Pit 99
New Lodge Pit 78
New Mills 90
New Red Marl 90
New Rock Colliery 72, 84, 92, 93, 95, 99, 100, 110, 159–160
New Rock Group 72, 73, 110
New Smith's Coal 84, 85, 90, 99, 158
New Vein 107
Newbury Colliery 72, 92, 93, 101
Newbury Veins 72, 101
Newnham 6, 13
Newton Brook 5
Newton St Loe 5, 8, 73, 99
  coal mining at 159
Nibley 102, 114
Night Vein 106
Nightingale Valley 52
Nine Inch Seam 119
Nine Inch Vein in Coal Measures 72, 116, 117
No. 9 Vein in Coal Measures 72
No. 8 Seam 119
No. 5 Vein (Farrington) 107, 108

No. 6 Vein (Farrington) 107, 110
No. 7 Vein (Farrington) 107
No. 8 Vein (Farrington) 106–107
No. 9 Vein (Farrington) 105, 106–107
No. 10 Vein (Farrington) 105–106, 110
Norman's Vein 83
North Braysdown Pit 117
North Hill 14, 17, 19
North Hill Fault 59
North Hill Pericline 41
North Shoots Coal 92
Northern Storm Water Interceptor Tunnel 51
Norton Hill Colliery 105, 107
Norton Malreward 3
Nunney 2, 135, 150

Oakhill 16, 152
Oatfield Batch 59
Oatfield Wood 59
ochres 153–154
Old Grove Colliery 107
Old Grove Collier Borehole 175
Old Mills Colliery 105, 107, 117
Old Mills Underground Borehole 105
Old Nibley Colliery 114
Old Pylemarsh Colliery 99
Old Red Sandstone 14–18, 19, 26, 52, 103
  surface relief 2, 3
Old Welton Colliery 107
Oldbury 5
Oldland Common 73, 76, 99, 100
Oldland Common Boreholes 171
olivine-basalt 11
Olveston 19, 43, 70, 104
Ordovician 10
Ornithella Beds 150
Over 19, 70, 95
Ox House Bottom 54
Oxbridge Pit 113

Palaeozoic rocks 5
Palate Bed 27–28
palynology
  Rhaetian 137
  Triassic 144–146
Parker Ground Pit 78
Parker Veins 78, 79, 80
Parkfield Colliery 102, 112, 114
Parkfield South Colliery 114
Parrot Vein 73, 99, 158
Patchway 96, 135, 142
Patchway Borehole 166
Patchway Tunnel 95, 140
Paulton Pits Boreholes 174
Peacock Coal 107
Peacock Vein 110, 111
Peart, The 75
Pen Hill 17, 19
Pen Hill Pericline 14
Pen Park Borehole 50
Pen Park Hole 7
  copper at 155
  lead ore at 154–155
Penarth Group 128, 129

Pennant Measures 67–68, 70, 71–73, 76, 95–104
 sections in 99, 100, 102
Pennant Sandstone 4
 building stone from 157
 conglomerates in 45
 iron ore in 153
Pennyquick Bottom 90
Pennywell Road Colliery 79, 80
Penpole Point 19, 104
 lead ore at 154
 limestone quarries at 156
Pensford 130, 132, 134
 coal mining at 159, 160
Pensford Basin 1, 3, 66, 68, 110–111, 117–118, 119, 124–127
Pensford Borehole 173
Pensford Colliery 110, 118, 119, 122, 124–126, 127, 159
 comparative sections in 109
Pensford Seams 110, 118, 119, 126, 127
Pensford Syncline 127
Perkins Course Coal 92, 93
Permo-Triassic 3
Perrink seam 73
Pigs Cheek Vein 83
Pill 133, 162
Pilning 104, 116
Pitcot 92, 93
Pitcot Colliery 101
Pleistocene 161–164
Pliensbachian 129
Porkskewett 65, 70, 95, 103, 135
Porkskewett Boreholes 65, 104, 166
Portbury 14, 16, 164
Portishead 5, 6, 14, 17, 19, 57, 67, 70, 95, 103, 104
Portishead Beds 14, 16, 17, 19, 26, 103
Portishead Boreholes 167
Portishead Down 57, 162
Portishead Point 57, 70
Portway Cutting 26
Portway Tunnel 16, 30, 35, 45, 178, 179
Pottershill 59
Potterswood 99
pottery and brick clay 158
Press's Quarry 46
Priddy 41
Primrose Vein 84, 85
Priston 5, 111
Priston Borehole 173
Priston Colliery 122–124
Priston Colliery Borehole 174
Privy Coal 92
Providence 43, 55
Publow 19, 66, 134
Publow Formation 67, 68, 73–74, 111, 126–127
Publow Leigh Seam 127
Pucklechurch 70, 90, 96, 102
Pucklechurch Boreholes 171
Pullastra Sandstones 140
Purton 10, 13, 130
Puxton Moor 1, 68
Pylemarsh 99
Pylle Hill 140, 142, 144

Quarry Pit 114
Quartz Conglomerate 12, 14, 17
Quartzitic Sandstone 43, 51, 55, 56, 62–65, 88, 91, 95
Quaternary 3
Queen Charlton 3, 111, 159

Radford Colliery Borehole 174
radiocarbon dating 165
Radstock 63, 105, 141,151
 building stone from 157
 vertical sections at 120
Radstock Basin 1, 5, 19, 42, 45, 64, 68, 70, 71–73, 91–93, 99, 105–110, 117, 119–126, 132, 135, 141, 142
 coal mining in 158, 159, 160
 isopachytes map 121
 surface features 2, 3
 section 117
Radstock Collieries Boreholes 174
Radstock Formation 67, 68, 119–124
 correlation of coals 122
 and Farmborough Fault 112
 section 118
Radstock Great Vein 111
Radstock Middle Vein 121
Radstock Old Pit 111
Radstock Slide 9
Rad Coal 73, 76
Rag Vein 99, 100, 113, 116
Ragged Seam 82
raised beaches 162
Ram Hill Pit 112, 113
Rangeworthy 70, 71, 73, 88–89, 96, 97
Rangeworthy Borehole 166
Red Ash Vein 74, 160
Red Axen Seam 91
Red Hill 130
Redcliffe 131
Redcliffe Bay 16
Redcliffe Sandstone 157
Redcliffe Sandstone Formation 128, 131–132
Redfield 99
Redfield Hill 99
Redhill 3, 43, 58
Redland 131, 134, 141, 142
Redwick 5
Regilbury 134
Rhaetic 90, 128–129, 131, 135
Rib Mudstone 41
Ridgehill Fault 58
Ridgeway Fault 70, 95
Ringing Pit 91
roadstone 155–157
Rock Seam 111
Rock Vein 71, 72, 84, 105, 107, 116
Rocks Wood 60
Rodent Earth (Pleistocene) 161
Rodford 102
Rodney Stoke 1, 164
Rodney Stoke Borehole 94
Rodney Stoke Fault 94
Rodway Hill 97
Rownham Hill 55, 56
Rownham Hill Coral Bed 43, 51, 56
Royal Oak Pit 78

Rudge Coal Seam 68
Rudge Pit 105
Rudge Pit Borehole 173
Rudge Vein 103, 105
Rush Hill 105
Rydons Pit 127

Saise Colliery 86
Salisbury Plain 8
Salter's Brook 126, 127
Saltford 3, 8
Salthouse Bay 56
Sandford Hill 143
Scrag Vein 99
Sea Walls 46, 48
seams
 principal 73
 see also veins and under individual names
Sedbury 5
Seminula Oolite 34, 37, 48, 50, 51, 52, 55
Seminula Pisolite 34, 48, 55
Serridge Engine Pit 113
17-inch Vein in Coal Measures 72
Severn Coal Basin 1, 5, 42, 65, 68, 70, 73, 95, 103, 104, 116
Severn River 4, 5, 37
 terraces 162
Severn Tunnel 6, 65, 70, 104, 132, 163, 164
Severnside 95
Severnside Borehole 104, 166
Severnside Evaporite Beds 130, 133, 155
shafts
 map of 69
 see also under individual collieries and pits
Sheep Wood 51
Shepton Mallet 130, 141, 147
Sherwood Sandstone Formation 128
Shirehampton 5, 19, 20, 49, 51–52
 limestone quarries at 156
 river terraces 162–163
Shirehampton Beds 17, 19, 26–28, 46, 52
Shirehampton Terrace 162
Shortwood Colliery 112, 114
Shortwood Hill 96
Sidelands Fault 55
Silurian 10–13
Silurian/Devonian 14–18
Sims Hill 133
Sinemurian 129
Siston 88, 90
Siston Brook 5
Siston Common Colliery 81, 82, 85, 90
Slade Bottom 91
Slate Vein (Soundwell) 79, 80
Slyving Vein 119, 121, 122
Small Coal Vein 101
Small's Quarry 60, 135
Smith Coal 88, 95
Smith's Coal 107, 158
Sneyd Park 28, 49
Sneyd Park Fish Bed 14, 17, 19, 26
Snuff Mills Park 97
Sodbury Seam 88

Somer, River 117
Somerdale Borehole 173
Somerset Coal Canal 7, 8, 159
Somerset Levels 3, 128, 164
Soundwell 70, 71
Soundwell Colliery 74, 78, 79, 80, 82–83, 84
Soundwell Seams 79, 90
Soundwell Smith Coal 79, 80, 86
South Liberty Colliery 74–76
South Shoots Coal 92
Southern Main Road quarry 148, 149
Southmead 19, 29, 37, 49, 50, 51, 140
Southside Wood 52
Speedwell Colliery 77, 78, 79, 80, 81, 82, 84, 86, 160
Speedwell Hard Vein 86
Speedwell Thrust 78
Sperrings Farm Borehole 103
sphalerite 142, 154
Spider Delf and Dog Veins 94, 95
Spring Cove 41
St Georges Park 99
St Mary Redcliffe 131, 149
St Maughans Group 14
St Michael's Hill 131
St Phillips March 98
St Vincent's rocks 6, 32
Standing Coal 72, 91–92
Stanton Drew 93, 134
  coal mining at 159
Stanton Prior 112
Stanton Wick 3, 112
  coal mining at 159
  glass sand from 157
Staple Hill 73, 96
Stapleton 5, 73, 96, 97, 131, 133
Starveall Colliery 73–76, 77
Stanton Drew 111
Stibbs Hill 99
Stinker Vein 112
Stinking Seam 96, 97
Stinking Vein 111, 113, 116
Stock Hill 111
Stockwell Hill Fault 102
Stockwood Borehole 173
Stoke Gifford 73, 87–88, 96, 133
  brickyards at 158
  building stone from 157
Stoke Gifford Boreholes 87, 88, 96, 166
Stoke Park 134, 135
Stoke Park Rock Bed 130, 132, 133, 135, 144
  celestite in 155
  mineralisation in 143
  sphalerite in 154
Ston Easton 8, 93–94
Ston Easton Boreholes 174
Stone Rag 91
Stony Vein (Soundwell) 79
Stowey 124
Strap Pit 91, 93, 94, 159
Strap Pit Borehole 94
Stratton-on-the-Fosse 70, 130, 132
Stratton-on-the-Fosse Borehole 175, 176
Streak Vein 107, 110
Stubbs Seam 84

stylolites 20, 63
Sub-Oolite Beds (Clifton Down Group) 37, 40, 42, 49, 55
Sudbrook 5, 70, 132
Sun Bed 141, 142
Supra-Pennant Measures 67–68, 73, 104–116
surface relief 2–4
Sutton Court 93, 111
Swallow Cliff 162
Swash Channel 6

Tanhouse 63
Tanhouse Limestone 27, 45
Tapwell Bridge 95
Tapwell Bridge Coal Seam 65
Tea Green Marl 128–129, 130, 135
Telychian Stage 11
Temple Cloud 3, 101, 130
  Mangotsfield Formation at 103
Temple Cloud Borehole 175
Temple Cloud Vein 103
Tenley 124
Terrace Gravel 164
terraces 162
  Bristol Avon 162–163
  Lower Severn 162
Thornbury 14, 16
Thornbury Beds 10, 13, 14, 16
Three Coal Vein 110
Thrubwell 3
Thurfer Vein 79, 82, 86, 87
Tickenham 5, 19, 24, 41, 57–58, 94, 103
  glass sand from 158
  Pleistocene deposits at 161
Tickenham Borehole 167
Timsbury 105, 111, 119
Tintern Sandstone Group 14, 17
Tites Point 10, 13
Titterstone Clee 23
Toarcian 129, 148
Tockington 131
Tog Hill 3, 149
Tommy Collier's Vein 107
Top Little Vein 122
Top Vein 107, 114, 116
Tortworth 10, 12, 13, 14, 15, 16, 135, 141
Tortworth Beds 11–12
Tortworth Inlier 10, 12
Tournaisian 20, 23, 25, 27, 42, 52
Tremadoc rocks 10
Triassic Conglomerate 144
  in Pleistocene deposits 161
Triassic 3, 5, 8, 55, 58, 60, 90, 93, 127, 128–146, 157
  base contour map of 138
  mineralisation and metasomatism 142–144
  sections through 50, 130
Trilobite Bed 34, 42, 48
Trolley Vein 118
Trough Vein 83
Trym, River 3, 5
  river terraces 162
Trym, Valley 53

Tubb's Bottom 112
Tuckingmill 127
tuff
  in Lower Palaeozoic 11, 17
  at Goblin Combe 46
Tump, The 144
Twerton 4
  Jurassic at 147
  river terraces 163
  sand and gravel from 157
Twerton Colliery 73, 90
Twerton Colliery Borehole 173
Twinhoe 147
Twinhoe Beds 147, 150, 151
Twinhoe Ridge 151
Two Foot seam 160
Tyning Colliery 119
Tyning Colliery Borehole 175
Tyning Farm, Codrington 39
Tyntesfield 55–56, 65, 94
Tytherington 37, 39, 43, 64, 70, 96, 130, 136, 137
  limestone quarries at 156
Tytherington Borehole 167

Upper Coal Measures 5, 43, 65, 67–68, 104
  classification 71–73
  distribution and thickness 68–71
  sections in 89, 100, 123
Upper Congyre Colliery 111
Upper Coral Bed 149, 150
Upper Cromhall Sandstone 27, 42, 43, 45, 48, 51, 55, 56, 58, 60, 61, 64, 95
Upper Five Coals Vein 82, 84, 85
Upper Inferior Oolite 129, 147, 149, 150
Upper Knole 19, 51
Upper Lias 129
Upper Old Red Sandstone 9, 10, 12, 15, 16–17, 63, 103
  isopachytes in 36
Upper Stanton Drew 159
Upper Trap 11
Upper Triassic 3
Upper Trigonia Grit 149, 150
Upper Whimsey Pit 113
Usk Anticline 14, 23

Vallis Limestone 27, 31, 41, 42
Vallis Vale 147, 150
Variscan orogeny 2, 3, 4
veins 65, 71
  see also seams and under individual names
Vimpennys Common 116
Viséan 20, 21, 23–25, 38, 42, 44, 51
Vobster 9, 91, 159
Vobster Breach Colliery 91
Vobster Quarry 64
Vobster Series 72
volcanic rocks 24, 41, 46, 59, 150

Wain's Hill 56–57
Wallsend Colliery 102
Walter's Pit 84
Walton-in-Gordano 57, 162

Walton Park 56–57
Wansdyke 3
Wanswell Green 13
Wapley 70, 73, 96
Wapley Borehole 144
Wapley Pit 90
Warkey Course 101
Warmley 73, 82, 84, 85, 90, 99
 brickyards at 158
Warmley Boreholes 171–172
Warmley Great Vein 82
Warren House (Broadfield Down) 58–59
Washingpool Farm Borehole 104
Watercress Farm Borehole 94
Weare 3
Webbs Heath 90
Wedmore 3, 135, 141
Wedmore Inlier 128, 135, 141
Wedmore Stone 141
Wellow branch of Somerset Coal Canal 8
Wells 94
Welton 150
Welton Colliery 119
Welton Hill Colliery 119
Wenlock 11, 12
West End (Grace's) Colliery 103
West Harptree 91, 93
West Harptree Borehole 173
West Leigh Court 144
West Mendip Basin 20, 21, 45
Westbury 161, 164
Westbury Anticline 14, 17, 51
Westbury Beds 129, 131, 135, 137, 140–141, 143
 sections in 142
Westbury Formation 129, 144, 146
Westbury-on-Severn 140
Westbury-on-Trym 19, 29, 48–51, 134, 136
 lead ore at 154
Westbury (Wilts) 1, 68, 91
Westbury (Wilts) Borehole 94, 142
Westclose Conglomerate 132, 134
Westerleigh 71, 88, 90, 132
Westerleigh Boreholes 72, 89, 90, 97–98, 144, 167, 172
Westerleigh Common 88, 97
Weston-in-Gordano 57, 156
Weston-super-Mare 24, 25, 41, 42
 lead and zinc ore at 155

Westphalian 66–127
 classification of 71–72
 distribution and thickness 68–71
 map of boreholes and shafts 69
 principal coal seam 73
Wethered Colliery 86
Whatley 150
Whimsey Pit 114
Whing seam 91
Whitchurch 3, 10, 66
White Axen seam 91, 93
White Hill 97
White Lias 3, 7, 128, 131, 141–142
 building stone from 157
White Oak Pit 94, 95
White Ox Mead (Head ?) 147, 150
Whitefaced Fault 73, 78, 79, 80, 85, 88, 97, 99, 102, 131
Whitehall Colliery 80, 82, 84
Whitehead Limestone 24, 37
Whites Hill 99
White's Top seam 94, 95, 103
Whitesill Common 97
Whitfield fault 10, 12, 13
Wick 1, 3, 5, 19, 20, 23, 37, 39, 42, 43, 63, 65, 70, 90–91
 earth pigments at 153
 iron ore at 153
 quarries at 157
Wick Fault 157
Wick Inlier 20, 90, 156–157
Wick quarry 62
Wicks Rocks Thrust 62
Wickwar 10, 14, 16, 19, 20, 23, 29, 37, 39, 88, 143
 limestone quarries at 156
Wigpool Syncline 23, 26
Willow Brook 5
Wills Hall 49
Willsbridge 142
Wilmot's Vein 72, 91
Winford 3, 19, 20, 43, 45, 63, 64, 65, 76–77, 124
 earth pigments at 153
 iron ore at 153
 quarries at 157
 haematitisation at 142
Winford Boreholes 73, 76, 95, 172
Winford Hospital borehole 60
Winterbourne 88–89, 102, 131
Winterbourne Borehole 72, 84, 88, 90, 96, 102, 167

Winterborne Marine Band 67, 71, 72, 76, 84, 87, 90, 95, 96, 97, 100, 103
Witchellia Bed 149
Withy Mills Colliery 73, 124
Withy Mills Colliery Borehole 175
Withy Mills Seam 119, 124, 126
Withy Mills Vein 72
Wolstonian 162
Wood Pit 114
Woodborough 164
Woodborough Colliery Borehole 175
Woodford Hill 134
Woodford Hill Sandstone 134, 143, 144
Woodhill 57
Woodhill Bay 6, 17
Woodhill Bay Conglomerate 17, 18
Woodhill Bay Fish Bed 14, 17
Woodspring 6, 20
Wooscombe Bottom 126
Worm Band 80, 81, 86
Wotton-under-Edge 3
Wraxall 52–56, 57–58, 94, 164
Wraxall Borehole 169
Wraxall Piece 54–55
Wrington 58, 68
Wrington Borehole 172
Wrington Hill Fault 58
Wrington Vale 1, 3, 19, 43, 95, 130, 134, 164
Wrington Warren 154
Writhington Collieries Boreholes 175
Writhington Colliery 121, 160
Wye, River 4, 5, 6

Yate 1, 20, 45, 63, 64, 70, 73, 96, 97, 132, 134
 celestite at 155
Yate Borehole 62, 64, 72, 84, 88, 167
Yate Deep Borehole 88, 89, 90, 98, 167
Yate Evaporite Bed 130, 135
Yate Hard seam 96
Yate Hard Vein 88, 90
Yate Little Vein 88
Yate Smith Coal 90
Yatton 19, 58, 164
Yatton Borehole 172
Yeo Valley 132
Young Wood Colliery 94, 95

zinc 154–155

BRITISH GEOLOGICAL SURVEY

Keyworth, Nottingham NG12 5GG
0602-363100

Murchison House, West Mains Road,
Edinburgh EH9 3LA   031-667 1000

London Information Office, Natural History Museum Earth Galleries, Exhibition Road, London SW7 2DE
071-589 4090

The full range of Survey publications is available through the Sales Desks at Keyworth and at Murchison House, Edinburgh, and in the BGS London Information Office in the Natural History Museum Earth Galleries. The adjacent bookshop stocks the more popular books for sale over the counter. Most BGS books and reports are listed in HMSO's Sectional List 45, and can be bought from HMSO and through HMSO agents and retailers. Maps are listed in the BGS Map Catalogue, and can be bought from Ordnance Survey agents as well as from BGS.

The British Geological Survey carries out the geological survey of Great Britain and Northern Ireland (the latter as an agency service for the government of Northern Ireland), and of the surrounding continental shelf, as well as its basic research projects. It also undertakes programmes of British technical aid in geology in developing countries as arranged by the Overseas Development Administration.

The British Geological Survey is a component body of the Natural Environment Research Council.

Maps and diagrams in this book use topography based on Ordnance Survey mapping

HMSO

HMSO publications are available from:

HMSO Publications Centre
(Mail, fax and telephone orders only)
PO Box 276, London SW8 5DT
Telephone orders 071-873 9090
General enquiries 071-873 0011
Queueing system in operation for both numbers
Fax orders 071-873 8200

HMSO Bookshops
49 High Holborn, London WC1V 6HB
  071-873 0011   Fax 071-873 8200
(Counter service only)
258 Broad Street, Birmingham B1 2HE
  021-643 3740   Fax 021-643 6510
33 Wine Street, Bristol BS1 2BQ
  (0272) 264306   Fax (0272) 294515
9 Princess Street, Manchester M60 8AS
  061-834 7201   Fax 061-833 0634
16 Arthur Street, Belfast BT1 4JY
  (0232) 238451   Fax (0232) 235401
71 Lothian Road, Edinburgh EH3 9AZ
  031-228 4181   Fax 031-229 2734

HMSO's Accredited Agents
(see Yellow Pages)

And through good booksellers